POLYMERS, PHOSPHORS, *and* VOLTAICS *for* RADIOISOTOPE MICROBATTERIES

POLYMERS, PHOSPHORS, *and* VOLTAICS *for* RADIOISOTOPE MICROBATTERIES

Edited by
Kenneth E. Bower
Yuri A. Barbanel'
Yuri G. Shreter
George W. Bohnert

CRC Press
Taylor & Francis Group
Boca Raton London New York

CRC Press is an imprint of the
Taylor & Francis Group, an **informa** business

CRC Press
Taylor & Francis Group
6000 Broken Sound Parkway NW, Suite 300
Boca Raton, FL 33487-2742

First issued in paperback 2019

ISBN-13: 978-0-8493-0915-1 (hbk)
ISBN-13: 978-0-367-39605-3 (pbk)

Library of Congress Card Number 2002017496

Library of Congress Cataloging-in-Publication Data

Polymers, phosphors, and voltaics for radioisotope microbatteries / edited by Kenneth E. Bower ...[et al.].
  p. cm.
Includes bibliographical references and index.
ISBN 0-8493-0915-8 (alk. paper)
1. Nuclear batteries--Materials. 2. Radioisotopes--Industrial applications. 3. Microelectromechanical systems. 4. Polymers--Electric properties. 5. Phosphors. I. Bower, Kenneth E.

TK2965.P65 2002
621.31′242—dc21
                2002017496

Visit the Taylor & Francis Web site at
http://www.taylorandfrancis.com

and the CRC Press Web site at
http://www.crcpress.com

Preface

This book is a remarkable testament to the value of collaboration on technical problems with broad social implications. More than 30 scientists in Russia and the U.S. have worked together over the last several years on original research into radioisotope battery design optimization.

Progress on this topic has the potential to revolutionize microelectronics by enabling emerging microelectromechanical systems (MEMS) and nanotechnology. Power-matched supplies would last decades, even centuries. No power cords, rectifiers, or transformers will be needed for a new generation of microdevices — only safe, direct, long-life, stable, integrated electric power from the highest energy density source available.

A key feature of this book is discussion of the materials of construction for miniaturized radioisotope power supplies. Progress in nuclear battery technology depends on characterization of functionally radiation-stable components. Substantial progress has been made to solve problems of using integrated radioisotope batteries for micro- and nanoelectronics. Each author has provided an authoritative assessment and indicated where development is needed. The first chapter's treatment of ionizing radiation serves to remind us that *all* materials are ionized in the pressence of 5.5 MeV decay particles. We seek materials that are *functionally* stable — the components perform the engineered task even after ionization.

This book is intended for MEMS designers, electrical and nuclear engineers, material scientists in general and polymer and phosphor chemists in particular, voltaic fabricators, radioactive material policy makers, and other interested nuclear industry professionals.

Significant technological progress depends today on coordinated interdisciplinary research. A large group of scientists and technologists was organized around the shared applied goal of building a better radioisotope battery that can enable MEMS technology. We have had the privilege of serving as project leaders and now as editors on this project. We are grateful for the diversity of talent and specialties focused on this work. We have optimism for the future of the radioisotope microbattery as a commercially significant development and hope that our collaboration provides part of a multinational solution to nuclear fission waste disposition.

Kenneth E. Bower, Charleston, Illinois

Yuri G. Shreter, St. Petersburg, Russia

Yuri A. Barbanel', St. Petersburg, Russia

George W. Bohnert, Kansas City, Missouri

Acknowledgment

This collaboration was sponsored by Initiatives for Proliferation Prevention, U.S. Department of Energy, and administered by the U.S. Civilian Research and Development Foundation. TRACE Photonics Inc. was supported by U.S. Army, Picatinny Arsenal engineers, G. Robert Haugeto and Mike Kajor through the Small Business Innovated Research (SBIR) Program, and Sensor Technology Development Fund venture capital partners Robert Knollenberg and Betty Ohannessian.

The Editors

Kenneth E. Bower earned his Ph.D. in chemistry in 1991 from the University of Akron, Ohio. He was a postdoctoral fellow at Los Alamos National Laboratory in chelating polymer synthesis and a team leader for actinide analytical chemistry sample management. Dr. Bower is research director for TRACE Photonics Incorporated, which is integrating the radioisotope microbattery into sensors and MEMS. He resides in Charleston, Illinois, a few blocks from "The Castle" of Eastern Illinois University.

Yuri G. Shreter was born in 1944 in Petropavlovsk. He earned his B.Sc. in physics and engineering from the Odessa Technical University (1967) and his Ph.D. (1971) and Doctor of Science (1993) from A.F. Ioffe Physico-Technical Institute in St. Petersburg. Dr. Shreter became a visiting Gauss professor in Gottingen University, Germany (1987-88) and a visiting professor at Cologne University and at UMIST, Manchester, United Kingdom (1991-92). Dr. Shreter is currently a leading scientist at A.F. Ioffe Physico-Technical Institute and a professor at St. Petersburg State Technical University. His research interests include material science of wide-band-gap semiconductors and devices.

Yuri A. Barbanel' was born in 1935 in Leningrad (now St. Petersburg). He graduated from the chemistry department of Leningrad State University in 1958 and earned his Ph.D. in 1964 and his D.Sc. in 1991. Dr. Barbanel' is currently a leading scientist at V.G. Khlopin Radium Institute (St. Petersburg). His research interests include optical spectra (including radioluminescence), energy level structure of the lanthanide and actinide ions in crystals, and the absorption spectra of the actinides and lanthanides in molten salts.

George W. Bohnert is a senior staff engineer in the materials engineering and business development section of Honeywell Federal Manufacturing & Technologies, a prime contractor for the U.S. Department of Energy. He received his B.S. in chemical engineering from the University of Missouri at Columbia in 1976. He joined Honeywell in 1981. His research areas include polymer processing, precision cleaning, process waste minimization, and investigation of alternative energy technologies. Mr. Bohnert is a coinventor of three U.S. patents. He resides on a small farm near Harrisonville, Missouri with his lovely wife and two children.

Contributors

Gennady P. Akulov
V.G. Khlopin Radium Institute
St. Petersburg, Russia

Vyatcheslav M. Andreev
Ioffe Physico-Technical Institute
St. Petersburg, Russia

Laura Yu. Barbanel'
St. Petersburg, Russia

Yuri A. Barbanel'
V.G. Khlopin Radium Institute
St. Petersburg, Russia

Natalia I. Bochkareva
Ioffe Physico-Technical Institute
St. Petersburg, Russia

George W. Bohnert
Honeywell FM&T
Kansas City, Missouri

Carl C. Bower
TRACE Photonics Inc.
St. Paul, Minnesota

Kenneth E. Bower
TRACE Photonics Inc.
Charleston, Illinois

Victor A. Bykov
V.G. Khlopin Radium Institute
St. Petersburg, Russia

Tatyana A. Iourre
Institute of Technology
St. Petersburg, Russia

Yuri L. Kaminski
V.G. Khlopin Radium Institute
St. Petersburg, Russia

Alexander G. Kavetsky
V.G. Khlopin Radium Institute
St. Petersburg, Russia

Natalia V. Klimova
Institute of Technology
St. Petersburg, Russia

Piotr M. Konovalov
V.G. Khlopin Radium Institute
St. Petersburg, Russia

Sergei P. Meleshkov
V.G. Khlopin Radium Institute
St. Petersburg, Russia

Sergei N. Nekhoroshkov
V.G. Khlopin Radium Institute
St. Petersburg, Russia

Alexander A. Panferov
V.G. Khlopin Radium Institute
St. Petersburg, Russia

Yuri T. Rebane
Ioffe Physico-Technical Institute
St. Petersburg, Russia

Ludmila I. Rudaya
Institute of Technology
St. Petersburg, Russia

Andrew F. Rutkiewic
TRACE Photonics Inc.
Albuquerque, New Mexico

Yuri G. Shreter
Ioffe Physico-Technical Institute
and St. Petersburg State Technical
  University
St. Petersburg, Russia

Valentina B. Shuman
Ioffe Physico-Technical Institute
St. Petersburg, Russia

Maxim M. Sychov
Institute of Technology
St. Petersburg, Russia

Shahid M. Yousaf
TRACE Photonics, Inc.
Charleston, Illinois

Symbol Designations

A	Activity of radioactive substance
A_{al}	Allowed activity of radioactive substance
A_m	Specific activity of radioactive substance
A_{mol}	Molar activity of radioactive substance
A_S	Surface activity of radioactive substance
A_v	Volumetric activity of radioactive substance
ARC	Antireflection coating in PV cells
B	Brightness
$b(\lambda)$	Relative spectral density of the radiant flux
c	Speed of light
Ci	Curie unit, $3.7 \bullet 10^{10}$ disintegrations per second
d	Substance density
D_{abs}	Absorbed dose
D_C	Cylinder diameter
D_{el}	Electron dose
D_g	Gelation dose
D_L	Linear dimension (thickness, etc.)
D_S	Sphere diameter
d_S	Surface density of a substance
E	Energy
E_A	Process activation energy
E_c	Conduction band of semiconductor
E_{av}	Average energy for creation of electron–hole pair
E_F	Fermi level in semiconductor
E_g	Bandgap (energy) of semiconductor
E_{lat}	Crystal lattice energy
E_T	Energy of activation of nonradiative transition
E_V	Valence band in semiconductor
FF	Fill factor of current-voltage characteristics
G	Radiation-chemical yield
g_c	Gel fraction
G_{el}	Electrical capacitance
h	Planck's constant
H_C	Cylinder height (length)
I	Current (strength)
I_{IC}	Ionization current
I_O	Saturation current in PV cell
i_O	Saturation current density in PV cell
I_{PC}	Photocurrent in PV cell
I_{sc}	Short-circuit current
i_{PC}	Photocurrent density in PV cell
k	Boltzmann's constant
k_l	Light concentration coefficient

K_m	Maximal luminous efficiency at the peak wavelength of 555 nm
K_T	Temperature instability coefficient of energy efficiency
L_L	Light output
L, L_n, L_p	Diffusion length of carriers in semiconductor
M, m	Mass
M_A	Atomic mass
M_M	Molecular mass
N	Number of particles (atoms, electrons, etc.)
N_A	Avogadro constant
N_a	Acceptor concentration in semiconductor
N_d	Donor concentration in semiconductor
n_β	Beta particle flux
n_i	Intrinsic carrier concentration in semiconductor
η_o	Backscatter coefficient
η	Efficiency of energy transfer, release, or conversion
η_{CL}	Energy efficiency of cathodoluminescence
η_d	Energy efficiency of direct conversion of energy
η_{ind}	Energy efficiency of indirect conversion of energy
$\eta_{l\text{-}el}$	Energy efficiency of conversion of light energy into electricity by a photovoltaic
η_q	Quenching coefficient of luminescence
η_{RL}	Energy efficiency of radioluminescence
η_{st}	Stokes losses coefficient
η_T	Thermalization coefficient
η_β	Efficiency of conversion of beta particle energy into available surface beta flux
$\eta_{\beta\text{-}el}$	Efficiency of conversion of beta particle energy into electricity by a betavoltaic
$\eta_{\beta\text{-}l}$	Efficiency of conversion of beta particle energy to light
OPV	Organic photovoltaic cell
PV	Photovoltaic cell
p	Pressure
P	Output electric power from PV cell
P_D	Dose rate
P_m	Maximal electrical power output of PV or betavoltaic cells
P_{sp}	Specific (per 1 Ci) power of radiation of radioactive substance
P_v	Specific (per unit volume) power
P_β	Power of the beta particle flux
Q	Carrier collection coefficient
Q_{int}	Internal quantum yield of PV cells
Q_{ext}	External quantum yield of PV cells
q	Electron charge
q_v	Frequency factor
R	Resistance
R_C	Cylinder radius
R'_m	Maximal path length of particles in a substance

R_o	Universal gas constant
r	Distance in surface density units
r_{op}	Primary electron backscattering coefficient
S	Surface area
s_c	Sol fraction
s_o	Surface recombination velocity of carriers in semiconductor
S_e	Energy radiosity
$S_e(\lambda)$	Spectral density of energy radiosity
S_V	Luminosity
T	Temperature
T'	Optical transmission
T''	Optical loss
$T_{1/2}$	Half-life
t, τ	Time
$U(\lambda)$	Spectral luminous efficiency function of monochromatic photopic light
V	Voltage
V_C	Potential difference in p–n junction of PV cell
v	Volume
V_{oc}	Open-circuit voltage
W	Average energy of ion pair formation in gas
w	Probability
w_r	Radiative recombination probability
w_T	Nonradiative recombination probability
$w(\varepsilon_\beta)$	Spectral response of beta particles
x, y	Chromaticity coordinates
Z	Atom nucleus charge (atomic number)
Z_{eff}	Efficient atomic number of substance
γ	Share of backscattered electrons
δ	Spectral band half-width
ε_{av}	Average kinetic energy of beta particles
ε_d	Dielectric permittivity
ε_{max}	Maximal kinetic energy of beta particles
ε_α	Kinetic energy of alpha particles
ε_β	Kinetic energy of beta particles
θ	Azimuthal angle
λ	Wavelength or radioactive decay constant
μ_L, μ_m	Linear and mass absorption coefficient of beta particles
v	Frequency
ρ	Distance
Φ_e,	Radiant flux in energy units
Φ_V	Light flux in photometric units
$\Phi_e(\lambda), \Phi_e(v)$	Spectral densities of radiant flux per unit wavelength and frequency ranges
ϕ	Polar angle
$\varphi(r)$	Dose function of the point beta emitter (source)

Ω	Solid angle
IR	Infrared
LET	Linear energy transfer
LPE	Liquid phase epitaxy
MSA	Minimal significant activity of radioisotope
MOCVD	Metal organic chemical vapor deposition
NMR	Nuclear magnetic resonance
RLS	Radioluminescent light source (RLSs – sources)
RLS-T	Tritium-based radioluminescent light source (RLSs-T – sources)
RTL	Radiothermoluminescence
UV	Ultraviolet
x	Mole fraction of component in (solid) solution

Contents

Chapter 3
Nonradioactive Materials for Nuclear Batteries 109
A.G. Kavetsky, Y.L. Kaminski, G.P. Akulov, and S.P. Meleshkov

Chapter 8
Wide-Band Semiconductors for Direct-Conversion Nuclear Batteries 365

Y.G. Shreter, Y.T. Rebane, and N.I. Bochkareva

Chapter 9

T.A. Iourre, L.I. Rudaya, and N.V. Klimova

Chapter 10
Radioisotope Microbattery Commercialization

K.E. Bower, A.F. Rutkiewic, C.C. Bower, and S. M. Yousaf

CHAPTER 1

Conversion of Radioactive Decay Energy to Electricity

A.G. Kavetsky, S.P. Meleshkov, and M.M. Sychov

CONTENTS

1.1   INTERACTION OF IONIZING RADIATION WITH MATTER

1.1.1   Types and Energy of Radioactive Decay

Radioactivity is a property of atoms of unstable isotopes that enables them to spontaneously transform into atoms of other isotopes, with emission of charged particles and quanta of electromagnetic radiation. Natural radioactivity was discovered by Becquerel in 1896 in the study of uranium salts; this marked the beginning of comprehensive examination and use of the phenomenon. A very interesting application is the use of energy released in radioactive decay for generation of electrical energy.

Spontaneous radioactive decay involves transformations of an unstable atomic nucleus leading to changes in its charge (Z), mass (M), and energy state. Several types of radioactive decay differ in the type of emitted particles. The most common types are alpha decay, electronic and positronic beta decay, and K-electron capture. Other kinds of radioactive decay do not play a significant role in practical applications of radioactivity.

Energy released in radioactive decay transforms into the kinetic energy of the daughter nucleus and emitted particles. This released energy is equal to the difference between the rest energy of the parent nucleus and the rest energy of the daughter nucleus and emitted particles. Kinetic energy of radioactive decay products (radioactive decay energy) can be converted into electricity. All types of radioactive decay obey the universal law of radioactive decay.

1.1.1.1   Radioactive Decay Law[1]

Variation of the number of radioactive atoms N in time t is proportional to $\exp(-\lambda t)$, where λ is the radioactive decay constant. This relationship follows from the assumption that the probability of decay of a nucleus of a given kind in a given period of time is constant. Indeed, dN minus the number of atomic nuclei decaying in the period from t to $t + $ dt is proportional to the time period dt and the number of nuclei N remaining by the time t:

$$dN = -\lambda N \, dt \qquad (1.1)$$

The term λ in Equation 1.1 is the radioactive decay constant characterizing the probability of decay in unit time. Integration of Equation 1.1 with respect to time from 0 to t, assuming the number of atoms at $t = 0$ is equal to N_o, gives

$$N = N_o \cdot \exp(-\lambda t) \qquad (1.2)$$

Equation 1.2 describes the statistical law of spontaneous radioactive decay of an isolated radionuclide. It is convenient to characterize the lifetime of a radioactive isotope by a period in which half of the initial number of its nuclei undergo decay. This period is termed the half-life, $T_{1/2}$.

$$T_{1/2} = \frac{\ln 2}{\lambda} \tag{1.3}$$

The most important characteristic of a radioactive substance is its activity, A. Activity is the number of nuclei of a given isotope decaying in unit time. As follows from Equation 1.1, A is the product of the decay constant and the number of radioactive nuclei in the sample:

$$A = -\frac{dN}{dt} = \lambda N \tag{1.4}$$

In the International System of Units, the dimension of activity is decay per second, becquerel (Bq). Another widely used activity unit is curie (Ci), equal to 3.7 $\times 10^{10}$ Bq.

Alpha decay is characteristic of natural and artificial radioactive isotopes with large atomic numbers.[1,2] For unstable atomic nuclei, it is accompanied by emission of alpha particles, i.e., double-ionized helium atoms. Alpha decay yields a daughter nucleus whose mass number is lower than that of the parent nucleus by four units and of the charge by two units. The alpha particles emitted in decay of a given nucleus can have the same energy or a set of discrete energies. When a radionuclide emits several groups of alpha particles with different discrete energies, the decay is accompanied by emission of gamma quanta of different discrete energies because the nuclei formed by alpha decay can occur in different energy states. Transitions of nuclei from the excited states to the ground state are accompanied by gamma emission, with the energy of the emitted quanta equal to the difference between the energies of the corresponding two groups of alpha particles (with correction for nucleus recoil energy).

The energy of alpha particles emitted in radioactive decay for the overwhelming majority of alpha-emitting nuclei ranges from 4 to 9 MeV, and the energy of the concomitant gamma quanta usually does not exceed 0.5 MeV. Alpha particles carry the major fraction of energy released in the decay. Only about 2% of the energy (for heavy radioactive nuclei) transforms into the kinetic recoil energy of the daughter nucleus.

Another type of radioactive decay is transformation of radioactive nuclei with preservation of their mass numbers and increase (electronic beta decay) or decrease (positronic beta decay, K-electron capture) of the charge of the nuclei formed.[1-3] The energy released in beta transformations of a radioactive nucleus ranges from 0.018 (^3H) to 16.4 MeV (^{12}N).

Electronic beta decay is characteristic of nuclei of both natural and artificial radioactive elements; it is accompanied by emission of an electron and an antineutrino, $\tilde{v}$. Owing to random distribution of decay energy between the two particles emitted in electronic decay, the energy spectrum of beta particles is continuous and covers the range from zero to the maximal energy of the beta particle. A typical example of electronic beta decay is the beta decay of tritium with a half-life equal to 12.34 years:

$$_1^3\mathrm{H} \xrightarrow{\text{12.34 years}} {}_2^3\mathrm{He} + \beta^- + \tilde{\nu}$$

Positronic beta decay is characteristic only of artificial radionuclides and is accompanied by transformation of one of the protons in the nucleus into the neutron with emission of positron and neutrino. The energy spectrum of positrons is continuous, as is that of beta particles. After escape from the nucleus, the positron unites with an electron to form two gamma quanta (0.51 MeV each).

K-electron capture is electron capture by electronic shells of the radionuclide. This competes with positronic beta decay. Capture from the closest electronic shell (K shell) is most probable, although capture from other shells (L, M, etc.) is also possible. This process is followed by electronic transitions to fill the vacancy formed in the electronic shells. Electronic transitions between shells of the forming atoms are accompanied by emission of characteristic x-ray radiation. Transition of an electron from an external electronic shell to the electronic vacancy can also occur without emission of x-ray quantum but with emission of another electron from the external (more remote than the nucleus) electronic shell (an Auger electron). The kinetic energy of the Auger electron is equal to the difference between the binding energy of the captured electron and that of the emitted electron. An example of an isotope that decays by K-electron capture is ^{55}Fe.

Electronic and positronic beta decay and K-electron capture (beta transformations), as well as alpha decay, can be accompanied by emission of gamma quanta of various discrete energies in cases when the daughter nucleus is formed in an excited state. Transition to the ground state is accompanied by emission of gamma quanta and can occur in several steps through intermediate excited levels. The energy of gamma quanta accompanying beta decay can reach 2.5 MeV.

1.1.2 Interaction of Ionizing Radiation with Matter

Ionizing radiation can interact with matter to give various effects, some of which offer a possibility of generating electrical energy. Ionizing radiation emitted in radioactive decay is a flux of charged particles or electromagnetic quanta. When passing through matter, ionizing radiation loses energy in elastic and nonelastic interactions with electrons and nuclei of atoms of the substance. In elastic scattering, the initial particles do not disappear, no new particles appear, and particles (e.g., nuclei) involved in the interaction do not change their internal energy. The total kinetic energy of particles participating in elastic interaction remains unchanged and is redistributed among these particles with changing of interacting particle motion directions. Nonelastic interaction is characterized by conversion (complete or partial) of the kinetic energy of the moving particle to other forms, e.g., to the excitation energy of atom or nucleus, radiation energy, and rest energy of newly formed particles.

In this section, the main concern is with the primary processes occurring in interaction of the ionizing radiation with matter. Secondary processes, such as luminescence, generation of electron–hole pairs, and radiolysis will be discussed in later sections of this book.

1.1.2.1 Interaction of Alpha Particles with Matter[1-3]

When passing through an exposed substance, alpha particles lose their energy in nonelastic and elastic scattering with electrons and elastic scattering with nuclei. The main mechanism of the energy loss by alpha particles is their nonelastic Coulombic interaction with substrate electrons, which causes either ionization or excitation of atoms of the substance (ionization stopping).

In each event of nonelastic scattering of alpha particle causing ionization of the atom, one or several electrons are knocked out. The part of knocked-out high-energy electrons whose energy exceeds the ionization potential of atoms (delta electrons) can cause secondary ionization. Their behavior and character of interaction with the matter are similar to those of high-energy electrons and beta particles. The alpha particle gradually exhausts its energy, mainly in ionization-stopping events, until its energy becomes comparable with the average energy of thermal motion of medium particles. Collisions of alpha particles with nuclei and deviations of alpha particles from the beam due to scattering on nuclei are rare and do not noticeably contribute to the energy loss because of the low ratio of the nucleus diameter to the atom diameter (ca. 10^{-4}).

The very low probability of elastic scattering of alpha particles on nuclei causes their trajectory to be nearly straight. The path length R', which is the distance traveled by the alpha particle in the substance, depends on its initial energy, $E_{0\alpha}$. Empirical tables and formulas of alpha particle path length in various substances are given in physical handbooks. For example, the path length of ^{226}Ra alpha particles ($E_{0\alpha} = 4.78$ MeV) is about 3.3 cm in air under normal conditions and about 33 μm in water.[2]

1.1.2.2 Interaction of Beta Radiation with Matter[1-5]

In interaction with matter, beta particles, delta electrons, and monoenergetic accelerated electrons consume and lose their kinetic energy in multiple elastic and nonelastic scattering events with atoms of the irradiated substance (ionization loss). Electrons with high kinetic energy can lose part of their energy by generating bremsstrahlung (radiation loss), which arises when an electron is decelerated in the Coulombic field of a nucleus. In each event of interaction of incident electron with matter, the change in its momentum is relatively large, which can result in significant deviations from the initial motion direction. As a result, the motion in a substance of electrons with kinetic energy less than 100 keV is chaotic and resembles diffusion rather than forward motion in the initial direction.

The total loss of electron kinetic energy as it passes through matter is a sum of ionization and radiation losses:

$$\left(-\frac{dE}{dx}\right) = \left(-\frac{dE}{dx}\right)_{ion} + \left(-\frac{dE}{dx}\right)_{rad} \qquad (1.5)$$

For nonrelativistic electrons (when $v/c \ll 1$, where v is the electron velocity and c is the light velocity), the specific ionization loss of kinetic energy of the incident electrons can be described by

$$\left(-\frac{dE}{dx}\right)_{ion} = \frac{4\pi q^4 ZN}{m_{oe}v^2} \ln \frac{m_{oe}v^2}{2J}$$ (1.6)

where q is the electron charge, m_{oe} is the electron rest mass, N is the number of atoms in 1 cm^3 of the substance, Z is the atomic number of the substance element, and J is the average ionization potential of the substance atoms. Since $N = N_A \, d/M_A$, where d is the substance density, N_A is the Avogadro constant, and M_A is the atomic mass of the substance, the specific ionization loss of kinetic energy of nonrelativistic electrons apparently increases with increasing density and atomic number (charge of atom nucleus) of the irradiated substance. At the same time, the specific ionization loss decreases with increasing kinetic energy of nonrelativistic electrons E, equal to $m_{oe} v^2/2$. The specific ionization loss of beta particles emitted by radionuclides used in nuclear batteries can be calculated by Equation 1.6; for alpha particles, similar equations can be used.

In accordance with classical electrodynamics, a decelerated electron stopping in the Coulombic field of an atomic nucleus with charge Z emits electromagnetic energy proportional to the acceleration squared. Since the Coulombic force is proportional to the product of charges of the interacting particles and acceleration is proportional to the force and inversely proportional to the particle mass, the energy emitted in the course of particle stopping is proportional to $(Z/M)^2$, where M is the particle mass. This dependence explains why the probability of energy emission by alpha particle during its stopping is lower by a factor of ca. 10^7 than in the case of the electron stopping. Bethe and Heitler found that the specific radiation loss of electron kinetic energy depends on the degree of shielding of nucleus with atomic electrons. The following relation was found valid for the examined ranges of electron kinetic energy:

$$\left(-\frac{dE}{dx}\right)_{rad} \sim Z^2 N E$$ (1.7)

According to Relation 1.7, radiation loss of electron kinetic energy increases in proportion to the squared charge Z of the nuclei of the irradiated substance, with the concentration of atoms N (and hence with the substance density), and with the electron kinetic energy.

The relation between radiation and ionization loss of electron kinetic energy can be estimated as

$$\frac{\left(-dE/dx\right)_{rad}}{\left(-dE/dx\right)_{ion}} \approx \frac{EZ}{800}$$ (1.8)

This equation shows that, for electrons with high kinetic energy (more than 0.5 MeV) in substances with high Z, the radiation loss is comparable to ionization loss. When determining the energy loss from relatively low kinetic energy electrons such as tritium beta particles, radiation loss is small compared to ionization loss. However, the amount of soft beta emitters can be estimated from the intensity of the bremsstrahlung, and health safety considerations should include the radiation loss, even from soft betas.

The true path length R' of electrons in a substance is determined from the total energy loss:

$$R' = \int_0^{E_0} \frac{dE}{dE/dx} \qquad (1.9)$$

where E_o is the initial electron energy. The true path length is the electron path length along a curvilinear trajectory. The projection of the true path length onto the initial direction of electron motion is termed the maximal path length, R'_m; this quantity can be determined experimentally. Beta particles of a given spectral distribution are commonly characterized by the maximal path length R'_m (or by the maximal depth of penetration of beta particles into substance) related to beta particles of the maximal energy ε_{max}. For aluminum, R'_m is calculated by the empirical equation

$$R'_m = 0.412\, \varepsilon_{max}^{(1.256 - 0.0954 \cdot \ln \varepsilon_{max})} \quad \text{for } 0.01 \leq \varepsilon_{max} \leq 3 \text{ MeV} \qquad (1.10)$$

In Equation 1.10, R'_m is expressed in grams per square centimeter and ε_{max} in MeV. In materials different from aluminum, R'_m can be calculated by the equation

$$R'_{m,x} = R'_{m,\text{Al}}\, \frac{(Z/M_A)_{\text{Al}}}{(Z/M_A)_x} \qquad (1.11)$$

where $R'_{m,x}$, $R'_{m,\text{Al}}$, $(Z/M_A)_x$, and $(Z/M_A)_{\text{Al}}$ are the maximal path lengths and charge-to-mass ratios for element x and aluminum, respectively.

For a continuous spectrum of beta particles with energy varying from practically zero to ε_{max}, dependence of the flux density of beta particles n_β on the substance layer thickness r is approximately exponential, where n_β° and n_β are the flux densities of beta particles in the incident beam. μ_m is the mass coefficient of beta particle absorption in cm^2/g.

$$n_\beta = n_\beta^\circ \exp(-\mu_m \cdot r) \qquad (1.12)$$

$$\mu_m = 15.5\, \varepsilon_{max}^{-1.41} \qquad (1.13)$$

1.1.2.3   Interaction of X-Ray and Gamma Radiation with Matter[1-5]

Interaction of hard electromagnetic quanta with matter is different from interaction of charged particles. First, hard electromagnetic quanta have zero rest mass and a velocity of light. Also, electromagnetic quanta have no charge and therefore are not subject to the action of long-range Coulombic forces. The probability of interaction of hard electromagnetic quanta with particles of a substance is considerably lower than for electrons and alpha particles, and the penetrating power of x-ray and gamma quanta is high.

Variation of gamma or x-ray quanta flux Φ passing through a substance of thickness D_L is described by an exponential function.

$$\Phi(D_L) = \Phi_0 \exp(-\mu D_L) \tag{1.14}$$

where μ is the linear extinction coefficient of the gamma or x-ray quanta flux in the given substance, cm^{-1}.

Interaction of x-ray and gamma quanta with matter involves significant effects: the photoelectric effect (photoeffect), Compton (noncoherent) scattering, and formation of electron–positron pairs.

The photoeffect is nonelastic interaction of gamma quanta (electromagnetic radiation quanta) with bound atomic electrons, in which the whole energy of the primary quantum is transferred to an electron of one of the atom electronic shells. As a result, the electron that took up the energy (photoelectron) is emitted by the atom with a kinetic energy equal to the difference between the energy of the primary quantum and the binding energy of this electron in the atom.

Compton scattering is elastic interaction (collision) of the incident quantum (treated as a particle) with an atomic electron. A high-energy gamma quantum can be fully absorbed in the Coulombic field of an atom nucleus or electron to generate an electron–positron pair. The major contribution to absorption of electromagnetic radiation of energy equal to hundredth and tenth fractions of a megaelectron-volt is made by the photoeffect and Compton effect. Interaction of hard photon radiation with matter results in generation of high-energy electrons whose interaction with the matter will give rise to secondary effects.

1.1.2.4   Dose and Dose Rate[6]

The energy transferred to a substance by ionizing radiation is quantitatively characterized by the absorbed radiation dose. Absorbed radiation dose D_{abs} is the ratio of the energy of charged particles or photons E, transferred by ionizing radiation in the volume element with mass m to that mass:

$$D_{abs} = \frac{E}{m} \tag{1.15}$$

The dimensional unit of the absorbed dose is gray (Gy), equal to 1 J of absorbed energy per kilogram of a substance. The rad unit is often used, where 1 rad = 0.01 Gy. The absorbed dose rate P_D is the dose absorbed in unit time:

$$P_D = \frac{D_{abs}}{t} \qquad (1.16)$$

The dimension of absorbed dose rate is gray per second (Gy/s), equal to watt per kilogram (W/kg) (1 Gy/s = 1 W/kg).

1.2 BASIC PRINCIPLES OF CONVERSION OF RADIOACTIVE DECAY ENERGY TO ELECTRICITY

Radioactive decay energy can be converted into electricity through conversion of kinetic energy of particles formed in radioactive decay to thermal energy, with subsequent conversion of the thermal energy to electrical energy. Alternatively, incidental electromagnetic radiation can be converted to thermal and electrical energy. The second way involves generation of the electrical energy without a thermal cycle for nuclear batteries of various types (direct-charge, direct-conversion, or indirect-conversion nuclear batteries).

1.2.1 Thermoelectric Converters

Devices generating electrical energy from radioactive decay energy based on a thermal cycle use radionuclide heat sources (RHSs), which are hermetically sealed containers or ampules that hold the radionuclide-containing material (radioisotope fuel) as a safe thermal source. Particles and electromagnetic radiation generated by the radionuclide decay are absorbed in the radioisotope fuel and structural material of the RHS fuel ampule, and they give off heat. The thermal power Q released at time t can be estimated as

$$Q(t) = 3.7 \cdot 10^{10} \cdot A_o \cdot \varepsilon_{av} \cdot \exp(-\lambda t) \qquad (1.17)$$

where A_o is the radionuclide activity in Ci at time $t = 0$, λ is the radioactive decay constant, and ε_{av} is the average energy of particles and quanta released in a decay event.

The characteristics of some $^{238}PuO_2$-based RHSs developed in the V. G. Khlopin Radium Institute, St. Petersburg, Russia, are listed in Table 1.1.

Of the many methods for conversion of thermal energy to electricity, the most suitable for RHSs are the dynamic (using the Renkin liquid–metal cycle or the Brighton gas cycle), thermoionic, and thermoelectric methods.[8,9]

The dynamic method for conversion of thermal energy to electricity is based on generation of electrical energy with a turbogenerator. The generator is driven by a circulating fluid in a closed circuit, which is evaporated (Renkin cycle) or heated

Table 1.1   Characteristics of some RHSs

RHS Designation	Thermal Power, W	Working Temperature, °C	Year of Development
Zemlya-1	22	400	1966
Zhizn'	300	920	1968
Sloi	2	400	1970
RHS	120	900	1971
Vysota	1000	950	1978
Pochva	20	200	1984
RHS-238-3, 7, 12	Series: 3, 7, 12	110	1986
Gemma-OKR	Series: 0.08–0.30	160	1987
RHS-238-0.22	0.22	180	1990

Source: Bartenev, S.A. et al., Radionuclides and articles thereof for science, technology, and medicine, in *V.G. Khlopin Radium Institute. On the 75th Anniversary,* Il'enko, E.I., Ed., St. Petersburg Institute of Nuclear Physics, 1997, 133 [Russian language].

(Brighton cycle) by a radionuclide fuel block.[8,9] The SNAP-1 unit developed in the 1950s in the U.S. utilized 3.5 MCi ^{144}Ce (A_m = 3183 Ci/g for 100% ^{144}Ce, $T_{1/2}$ = 284 days) as the heat source, generated 500 W of electrical power at 115 V, and had a conversion efficiency of about 10%. The theoretical efficiency limit for dynamic energy conversion is 15% for electrical output greater than 1 kW.[9] However, dependence on many moving parts, necessity of using huge quantities of radioisotope fuel, and low conversion efficiency for electrical power less than 1 kW severely restricted possible applications.[8,9]

The thermoionic method for conversion of thermal energy to electricity is based on the thermoelectron emission phenomenon. When heated to a high temperature (up to 1700°C), a cathode emits electrons that pass through alkali metal (cesium) vapor to eliminate space charge and are collected on an anode kept at a considerably lower temperature (up to 700°C).[8] The theoretical efficiency of thermoionic devices is predicted to reach 20%, with short-circuit current density as high as 100 A/cm^2, 0.7 V between converter terminals at maximal power, and continuous operation for 20,000 hours.[8] The SNAP-13 radionuclide thermoionic generator containing the radioisotope ^{242}Cm generated 12.5 W.[9] Energy conversion efficiency was not optimal because of inadequate heat insulation at very high working temperatures. At these temperatures, stringent requirements are imposed on the strength and corrosion resistance of the fuel capsule, and the presence of cesium vapor further constrains material choices for emitter, collector, and insulators.

Dynamic and thermoionic methods cannot compete with the practical thermoelectric method. In 1929, Ioffe[8] was the first to propose thermoelectric converters for generation of electrical energy. Development of radioisotope thermoelectric generators (RTGs) was initiated in the U.S. in the early 1950s.[7,9]

The thermoelectric method for conversion of thermal energy to electricity is based on the thermoelectromotive force arising from a temperature gradient between two branches of an electric circuit composed of different conductors or semiconductors. Although designs of working RTGs are diverse,[8-10] their principal scheme is similar. Figure 1.1 shows the principal scheme of an RTG with external electrical

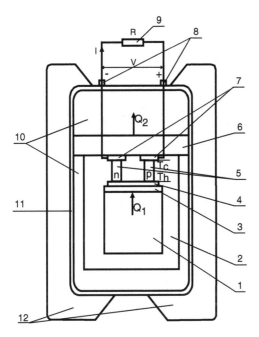

Figure 1.1   Principal scheme of RITEG: (1) RHS, (2) thermal insulation, (3) hot heat conductor, (4) commutating plate of the hot junctions, (5) semiconductor branches with different types of conductivity, (6) cold heat conductor, (7) commutating plates of cold junctions, (8) power points, (9) external electric resistance, (10) biological shield, (11) casing, and (12) cooling ribs. Designations: T_h, T_c are the temperatures of hot and cold junctions, respectively; Q_1, Q_2 are heat power emitting by RHS and dissipated heat power, respectively.

load. For optimal utilization of thermal power, several RHSs are arranged in a fuel container in the center of the RTG. The hot junctions of the thermocouples commutated in blocks are in thermal contact with the side or butt surfaces of the fuel container.

The cold junctions of the thermocouples are cooled by heat removal through the heat conductor, casing, and cooling ribs of the RTG, which also includes devices for thermal and electric control (not shown in Figure 1.1). These devices are intended for stabilization of the electric parameters of the RTG at a preset working level, since the generated thermal and electric power decrease in the course of operation according to radioactive decay law. To reduce the dose rate of ionizing radiation to a safe level, RTGs are equipped with a biological shield; its material and design depend on the kind and activity of radioisotope fuel.

RTGs are used in various autonomous devices. Among such devices are electric cardiostimulators, autonomous power sources for optical and radio beacons, meteorological stations, deep-sea buoys, and spacecraft electronics. The main characteristics of RTGs are listed in Table 1.2. The electric power generated by RTGs ranges from 10^{-3} to 10^2 W; the efficiency of energy conversion is up to 6%.[8-10] RTGs often utilize the radioisotope ^{90}Sr because of its relatively low cost and availability[11]; ^{238}PuO$_2$ is preferred in spacecraft RTGs and for use in electric cardiostimulators.

Table 1.2 Characterization of RITEGs

Designation, Country	Power, W — Thermal	Power, W — Electrical	Voltage, V	Efficiency, %	Radionuclide	Fuel Loading, Ci (g)	Service Life, Years	Mass, kg	Ref.
SNAP-3B7, U.S.	52	2.7	3.5	5.2	^{238}Pu	1,600	5	2.1	9, 10
SNAP-7B, U.S.	1440	68	12	4.7	^{90}Sr	225,000	10	2090	9
SNAP-7C, U.S.	256	11.6	5	4.5	^{90}Sr	40,000	10	850	9
SNAP-11, U.S.	396	19	3	4.8	^{242}Cm	(6.2)	0.5	7.55	9
SNAP-17, U.S.	—	30	—	—	^{90}Sr	—	5–10	11.4 (without protection)	9
SNAP-27, U.S.	—	63	—	—	^{238}Pu	—	1	14	9
RTG-3, U.S.	—	1	—	—	^{238}Pu	—	20	4.4	9
RIPPLE-1, GB	—	0.075	—	1.71	^{90}Sr	—	—	600	9
Beta-3, USSR	265	12	12	4.5	^{90}Sr	40,000	10	250	8
Beta-h, USSR	208	10	6	4.8	^{90}Sr	31,000	10	156	8
G-90-60/40, USSR	1650	60	40	3.6	^{90}Sr	250,000	10	1200	8
Ritm, USSR	0.2	10^{-3}	1	0.5	^{238}Pu	—	10	0.050	8

The major factor restricting application of RTGs is that they require large amounts of radiotoxic nuclides such as ^{90}Sr and ^{238}Pu. Small quantities of the isotopes do not generate sufficient thermal gradients. On the other hand, the use of pure beta emitters with energy less than 200 to 300 keV is relatively safe. Even large amounts of radionuclides such as tritium and ^{63}Ni do not require heavy biological shields. Although the power released in beta decay of these radionuclides is insufficient for their use in RTGs, their beta radiation can be used for energy generation in nuclear batteries.

1.2.2 Direct-Charge Nuclear Batteries

The operational principle of direct-charge nuclear batteries is based on the fact that the voltage across the battery electrodes (emitter and collector) is provided by direct collection of charged particles on one of the electrodes (collector). Direct-charge nuclear batteries allow high voltages (up to hundreds of kilovolts) to be obtained at small currents (nanoamperes) determined by the rate of the radionuclide decay. The electricity is discharged by close of the circuit through a working load.

In the simplest case, such a nuclear battery consists of two concentric, coaxial, or parallel electrode surfaces insulated from each other and separated by an evacuated space or a space filled with a dielectric.[8,9] A radioactive substance emitting charged particles can be applied as a thin layer on the surface of one of the electrodes (emitter)[12,13] or, if it is gaseous (tritium, ^{85}Kr), placed in the sealed interelectrode space of the battery.[9,14] Some of the charged particles formed by the radioactive decay and ejected toward the collector are collected on its surface. The charge transfer to the collector at the open electric circuit can continue (in the ideal case) until voltage across the electrodes reaches the value close to the maximal kinetic energy of the charged particles emitted by the radioactive substance.

To reach the collector, the charged particles must overcome the electrostatic field of like charge already built up. Very high voltage is limited by the internal resistance of the nuclear battery components, even in the open-circuit state. Charge is lost by leakage through the insulator surfaces (in the case of evacuated interelectrode space) or through the surface of the dielectric separating the electrodes. Therefore, the maximal voltage V_{oc} generated at open circuit depends on the energy of the charged particles emitted by the radioactive substance and the nature of the dielectric separating the electrodes. V_{oc} can be calculated by Equation 1.18[9]:

$$V_{oc} = R_i I \tag{1.18}$$

where R_i is the internal resistance of the battery and I is the charge current due to the radionuclide decay.

When the battery electrodes are closed through a loading resistance R_e, the current I passes and the voltage V decreases to the value given by Equation 1.19.[9]

$$V = R_{sum} I = \frac{R_e R_i I}{R_e + R_i} \tag{1.19}$$

In this case, the power P of the nuclear battery whose emitter gives off charged particles in an angle of 2π can be estimated by Equation 1.20[8]:

$$P = I \cdot V = \eta \, D_L \, P_v \, S \tag{1.20}$$

where I and V are the working current and voltage, respectively, η is the efficiency of the nuclear battery, D_L is the radionuclide layer quantity, P_v is the specific (per unit volume) power of the radionuclide, and S is the area of the emitting surface of the emitter.

Equation 1.20 shows that, to increase the power of direct-charge nuclear batteries, it is necessary to increase the working surface area of the electrodes and battery efficiency. This can be done by choosing the radionuclide layer thickness and the emitter thickness appropriately, so absorption of charged particles in these layers is minimal and emission of charged particles from the emitter surface occurs in a solid angle of 4π at maximal current densities.

The first nuclear battery operating according to this principle was suggested by Moseley in 1913.[15] For the emitter, he used a thin-walled spherical quartz ampule filled with radium. The ampule walls retained alpha particles but transmitted beta particles. This ampule was concentrically fixed with a thin quartz rod inside a sphere, with the silver-plated inner surface serving as collector of beta particles. After evacuation of the space between the sphere surfaces, an open-circuit voltage of 150 kV was obtained corresponding to electric breakdown on the insulator surface. The current at electrode closure was 10^{-11} A.

In a vacuum nuclear battery based on ^{90}Sr and ^{90}Y and developed by Linder,[12] the emitter of beta particles was a thin-walled (about 20 μm) complex-shaped tubular structure with spherical ends. The inner surface of this structure was coated with a layer of the radioisotope. The output open-circuit voltage of this battery reached 365 kV and the short-circuit current was about 1 nA. The efficiency of Linder beta radiation utilization (relative to the total amount of beta particles formed in radio-active decay) was about 75%, while the efficiency of Moseley's battery was 8%.[8]

As a practical example of a nuclear battery utilizing alpha emitters for generating high voltage across the electrodes, a design resembling a triode has been constructed and characterized.[16] The design includes a cylindrical emitter coated with ^{210}Po on its external surface, coaxially fixed inside a cylinder of a larger diameter (collector) and separated from the collector by a control grid. A negative potential of several hundred volts, fed to the control grid, suppresses the current of secondary electrons arising from interaction of alpha particles emitted by ^{210}Po with the emitter matter. The open-circuit voltage across the electrodes of this nuclear battery was 50 kV at a control grid voltage of −800 V and residual pressure in the interelectrode space of about 0.1 Pa.[8] The characteristics of other direct-charge batteries are listed in Table 1.3.[8,9]

Direct-charge nuclear batteries produce a high voltage (tens and hundreds of kV) and operate in the pulse mode at the engineered breakdown target; the electric power generated by them ranges from micro- to milliwatt, since the current is proportional to the flux of charged particles and does not exceed fractions of milliampere. The

Table 1.3   Characteristics of Some Direct-Charge Nuclear Batteries

Radionuclide	Activity, Ci	Interelectrode Space	emf, V	I_{sc}, A	Mass, kg	Service Life, Years
^{3}H	0.2	Vacuum	1200	$5 \cdot 10^{-10}$	0.02	Weeks
^{85}Kr	0.04	Polystyrene	3000	10^{-10}	0.03	5–10
^{85}Kr	0.3	Polystyrene	5000	10^{-9}	0.03	5–10
^{85}Sr	0.02	Polystyrene	2000	$5 \cdot 10^{-11}$	0.15	>5

Source: Corliss, W.R. and Harvey, D.G., *Radioisotopic Power Generation*, Prentice-Hall, Inc., Englewood Cliffs, NJ, 1964.

extremely high efficiency of this approach suggests the value of designing the electronic load around this pulse mode capacitor.

1.2.3   Direct-Conversion Nuclear Batteries

Research on direct conversion of radioisotope decay energy followed several lines,[9] including nuclear batteries with contact voltage electrodes. In such batteries, any kind of radiation (alpha, beta, or gamma) causes volumetric ionization of the gas filling the space between two metal electrodes that have different work functions. The electrode contact potential difference creates an electric field carrying electrons and positively charged ions in opposite directions. Upon switching in an external load, an electric current passes in the circuit, depending on the kind and intensity of ionizing radiation as well as on the nature and pressure of the interelectrode gas, electrode material, etc. These nuclear batteries using 10 mCi ^{90}Sr created a voltage of about 1 V with short-circuit current of $4 \cdot 10^{-10}$ A.[9] However, the energy conversion efficiency for such batteries was low (0.5%), mainly because of high average energy of ion pair formation in the gas (about 30 eV).

Much greater promise is offered by beta flux irradiation of semiconductor elements of different conductivity types (*p–n* or *p–i–n* junctions). This is based on separation of the electron–hole pairs originating on exposure of the semiconductor materials by an electric field created by *p* and *n* layers of the *p–n* or *p–i–n* junctions. As a result, the *n*-region charges negatively and the *p*-region charges positively. As every beta particle creates in a semiconductor material up to several tens of thousands of electron–hole pairs, *p–n* junction–based devices convert a small number of high-energy beta particles to a much greater current of low-energy electrons. However, not all electron–hole pairs created by beta radiation are involved in the current generation in the external circuit. The factors responsible for electron–hole pair loss are analyzed in Section 7.2 in Chapter 7.

The initial stage in conversion of the ionizing radiation energy (subsequent discussion will concern only beta radiation) to electrical energy consists in the outlet of the beta flux from the radionuclide-containing substance. The betas of the radionuclides typically employed in nuclear batteries have a fairly low energy whose portion will be lost by absorption in a carrier substance. Where the efficiency of conversion of the total beta energy to a beta flux energy at the source surface is η_β (the remainder being self absorbed) and the efficiency of conversion of the beta flux energy to electrical energy is $\eta_{\beta\text{-}el}$, the efficiency of the direct energy conversion is equal to the product

$$\eta_d = \eta_\beta \cdot \eta_{\beta-el} \qquad (1.21)$$

The efficiency of the current generation in the external circuit $\eta_{\beta\text{-}el}$ can be defined as the ratio of the electrical power generated under optimum external load to the beta flux power absorbed in the semiconductor. (This definition is not the same as the ratio of electrical power to the activity of the source, which is included in the term η_β.) Owing to the loss of electron–hole pairs, the $\eta_{\beta\text{-}el}$ parameter is much less than unity. The dependence of $\eta_{\beta\text{-}el}$ on the band gap energy of the semiconductor has been theoretically calculated.[17] This calculation shows that for semiconductors with the band gap energy E_g of 1.9 eV (AlGaAs), $\eta_{\beta\text{-}el}$ can reach 20 to 22%, and for those with 1.1 eV (Si), $\eta_{\beta\text{-}el}$ can reach 13 to 14%. At low excitation levels characteristic of a semiconductor exposed to tritium beta particles, the energy conversion efficiency $\eta_{\beta\text{-}el}$ is equal to 15% and 7 to 8% for AlGaAs- and Si-based semiconductors, respectively. The practically achievable $\eta_{\beta\text{-}el}$ values for selected betavoltaics are given in Table 1.4.

A typical scheme of a device generating electrical energy via exposure of a semiconductor converter to ionizing radiation is shown in Figure 1.2. It includes a source of radioactive radiation and a converter. Different designs utilize diverse sources of ionizing radiation and betavoltaic converters. The performance characteristics of selected devices are presented in Table 1.4. These refer, for the most part, to prototype models of direct-conversion betavoltaics. Commercial "beta cell" (McDonnell- Douglas) batteries are included in Table 1.4. The beta cell design is shown in Figure 1.3a, and its current-voltage characteristic is given in Figure 1.3b.

Table 1.4 also presents the efficiencies of conversion of energy of the beta flux from gaseous tritium to electrical energy by betavoltaics such as GaP, $Al_{0.1}Ga_{0.9}As$, and amorphous silicon (items 5, 6, and 7). The same table presents the characteristics of two models of direct-conversion nuclear batteries with solid-state tritium-based beta sources: those with titanium-tritide beta source (item 9) and with tritium incorporated into the i-layer of the n–i–p silicon converter (item 8). Table 1.4 shows that the greatest efficiency is exhibited by wide-band gap betavoltaics based on GaP and AlGaAs ($\eta_{\beta\text{-}el}$ of 5 to 6%).

Thus, nuclear batteries employing semiconductors converting decay energy to electrical energy via p–n junction significantly surpass in conversion efficiency those employing the contact voltage. Their conversion efficiency can reach several percent. For tritium-based nuclear battery models, the current density generated by the betavoltaic can reach 1 $\mu A/cm^2$, and for the open-circuit voltage of serially connected betavoltaics, it can reach several volts.

1.2.4   Indirect-Conversion Nuclear Batteries

Besides research into direct conversion of the energy of ionizing radiation into electrical energy, research into indirect energy conversion has also been performed

Table 1.4 Characteristics of Direct-Conversion Betavoltaics

No.	Battery Units		Battery Parameters					
	Radiation Source	Converter	P_m, μW	$\left(\dfrac{dP_m}{dS}\right)$: W/cm²	V_{oc}, V	I_{sc}, μA (i_{sc}, μA/cm²)	Diameter × height mm × mm	η_d, % ($\eta_{\beta-el}$, %)	Ref.
1	⁹⁰Sr - ⁹⁰Y, A = 50 mCi	Silicon, n–p	0.8	—	—	—	—	0.94	8
2	¹⁴⁷Pm, A_m = 680 Ci/g	Silicon, n–p, junction depth 3 μm	—	—	—	(10)	—	(1.5)	8
3	¹⁴⁷Pm (Pm_2O_3)	Silicon n–p (package of converters)	43		1.79	44	1.2 × 0.78	1.04	8
4	¹⁴⁷Pm (Pm_2O_3)	Silicon n–p (package of converters)	212		4.75	77	1.32 × 1.57	0.84	8
5	Tritium, gas, p = 1.03 MPa, D_L = 2 cm	GaP	(0.68)		1.05	(1)	—	(6)	18
6	Tritium, gas, p = 0.206 MPa	$Al_{0.1}Ga_{0.9}As$, Converter at the center of a hemisphere with D_S = 2 cN	(0.4)		0.5	(1)	—	(5)	19
7	Tritium, gas	Amorphous silicon, n–i–p	0.129		0.44	0.58	—	(1.2)	20
8	Tritium, in Si-³H, 10 at.% of tritium (5·10²¹ at/cm³), at D_L = 1 μm, A_S = 0.024 Ci/cm²	Amorphous silicon, n–i–p, tritium in a composition of the i-layer.	(6·10⁻⁵)		0.089	(9·10⁻⁴)	—	0.007	21
9	Tritium, T³H₂, A_S = 0.22 Ci/cm²	$Al_{0.25}Ga_{0.75}As$	(0.015–0.027)		0.6	(0.040–0.058)	—	(3–5)	19

Note: Designations: P_m is the electrical power released under optimal load; V_{oc} is the open-circuit voltage; I_{sc} is the short-circuit current; i_{sc} is the short-circuit current density; $\eta_{\beta-el}$ is the efficiency of conversion of the beta particle energy to electrical energy by the betavoltaic; η_d is the direct conversion efficiency.

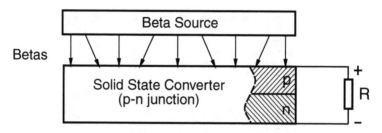

Figure 1.2 Scheme of direct-conversion betavoltaic.

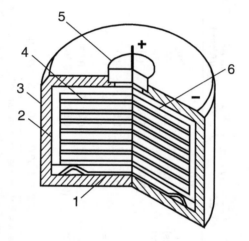

Figure 1.3a Betacel battery design: 1) spring; 2) insulating capsule; 3) case; 4) radiation source; 5) sealed contact; 6) silicon converter. (From Kodyukov, V.M. et al., *Radioisotope Sources of Electrical Energy*, Fradkin, G.M., Ed., Atomizdat, Moscow, 1978 [in Russian].)

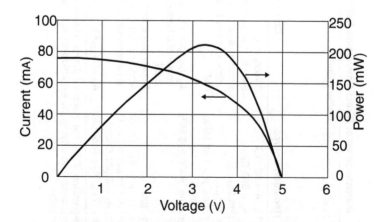

Figure 1.3b Current-voltage characteristic of betacel battery. (From Kodyukov, V.M. et al., *Radioisotope Sources of Electrical Energy*, Fradkin, G.M., Ed., Atomizdat, Moscow, 1978 [in Russian].)

since the 1950s. This method consists of conversion by radioluminescent materials of the energy released in the decay to energy of electromagnetic (light) radiation to be further converted to electrical energy by a photovoltaic. This is a two-stage conversion: radioactive radiation → light → electrical energy. This method may appear less efficient than the one-stage direct conversion: radioactive radiation → electrical energy. However, it has advantages in reduced radiation influence and in radiation protection of the sensitive photovoltaic element.

In both schemes, the initial stage of the energy conversion consists in the outlet of the beta flux from the radionuclide-containing carrier substance to its surface with conversion efficiency η_β. Conversion efficiency of beta particle energy to light energy is defined as $\eta_{\beta\text{-}l}$, and the efficiency of conversion of the light energy to electrical energy by a photovoltaic is defined as $\eta_{l\text{-}el}$. Hence, the indirect energy conversion efficiency η_{ind} can be defined by:

$$\eta_{ind} = \eta_\beta \cdot \eta_{\beta-l} \cdot \eta_{l-el} \tag{1.22}$$

With appropriately chosen beta source designs, fairly high η_β values can be achieved. For example, in certain modifications of tritium gas–filled radioluminescent light sources, η_β is 0.84 (see Section 2.2.4.3 in Chapter 2). The efficiency of conversion of radioactive radiation to light by a luminescent material $\eta_{\beta\text{-}l}$ can reach 0.25 (see Section 3.1.1.1 in Chapter 3) and that of conversion of light energy to electrical energy on illumination of semiconductor converters to radioluminescent light sources can reach up to 0.35 (see Chapter 7). Hence, the conversion efficiency of indirect conversion η_{ind} can attain 0.07. At the same time, Table 1.4 shows that, for direct-conversion breadboard batteries, $\eta_{\beta\text{-}el}$ does not exceed 0.06 and η_d slightly exceeds 0.01. This suggests that optimization development for indirect conversion schemes is justified. Constructed prototypes of indirect conversion batteries are comparable in conversion efficiency to models of direct-conversion batteries.

Conceptual designs of betavoltaic batteries differ from each other in the type of the radioluminescent light source, radionuclide, and photovoltaic employed. The radionuclides most frequently used in such batteries are [147]Pm and [3]H. Early designs of the batteries utilized silicon, selenium, and cadmium sulfide–based photoelectric cells. Recently, photoelectric cells based on A^3B^5 compounds have been preferentially used.

The radioluminescent light source can be represented by:

- A mixture of powdered radionuclide-containing substance and a luminescent material[22]
- A powdered phosphor with a radionuclide incorporated in its crystal lattice[8]
- A mixture of gaseous radionuclides ([3]H$_2$ or [85]Kr) with inert gases (Ar, Kr, Xe) or inert gas–mercury vapor mixtures[22,23]
- Dusty solid particles containing alpha- or beta-emitting radionuclide, homogeneously dispersed in an inert gas (Xe); aerosols[11]
- A hermetic glass capsule of any shape whose inner surface is coated with a phosphor and whose cavity is filled with gaseous tritium[22,24-27]

- A panel comprising empty microspheres, filled with phosphor particles and gaseous tritium at elevated pressure[28]
- An aerogel–phosphor composition saturated with tritium[29]
- Other compositions, e.g., a mixture of an organic tritium–containing compound with an organic luminescent material (see section 2.3.6 in Chapter 2)

Electrical energy generation efficiency for any design of indirect conversion battery depends on how well matched the maxima of the spectral response curve of the semiconductor converter (photovoltaic) is to the emission maxima of the luminescent material.[8,11]

Selected schemes of indirect-conversion betavoltaic batteries are shown in Figures 1.4a to 1.4d. Table 1.5 presents characteristics of the betavoltaic batteries employing various light sources and photovoltaic converters, including activity of the radionuclide A, maximal electrical power P_m, volume of the battery v, specific (per unit activity) electrical power Q, and energy conversion efficiency η_{ind}. The parameter Q characterizes the efficiency of the use of the radionuclide activity and is defined as the P_m-to-A ratio. It is related to the energy conversion efficiency as

$$Q = P_{sp}\eta_{ind} \qquad (1.23)$$

where P_{sp} is the specific power released in the decay of 1 Ci of the radionuclide (data in Table 1.5 are initial values of the parameters of interest; the size of the battery is given without structural and enclosure units).

The [147]Pm-based betavoltaic battery whose configuration is shown in Table 1.5 (item 1) was manufactured in the late 1950s[8] as the Elgin–Kidde atomic battery. It was characterized by a fairly high promethium-147 quantity and a service life of only 3 years. Thus, it had limited application.

The parameters of the batteries with a gas–dust mixture of the radioisotope-containing substance and a noble gas were calculated theoretically and presented in items 2 to 7 in Table 1.5. In this battery design, the decay energy is utilized efficiently. However, realization of such a device requires solving a number of problems. The first is to create a homogeneous stable gas–dust mixture and maintain it in this condition for a fairly long time (10+ years). The second problem is to create a wide-band-gap photovoltaic to convert vacuum UV radiation energy to electrical energy with high (up to 50%) efficiency. Notably, absorption of ionizing radiation by xenon causes radioluminescence with λ_{max} = 172 nm. Such a converter can be based on doped diamond or aluminum nitride.[11] According to Baranov et al.,[11] prerequisites to finding a solution to these problems and creating this nuclear battery design exist.

The performance characteristics (radionuclide activity A, maximal electrical power P_m, and volume v) of the batteries with an aerogel–phosphor composition as the light source (items 8 and 9 in Table 1.5) were also calculated theoretically.[30] Such light sources with energy radiosity of 23 μW/cm^2 exhibit very unstable performance characteristics, owing to the high specific content of tritium.[30] Their radiosity decreases tenfold within 250 days.[29] The aerogel–phosphor composition,

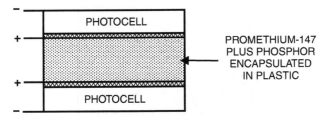

Figure 1.4a Phosphor + promethium oxide mixture–filled light source–based battery. (From Walko, R.J. et al., Electronic and photonic power applications, in *Radioluminescent Lighting Technology. Technology Transfer Conference Proceedings*, U.S. DOE, Annapolis, MD, 1990, 13-1.)

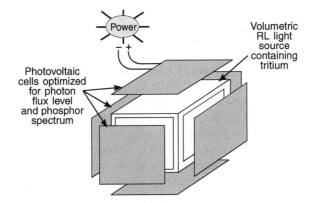

Figure 1.4b Aerogel–phosphor composition–filled (volumetric) light source–based battery. (From Walko, R.J. et al., Electronic and photonic power applications, in *Radioluminescent Lighting Technology. Technology Transfer Conference Proceedings*, U.S. DOE, Annapolis, MD, 1990, 13-1.)

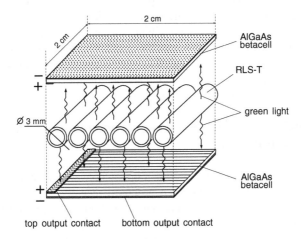

Figure 1.4c Tritium gas–filled light source–based battery.

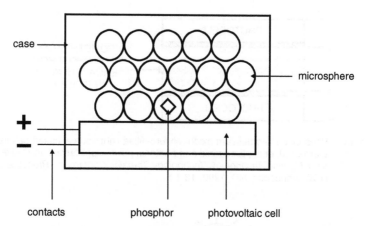

Figure 1.4d Self-luminous microspheres–containing light source–based battery. (From Rivenburg, R.C. et al., U.S. Patent 5,443,657, 1995.)

saturated with tritium at a pressure of 1 rather than 100 atm (as in the case above), is characterized by about 100 times less tritium and 50 to 100 times less radiosity.

At the same time, according to Renschler et al.,[29] radiosity of the light source decreases over 4 years proportionally to decreasing tritium activity. Apparently, with such light sources it will be possible to manufacture a battery with specific (per unit activity) electrical power, Q, of 0.5 µW/Ci. Although the specific power of this battery would be 100 times smaller than that of the light source with energy radiosity of 23 µW/cm^2, the fact that its specific power degrades only proportionally to the tritium decay makes such an option preferable.

The characteristics of actual models of batteries employing tritium gas–filled light sources are presented in Table 1.5 (item 10). The electric power of such batteries can be improved by no more than two- to threefold by optimizing the design and components employed, e.g., by utilizing converters (photovoltaics) with higher conversion efficiency, $\eta_{l\text{-}el}$. The batteries employing microsphere illuminators are relatively powerful and miniature. At the same time, lack of data on loading activity and long-term luminosity prevents unambiguous characterization of this device.

Table 1.5 suggests a relatively high efficiency of indirect conversion of radioactive decay energy to electricity compared to other conversion methods. The potential of indirect conversion will be unambiguously demonstrated only when better prototypes of the sort described are built and tested.

1.2.5 Light-Concentration Schemes for Indirect-Conversion Nuclear Batteries

In spite of the long history of the indirect conversion approach to nuclear batteries, some new designs have been proposed recently. Waveguide-based light concentration schemes look very promising and will be discussed. One of the major barriers to nuclear battery commercialization is self-absorption of the ionizing radiation in direct- and indirect-conversion schemes requiring large semiconductor

Table 1.5 Parameters of Indirect-Conversion Betavoltaic Batteries

	Battery Units			Battery Parameters					
No.	Isotope	Light Source	Photovoltaic	W, Ci	g_m, μW	v, cm³ (S^a, cm²)	Q, μW/Ci	η_{ind}, %	Ref.
1	^{147}Pm	Luminophore based on CdS with ^{147}Pm incorporated into the lattice	Silicon	3.3	12	—	3.64	1	8
2	^{90}Sr	Xenon containing a dust of radioactive material	Doped diamond, aluminum nitride	$6 \cdot 10^6$	10^{10}	$10^6 - 10^7$	$1.7 \cdot 10^3$	25	11
3	^{208}Po			$1.3 \cdot 10^6$			$7.7 \cdot 10^3$	25	
4	^{228}Ra			$3.5 \cdot 10^5$			$2.8 \cdot 10^4$	—	
5	^{228}Th			$2.2 \cdot 10^5$			$4.5 \cdot 10^4$	—	
6	^{238}U			$1.9 \cdot 10^5$			$5.3 \cdot 10^4$	—	
7	^{238}Pu			$1.2 \cdot 10^6$			$8.3 \cdot 10^3$	25	
8	^{3}H	Aerogel–phosphor (ZnS) composition saturated with tritium	A³B⁵ (GaAs, AlGaAs, GaAsP, GaP)	3200	2000	32	0.625	1.8	30
9	^{3}H	Aerogel–phosphor (ZnS) composition saturated with tritium	a-SiH	5800	2000	60	0.345	1	30
10	^{3}H	Gas-filled tritium light sources	AlGaAs	0.5 / 1.2	0.09 / 0.15	(3.4) / (3.4)	0.18 / 0.12	0.5 / 0.4	See Section 7.2.8 (Chap. 7)
11	^{3}H	Panel comprising microspheres 0.025 cm in diameter	AlGaAs	—	50	(7.5)	—	—	28

a S is the photovoltaic converter area.

surfaces to collect tiny radiation flux. Since semiconductor cost dominates battery cost, the large semiconductor surface area increases the cost-to-power ratio. Direct conversion also has a central problem of semiconductor radiation damage, which suggests the value of indirect conversion designs in which the semiconductor is not exposed to highly ionizing radiation.

Use of light concentration schemes based on waveguides improves indirect design by increasing the intensity of light incident on the photovoltaic cell. This greatly improves the efficiency of the photovoltaic, which is essential for miniaturization and cost reduction. Multilayer structures needed for practical energy output are cheaper to fabricate when one of the repeating components, in this case the photovoltaic cell, is removed from the structure and placed outside. Such a design better protects the photovoltaic from radioisotope diffusion, which can radiolytically and chemically damage the semiconductor.

Use of waveguide-based light concentration is well known in the technology of solar cells.[31] Large-area flat waveguides luminesce under excitation by sun radiation. Concentrated light is emitted at the edge of the waveguide and coupled to the photovoltaic; the effect is to increase the light flux at the surface of the photovoltaic. The same idea may be used for a radioluminescent light source in which a waveguide luminesces under exposure to ionizing radiation.

1.2.5.1 Nuclear Battery Design

Figure 1.5 shows a nuclear battery design using waveguide principles.[32] Hermetic sealing of the radioisotope with a tritium getter such as 1,4-bis(phenylethynyl)benzene (DEB) prevents accidental isotope leakage. Scintillation glass waveguides are coated with thin metal mirrors (or a high difference refractive index barrier) and a radioisotope or its compound. Waveguides may be in the form of fibers or plates. The isotope or its compound generates beta radiation; the beta particles penetrate the scintillating waveguide to generate photons that are piped to the emitting edge. A radiation-hard borosilicate glass window is optically coupled between the waveguide and photovoltaic, preventing radioisotope diffusion to the semiconductor. The spectral response of the photovoltaic cell should be matched to emission of the scintillating material. The radioisotope may be advantageously incorporated into the waveguide; for example, promethium oxide may be incorporated into the glass.

Advantages of the nuclear battery using scintillating waveguides are:

- The semiconductor is not exposed to ionizing radiation.
- Photovoltaic cell conversion efficiency is increased by an increase of light flux.
- Smaller photovoltaic cells offer miniaturization and cost reductions.
- Waveguides may be fabricated in device-compatible shapes with thickness equal to the penetration depth of isotope betas, thereby allowing further miniaturization.
- Emission spectra of radioluminescent material can be tuned to better match spectral sensitivity of the photovoltaic cell, greatly increasing efficiency.
- There is a possibility of incorporating the isotope into the scintillating material to improve power density and efficiency.

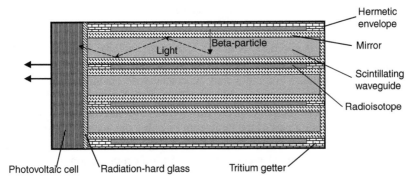

Figure 1.5   Schematic view of the radioluminescent waveguide–based light concentration battery design.

Scintillation glass will work if the radioluminescent light is successfully piped to the emitting edge of the waveguide, which is possible when several conditions are fulfilled. Material for the waveguide should not absorb or scatter the radioluminescent light; this light must be reflected on the interface between waveguide and the surrounding media. The first option is to coat the waveguide with a metal mirror, which should have high reflectivity. The problem is that even a thin film of the metal will absorb part of the useful beta radiation. To minimize this effect, metal with low Z (atomic number) is needed. Aluminum meets both requirements. Another option is to put the waveguide having refraction coefficient n_1 into the surrounding media with lower refraction coefficient n_2. Then part of the radioluminescent light emitted at the angle lower than critical will undergo multiple total internal reflection and be piped to the edge. (See Figure 1.6.)

Light emitted within the solid angle 2Φ escapes the waveguide and is therefore lost for useful utilization. The value of critical angle is defined as follows:

$$\Phi = \text{arc sin}\left(n_2/n_1\right) \tag{1.24}$$

For example, if the waveguide is made of glass or plastic with $n_1 = 1.6$ and surrounded by gas with $n_2 = 1$, then $\Phi = 39°$. Trapping efficiency, η_{trap}, is the efficiency of light transferred through the waveguide[31]:

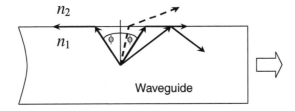

Figure 1.6   Scheme of light propagation in the waveguide.

$$\eta_{trap} = \cos \Phi \qquad (1.25)$$

In this example, $\eta_{trap} = 0.78$. If $n_1 = 1.7$, then $\eta_{trap} = 0.81$. Thus, about 80% of radioluminescent light may be trapped in the waveguide. Instead of use with gas as the surrounding media, waveguides may be coated with cladding with a very low n_1 value. For example, silica aerogel has a refraction index close to 1.[33]

When light reaches the edge of the waveguide, total internal reflection plays a negative role, preventing part of the light from escaping the waveguide and reaching the photovoltaic. For better output, the emitting edge must be optically coupled to the photovoltaic with optical grease or glue. Antireflection coatings may also be useful.

1.2.5.2 Suitable Radioluminescent Materials

Light concentration schemes require efficient scintillators. Among the most efficient radioluminescent materials known are A_2B_6-based phosphors. Hamil and co-authors[34] suggested use of ZnS radioluminescent phosphor waveguides in a form of thin sheets or long whiskers mounted in a sealed envelope filled with tritium gas. However, A_2B_6 compounds are not very transparent for the emitted light, and self-absorption is a problem preventing concentration since, in the waveguide, light must travel a significant distance. To overcome the problem of light self-absorption, waveguiding plates have been tried; these are made of glass or other transparent material on which vapor-deposited thin films of radioluminescent phosphor have been deposited.[35] The distance that light can pass is increased, since most of the distance to the voltaic is in the glass.

To further decrease absorption losses, phosphor with better transparency to its own light has been used: cerium and europium doped-calcium sulfide. Use of comparatively thick plates will limit miniaturization and concentration possibilities. Using tritium gas will not allow very high specific power unless elevated pressures are used, which is not safe or practical.

Another class of efficient radioluminescent materials is alkali halides such as NaI(Tl) and CsI(Tl). This class of materials possesses low light absorption. Designs of radioluminescent light sources have been based on polished CsI(Tl) planar waveguides.[36] The authors fabricated a "sandwich" of monocrystal plates 35×60 mm in size and 0.4 to 0.5 mm thick. Between them, ^{35}S was dispersed in poly-vinylalcohol. However, fabrication of thin alkali halide waveguides by polishing is costly and time consuming. Since the waveguide thickness must be comparable to the penetration depth of betas (or alphas if alpha emitter is used), use of tritium would require micrometer layers. An additional disadvantage of alkali halides is that they are hygroscopic and would require moisture protection. It is possible to utilize plastic scintillators that are used in the detection of penetration radiation, but their radiation stability is poor.

Scintillating glass is nearly ideal as a waveguiding material for this application. Glasses have good optical properties and have been used in detection and visualization of penetration radiation. Li-containing glasses are used for the detection of neutrons. Bicron GS1 glass is suggested for detection of weak beta emitters like ^{14}C.[37] Threshold energy for the glass radioluminescence is above an excitation energy of approximately 2.5 eV.[38] Radioluminescent glasses are also widely used for the detection and visualization of x-rays.[39]

Use of Tb-doped radioluminescent glass fibers as electron scintillators has been described.[40] Fiber thickness was 6 to 16 µm and electron energies were 100 to 400 keV; the experiment proved that glass is an appropriate candidate for radioluminescent light concentration scheme. For improvement of radioluminescence efficiency, Tb-doped glasses may be co-doped with gadolinium oxysulfide.[41] Radioluminescence efficiency of glasses under the beta exposure reported so far does not exceed 2 to 3%; however, recent significant progress allows one to hope that it will be further increased. One promising way may be to incorporate nanoparticles of efficient radioluminescent materials into the porous glass; for example, ZnS nanoparticles have been incorporated into glass.[42]

Optical properties of glasses are quite good, but exposure to ionizing radiation glasses darkens glass due to formation of color centers.[43] As a result, the optical absorption edge of the glass, typically in the range of 300 nm and far from the glass emission band, moves toward longer wavelengths. The radiation stability of the glass may be significantly improved by doping with cerium dioxide. Addition of up to 2% by weight cerium dioxide significantly decreases the rate of the degradation process because cerium easily changes its valence. The Ce^{4+} ion has a valence electron that reacts readily with free electrons formed as result of radiation:

$$Ce^{4+} + e^- \rightarrow Ce^{3+} \qquad (1.26)$$

This prevents formation of the color center. Three-valence cerium in turn reacts with holes, preventing formation of the hole-based coloring center.[44] Another improvement of radiation stability of the glass may be achieved if the position of the radioluminescence emission maximum is far from the absorption edge. For example, 3 million rad x-ray exposure of the Type IQI 301 Tb-doped silicate glass caused only a several-percent increase of the adsorption of the 550-nm light emitted by terbium ion.[45] Annealing at 375°C for 4 h fully restored the glass.

An additional advantage of luminescent glasses is that they may be RF-magnetron sputtered to form flat waveguides that may be very useful for fabricating multilayer waveguide/isotope configurations.[46]

1.2.5.3 Experiments with Scintillating Glass

For the practical nuclear battery, choice of the isotope is critical. Radioisotopes should possess intermediate decay rates with the necessary useful life, high specific activity, a minimum of gamma radiation with high penetration depth and requiring complex protection; low cost, availability, acceptable regulatory and safety restrictions, and a convenient fabrication technology. Alpha emitters provide the highest energy density but degrade crystalline and amorphous materials. Tritium is an acceptable isotope due to moderate decay time, availability, and low cost. The soft betas may be used without extraordinary radiation shielding. Therefore, for these experiments, tritium was used. Gaseous tritium limits miniaturization, so bound isotope is preferable, even though it increases radiation source self-absorption. A useful solid form is titanium tritide, obtained by saturation of a fresh thin layer of titanium with tritium with annealing at 450 to 500°C. Glass withstands the high temperatures needed for deposition of titanium tritide.

In the experiments, 1-mm-thick LKH-6 Tb-activated scintillating glass flat samples obtained from Collimated Holes Inc. were used. This glass has improved radioluminescence efficiency and radiation stability due to additional doping with Gd and Ce. To validate the stability of the glass with tritium exposure, a glass disk was sandwiched between two titanium tritide sources and the light output was measured for 31 days with an International Light 1700 Photometer and SHD033 detector (silicon photodiode).

Light output from the scintillation glass pinned between tritiated titanium was stable within the accuracy of the measuring device. (See Figure 1.7.) To monitor the optical properties of the glass, UV/VIS spectra were taken before and after a 31-day exposure. No change was seen in the absolute transmittance or position of the absorption edge. (See Figure 1.8.) The absorption coefficient of the glass was found to be about 0.06 cm^{-1} for the wavelength of emission peak — enough to pipe photons efficiently several centimeters in a waveguide. The glass is therefore compatible with tritium beta radiation, making a practical radioluminescent source.

Photoluminescent properties of glass were studied for better matching of the photovoltaic to glass emission. The maximum on the excitation spectrum of the glass is 270 nm, so this wavelength was used for excitation during collection of the photoluminescence spectrum. The main emission peak shown in Figure 1.9 is at 542 nm, which is typical for Tb^{3+} ion photoluminescence. It is assumed that the radioluminescent spectrum has a similar appearance to the photoluminescent spectrum.

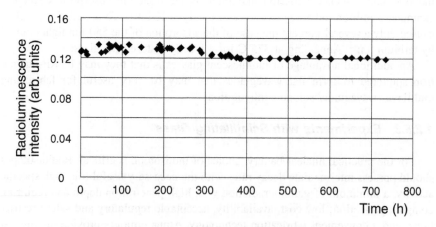

Figure 1.7   Relative light output of scintillation glass exposed to tritium.

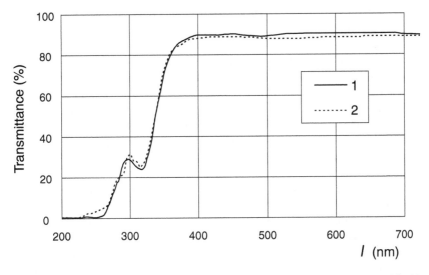

Figure 1.8 UV/VIS spectra of the radioluminescent glass before (1) and after (2) tritium exposure.

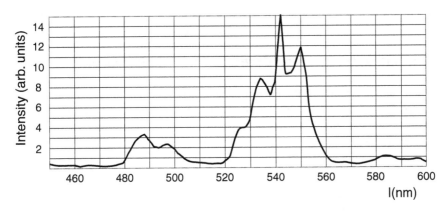

Figure 1.9 Emission spectrum of LKH-6 type glass under 270-nm excitation.

In one experiment with photovoltaics, polished glass was cut into four 21 × 5 mm thick pieces and assembled into a stack. Each piece was sandwiched between two titanium tritide sources on a steel substrate of ~100-μm thickness. A highly sensitive (low-light threshold) AlGaAs photovoltaic cell with a 4 × 4 mm aperture was utilized. The description of the low-light A_3B_5 photovoltaic cells is provided in Chapter 7. The cell's response curve matched the emission maximum of the glass and the cell had a low leakage current of 10^{-12} A/cm² at U = 10 mV.

At the photocell, a 10 nA/cm² current was measured. (Due to source self-absorption, titanium tritide provides one-tenth the power flux of tritium gas used in conventional tritium bulbs, for which hundreds of nA/cm² are typically obtained.[47])

The open circuit voltage was 220 mV, giving a total power output of about 2nW/cm² for the assembly.

Emission from the flat glass samples did not follow the cosine law. Angular distribution of the emission from the 5 × 1 mm edge of the 21 × 5 × 1 mm glass sample was studied. The edge around which rotation took place was 5 mm. (See Figure 1.10.)

The maximum intensity of emission is not at the zero angle but at 30 to 35°. This effect is attributed to the beta absorption range in the glass of no more than 3 μm, while the glass waveguide was 1000 μm (1 mm) thick. The brightest emissions emanated from very thin layers near the exposed surfaces.

For commercial uses, the product would require microwatt output. No mirrors were used in these experiments because it is believed that mirrors, optical coupling of the radioluminescent source to the cell, and reduction of waveguide thickness to the beta range (several micrometers instead of current 1000 μm) would increase the light output two orders of magnitude. These optimization steps are expected to provide 1 μA/cm².[47] Incorporation of the radioisotope directly into the scintillating glass is worth consideration since it minimizes self-absorption and immobilizes the radioisotope at the same time. Promethium-147 seems to be suitable for this purpose; however, radiation stability of the glass under the more energetic isotope must be studied. This light-concentrating radioluminescent light source concept may be used for power generation and in light sources for low-intensity lighting as well as in microelectronics where self-sustained light sources are beneficial.

1.2.6   Indirect-Conversion Based on Thin-Film Phosphors

For miniaturization of the photon battery, it is beneficial to use titanium tritide sources for generation of radioluminescent light instead of the tritium gas. This approach requires either powder or thin-film phosphor deposited on a transparent substrate to be exposed to beta flux. Light may then be converted into electricity by

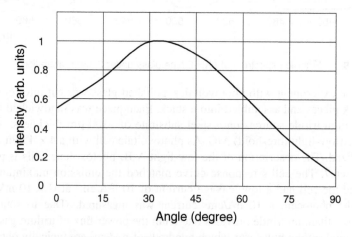

Figure 1.10   Angular distribution of light emitted from 5 × 1 mm edge of the 21 × 5 × 1 mm glass sample.

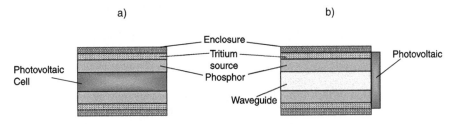

Figure 1.11   Schematic view of repeating unit of photon battery designs: 1 — tritium source; 2 — reflecting layer; 3 — phosphor layer; 4 — WLS waveguide; 5 — photovoltaic cell.

a photovoltaic cell deposited on the transparent substrate (Figure 1.11a) to allow double-sided illumination. For example, an amorphous silicon cell may be deposited on a polymer or glass substrate. Alternatively, for saving space and cost and to increase efficiency by increasing light flux, radioluminescent light may be waveguided by conventional or high-efficiency wavelength shifting (WLS) glass or plastic (Figure 1.11b).

Thin-film phosphors with thickness equal to the range of beta particles have the advantage of reduced diffusive light scattering; they are extensively used in thin-film electroluminescent (TFEL) devices. The most common material is ZnS:Mn, but new CaS, SrS, gallates, and thiogallates have been developed.[48] Thin cathodoluminescent films are gaining increasing acceptance for use in field emission displays (FEDs) and plasma display panels (PDPs).[49,50]

It is known that the efficiency of cathodoluminescent phosphors rises when current density decreases, which is favorable for low-current excitation by isotope. However, efficiency decreases rapidly below a certain threshold value, U_{th}.[51] FED phosphors are optimized to have high luminescence efficiency when excited by low-energy electrons, constituting a significant part of tritium's radiation.

Thin-film FED phosphors have the following advantages:

- Thin-film technology is beneficial for miniaturization of indirect-conversion designs.
- Phosphor layers may be deposited directly on the photovoltaic cell, even during the fabrication process. A_2B_6 compounds are used as antireflection coatings for solar cells, so the deposition technology is available.
- Losses from radioisotope self-absorption can be reduced when radioactive components are used in very thin layers.
- There is no diffusive light scattering.
- No binder is used to attach the phosphor; therefore, binder degradation is absent.
- Unlike powder phosphors, light concentration schemes are easy to design.
- FED phosphors are optimized for low-energy electrons.

The thin-film approach has some special problems:

- Losses due to total internal reflection requires optimization of light coupling to the voltaic. The problem might be minimized if the phosphor film is deposited directly on the photovoltaic cell, which is typically made of high refractive index material. In addition, it is possible to grow thin films with columnar structures to improve light output.[49]

- Since the miniaturized photon battery is not likely to be held under vacuum, electron-stimulated surface chemical reactions (ESSCRs) may occur, e.g., formation of an oxide layer on ZnS phosphor that reduces phosphor efficiency.[50]

For deposition of thin-film phosphors, many techniques may be used, including sputtering, atomic layer chemical vapor deposition (ALCVD), metallorganic CVD (MOCVD), and dip coating.[54-56] ALCVD is the most promising technique. In this method, the substrate is repeatedly treated with phosphor precursors. If ZnS:Cu phosphor is desired, for example, $ZnCl_2$, CuCl, and H_2S are used in reactive stoichiometry. Repeated treatment cycles allow very precise composition and thickness control of the phosphor layer with improved stoichiometry and reduced defects.[54]

Powder and thin-film phosphors were experimentally compared. Samples of glass slides coated with powder phosphors were prepared; glass was cleaned, treated with a 5% weight solution of phosphoric acid in acetone, and then dusted with phosphor powder. FK-106z, a ZnS:Cu cathodoluminescent phosphor, was used. After removing loose powder, samples were baked at 200°C for 1 h. This treatment improved phosphor adherence to the glass substrate.

To exclude photoluminescence influence, samples were preconditioned in darkness. Titanium tritide source with specific activity 0.47 μW/cm^2 was utilized to excite samples and brightness was measured with an IL 1700 photometer coupled with a SHD033 detector. Samples were monitored for 30 min to exclude nonequilibrium effects.

The dependence of the brightness on phosphor layer thickness is plotted in Figure 1.12. Data show that maximum brightness of 14 mcd/m^2 is achieved at 3.5 mg/cm^2 phosphor layer thickness.

Rare-earth-activated strontium aluminum borate NP 2820 (Nichia) and ZnS:Cu FKP-03K (Luminophor Corp.) persistent phosphors, as well as ZnS:Cu EL728 (Sylvania) AC electroluminescent phosphor, were also tested for radioluminescence. Tested layer thickness was the same as for FK-106z. Samples showed 3, 25, and 15 mcd/m^2, respectively. Long afterglow and electroluminescent properties are not necessarily useful for betaluminescence.

For estimation of conversion efficiency, radioluminescence spectra of studied phosphors were acquired with the use of a SDL-2 fluorometer (LOMO Corp.). See Figure 1.13.

The same titanium tritide source was used to excite radioluminescence. Nichia NP 2820 phosphor, a rare-earth-activated ion afterglow phosphor, showed insufficient radioluminescence to obtain a resolved spectrum.

Data obtained allowed efficiency estimation of radioluminescence (RL) in lumens per watt, η'_{RL}, and in percent, η_{RL}. The following formulas were used:

$$\eta'_{RL} = 0.1\,\pi\,B/P_s \qquad (1.27)$$

$$\eta_{RL} = 10\,\pi\,B/\left(K_m Z P_s\right) \qquad (1.28)$$

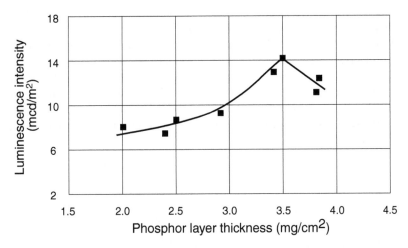

Figure 1.12   Correlation between phosphor layer thickness and brightness.

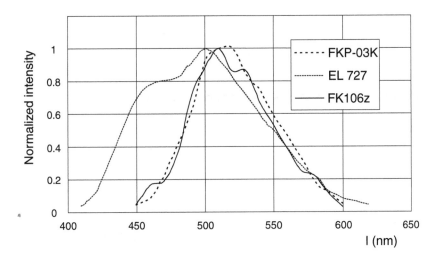

Figure 1.13   Radioluminescence spectra of powder phosphors.

Here B is brightness in mcd/m²; P_s is the specific power of the beta particle flux in μW/cm²; and K_m is the maximum light efficiency in 683 lm/W.

$$Z = \frac{\int\limits_{380}^{780} U(\lambda) \cdot b(\lambda)\mathrm{d}\lambda}{\int\limits_{0}^{\infty} b(\lambda)\mathrm{d}\lambda} \qquad (1.29)$$

where $b(\lambda)$ is the relative spectral density of radioluminescence flux (a.u.) and $U(\lambda)$ is the relative spectral density of photopic light.

Thin film ZnS:Mn, ZnS, Mn, Cu, and ZnS:Cu samples were obtained from Ghent University, Belgium, courtesy of Neyts and Hikavyy. Deposition had been performed by ALCVD technique on the glass substrates. Excitation off the samples, optical measurements, and calculations were performed in the same manner as for powder phosphor samples. Figure 1.14 shows the normalized radioluminescence spectra of tested samples, while Figure 1.15 shows the effect of phosphor thicknesses below 1 μm.

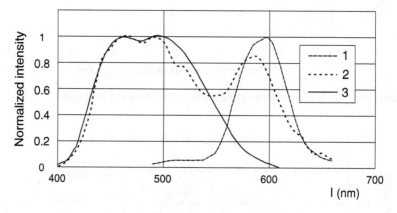

Figure 1.14   Thin-film phosphors radioluminescence spectra.

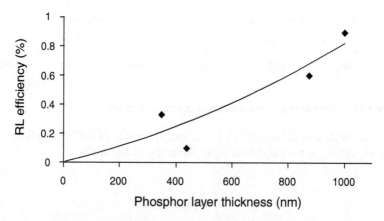

Figure 1.15   Experimentally determined radioluminescent efficiency dependence on phosphor thickness under 1 μm.

ZnS:Mn is the most efficient phosphor tested, although rigorous comparisons find ZnS to be the most efficient tritium radiophosphor (Chapter 3). Optimal thickness must be determined experimentally, but it is a compromise between maximum capture of beta particles and minimal light self-absorption.

While the brightness obtained is lower than that of powder phosphors in similar conditions, a potential exists for greater improvement. In these experiments, no mirror was deposited on the thin film. Such a mirror would absorb part of the useful beta radiation, but increase in the light output would compensate the reduced energy flux. Since low Z material is required to minimize absorption of betas, it may also reduce the backscattering effect.

The value of backscattering, η_0, may be estimated with the following formula[57]:

$$\eta_0 = (1/6)\ln Z_m - (1/4) \tag{1.30}$$

Here Z_m is the mean atomic number of the solid. For zinc sulfide, $\eta_0 = 0.27$, while for aluminum (which may be used as a mirror), $\eta_0 = 0.18$.

For the absorption of all beta particles incident on the film, thickness must be comparable with the penetration depth of emitted radioisotope particles. Penetration depth R (nm) of beta particles with energies E more than 10 keV in the material with density d (g/cm³) may be estimated by the following empirical formula[58]:

$$R = 45\, d^{-0.9} E^{1.7} \tag{1.31}$$

Therefore, for tritium betas with maximum energy of 18.6 keV and ZnS-based thin-film phosphor, the maximum penetration depth will be ~1.8 μm, which is deeper than the tested phosphor thickness. On the other hand, thicker film absorbs more emitted light.

Finally, light coupling must be improved. There are double losses due to total internal reflection at both phosphor–glass and glass–air interfaces. If the phosphor is deposited directly on the surface of the semiconductor, total internal reflection from the phosphor–photovoltaic interface is reduced, since photovoltaics typically have an equal or higher refractive index. Internal reflection from the opposite side of phosphor film beneficially increases light output in the direction toward the photovoltaic. If waveguiding is utilized, a transparent waveguiding plate may be two-side coated with thin phosphor films. The waveguide must have the highest possible refraction index to prevent total internal reflection of the light from the phosphor through the phosphor–waveguide interface. On the outer phosphor surface, a mirror is necessary for the waveguiding effect to be improved. Note that lithium tritide is white and provides higher energy flux than titanium tritide; it might also serve as its own light reflector.[59]

Study of thin-film phosphor radioluminescence will result in its utilization in light sources and batteries. Even unoptimized thin-film samples show results comparable to current powder-coat phosphors. For further improvement, sample thickness must be optimized, mirrors should be used, phosphor composition must be improved with respect to radioluminescence efficiency, and light self-absorption must be reduced.

REFERENCES

1. Mukhin, K.N., *Experimental Nuclear Physics. Atom Nucleus Physics*, Vol. I., 3rd ed., Atomizdat, Moscow, 1974 [Russian language]; Knoll, G.F., *Radiation Detection and Measurement*, 2nd ed. John Wiley & Sons, New York, 1989.

2. Babichev, A.P. et al., *Physical Values: Handbook*, Grigor'ev, I.S. and Meilikhov, E.Z., Eds., Energoatomizdat, Moscow, 1991 [Russian language]; Browne, E. and Firestone, R.B., *Table of Radioactive Isotopes*, Wiley-Interscience, New York, 1986.

3. Klimov, A.N., *Nuclear Physics and Nuclear Reactors*, Atomizdat, Moscow, 1971 [Russian language]; Ehmann, W.D. and Vance, D.E., *Radiochemistry and Nuclear Methods of Analysis*, Wiley-Interscience, New York, 1991.

4. Shirokov, Yu. M. and Yudin, N.P., *Nuclear Physics*, 2nd ed., Nauka, Moscow, 1980 [Russian language].

5. Henley, E.J. and Johnson, E.R., *The Chemistry and Physics of High Energy Reactions*, University Press, Washington, D.C., 1969.

6. Den'gub, V.M. and Smirnov, V.G., *Units of Quantities, Dictionary-Reference Book*, Izd. Stand., Moscow, 1990 [Russian language].

7. Bartenev, S.A. et al., Radionuclides and articles thereof for science, technology, and medicine, in *V. G. Khlopin Radium Institute. On the 75th Anniversary*, Il'enko, E.I., Ed., St. Petersburg Institute of Nuclear Physics, RAS, St. Petersburg, 1997, 133 [Russian language].

8. Kodyukov, V.M. et al., in *Radioisotope Sources of Electrical Energy*, Fradkin, G.M., Ed., Atomizdat, Moscow, 1978 [Russian language].

9. Corliss, W.R. and Harvey, D.G., *Radioisotopic Power Generation*, Mir, Moscow, 1967 [Russian language].

10. Sarkisov, A.A., Yakimov, V.A., and Kaplar, E.P., in *Thermoelectric Generators with Nuclear Heat Sources*, Sarkisov, A.A., Ed., Energoatomizdat, Moscow, 1987 [Russian language].

11. Baranov, V. et al., Radioactive isotopes as energy sources in photovoltaic nuclear battery based on plasma-dusty structures, in *Isotopes*, Baranov, V. Yu., Ed., IzdAT, Moscow, 2000, chap. 17 [Russian language].

12. Linder, E.G. and Christian, S.M., The use of radioactive material for generation of high-voltage, *J. Appl. Phys.*, 23, 1213, 1952.

13. Coleman, I.N., Radioisotopic high potential low-current sources, *Nucleonics*, 11, 42, 1953.

14. Windle, W.E., Microwatt radioisotope energy converters, *IEEE Trans. Aerospace*, 2, 646, 1964.

15. Moseley, H.G.J. and Harling, J., The attainment of high potentials by the use of radium, *Proc. R. Soc.*, 88A, 471, 1913.

16. Anno, J.N., A direct-energy-conversion device using alpha particles, *Nucl. News*, 5, 3, 1962.

17. Olsen, L.S., Review of betavoltaic energy conversion, in *Proc. XII Space Photovoltaic Res. Technol. Conf.*, Cleveland, OH, 256, 1992.

18. Walko, R.J. et al., Tritium-fueled beta cells, *IEEE Proc. Intersociety Energy Conversion Engineering Conf.*, 6, 135, 1997.

19. Andreev, V.M. et al., Tritium-powered beta cells based on $Al_xGa_{1-x}As$, *Proc. 28th IEEE PVSC*, Anchorage, AK, Sept. 15–22, 2000.

20. Deus, S., Alpha- and betavoltaic cells based on amorphous silicon, *Proc. 28th IEEE PVSC*, Anchorage, AK, Sept. 15–22, 2000.

21. Kostesky, T. et al., Tritiated amorphous silicon films and devices, *J. Vac. Sci. Technol., A*, 16, 893, 1998.
22. Mikhal'chenko, A.G., *Radioluminescent Emitters*, Energoatomizdat, Moscow, 1988 [Russian language].
23. Mikhal'chenko, A.G., Zhurikh, M. Yu., and Chumak, N.V., Radioluminescent emitters and radiation-chemical processes in them, in *Radiation-Chemical Transformations in Inorganic and Organic Materials*, Vasil'ev, I.A., Ed., Izd. St. Petersburg Technol. Inst. (Techn. Univ.), St. Petersburg, (2000), 18 [Russian language].
24. Nuclear Lamps, Defence Standard 62-4/Issue 2, Ministry of Defence of G.B., 1972.
25. American National Standard N540; Classification of Radioactive Light Sources (NBS Handbook 116), U.S. Government Printing Office, Washington, 1976.
26. SH3.R05 and SH3.R06 Type Radioluminescent Light Sources. Technical Specifications, TU 95 2639–2697, 1997 [Russian language].
27. SH3.R07 Type Radioluminescent Light Sources. Technical Specifications, TU 95 2681–2698, 1998 [Russian language].
28. Rivenburg, R.C. et al., U.S. Patent 5,443,657, 1995.
29. Renschler, C.L. et al., Solid state radioluminescent lighting, *Radiat. Phys. Chem.*, 44, 629, 1994.
30. Walko, R.J. et al., Electronic and photonic power applications, in *Radioluminescent Lighting Technology Transfer Conference Proceedings*, U.S. DOE, Annapolis, MD, 1990, 13-1.
31. Reisfeld, R. and Jorgensen, C.K., Luminescent solar concentrators for energy conversion, in *Structure and Bonding* 49, Springer-Verlag, Berlin, Heidelberg, 1982, 2–36.
32. Sychov, M.M. et al., Radioluminescent glass based light and power source, *Abstract Book of 103 American Ceramic Society Annual Meeting*, Indianapolis, 2001, 34.
33. Sprehn, G.A. et al., Aerogel-clad optical fiber, US Patent 5,684,907, 1994.
34. Hamil, R.A. et. al., Luminescent light source for laser pumping and laser system containing same, US Patent 5,313,485, 1992.
35. Soltani, P.K., Wrigley, C.Y., and Storti, G.M., High-luminance radioluminescent lamp, US Patent 4,855,879, 1989.
36. Trykov, O.A. et.al., High-intensity autonomic light sources, Obninsk, FEI, 1998 [Russian language].
37. Bicron Technical Data Sheet, *www.bicron.com*.
38. Ellis, J., Glass and phosphor scintillators for x-ray imaging, *SCIFI-97*, AIP Woodbury, New York, 39–46, 1997.
39. Fan, G.Y., Bueno, C., and Dunkelberger, D., Performance Characteristics of Radioluminescent Fiber Optics as Electron Scintillators, *J. Electron Microsc.*, 42(6) 419–423, 1993.
40. Buchanan, R.A. and Bueno, C., Terbium activated silicate luminescent glasses for use in converting x-ray radiation into visible radiation, US Patent 5,122,671, 1991.
41. Huston, A.L. and Justus, B.L., Glass matrix doped with activated luminescent nanocrystalline particles, US Patent 5,585,640, 1995.
42. Blasse, G. and Bril, A., Characteristic luminescence, *Philips Tech. Rev.*, 31(10), 304, 1970.
43. Swallow, A.J., *Radiation Chemistry*, Longman, New York, 1973.
44. Pikaev, A.K., *Modern Radiation Chemistry*, Nauka, Moscow, 1987 [Russian language].
45. Industrial Quality Inc., Technical Datasheet, *www.indqual.com*.

46. Bruce, A.J., Grodkiewicz, W.H., and Shmulovich, J., Er^{3+}-doped soda-lime glasses and planar waveguide devices, *Synthesis and Application of Lanthanide-doped Materials*, Potter, B.G. and Bruce, A.J., Eds., American Chemical Society, Washington D.C., 69–76, 1995.

47. Andreev, V.M., Advanced beta cell and low-intensity photovoltaic cells and arrays based on III-V compounds, *Ioffe Institute Report*, Ioffe Institute, St. Petersburg, Russia, 2000.

48. Muller, G.O., The new electroluminescence concept – thin films, in *Solid State Luminescence*, Chapman & Hall, London, 1993, 133–157.

49. Chakhovskoi, A.G., Characterisation of novel powder and thin film RGB phosphors, *Tech. Dig. 9th IVMC Conf.*, St. Petersburg, Russia, 614–618, 1996.

50. Bondar, V.D., Energy-controlled variable color thin film luminescent screens, *Tech. Dig. 1st Int. Conf. Display Phosphors*, San Diego, 287–290, 1995.

51. Shea, L.E. and Walko, R.J., Development of standards for characterization of cathodoluminescence efficiency, *Proc. SPIE*, 3636, 105–115, 1999.

52. Givargizov, E.I., Phosphors with columnar structure, *SID'99 Abstract Book*, San Diego, 301–304, 1999.

53. Swart, H.C. and Hillie, K.T., Degradation of ZnS FED phosphors, *Surf. Interface Anal.*, 30, 1, 383–386, 2000.

54. Sanders, B.W., Atomic layer epitaxy of phosphor thin films, in *Solid State Luminescence*, Chapman & Hall, London, 293–311, 1993.

55. Yu, J.E., Characterization of ZnS layers growth by MOCVD for thin film electroluminescent devices, *MRS Symposium Proc.*, 242, 215–220, 1992.

56. Minami, T., EL devices using Ga_2O_3 phosphor thin films prepared by dip-coating, *SID'99 Abstract Book*, San Diego, 105–108, 1999.

57. *Phosphor Handbook*, Shionoya, S. and Yen, W.M., Ed., CRC Press, New York, 115, 1999.

58. Gaber, M. and Fitting, H.J., *Phis. Stat. Sol.*, (a) 85, 195, 1984.

59. Kherani, N.P. and Shmayda, W.T., Radioluminescence using metal tritides, *Int. Symp. Metal-Hydrogen Sys.*, Uppsala, Sweden, 1992.

CHAPTER 2

Radioactive Materials, Ionizing Radiation Sources, and Radioluminescent Light Sources for Nuclear Batteries

A.G. Kavetsky, S.N. Nekhoroshkov, S.P. Meleshkov,
Y.L. Kaminski, and G.P. Akulov

CONTENTS

39

2.1   TRITIUM-CONTAINING RADIOACTIVE MATERIALS

2.1.1   General Information and Requirements for Radioisotopes

Use of energy released by natural radioactive decay for heat, light, or electricity is an attractive concept. Devices utilizing this energy source could operate for prolonged periods in completely autonomous mode; however, generating electrical power greater than 1 W requires huge quantities of radioactive material. For example, the thermoelectric generator SNAP-7B, with an output of 68 W, contained 225,000 Ci of ^{90}Sr.[1] Accumulation of radioactive material poses a potential hazard to the environment and human health.

The least hazardous sources are beta-emitting radioactive isotopes with energy under 200 to 300 keV and stable isotope decay products. Such radionuclides pose relatively low radiation hazard. Even in significant amounts, they typically do not require heavy biological protection. It is expedient that energy sources in the devices for light or electricity generation be "pure" beta emitters with half-lives of over several years. In this case, the beta flux power decline will be relatively insignificant, and the engineered lifetime of the device can be many years. These types of isotopes do not generate much heat and therefore do not make good thermoelectric sources.

Among about 2000 radioactive isotopes, only a few pure beta emitters are characterized by half-lives of over several months with stable isotopes as decay products. These include ^{3}H, ^{14}C, ^{45}Ca, and ^{63}Ni. The list of candidates can be supplemented with ^{55}Fe decaying by electron capture and emitting x-rays and Auger electrons, ^{147}Pm emitting not only beta but also soft gamma rays, and ^{204}Tl emitting beta radiation with fairly high energy.

Selected characteristics of these isotopes are presented in Table 2.1, which contains the half-life, $T_{1/2}$, of the isotope and radiation type. It also presents the values of average energy ε_{av} and maximal energy ε_{max} of beta particles,[2] as well as those of the specific (per 1 Ci) power P_{sp}, the minimal significant activity of the isotope (MSA), and the radioactive hazard group. The ^{55}Fe isotope is also characterized by Auger electron energy and x-ray energy.

The parameter P_{sp} is calculated as

$$P_{sp} = 3.7 \cdot 10^{10} \cdot \int_0^{\varepsilon_{max}} w(\varepsilon_\beta) \cdot \varepsilon_\beta \cdot d\varepsilon_\beta = 3.7 \cdot 10^{10} \cdot \varepsilon_{av} \qquad (2.1)$$

where ε_β is the kinetic energy of beta particles, $w(\varepsilon_\beta)$ is the spectral distribution, and the factor $3.7 \cdot 10^{10}$ characterizes the number of decay events of the unit activity, 1 Ci. For ε_{av} measured in kiloelectron volts and P_{sp} in microwatts/curie, one obtains

$$P_{sp} = 5.92 \cdot \varepsilon_{av} \qquad (2.2)$$

MSA and the radiation hazard group characterize the radiation hazard posed by the isotope in accordance with hygienic standards in Russia.[3,4] The radionuclides

Table 2.1   Characteristics of Isotopes

Isotope	^{3}H	^{14}H	^{45}Ca	^{55}Fe	^{63}Ni	^{147}Pm	^{204}Tl
$T_{1/2}$, year	12.34	5710	0.44	2.72	100	2.7	3.8
Radiation type	β⁻	β⁻	β⁻	e⁻, X	β⁻	β⁻, weak γ	β⁻, X
ε_{av}, keV	5.7	49	77	e⁻ – 5.2	17.6	62	243
ε_{max}, keV	18.6	156	257	X – 5.9	67	224	763
P_{sp}, μW/Ci	34	290	456	e⁻ – 18 X – 10	100	367	1440
MSA, Bq	10^9	10^7	10^7	10^6	10^8	10^7	10^4
Radiation hazard group	D	C	C	C	D	C	B

Sources: From Khol'nov, Yu.V. et al., *Characteristics of Radiation Emitted by Radioactive Nuclides Employed in the People's Economy*, Atomizdat, Moscow, 1980 [in Russian], and *Principal Organic Rules of Ensuring Radiation Safety*, RF Minzdrav, Moscow, 1985 [in Russian].

are sorted into four radiation hazard groups: A, B, C, and D, of which group A poses the greatest hazard.

The radionuclide preferred for light generation is tritium, with its favorable combination of nuclear-physical and hygienic characteristics. Table 2.1 shows that tritium betas have energies no greater than 18.6 keV, which approximates that of cathode rays utilized in TV sets. The maximal path lengths in solids for electrons with this energy level do not exceed several micrometers. Therefore, use of tritium, like cathode rays, does not require heavy biological protection. At the same time, tritium is energetic enough and its lifetime long enough for many practical applications.

Tritium-containing substances can be utilized in radioluminescent light sources or nuclear batteries in diversified chemical forms:

- Gaseous tritium
- Metal tritides, in particular, titanium tritide
- Tritium-loaded silicon
- Tritium-containing zeolites and aerogels
- Selected organic tritium–containing compounds, including fullerenes and nanotubes

These tritium-containing substances exhibit relatively high specific activity, low tritium desorption rate, and comparatively high radiation stability under beta irradiation.

2.1.2 Tritium Gas

The chemical properties of tritium gas are similar to those of hydrogen gas; its specific activity is 9700 Ci/g.[5] The density of gaseous tritium can be estimated as $d = 2.68 \cdot 10^{-4}$ g/cm^3 by dividing the mass of 1 mol of the tritium diatomic molecule, 6 g, by the volume of 1 mol of ideal gas at standard condition, 22,400 cm^3. Generating gaseous tritium in the laboratory typically involves desorption from uranium tritide at temperatures above 400°C.[6] At temperatures below 200°C, tritium is adsorbed by uranium, and at 20°C the residual pressure of tritium after adsorption by uranium is about 0.13 Pa. This corresponds to the equilibrium concentration of tritium above uranium[5,7] of $3.5 \cdot 10^{-6}$ Ci/cm^3. Figure 2.1 presents a typical apparatus for uranium tritide storage (uranium bed).[8]

Handling gaseous tritium in amounts of up to 50,000 Ci employs variously designed apparatus, like that schematically shown in Figure 2.2.[8] This comprises a system of stainless-steel manifolds and valves hermetically connected to uranium tritide storage facilities, calibrated tanks for precisely metering gaseous tritium, circulation pumps, a system for creating and maintaining vacuum, and a variety of analytical instruments. The facility shown in Figure 2.2 is the basic framework for various manipulations with tritium that can be supplemented with other units and systems needed for manufacturing gas-filled radioluminescence light sources (RLSs-T).

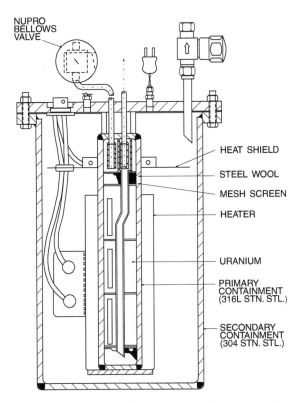

Figure 2.1 Apparatus for uranium tritide storage (uranium bed) designed at the Mound Laboratories. (From Bokwa, S.R. et al., Handling high-specific-activity tritium, *Nucl. Instrum. Methods Phys. Res., Sect. A*, 257, 52, 1987.)

2.1.3 Inorganic Compounds of Tritium

2.1.3.1 Titanium Tritide

Practical use of soft beta radiation often utilizes metal tritides, mostly titanium tritide.[5,7,9] Titanium tritide has a fairly high specific activity, up to 1100 Ci/g.[5] It is stable in air[5] (unlike a number of other tritides that can behave pyrophorically) and exhibits water resistance up to 100°C. However, under normal temperature and pressure, tritium is released from titanium tritide at a rate of $2.5 \cdot 10^{-6}$ Ci/h per curie of adsorbed tritium. This corresponds to about 0.2 Ci of tritium desorbed within 10 years. Along with the radioactive decay of tritium, desorption also reduces the specific beta flux power. Nevertheless, titanium tritide is considered fairly suitable as a solid beta source.

Steps in preparing titanium tritide thin films are as follow:

1. Preparing metal tritide sources starts with making a clean substrate. A plate made of nickel, copper, silver, molybdenum, tungsten, tantalum, platinum, or other corrosion-resistant material intended as a substrate is polished, and its surface is chemically decontaminated from oxide films.[7]

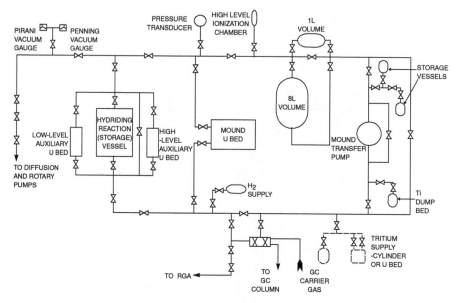

Figure 2.2 Tritium-handling apparatus. (From Bokwa, S.R. et al., Handling high-specific-activity tritium, *Nucl. Instrum. Methods Phys. Res., Sect. A*, 257, 52, 1987.)

2. The plate is degassed in a vacuum chamber at high temperature and pressure less than $6.6 \cdot 10^{-3}$ Pa.
3. This is followed by deposition of a thin (several hundreds of nanometers to several micrometers) titanium layer by vacuum evaporation. Titanium is evaporated either by induction[7] or by electron bombardment,[7,10] with the pressure in the evaporation chamber no greater than $6.6 \cdot 10^{-4}$ Pa.
4. The titanium to be evaporated is degassed under continuous evacuation at a high temperature (up to 1100°C) and pressure no greater than $2.6 \cdot 10^{-3}$ Pa. The resulting titanium coating can adsorb up to two tritium atoms per titanium atom.
5. After depositing the titanium layer onto the substrate, the latter is transferred under an inert atmosphere to a vacuum chamber for tritium loading.
6. The chamber is evacuated below $1 \cdot 10^{-6}$ Pa, and the temperature in the chamber is raised to the optimum for tritium adsorption, 350°C.[5,7,10]
7. The chamber is then isolated from the high-vacuum pumps and tritium gas is introduced into it from the uranium trap (bed) to a pressure greater than 66 kPa.
8. When tritium adsorption by titanium is complete, the substrate is cooled to room temperature and the remaining tritium gas is collected into the apparatus for uranium tritide storage (uranium bed).

Repeat heating to 350°C, introducing tritium and adsorption as needed to load the film. After final cooling of the substrate to room temperature, tritium gas is returned to the uranium trap and the newly prepared beta source is taken out of the vacuum chamber. The titanium-to-tritium atomic ratio in the source prepared this way is typically 1:1.5. This ratio can be increased by dynamic loading of titanium with tritium. In this case,[7] titanium is preheated in a tritium atmosphere to 700 to 800°C and then cooled to room temperature, which results in a titanium-to-tritium atomic ratio of 1:2. Beta sources thus prepared have a specific activity of up to 1100 Ci/g, the volume of the gaseous tritium bound to 1 g of titanium equal to 470 cm^3.

In certain cases, the tritiated titanium layers can be coated with protective films such as silicon dioxide or cellulose nitrate.[10] This reduces tritium leakage and radiation handling hazard.

2.1.3.2 *Tritiated Silicon*

Tritiated silicon, both porous crystalline (TPS) and amorphous (a-Si:^{3}H), is suitable as a beta source in betavoltaics converting the radioactive decay energy to electrical energy.[11-15] TPS has been used as an independent light source owing to luminescence excited by tritium beta particles.[12-14]

TPS is prepared by electrolytic etching of crystalline silicon in a hydrofluoric acid solution until a transparent sample with a total porosity of 50 to 90% is obtained.[12,13] The resulting porous silicon is characterized by long narrow pores with a size ranging primarily from 1 to 100 nm. Interaction with solvent renders the inside pore surface covered with hydride species of various composition such as $\equiv$SiH, =SiH$_2$ (predominantly), and $-$SiH$_3$. Porous silicon is placed in a medium such as H^{3}H–^{3}H$_2$ gas mixture diluted with He or H$_2$. Isotopic exchange of tritium atoms from the gas phase with hydrogen atoms on the pore surface of porous silicon proceeds at room temperature to equilibrium. The amount of tritium reacted with the porous silicon surface can be estimated from change in the concentration of tritium occurring in the gas phase.

TPS can be prepared by electrolytic dissolution of bulk silicon sample in a mixture of fluorine hydride (HF) and fluorine tritide (^{3}HF).[14] The amount of tritium in the pore interior of TPS prepared in the manner described depends on the HF-to-^{3}HF ratio in the initial electrolyte.

TPS can contain up to 10 Ci of tritium in 1 cm^3, assuming 10% conversion of the surface =SiH$_2$ groups to =SiH^{3}H groups. Tam estimated that a TPS film 10 μm in thickness and 1 cm^2 in area contains 10 mCi of tritium and has a brightness of 1.16 cd/m^2.[12-14]

Amorphous Si:^{3}H can be prepared by the same procedures employed for preparing hydrogenated amorphous silicon (a-Si:H).[11,15] These include glow-discharge dissociation of silane SiH$_4$, reactive sputtering or evaporation of silicon in a tritium atmosphere, thermal chemical vapor deposition using SiH$_4$, and electron cyclotron resonance plasma deposition from SiH$_4$. To prepare mechanically stable a-Si:H films free of flaking or blistering, glow discharge in SiH$_4$ is used that is ignited in a direct current (DC) saddle-field plasma chamber.[16-18]

Hydrogen (or tritium) incorporation into silicon can be controlled through deposition conditions. This procedure is preferred for preparing thin a-Si:^{3}H films.[11,15,19] By alternative and combined supply of tritium, silane, phosphine, and diborane into the discharge space, a *p–i–n* junction comprising an a-Si:H *p*-layer, an a-Si:^{3}H *i*-layer, and an a-Si:H *n*-layer can be conceivably fabricated as a source-integrated betavoltaic.[11]

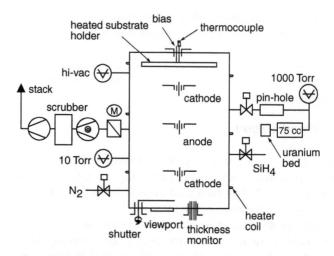

Figure 2.3   Schematic of the direct current saddle-field deposition system. (Adapted from Kostesky, T. et al., Tritiated amorphous silicon films and devices, *J. Vac. Sci. Technol., A*, 16, 893, 1998.)

Figure 2.3 presents the schematic of an apparatus for preparing a-Si:^{3}H by the DC saddle-field glow discharge deposition method.[18] Tritium, which is stored as uranium tritide, and silane (see Figure 2.3) are introduced through individual mass flow controllers into a preliminarily evacuated stainless-steel chamber being heated. The chamber contains five electrodes: three of them are stainless-steel mesh electrodes, to which a high voltage (hundreds of volts) is applied. For preparing a-Si:^{3}H on substrate inside the chamber, the saddle-field discharge is created between the electrodes. A chamber flange supports the substrate holder. The substrate onto which a-Si:^{3}H is deposited can be heated to 500°C. The process is run at a pressure of 6.6 to 21.1 Pa.[18] The resulting a-Si:^{3}H contains up to 10 at.% of tritium,[18] equivalent to specific activity of about 100 Ci/g.

The authors of the preparation procedures for TPS[12-14] and a-Si:^{3}H[11,15,18] noted high stability due to chemical binding of tritium with silicon. However, directly measured rates of tritium leakage from tritiated silicon are not reported. Thus, the radiation hazard posed by tritiated silicon cannot yet be inferred.

2.1.3.3   Tritium-Containing Zeolites and Aerogels

The main limitation of the tritium-based beta sources considered previously is loss of the beta energy directly in the tritium-containing material — self-absorption. This limitation can be overcome by bringing the beta emission source closer to the irradiation object, which will minimize energy losses for betas when transferred to the phosphor. One can manufacture tritium-excited RLSs, with intimate tritium-containing materials and phosphors, based on a variety of compounds and procedures. The example is tritiated porous silicon, already considered. Next, preparation procedures for other RLSs-T are discussed, namely, silicon aerogels and zeolites containing phosphors and tritium.

Preparing RLSs-T with finely ground phosphor (particle size within 1 to 20 μm) dispersed throughout a porous inorganic gel (aerogel or xerogel) with a pore size within 1 to 500 nm and surrounded with tritium on gel pore surfaces has been described.[20-23] Phosphor, especially ZnS:Cu, is introduced into the initial solution (sol) consisting of a polymerizing precursor (oxides and alkoxides of metals and silicon) and a solvent (ethanol or methanol). The phosphor is dispersed in the solution with ultrasound, which reduces phosphor particle agglomerates. In the case of silicon aerogel, the amount of the phosphor can vary between 0.005 and 1.0 g/ml of solution. Gelling is initiated by introducing a promoter into the sol-phosphor dispersion. NH_4OH with a concentration between $2.4 \cdot 10^{-4}$ and $3.3 \cdot 10^{-3}$ mol/l is a preferred promoter.

In the course of gelling, the sol–phosphor mixture is homogenized by mechanical stirring, thus preventing precipitation of the phosphor particles. Before attaining the gel point, the mechanical stirring is stopped. When gelling is complete, the silicon gel–phosphor system is maintained at 50°C up to 2 days to strengthen the polymer network. The aerogel is prepared from the gel by replacing the solvent (ethanol) with liquid CO_2. The latter is removed at a temperature of 36°C and a pressure of 8.6 MPa by passing through the critical point. To remove the water and organic impurities adsorbed on the surface of the silicon aerogel–phosphor sample, the latter is heated in a vacuum at 150 to 400°C. Tritium is loaded into the aerogel–phosphor composition using tritium gas by isotopic exchange in several stages until a vapor pressure of 0.14 MPa is reached or using tritiated water vapors by adsorption over several days.[23] A tritiated silicon aerogel sample prepared by this procedure had a volume activity of 20 Ci/cm^3 and initial brightness of 3.1 cd/m^2.

Another way of tritiating the aerogel–phosphor matrix uses introduction of a solution of tritiated di(phenylethyl)benzene (TDEB) in ethylbenzene or other organic solvent[21,23] into the aerogel pores to be followed by distilling off the solvent. One sample prepared by this procedure had volume activity of 32 Ci/cm^3 and initial brightness of 0.8 fL (2.7 cd/m^2), while another with an initial volume activity of 150 Ci/cm^3 had an initial brightness of 1.2 fL (4.1 cd/m^2). The degradation of these highly loaded samples was rapid.

Zeolites containing ions of rare-earth elements loaded with tritiated water vapor have been prepared.[24,25] Zeolites 4A ($Na_2O:Al_2O_3:2SiO_2$) and 13X ($Na_2O:Al_2O_3:2.5SiO_2$) were maintained in nitrate solutions of rare-earth elements to exchange the sodium ions with ions of europium, terbium, and other rare-earth elements. The zeolites were dried at 400°C in a vacuum. A glass ampule containing the zeolites was connected to an ampule with tritiated water. The zeolite acted as a "sponge" for tritiated water vapors. To accelerate adsorption of water vapors, the sample was cooled to 77 K for 2 min, followed by heating to room temperature. A 52-mg Eu:X zeolite sample containing 36 Ci of tritiated water and emitting in a narrow spectral band λ_{max} = 620 nm had a brightness of 1.7 cd/m^2.

2.1.4   Organic Compounds of Tritium

The most widely used organic tritium compounds in beta sources and radiolu-minescent light sources are tritiated polymers. For this purpose, polystyrene, polyethylene, polyacrylates, polycarbonates, polyurethanes, polyamides, and polysiloxanes have been suggested.[26] Chemical aspects of tritium labeling of organic polymers hold interest for many applications, so this subject is discussed in greater detail.

Polystyrene. The most convenient procedure for preparing [^{3}H]polystyrene consists in polymerization of [^{3}H]styrene. [7,8-^{3}H$_2$]styrene can be synthesized by catalytic hydrogenation of phenylacetylene with gaseous tritium.[26-29]

$$C_6H_5-C\equiv CH \rightarrow C_6H_5-C^3H=C^3HH$$

[7-^{3}H]styrene is prepared by catalytic reduction of acetophenone with gaseous tritium, followed by dehydration of methylphenylcarbinol.[30]

$$C_6H_5-CO-CH_3 \rightarrow C_6H_5-C^3H(OH)-CH_3 \rightarrow C_6H_5-C^3H=CH_2$$

[4-^{3}H]styrene (i.e., styrene with a label in the benzene ring) can be prepared by catalytic dehalogenation of p-iodophenylmethylcarbinol, with subsequent dehydration of methylphenylcarbinol[31]:

$$p\text{-}I\text{-}C_6H_4\text{-}CH(OH)CH_3 \rightarrow p\text{-}^3H\text{-}C_6H_4\text{-}CH(OH)CH_3 \rightarrow p\text{-}^3H\text{-}C_6H_4\text{-}CH=CH_2$$

Polysiloxanes. Vinyltrichlorosilane is reacted with ethoxide to give vinyltriethoxysilane, hydrogenated with gaseous tritium and hydrolyzed to a siloxane polymer[32-33]:

$$CH_2=CH-SiCl_3 \rightarrow CH_2=CH-Si(OC_2H_5)_3 \rightarrow CH_2{}^3H-CH^3H-Si(OC_2H_5)_3 \rightarrow$$

$$\rightarrow \sim O-Si(OC_2H_5)(CH^3H-CH_2{}^3H)\sim$$

Also possible is hydrogenation of vinylchlorosilane to [^{3}H]ethylchlorosilane, which is then mixed with phenyltrichlorosilane and hydrolyzed to arylalkylsiloxane.[33-34]

Poly(methyl methacrylate). Poly(methyl methacrylate) is subjected to isotopic exchange with gaseous tritium by the modified Wilzbach procedure[35] or synthesized from the labeled monomer. Methacrylic acid is subjected to isotopic exchange with tritiated water and esterified with diazomethane and the resulting ester polymerized[26]:

$$CH_2=C(CH_3)COOH \rightarrow CH_2=C(CH_3)COO^3H \rightarrow CH_2=C(CH_3)COO-CH_2{}^3H \rightarrow$$

$$\rightarrow \sim CH_2-C(CH_3)(COOCH_2{}^3H)\sim$$

Polymethylacrylate. Reaction of tritiated water with calcium carbide yields labeled acetylene, which is further carbonylated over nickel tetracarbonyl in the presence of an alcohol. The resulting ester is polymerized[36]:

$$^3H_2O \rightarrow C^3H\equiv C^3H \rightarrow C^3HH=C^3H-COOCH_3 \rightarrow \sim C^3HH-C^3H(COOCH_3)\sim$$

Polyamides. Muconic acid is catalytically hydrogenated with gaseous tritium to adipic acid, which is then condensed with hexamethylenediamine, thus giving tritium-labeled Nylon 6,6[26]:

$$HOOC-(CH=CH)_2COOH \rightarrow HOOC-(CH^3H)_4-COOH \rightarrow$$

$$\sim CO-(CH^3H)_4-CO-NH-(CH_2)_6-NH\sim$$

Fumaric acid is catalytically hydrogenated to succinic acid and in a similar way condensed with hexamethylenediamine,[28] which gives polyamide of the following structure:

$$\sim CO-(CH^3H)_2-CO-NH-(CH_2)_6-NH\sim$$

Polyurethanes. 1,4-butinediol is catalytically hydrogenated with tritium to butanediol and treated with diisocyanate[37]:

$$HOCH_2C\equiv C-CH_2OH \rightarrow HOCH_2-(C^3H_2)_2-CH_2OH \rightarrow$$

$$\sim CH_2-(C^3H_2)_2-CH_2O-CO-NH-(CH_2)_6-NH-COO\sim$$

A tritium label can be introduced in the last stage by catalytic hydrogenation of the unsaturated polymer[38]:

$$\sim CH_2C\equiv C-CH_2O-CO-NH-(CH_2)_6-NH-COO\sim$$

$$\downarrow$$

$$\sim CH_2-(C^3H_2)_2-CH_2O-CO-NH-(CH_2)_6-NH-COO\sim$$

Also described is synthesis of other polyurethanes from polyols easily hydrogenated with tritium at the phenyl ring.[39]

Polyethylene. Acetylene and its lowest homologs can be catalytically hydrogenated with gaseous tritium to ethylene and its homologs that are then polymerized[28]:

$$CH\equiv CH \rightarrow C^3HH=C^3HH \rightarrow \sim C^3HH-C^3HH\sim$$

Labeled polyethylene can also be obtained by hydrogenation of poly-1,3-butadiene or polyacetylene with gaseous tritium[40]:

$$\sim CH=CH\sim \rightarrow \sim C^3HH-C^3HH\sim$$

Poly(vinyl acetate). Reaction of acetylene with acetic acid tritiated at the carboxy group in the presence of the mercury catalyst gives labeled vinyl acetate, which is further polymerized[41]:

$$HC \equiv CH + {}^3HO(O)CCH_3 \rightarrow \rightarrow \sim C^3HH-CH(OCOCH_3)\sim$$

The specific activity of most polymers was typically no greater than 10 to 50 Ci/g, depending on the synthesis procedure and structure of the labeled polymer. One of the largest specific activities was that of [${}^3H$]polyethylene, namely, 600 Ci/g.[40] Considerable values of the specific activities were reported for [7,8-${}^3H_2$]polystyrene (500 Ci/g[28] and even 1300 Ci/g[27]), the highest possible theoretical value being 560 Ci/g with two tritium atoms incorporated into the styrene molecule. This suggests that a specific activity of polystyrene in excess of 560 Ci/g can be achieved via isotopic exchange reactions of the hydrogen atoms of the vinyl group and the aromatic ring of styrene in excess gaseous tritium in parallel with hydrogenation. Similar reactions were studied in detail using ${}^3H$ NMR.[42]

Tritium-labeled polymers are less stable in storage when in solid or liquid state compared to their solutions (e.g., solutions of [${}^3H$]polystyrene in benzene, [${}^3H$]polysiloxane in ethanol, and [${}^3H$]polyurethane in dimethylformamide[38,43]). The [${}^3H$]polystyrene manufactured in the former USSR as a benzene solution had a specific activity of up to 50 Ci/g polymer, and that in solid state, up to 2.5 Ci/g.[43]

Polymer films obtained by evaporation of such solutions are valuable as tritium carriers for devices converting β-decay energy to electric energy, calibration of β-spectrometers, and neutralization of static electricity. For example,[44] films prepared by evaporation of a benzene solution of [7,8-${}^3H_2$]polystyrene with a specific activity of 31 Ci/g were $(2 \text{ to } 5) \cdot 10^{-4}$ g/cm^2 thick and currents attained in an ionizing chamber were on an order of $(2 \text{ to } 3) \cdot 10^{-10}$ A/cm^2. The tritium desorption rate from these films immediately after preparation was $(0.6 \text{ to } 2.2) \cdot 10^{-8}$ Ci/(cm^2·h), and after 6 months it increased by a factor of 1.4 to 4.7.[44]

Jensen et al.[45,46] and Nelson et al.[47] assumed that polystyrene containing tritium in the aromatic ring is more stable than that labeled at the side chain; they modeled the reaction of tritium incorporation into the aromatic ring by isotopic exchange in the $CH_2Cl_2 + TiCl_4 + {}^3HCl$ medium. To this end, they employed data obtained in a model experiment with deuterium. These showed that five hydrogen atoms in the aromatic ring can be replaced by deuterium atoms. As recalculated for tritium, this would produce specific activity of about 1300 Ci/g. Practical achievement of this concentration has not been reported.

Thus, tritium-containing polymers exhibit a fairly diverse chemistry. A special group is formed by transparent nonradioactive polymers containing dissolved low-molecular-weight compounds with super-high specific activity as tritium carriers. This group will be considered in Section 8.3.4 in Chapter 8.

A compact storage form of hydrogen has been sought for fuel cell feed. Substances are needed that absorb high concentrations of hydrogen, on the one hand, and then release them under controlled conditions. A possible form of hydrogen storage of this kind can be found in metal hydrides and, in particular, metal tritides as applied to tritium (see Section 2.1.3). Among organic compounds suitable for hydrogen storage, fullerenes and nanotubes show promise. Therefore, the major properties of these compounds and tentative estimates of their promise as tritium carriers will be briefly discussed.

Fullerenes were discovered in 1985 in carbon vapors from laser-induced graphite evaporation. In 1990, a preparative method of making fullerenes by electric-arc evaporation of graphite electrodes in a helium atmosphere was discovered. Fullerenes constiute a family of polyhedral molecules composed solely of carbon atoms. The principal feature distinguishing fullerenes from other allotropic modifications of carbon (infinitely extended systems such as diamond, graphite, and carbine) is their spatially closed structure. The most widely distributed species is C_{60} fullerene, characterized by two kinds of carbon–carbon bonds, namely, double C=C bonds shared by hexagons and ordinary C–C bonds shared by a hexagon and a pentagon (Figure 2.4).[48]

The system of conjugated (though strained) double bonds makes fullerenes similar in chemical behavior to polyalkenes. Fullerenes undergo addition reactions but not substitution reactions, since they have nothing to be substituted. A unique feature of fullerenes is their ability to dissolve in hydrocarbons (especially aromatic) and form carbon films upon solvent evaporation. Another feature of fullerenes is the occurrence of an interior cavity with a size sufficient for accommodating one or several "guest" atoms. Even atoms of inert gases can be incorporated into the cavity of the host fullerene at high temperature and pressure.[48] Theoretical calculations[49] suggest that hydrogen is a suitable fullerene guest.

The compound $^3H@C_{60}$ with one guest tritium atom per fullerene molecule can be synthesized by two procedures. In the first, the lithium salt of fullerene is exposed to neutron irradiation in a nuclear reactor.[50] The tritium formed by the reaction

$$^6Li + {}^1n \rightarrow {}^3H + {}^4He + 4.8 \text{ MeV}$$

occupies the guest niche. The second procedure[51] consists in preparation of the $^3He@C_{60}$ compound from fullerene and gaseous helium at high temperature and pressure. Neutron flux converts ^3He to tritium by the reaction

$$^3He + {}^1n \rightarrow {}^3H + {}^1H + 760 \text{ keV}$$

In the former case, trace amounts of product are formed, and in the latter, yields are significantly higher though insufficient for practical application. Neither of these procedures of tritium incorporation into fullerenes yields products with high specific activity.

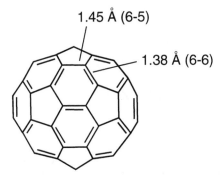

1.45 Å (6-5)

1.38 Å (6-6)

Figure 2.4   Structure of the C_{60} fullerene molecule. (From Gol'dshleger, N.F. and Moravskii, A.P., Fullerene hydrides: preparation, properties, and structure, *Usp. Khim.*, 66, 353, 1997 [in Russian]).

The most important property of fullerenes in this respect is their ability to add hydrogen at double bonds to form fullerene hydrides.[52,53] Fullerenes can also be hydrogenated by a wide variety of reducing agents, such as gaseous hydrogen; a catalyst is not necessary for these reductions. For example, fullerene C_{60} reacts with high-purity hydrogen at 300 to 673 K and 10 to 25 atm, yielding crystalline hydrides of fullerene containing 10 to 30 hydrogen atoms per molecule. These hydrides release hydrogen at 800 K, thus regenerating the initial fullerene.[54] Catalytic hydrogenation in the presence of transition metals (450 K, 100 atm) gives a mixture of $C_{60}H_{36}$ and $C_{60}H_{18}$. Catalytic hydrogenation in toluene (550 K, 130 atm) yields highly hydrogenated fullerenes (up to $C_{60}H_{50}$). The $C_{60}H_{60}$ hydride has not yet been prepared.[52,53]

Most theoretical calculations show that the greatest thermodynamic stability is characteristic of $C_{60}H_{36}$ hydride. A probable reason is that the 12 double bonds remaining in the hydride molecule form four benzenoid rings sited at the vertices of a tetrahedron (Figure 2.5a). (However, the planar aromatic rings are responsible for added steric strain). Some calculations show that the $C_{60}H_{36}$ hydride contains two benzenoid rings and six partially isolated double bonds (Figure 2.5b).[55] Different synthesis procedures may yield $C_{60}H_{36}$ hydrides with different structures.[52,53]

Solid hydrides of fullerenes can generally differ significantly in resistance to atmospheric oxygen and light, even if they have identical empirical formulas but differ in synthesis procedures. The most stable fullerene hydride species are those prepared in a hydrogen atmosphere at high temperature without catalyst, but these hydrides slowly degrade in solutions, too.[52,53]

The all-tritiated hydride $C_{60}H_{36}$ would have a specific activity of about 1250 Ci/g. Its preparation has not been reported, probably due to difficulties in handling gaseous tritium at high temperatures and pressures. Nevertheless, the use of fullerene tritide as a tritium carrier does not seem especially promising due to low chemical stability and radiation stability (see Section 5.3.4). Ultimate conclusions can be drawn only after direct verification.

Carbon nanotubes were discovered in 1991[56] and have already generated a large number of publications devoted to methods of preparation (by electric-arc

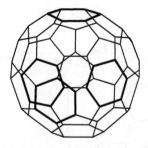

Figure 2.5   Assumed structure of the most stable form of fullerene hydride, $C_{60}H_{36}$. (From Dunlap, B.I., Brenner, D.W., and Schriver, G.W., Symmetric isomers of $C_{60}H_{36}$, *J. Phys. Chem.*, 98, 1756, 1994.)

and laser-induced evaporation, pyrolysis of hydrocarbons, decomposition of metal carbides, etc.), study of their properties, and possible applications. The great interest aroused by this unique material is due, above all, to its unusual structure.[57-59]

Nanotubes are cylindrically shaped structures whose walls are formed by a folded graphite plane, i.e., by a surface faced with regular hexagons with carbon atoms, linked by ordinary bonds, at vertices. The ends of closed nanotubes are covered by fullerene hemispheres containing regular pentagons along with regular hexagons (Figure 2.6).[58] The length of nanotubes is several tens of micrometers, which exceeds by several orders of magnitude their diameter (typically several nanometers). In addition to single-walled nanotubes, structures consisting of several graphite layers, enclosed within one another or wound on a common axis, have been described. The interior channels of the tubes are several nanometers in diameter, which favors efficient capillary drawing-in of liquids and gases.

Hydrogen is incorporated in nanotubes by exposing open nanotubes in platinum foil to hydrogen atmosphere at 300 mm Hg and 273 K (0°C) for 10 min, followed by cooling to 90 K (–183°C) and evacuation. The hydrogen is desorbed upon raising the temperature to 223 K (–50°C) under high-vacuum conditions (10^{-8} mm Hg). The temperature-dependant desorption spectrum exhibits a peak at 288 K (15°C).[60] Because hydrogen can be condensed in nanotubes to extremely high density, they are very promising for hydrogen fuel storage. Indeed, the gravimetric density of hydrogen stored in nanotubes is 5 to 10% (wt), greater than even metal hydrides.

Comparison of hydrogen content in a series of materials (metal hydrides, polymers, liquid hydrogen, hydrogenated nanotubes) reveals clear superiority of hydrogenated nanotubes.[60] The volumetric content of hydrogen (kg/m^3) in nanotubes exceeds by a factor of five that expected, assuming a smallest neighbor distance of 0.35 nm with proximity to the tube walls of 0.29 nm. This high hydrogen density is attributed to the attraction potential of the pore walls (the capillary effect). There are published data of up to 20 wt% hydrogen sorption by nanotubes[61] and up to 67 wt% on carbon nanofibers.[62]

The simulation of protium–tritium mixture sorption suggests preferential preparation of tritium-containing nanotubes.[63] Wang et al. conclude that, at certain pressure, temperature, and size, nanotubes can exhibit sorption of tritium by a factor of 1000 more effectively than with protium. Figure 2.7 presents results of simulation

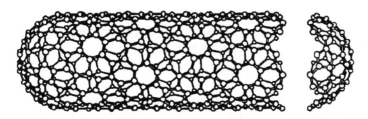

Figure 2.6   Structure of closed carbon nanotubes. (From Dresselhaus, M.S., Down the straight and narrow, *Nature*, 358, 195, 1992.)

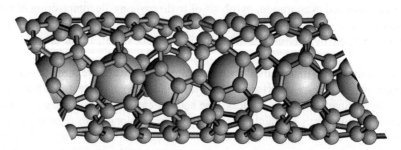

Figure 2.7   Tritium adsorption inside carbon nanotubes. (From Johnson, J.K., *http://www.engrng.pitt.edu/~chewww/johnson.html*.)

of tritium sorption in nanotubes,[64] suggesting a method of isotope fractionation termed "quantum sieving."

The isotope effect should favor gaseous tritium over protium in procedures for incorporating hydrogen into nanotubes; however, for lack of information about radiation resistance and optical properties of the nanotubes and the hydrogen-loaded tubes, the suitability of this material as a tritium carrier for nuclear batteries and RLSs-T application cannot yet be judged. Radiochemical investigations using tritium in place of protium would also provide data for these more general nanotube hydrogen-storage applications. Handling carbon nanomaterials has been complicated by the lack of reliable and reproducible methods of their preparation in the parent form, as well as by lack of generally accepted procedures of synthesis standardization.[53] This is more pertinent for hydrogenated materials and even more so for tritiated materials.

While metal and fullerene hydrides can be treated (to a good approximation) as stoichiometric compounds, this is not true for hydrogenated nanotubes. Any attempt at directly transferring quantitative data for protium adsorption by nanotubes to tritium adsorption has been of doubtful success, since published data do not clarify whether the carbon-to-hydrogen molar ratio or the weight amount of hydrogen is taken as invariant. Assuming 10% by weight of hydrogen is retained in nanotubes on changing from protium to tritium, the specific activity of such tritium-containing nanotubes would be 1000 Ci/g. If one takes the carbon-to-hydrogen molar ratio as invariant, the specific activity increases threefold.

2.2 SPECTRAL AND POWER CHARACTERISTICS OF BETA SOURCES

Selection of tritium-based beta sources for betavoltaic converter or radioluminescence light sources is to be governed by their principal parameters. This is, above all, the specific beta-flux power dP_β/dS, W/cm², attainable with a beta source based on a tritium-containing material. dP_β/dS is the integral with respect to beta energy of the product of the specific flux of betas n_β with the energy ε_β reaching the emitting surface by their energy ε_β (see Equation 2.21).

Another characteristic essential for tritium-based beta source is its spectrum $n_\beta(\varepsilon_\beta)$. Betas emitted by a radionuclide have a continuous energy distribution within the range of zero to the maximal value ε_{max}. This pure spectrum can be calculated on the basis of beta-decay theory. The absorption coefficient of betas in the material is strongly dependent on their energy. Therefore, the difference in the absorption of betas with different energies that propagate through the volume to the surface of the radionuclide-containing material will distort the initial energy distribution pattern. Thus, the spectrum of any specific beta source will be slightly different from that described by the beta-decay theory.

Any tritium-based beta source has an optimal layer thickness from which betas reach the working surface of the source. The optimal thickness approximates the maximal path length of betas. In the case of tritium, betas are characterized by the path length (in surface density units) on an order of 0.7 mg/cm^2.[5] This suggests that the "working layer" thickness for a solid tritium-containing substance is 1 to 3 μm. The activity utilized in a beta source will be determined by the specific activity, density, and layer thickness of radioactive substance. Use of radioactive material in amounts exceeding the functionally minimum are costly and create unnecessary handling problems. A precise estimation of the optimal thickness of the radioactive material layer is therefore needed.

2.2.1 Energy Distribution Function for Tritium Beta Particles (Beta Spectrum of Tritium)

To calculate the specific power of the beta flux reaching the surface of a tritium-containing material, one needs the distribution function (number) of betas emitted per unit time in the decay of tritium atoms n_β over their energy ε_β, i.e., the beta spectrum. The theory of beta decay of nuclei[65] describes the distribution of betas $w(\varepsilon_\beta)d\varepsilon_\beta$ as

$$w\left(\varepsilon_\beta\right)d\varepsilon_\beta = C_1 \cdot E' \cdot \left(E'_m - E'\right)^2 \cdot \left(E'^2 - 1\right) \cdot F(E', Z)d\varepsilon_\beta \qquad (2.3)$$

where C_1 is the constant incorporating the normalizing factors and parameters of the theory

$$E' = 1 + \frac{\varepsilon_\beta}{m_{oe} \cdot c^2} \qquad (2.4)$$

$$E'_m = 1 + \frac{\varepsilon_{max}}{m_{oe} \cdot c^2} \qquad (2.5)$$

$m_{oe} \cdot c^2$ is the rest energy of betas, namely, 511 keV,[66] and the factor $F(E', Z)$ takes into account interaction of betas with the Coulomb field of the daughter nucleus. In the case of tritium, the charge of the daughter nucleus Z (^{3}He) is equal to 2, and the factor $F(E',Z)$, according to Oliver et al.,[67] can be represented as

$$F(E',Z) = \frac{2 \cdot \pi \cdot Z \cdot \alpha}{\sqrt{E'^2 - 1}} \cdot \frac{1.0002037 \cdot E' - 0.001427 \cdot \sqrt{E'^2 - 1}}{1 - \exp\left(-2 \cdot \pi \cdot \alpha \cdot E'/\sqrt{E'^2 - 1}\right)} \tag{2.6}$$

where α is the fine-structure constant, $1/137$.[66]

Lewis[68] measured the beta spectrum of tritium using a lithium-drifted silicon detector with tritons implanted in the depletion layer. The spectrum proved to be virtually unaffected by their absorption in the material. In this case, the number of the tritium betas detected per unit time n_β with energy ε_β is proportional to the probability that they form in the beta decay of the nuclei, i.e., to $w(\varepsilon_\beta)$. Figure 2.8 presents the beta spectrum measured in this manner. This spectrum can be analyzed using Equation 2.3. This requires plotting

$$\sqrt{n_\beta(\varepsilon_\beta) \Big/ \left[F(E',Z) \cdot \left(E'^2 - 1\right)^{1/2} \right]}$$

(to be designated as F–C) vs. ε_β (Fermi–Curie plot). The resulting linear dependence (see Figure 2.9) suggests that the experimentally observed beta spectrum of tritium is adequately described by the equations of the beta-decay theory.

For Equation 2.3 to be used for calculating the spectrum for the surface of a tritium-containing material, the constant C_1 will be determined from the condition of normalizing the spectrum to unity. In this case, $C_1 = 681$. This value of the parameter C_1 for the energy distribution function of tritium betas is subsequently used.

2.2.2   Spectral Characteristics of Tritium-Based Beta Sources

The spectral response of the beta flux from the tritium-containing substance will be calculated using empirical data on the change in the number of electrons (betas), n_β, with the initial energy ε_β^o in dependence with the path length in the substance and empirical data on variation of the energy of electrons (betas) in dependence with the path length in the substance. This approach implies characterization of the absorption of betas by the absorption coefficient $\chi(\varepsilon_\beta^o, \rho)$ and describes their energy ε_β as a function of the initial energy and the path length, $f(\varepsilon_\beta^o, \rho)$.

A beta source will be considered based on a radionuclide-containing material. On the surface of this source, an elementary area dS will be singled out, and in the bulk of the source, an elementary volume dv will be singled out. The device with the volume dv (see Figure 2.10) is assumed to emit betas isotropically every second into the solid angle 4π, dn_v for which holds

$$dn_v = 3.7 \cdot 10^{10} A_m d \, dv \tag{2.7}$$

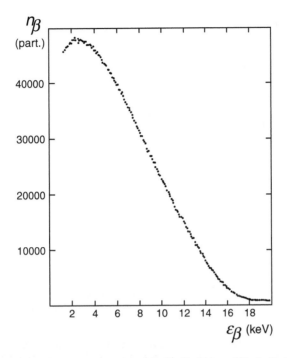

Figure 2.8   Beta spectrum of tritium measured with lithium-drifted silicon detector, with tritons implanted in the depletion layer. (Adapted from Lewis, V.E., Beta decay of tritium, *Nucl. Phys.*, *A*, 151, 120, 1970.)

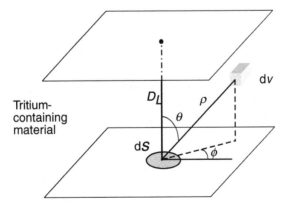

Figure 2.9   Fermi–Curie plot corresponding to the beta spectrum presented in Figure 2.8. (From Lewis, V.E., Beta decay of tritium, *Nucl. Phys.*, *A*, 151, 120, 1970.)

where $3.7 \cdot 10^{10}$ is the dimensional coefficient for activity conversion from curie into becquerel (number of decay events per second), A_m is the specific activity of the substance in Ci/g, and d is the density of the substance in g/cm^3. The number of betas going in the direction of the elementary area dS from the elementary volume

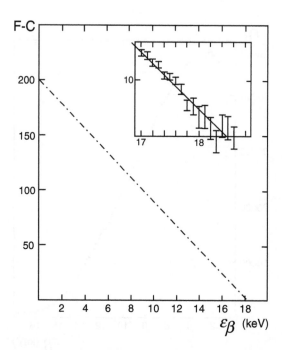

Figure 2.10   Scheme visualizing calculation of the spectra and the specific power of the beta flux from the volume to the surface of the planar layer of the tritium-containing material.

dv, separated from dS by the distance ρ, is $dn_{v\text{-}s}$. It is equal to the product of dn_v by the ratio between the solid angle, at which dS is "seen" from dv, and the whole solid angle 4π:

$$dn_{v\text{-}s} = 3.7 \cdot 10^{10} A_m d \cdot \frac{\cos\theta \cdot dS}{4\pi\rho^2} dv \qquad (2.8)$$

The energy distribution function for emitted betas $w(\varepsilon_\beta^o)$ can be described by Equation 2.3. The differential of the energy distribution of the beta flux $dn_\beta(\varepsilon_\beta)$ coming from the elementary volume dv in the substance layer to the elementary area dS on its surface is determined as

$$dn_\beta\left(\varepsilon_\beta\right) = w\left(\varepsilon_\beta^o\right) \cdot \chi\left(\varepsilon_\beta^o, \rho\right) \cdot dn_{v\text{-}s} = \frac{\cos\theta \cdot dS}{4\pi \cdot \rho^2} \cdot 3.7 \cdot 10^{10} A_m \cdot d \cdot w\left(\varepsilon_\beta^o\right) \cdot \chi\left(\varepsilon_\beta^o, \rho\right) \cdot dv \qquad (2.9)$$

where θ is the angle made by the normal to the area dS and the dS-dv direction and ε_β^o is the initial energy of betas when escaping dv. The initial energy ε_β^o of betas with the energy ε_β on the surface dS as the inverse function to $f(\varepsilon_\beta^o, \rho)$, is

$$\varepsilon_\beta^o = f^{-1}\left(\varepsilon_\beta, \rho\right) \qquad (2.10)$$

The $n_\beta(\varepsilon_\beta)$ distribution function can be calculated by integrating Equation 2.9 with respect to volume. A spherical coordinate system will be introduced with

- The origin in dS
- The azimuthal angle θ measured from the normal to the area dS
- The polar angle ϕ lying in the plane of the surface incorporating the dS area
- For the volume element dv separated from the dS area by the distance ρ holds in this coordinate system

$$dv = \rho^2 \sin\theta \, d\rho \, d\theta \, d\phi \qquad (2.11)$$

Substitute dv into Equation 2.9 according to Equation 2.11 and then integrate with respect to the angle θ between arccos (D_L/ρ) and $\pi/2$ (D_L is the thickness of the material layer) and with respect to ϕ between 0 and 2π. The energy distribution of the beta flux reaching the surface dS can be characterized by

$$n_\beta(\varepsilon_\beta) = \frac{3.7 \cdot 10^{10} \cdot A_m \cdot d}{8} \int_0^\infty \left\{ 1 + \cos\left[2 \arccos\left(\frac{D_L}{\rho} \right) \right] \right\} \cdot \chi(\varepsilon_\beta^o, \rho) \cdot w(\varepsilon_\beta^o) d\rho \qquad (2.12)$$

It is difficult to derive an explicit state of expression for the $\varepsilon_\beta^o = f^{-1}(\varepsilon_\beta, \rho)$ function using empirical Equations 2.17 and 2.19 that describe energy lost by electrons as a function of their path length in a substance and of the initial energy. This difficulty can be avoided by replacing the variable ρ in the integrand of Equation 2.12 by ε_β^o. In this case, Equation 2.12 takes the form

$$n_\beta(\varepsilon_\beta) = \frac{3.7 \cdot 10^{10} \cdot A_m \cdot d}{8} \int_{\varepsilon_\beta}^{\varepsilon_{max}} \left\{ 1 + \cos\left[2 \arccos\left(\frac{D_L}{\rho(\varepsilon_\beta, \varepsilon_\beta^o)} \right) \right] \right\} \cdot$$

$$\chi\left[\varepsilon_\beta^o, \rho(\varepsilon_\beta, \varepsilon_\beta^o) \right] \cdot w(\varepsilon_\beta^o) \cdot \frac{d\rho(\varepsilon_\beta, \varepsilon_\beta^o)}{d\varepsilon_\beta^o} d\varepsilon_\beta^o \qquad (2.13)$$

Here, ε_{max} is the maximal energy of betas in the spectrum and $\rho(\varepsilon_\beta, \varepsilon_\beta^o)$ is the path length of electrons (betas) in a substance corresponding to its energy change from the initial value ε_β^o to ε_β. The equations for $\rho(\varepsilon_\beta, \varepsilon_\beta^o)$ are directly derived from empirical Equations 2.17 and 2.19 for $\varepsilon_\beta = f(\varepsilon_\beta^o, \rho)$.

Next, the spectral response of the beta flux on the surface of a tritium-containing substance using titanium tritide can be calculated as an example. Calculations of the beta spectrum on the titanium tritide surface by Equations 2.12 and 2.13 require the $\chi(\varepsilon_\beta^o, \rho)$ and $f(\varepsilon_\beta^o, \rho)$ dependence for titanium. A number of works[70-73] contain data on the dependence for the coefficient of absorption of fast monoenergetic electrons by films of various substances (in particular, titanium). Also, those works present the energy of electrons passing through titanium layers as a function of the initial electron energy (within the range of tritium beta energies) and the beta path length ρ. The corresponding expressions have the form[70,71]

$$\chi\left(\varepsilon_\beta^o,\rho\right)=\exp\left\{-4.605\cdot\left[\frac{\rho}{R'\left(\varepsilon_\beta^o\right)}\right]^2\right\} \qquad (2.14)$$

where $R'(\varepsilon_\beta^o)$ is the path length in the material for betas with initial energy ε_β^o:

$$R'\left(\varepsilon_\beta^o\right)=90\cdot d^{-0.8}\cdot\left(\varepsilon_\beta^o\right)^{1.3} \text{ at } \varepsilon_\beta^o<10 \text{ keV} \qquad (2.15)$$

$$R'\left(\varepsilon_\beta^o\right)=45\cdot d^{-0.9}\cdot\left(\varepsilon_\beta^o\right)^{1.7} \text{ at } \varepsilon_\beta^o>10 \text{ keV} \qquad (2.16)$$

$$\varepsilon_\beta=f\left(\varepsilon_\beta^o,\rho\right)=\varepsilon_\beta^o\cdot\exp\left[-\frac{\rho}{R'\left(\varepsilon_\beta^o\right)}\right] \qquad (2.17)$$

Makhov[72,73] presents the following equations:

$$\chi\left(\varepsilon_\beta^o,\rho\right)=\exp\left\{-\left[\frac{\rho\cdot d}{50\cdot\left(\varepsilon_\beta^o\right)^{1.56}}\right]^2\right\} \qquad (2.18)$$

$$\varepsilon_\beta=f\left(\varepsilon_\beta^o,\rho\right)=\varepsilon_\beta^o\cdot\exp\left\{-0.95\cdot\left(\frac{\rho\cdot d}{50\cdot\left(\varepsilon_\beta^o\right)^{1.56}}\right)^{0.9}\right\} \qquad (2.19)$$

In Equations 2.14 to 2.19, ρ is expressed in nanometers and ε_β^o is expressed in kiloelectron volts. Equations 2.17 and 2.19 give expressions for $\rho(\varepsilon_\beta,\ \varepsilon_\beta^o)$ and $\dfrac{d\rho(\varepsilon_\beta,\ \varepsilon_\beta^o)}{d\varepsilon_\beta^o}$.

Figures 2.11a and 2.11b show spectral response of the specific beta flux on the surface of the titanium tritide layers with different thicknesses. The spectra were calculated by Equation 2.13 using expressions for $\chi(\varepsilon_\beta^o,\ \rho)$ and $f(\varepsilon_\beta^o,\ \rho)$ from Fitting,[70] Gaber and Fitting,[71] and Makhov.[72,73] Spectral curve intensity tends to increase with increasing titanium tritide layer thickness up to about 600 nm, whereupon the spectrum remains virtually unchanged, preserving the values shown in Figures 2.12a and 2.12b for $D_L=1000$ nm. Figures 2.12a and 2.12b present the $w(\varepsilon_\beta)$

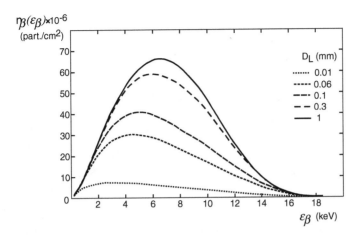

Figure. 2.11a Spectral distribution of the specific beta flux on the surface of titanium tritide layers of different thicknesses. Calculations with equations for $\chi(\varepsilon_\beta^o, \rho)$ and $f(\varepsilon_\beta^o, \rho)$. (From Fitting, H.-J., *Phys. Status Solidi*, A, 26, 525, 1974, and Gaber, M. and Fitting, H.-J., *Phys. Status Solidi*, A, 85, 195, 1984.)

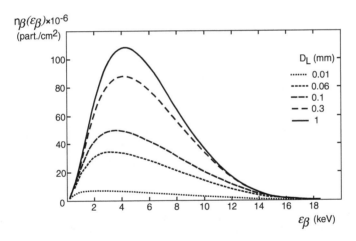

Figure 2.11b Spectral distribution of the specific beta flux on the surface of titanium tritide layers of different thicknesses. Calculations with equations for $\chi(\varepsilon_\beta^o, \rho)$ and $f(\varepsilon_\beta^o, \rho)$. (From Makhov, A.F., *Fiz. Tverd. Tela*, 11, 216, 1960 [in Russian], and Makhov, A.F., *Fiz. Tverd. Tela*, 11, 2161, 1960 [in Russian].)

vs. ε_β normalized spectral response curves for differing titanium tritide layers. The same figures show the initial distribution pattern for tritium betas over the energy $w(\varepsilon_\beta^o)d\varepsilon_\beta^o$ (see Section 2.2.1). Comparison of the initial and final beta spectra reveals only minor changes in the spectral responses: with increasing layer thickness, the spectrum is "depleted" in low- and high-energetic betas; correspondingly, the average beta energy as calculated by the equation

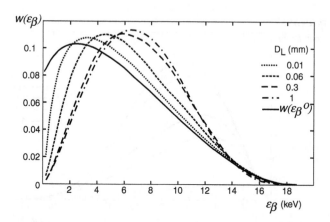

Figure 2.12a   Relative spectral distribution of the specific beta flux on the surface of titanium tritide layers of different thicknesses. Calculations with equations for $\chi(\varepsilon_\beta^o, \rho)$ and $f(\varepsilon_\beta^o, \rho)$. (From Fitting, H.-J., *Phys. Status Solidi*, A, 26, 525, 1974, and Gaber, M. and Fitting, H.-J., *Phys. Status Solidi*, A, 85, 195, 1984.)

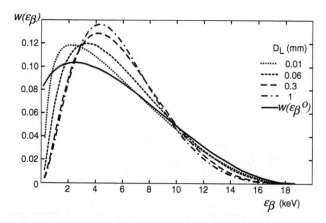

Figure 2.12b   Relative spectral distribution of the specific beta flux on the surface of titanium tritide layers of different thicknesses. Calculations with equations for $\chi(\varepsilon_\beta^o, \rho)$ and $f(\varepsilon_\beta^o, \rho)$. (From Makhov, A.F., *Fiz. Tverd. Tela*, 11, 216, 1960 [in Russian], and Makhov, A.F., *Fiz. Tverd. Tela*, 11, 2161, 1960 [in Russian].)

$$\varepsilon_{av} = \frac{\displaystyle\int_0^{\varepsilon_{max}} n_\beta(\varepsilon_\beta) \cdot \varepsilon_\beta \, d\varepsilon_\beta}{\displaystyle\int_0^{\varepsilon_{max}} n_\beta(\varepsilon_\beta) \, d\varepsilon_\beta} \qquad (2.20)$$

varies in the former case within 5.7 to 7.2 keV and in the latter within 5.7 to 5.9 keV.

2.2.3   Flux Power for Beta Sources

2.2.3.1   Calculations Based on the Electron Absorption Coefficient

In Section 2.2.2, the absorption coefficient and the energy of electrons as functions of the path length and initial energy, $\chi(\varepsilon_\beta^o, \rho)$ and $\varepsilon_\beta = f(\varepsilon_\beta^o, \rho)$, respectively, were used for calculating the spectral response of the beta flux escaping the titanium tritide layer $n_\beta(\varepsilon_\beta)$. The latter also enables calculation of the specific beta flux power dP_β/dS by integrating the product $n_\beta(\varepsilon_\beta) \cdot \varepsilon_\beta$ with respect to ε_β:

$$\frac{dP_\beta}{dS} = \int\limits_0^{\varepsilon_{max}} n_\beta\left(\varepsilon_\beta\right) \cdot \varepsilon_\beta d\varepsilon_\beta \tag{2.21}$$

where $n_\beta(\varepsilon_\beta)$ is defined by Equation 2.13. It is convenient to use the MathCAD 7/2000 program package to calculate Equation 2.21.

As expected, calculations of the dP_β/dS parameter by Equation 2.21 for titanium tritide layers of different thicknesses D_L show that (dP_β/dS) (D_L) plots attain saturation with increasing D_L. Correspondingly, the saturation-specific beta-flux power is about 0.63 μW/cm^2 with Equations 2.14 and 2.17 and about 0.72 μW/cm^2 with Equations 2.18 and 2.19. The layer thickness $D_{0.99}$, at which the specific beta-flux power was 99% saturated, was about 600 nm in both cases.

2.2.3.2   Calculations Based on the Normalized Function of Electron Energy Loss

The specific power of the beta flux reaching the surface of the tritium-containing material can also be determined using dependence of the energy loss per unit path length in the material layer $(d\varepsilon/d\rho)$ for monoenergetic electrons passing through this layer on the electron energy ε. This dependence can be generally represented by the well-known Bethe–Born relationship.[74] Practically, it is more convenient to take advantage of the normalized function describing the energy loss of electrons as dependent on their path length in the material layer[75]:

$$g(y) = \frac{R'\left(\varepsilon^o\right)}{\varepsilon^o} \cdot \frac{d\varepsilon}{d\rho} \tag{2.22}$$

where $y = \dfrac{\rho \cdot d}{R'(\varepsilon^o)}$, ε^o is the initial electron energy, and $R'(\varepsilon^o)$ is the electron path length in the material with density d. The path length for solids (in μg/cm^2) at electron energies between 5 and 25 keV can be represented as[75]

$$R'\left(\varepsilon^o\right) = k \cdot \left(\varepsilon^o\right)^{1.75} \tag{2.23}$$

For the energy range indicated, $k = 4.28^{76}$ with ε^o in kiloelectron volts. The parameter $R'(\varepsilon^o)$ calculated by the Equation 2.23 for the maximal tritium beta energy of 18.6 keV proved to be 713 µg/cm². This approximates the empirically determined maximal path length of tritium betas in the material of 700 µg/cm².[5]

Electron flux energy (relative to the initial energy) at depth y' can be determined as[75]

$$\phi(y') = 1 - \int_0^{y'} g(y)dy \qquad (2.24)$$

The normalized function of the energy loss in the material layer can be represented as the polynomial[75]

$$g(y) = 0.60 + 6.21y - 12.40y^2 + 5.69y^3 \qquad (2.25)$$

The equation describing the energy of a monoenergetic electron flux as dependent on the path length in a substance can be represented as

$$\phi(y') = 1 - 0.6 \cdot y' - 3.105 \cdot y'^2 + 4.13 \cdot y'^3 - 1.42 \cdot y'^4 \qquad (2.26)$$

Treating the beta flux escaping from the elementary volume dv of a tritium-containing substance as consisting of monoenergetic electron fluxes with the initial energy ε_β^o, the contribution from each monoenergetic electron flux is proportional to its share in initial spectral response of tritium betas $w(\varepsilon_\beta^o)$ (see Section 2.2.1). With previously proposed designations for each volume element of the tritium-containing substance dv emitting a beta flux toward the elementary area dS singled out on its surface, the specific power of the beta flux dP_β through dS can be calculated as

$$\frac{dP_\beta}{dS} = \int_{\varepsilon_\beta^o} \int_v \varepsilon_\beta^o \cdot 3.7 \cdot 10^{10} \cdot A_m \cdot d \cdot w\left(\varepsilon_\beta^o\right) \cdot \frac{\cos\theta}{4\pi\rho^2} \cdot \phi(y') \cdot d\varepsilon_\beta^o dv =$$

$$\qquad (2.27)$$

$$= 4.6 \cdot 10^9 \cdot A_m \cdot d \int_0^{\varepsilon_{max}} \varepsilon_\beta^o \cdot w\left(\varepsilon_\beta^o\right) \cdot \int_0^\infty \left\{ 1 + \cos\left[\arccos\left(\frac{D_L}{\rho}\right)\right] \right\} \cdot \phi(y') d\rho d\varepsilon_\beta^o$$

Here, D_L is the tritium-containing substance layer thickness, $y' = \dfrac{\rho \cdot d}{R'(\varepsilon_\beta^o)}$, d is the tritium-containing substance density, and integration with respect to energy is from the initial energies of betas in the flux to zero.

Calculations of dP_β/dS by Equation 2.27 for titanium tritide layers of different thicknesses show that, as in the previous case (see Section 2.2.3), $(dP_\beta/dS)(D_L)$ attains saturation with increasing D_L. In this case, the maximal specific beta-flux power $(dP_\beta/dS)_{sat}$ is equal to 0.82 μW/cm^2, where $D_{0.99}$ is about 600 nm.

In calculations of specific beta-flux power based on the normalized energy loss function for monoenergetic electrons $g(y)$ and on the absorption coefficient of monoenergetic electrons $\chi(\varepsilon_\beta^0, \rho)$, the beta-particle flux is treated as the sum of monoenergetic electron fluxes. This description presumes a linear trajectory of the flux electrons, which in reality is known to be zigzag, because of the scattering of electrons (and betas) on electrons of atoms of the medium. Thus, the $(dP_\beta/dS)(D_L)$ patterns can be determined more precisely if one knows how the specific power of the entire beta flux varies in passing through the substance. Such patterns are usually expressed by the dose function of the point beta source.

2.2.3.3 Calculations Based on the Point Beta Source Dose Function

Decrease in the specific beta-flux power due to interaction with the medium can be described in terms of the concept of the dose function of the point beta emitter, $\varphi(r)$. This function describes variation of the energy absorbed by a substance per decay event with distance r from the emitter,[77] expressed in surface density units. To use $\varphi(r)$ for determining the specific beta-flux power as a function of the distance passed, one apparently must determine beta-flux energy per decay event (one beta in the flux) as a function of path length. For determining the power of the flux of betas coming from dv to dS, dP_S, this function should be multiplied by the number of betas coming per unit time from dv to dS. The dependence of flux energy $E_1(\rho)$ per one beta particle can then be expressed via $\varphi(r)$ as

$$E_1(\rho) = E_1(0) - \int_0^{\rho \cdot d} 4\pi r^2 \cdot \varphi(r)dr \qquad (2.28)$$

Here, $E_1(0)$ is the mean initial energy of the flux per particle, which is, evidently, equal to ε_{av}. Correspondingly, the equation for dP_S takes the form

$$dP_S = dn_{v-S} \cdot \left[\varepsilon_{av} - \int_0^{\rho \cdot d} 4\pi r^2 \cdot \varphi(r)dr \right] \qquad (2.29)$$

By substituting Equation 2.8 into Equation 2.29 and integrating the resulting equation with respect to the volume, the specific (per unit surface) power dP_β/dS of the flux of betas coming from all the emitter volume to surface dS is obtained:

$$\frac{dP_\beta}{dS} = \frac{3.7 \cdot 10^{10} \cdot A_m \cdot d}{4\pi} \cdot \int_v \frac{\cos\theta}{\rho^2} \left[\varepsilon_{av} - \int_0^{\rho \cdot d} 4\pi r^2 \cdot \varphi(r) dr \right] dv \qquad (2.30)$$

It should be noted that the specific power of the beta flux is calculated with the angular distribution in the spatial angle 2π taken as isotropic.

The equations for dose function of the point beta emitter are derived from analysis of the experimental data. Depending on the spectrum of beta radiation and the absorbing medium, various kinds of $\varphi(r)$ functions can be used. The simplest approximation can be expressed in the form[77]

$$\varphi(r) = a \cdot \frac{\exp(-\mu_m \cdot r)}{r^2} \qquad (2.31)$$

where a is the factor depending on the measurement units chosen and μ_m is the mass absorption coefficient of betas. The factor a can be determined in view of

$$\int_0^\infty 4\pi r^2 \cdot \varphi(r) dr = \varepsilon_{av} \qquad (2.32)$$

Hence,

$$a = \frac{\mu_m \cdot \varepsilon_{av}}{4\pi} \qquad (2.33)$$

This elementary approximation can be replaced by more precise empirical equations. For nuclides with ε_{max} ranging from 167 keV (^{35}S) to 2.24 MeV (^{90}Y), the so-called Lovinger function can be utilized.[77] For radionuclides characterized in Table 2.1, the integrand in Equations 2.28 to 2.30 can be calculated by the empirical equation[78]

$$4\pi r^2 \varphi(r) = W(r) = 0.25 \cdot W_0 \cdot \exp(-10 \cdot r \cdot v') + 0.75 \cdot W_0 \cdot \exp(-2 \cdot r \cdot v') +$$
$$+ (\varepsilon_{av} \cdot v' - 0.4 \cdot W_0) \cdot r \cdot v' \cdot \exp(-r \cdot v') \qquad (2.34)$$

with $W(r)$ in units keV·cm^2/mg. The values for the parameters of Equation 2.34 for various nuclides are given in Table 2.2. The ε_{av} values presented in Table 2.2 differ insignificantly from those in Table 2.1. According to Bochkarev et al.,[78] these specific ε_{av} values are to be substituted in Equation 2.34. Also, Table 2.2 contains the values

Table 2.2   Parameters for the Dose Function of the Point Emitter

Isotope	³H	¹⁴C	⁴⁵Ca	⁶³Ni	¹⁴⁷Pm	²⁰⁴Tl
ε_{av}, keV	5.7	47	76	17.6	64	238
W_0, keV·cm²/mg	56.6	13.8	12.7	30.6	13.8	6
v', cm²/mg	15	0.34	0.15	1.48	0.19	0.027
R'_m, mg/cm²	0.72	30	64	7	54	310

Source: Bochkarev, V.V. et al., Distribution of absorbed energy from a point beta source in a tissue-equivalent medium, *Int. J. Appl. Radiat. Iso.*, 23, 493, 1972.

of the maximal path length of betas in the substance, with R'_m expressed in surface density units.

Next to be calculated are the specific power and efficiency of the outlet to the surface for the beta flux in sources with carrier substances containing various beta-active nuclides and shape of thin layer, cylinder, or sphere. A spherical coordinate system will be introduced such that:

- It originates in the dS area on the surface of the emitter.
- The axis from which the azimuthal angle θ is measured is perpendicular to the area dS.
- The plane in which the polar angle ϕ is measured is tangential to the emitter surface in dS.

2.2.3.4   Calculations Based on the Point Beta Source Dose Function for Thin-Layer Sources

Consider a source in the form of a planar layer with thickness D_L. Equation 2.30 can then be represented as

$$\frac{dP_\beta}{dS} = \frac{3.7 \cdot 10^{10} \cdot A_m \cdot d}{2} \cdot \int_0^{\pi/2} \cos\theta \cdot \sin\theta \cdot \int_0^{D_L/\cos\theta} \left[\varepsilon_{av} - \int_0^{\rho \cdot d} 4\pi r^2 \cdot \varphi(r) dr \right] d\rho d\theta \quad (2.35)$$

With $\varphi(r)$ in the form of Equation 2.31 and using Equations 2.1 and 2.33, Equation 2.35 takes the form:

$$\frac{dP_\beta}{dS} = \frac{A_m \cdot P_{sp}}{2\mu_m} \cdot \int_0^1 z \left[1 - \exp\left(\frac{\mu_m \cdot d \cdot D_L}{z} \right) \right] dz \quad (2.36)$$

where z is the integration parameter.

The parameter η_β characterizes the efficiency of conversion of beta-particle energy in the decay into the energy of the beta flux led out from the radionuclide-containing carrier substance to its surface. It can be calculated as the ratio between the energy of the flux passing through the entire surface of the layer (on both sides) S to the energy released in the volume v of the layer:

$$\eta_\beta = \frac{\dfrac{dP_\beta}{dS} \cdot S}{A_m \cdot P_{sp} \cdot d \cdot v} = \frac{1}{\mu_m \cdot d \cdot D_L} \int_0^1 z \left[1 - \exp\left(\frac{\mu_m \cdot d \cdot D_L}{z}\right)\right] dz \qquad (2.37)$$

Figure 2.13 presents, by way of example, the dependence of dP_β/dS and that of η_β on the layer thickness for a planar titanium tritide–based source as calculated by Equations 2.36 and 2.37. The parameters for titanium tritide are given in Table 2.3. The parameter μ_m, cm^2/g, was calculated by empirical equation[66,79] (ε_{max} is expressed in MeV units):

$$\mu_m = 15.5 \cdot \varepsilon_{max}^{-1.41} \qquad (2.38)$$

These relations describe trends for beta-flux specific power on the surface of planar sources.

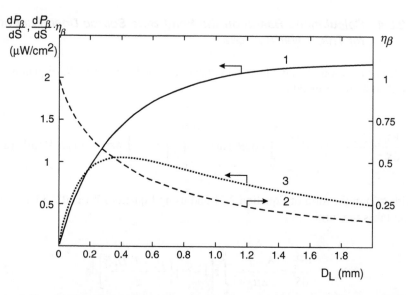

Figure 2.13   (1) Specific beta-flux power, (2) efficiency of conversion of the beta energy in the decay to the beta-flux energy, and (3) product of the specific beta-flux power and efficiency as functions of the layer thickness of titanium tritide–based source (calculated by Equations 2.36 and 2.37).

Table 2.3 Calculated Characteristics of Beta Sources

Isotope	Carrier	Carrier Density, g/cm³	A_m, Ci/g	$\left(\dfrac{dP_\beta}{dS}\right)_{sat}$, $\dfrac{W}{cm^2}$	$D_{0.99}$, μm	$\left(\dfrac{dP_\beta}{dS}\right)_{sat} \cdot A_m^{-1}$, μW·g/(Ci·cm²)	$\eta_\beta(D_{0.99})$, %	Ref.
³H	Ti³H₂	3.91	1100	0.99	0.77	0.0009	17	5
	Li³H	0.816	2910	2.6	3.7	0.0009	17	5
	Zr³H₂	5.67	600	0.54	0.53	0.0009	17	5
	α-SiH (³H₂)	2.0	250	0.22	1.5	0.0009	17	17, 18
	Tritiated organic compounds	1.0	100–1000	0.09–0.9	3.0	0.0009	17	See Section 2.1.4
¹⁴C	PMMA[a]	1.2	0.05	0.015	110	0.30	16	83
⁴⁵Ca	Ca (metal)	1.5	0.05	0.05	200	1.0	14	84
⁵⁵Fe[b]	Fe (metal)	7.8	200	3.4	30	0.017	14	85
⁶³Ni	Ni (metal)	8.9	10	0.21	3	0.021	16	86
¹⁴⁷Pm	Pm₂O₃ in enamel	3.0	50	15	35	0.30	15	c
²⁰⁴Tl	Tl₂O₃	9.0	75	1350	200	18	14	87

[a] PMMA is poly(methyl) methacrylate.
[b] All values for x-rays.
[c] Calculated on the basis of the specific activity of Pm₂O₃ and enamel composition.

From Henley, E.J. and Johnson, E.R., *The Chemistry and Physics of High Energy Reactions*, University Press, Washington, D.C., 1969.

First, with increasing D_L the plot of the specific power of the beta flux vs. the layer thickness attains saturation. In what follows, the layer thickness corresponding to the maximal level, $(dP_\beta/dS)_{sat}$, will be taken as equal to $D_{0.99}$, for which $dP_\beta/dS = 0.99 \, (dP_\beta/dS)_{sat}$. Second, the basic parameter governing $(dP_\beta/dS)_{sat}$ for sources with various carrier substances containing the same isotope is the specific activity of the substance. This follows from the fact that the mass absorption coefficient of betas μ_m is virtually independent of the atomic number Z of the element in the composition of the carrier substance.[79] The parameter μ_m depends on the ratio between the atomic number of the element Z in the composition of the carrier substance and its atomic mass M_A. The Z/M_A ratio is virtually independent of the atomic number of the element (especially in the middle of the periodic system) ($Z/M_A \approx 0.5$ for $Z = 2$, and $Z/M_A \approx 0.4$ for $Z = 89$). Third, the parameter $D_{0.99}$ depends on the density of the substance. Fourth, the efficiency of conversion of beta-particle energy in the decay to the energy of the flux of betas escaping from the substance tends to decrease monotonically with increasing layer thickness.

To choose the beta emitter with the best combination of specific beta flux power dP_β/dS and conversion efficiency η_β, one must elucidate how the product of the efficiency η_β and dP_β/dS varies with the layer thickness. The corresponding dependence is presented in Figure 2.13. Designate the parameter D_L corresponding to the $dP_\beta/dS \cdot \eta_\beta$ product maximum as D_{opt} and the dP_β/dS and η_β parameters corresponding to D_{opt} as $(dP_\beta/dS)_{opt}$ and $\eta_\beta(D_{opt})$, respectively. To improve the reliability of the $(dP_\beta/dS)_{sat}$, $D_{0.99}$, $(dP_\beta/dS)_{opt}$, D_{opt}, $\eta_\beta(D_{0.99})$, and $\eta_\beta(D_{opt})$ values relative to those estimated by Equations 2.36 and 2.37, one can utilize the dose function described by Equation 2.34.

2.2.3.5 Calculations Based on the Point Beta Source Dose Function for Cylindrical Sources

Specific power of the beta flux coming to the inside surface of a cylindrically shaped source is calculated for analyzing energy transfer from gaseous tritium to the radioluminescent phosphor layer. Such light sources, often with cylindrical shapes, are suitable for indirect-conversion device application.

For the lateral surface of the cylinder, Equation 2.30 takes the form

$$\frac{dP_\beta}{dS} = \frac{3.7 \cdot 10^{10} \cdot A_m \cdot d}{4\pi} \cdot \int_0^{2\pi}\int_0^{\pi/2} \cos\theta \cdot \sin\theta \cdot \int_0^{C(D_C,\theta,\phi)} \left[\varepsilon_{av} - \int_0^{\rho \cdot d} 4\pi r^2 \cdot \varphi(r) dr \right] d\rho d\theta d\varphi$$

(2.39)

where D_C is the diameter of the cylinder and

$$C(D_C, \theta, \phi) = \min\left(\frac{D_C \cos\theta}{\sin^2\theta \sin^2\phi + \cos^2\theta}, \frac{10D_C}{\sin\theta \cos\phi} \right).$$

Equation 2.39 gives the specific power of the beta flux at the center of the lateral surface of the cylinder. The cylinder length is taken as $20D_C$. With the dose function in the form of Equation 2.34 substituted into Equation 2.39, the integrals with respect to dr and $d\rho$ can be treated analytically. This yields cumbersome equations that are not presented here. The integration can be performed using the MathCAD 7 symbolical transformation commands. Integration with respect to $d\theta$ and $d\phi$ can be carried out numerically in the same program package.

Similar calculations were carried out for the butt surface of the cylindrical gas cavity. The corresponding equation has the form

$$\frac{dP_\beta}{dS} = \frac{3.7 \cdot 10^{10} \cdot A_m \cdot d}{2\pi} \cdot \int_0^{\pi/2} \cos\theta \cdot \sin\theta \cdot \int_0^{C(R_C,H_C,\theta)} \left[\varepsilon_{av} - \int_0^{\rho \cdot d} 4\pi r^2 \cdot \varphi(r)dr \right] d\rho d\theta$$

(2.40)

where R_C is the radius of the cylinder and H_C is the height (length) of the cylinder.

$$C(R_C,H_C,\theta) = \frac{H_C}{\cos\theta}, \text{ if } 0 \le \theta \le \arccos\left(\frac{H_C}{\sqrt{R_C^2 + H_C^2}}\right), \text{ and } C(R_C,H_C\theta) = \frac{R_C}{\sin\theta}, \text{ if }$$

$$\arccos\left(\frac{H_C}{\sqrt{R_C^2 + H_C^2}}\right) < \theta \le \frac{\pi}{2}.$$

2.2.3.6 Calculations Based on the Point Beta Source Dose Function for Spherical Sources

Calculating specific power of the beta flux and efficiency of the energy transfer from the volume of a spherical source containing a radioactive substance to its surface is essential for analyzing the efficiency of the energy transfer from a finely ground radioactive substance to a phosphor in which the radioactive substance is uniformly distributed.[20] Furthermore, such a calculation is necessary for determining the specific power transferred from the gas volume to its enclosing surface for spherical light sources containing gaseous tritium. When the gaseous medium contains another gas along with tritium (e.g., xenon), the absorption of the tritium beta radiation by xenon is responsible for emission with a wavelength of approximately 172 nm.[80] The emitted light is virtually not absorbed by the gas mixture and is completely absorbed by the substance (e.g., phosphor) coating the inner surface of the gas-enclosing sphere. At a certain diameter of the sphere, the specific power of the emitted flux with a wavelength of approximately 172 nm to the sphere surface will evidently exceed that of the beta flux to the surface from the volume of a sphere with identical diameter and with tritium content identical to that in the mixture.

Quantitative assessments imply knowledge of variation of dP_β/dS and η_β with diameter for spherical sources differing in the gas mixture composition. In the case of a sphere, Equation 2.30 takes the form

$$\frac{dP_\beta}{dS} = \frac{3.7 \cdot 10^{10} \cdot A_m \cdot d}{2} \cdot \int_0^{\pi/2} \cos\theta \cdot \sin\theta \cdot \int_0^{D_S \cdot \cos\theta} \left[\varepsilon_{av} - \int_0^{\rho \cdot d} 4\pi r^2 \cdot \varphi(r) dr \right] d\rho d\theta \quad (2.41)$$

where D_S is the sphere diameter. With the dose function in the form of Equation 2.31, Equation 2.41 can be transformed to

$$\frac{dP_\beta}{dS} = \frac{A_m \cdot P_{sp}}{4 \cdot \mu_m} \left\{ 1 + \frac{2}{(\mu_m \cdot d)^2 \cdot D_S^2} \left[(1 + \mu_m \cdot d \cdot D_S) \cdot \exp(-\mu_m \cdot d \cdot D_S) - 1 \right] \right\} \quad (2.42)$$

The efficiency of the energy transfer to the surface of the sphere will be equal to

$$\eta_\beta = \frac{6}{4 \cdot \mu_m \cdot d \cdot D_S} \left\{ 1 + \frac{2}{(\mu_m \cdot d)^2 \cdot D_S^2} \left[(1 + \mu_m \cdot d \cdot D_S) \cdot \exp(-\mu_m \cdot d \cdot D_S) - 1 \right] \right\}$$

$$(2.43)$$

With the dose function in the form of Equation 2.34, calculations by Equation 2.41 can easily be performed with a widely used software package. In this case, Equation 2.43 for calculating η_β takes the form

$$\eta_\beta = \frac{\dfrac{dP_\beta}{dS} 6}{A_m \cdot d \cdot D_S \cdot P_{sp}} \quad (2.44)$$

2.2.4 Results of Calculations of the Beta-Flux Power for Beta Sources

2.2.4.1 Specific Beta-Flux Power as a Function of Radioactive Substance Layer Thickness

Figure 2.14 presents plots of the specific power of the beta flux dP_β/dS coming to the surface of the titanium tritide layer of thickness D_L. These plots (curves 1, 2, 4, and 5) were obtained by the various methods described in Sections 2.2.3.1 to 2.2.3.4. As mentioned earlier, dP_β/dS tends to grow monotonically with increasing thickness of the titanium tritide layer, asymptotically approaching the saturation level $(dP_\beta/dS)_{sat}$. The latter was estimated by the previously described procedures at 0.72 to 0.99 µW/cm².

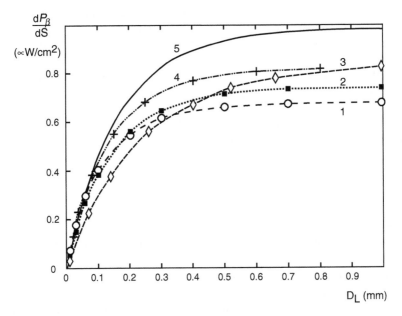

Figure 2.14   Specific beta-flux power on the titanium tritide surface as a function of the layer thickness. Calculations by (1) Equation 2.21 with $\chi(\varepsilon_\beta^o, \rho)$ and $f(\varepsilon_\beta^o, \rho)$. (From Fitting, H.-J., *Phys. Status Solidi*, A, 26, 525, 1974, and Gaber, M. and Fitting, H.-J., *Phys. Status Solidi*, A, 85, 195, 1984.), (2) Equation 2.21 with $\chi(\varepsilon_\beta^o, \rho)$ and $f(\varepsilon_\beta^o, \rho)$. (From Makhov, A.F., *Fiz. Tverd. Tela*, 11, 216, 1960 [in Russian], and Makhov, A.F., *Fiz. Tverd, Tela*, 11, 2161, 1960 [in Russian].), (3) Monte Carlo method. (From Tompkins, J.A. et al., in *Radioluminescent Lighting Technology. Technology Transfer Conference Proceedings*, U.S. DOE, Annapolis, MD, 1990, 5-1.), (4) Equation 2.27, and (5) Equation 2.30 with the dose function of the point beta emitter according to Equation 2.34.

The layer thickness $D_{0.99}$, for which $dP_\beta/dS = 0.99 \cdot (dP_\beta/dS)_{sat}$, was estimated by various methods to be 0.6 to 0.8 µm. These results will be compared with measured values and those calculated by other methods.

Practical titanium tritide–based beta sources are metallic (tungsten, molybdenum, etc.) substrates coated with a tritium-loaded titanium layer. One modification of such beta sources has the following characteristics[7]:

- Titanium tritide layer thickness of about 0.8 µm
- Coefficient of saturation of titanium with tritium (the number of tritium atoms per titanium atom) of two
- Specific surface activity of 0.35 Ci/cm^2
- Ionization current per unit area of the source (current generated by the unit surface area of the beta source in the ionization chamber) of $(1.5 \text{ to } 3) \cdot 10^{-8}$ A/cm^2

According to Lomonosov and Soshin,[81] the ionization current I_{IC} can be calculated as

$$I_{IC} = n_\beta \varepsilon_{av} \, q/W \tag{2.45}$$

where n_β is the beta flux passing through the sensitive volume of the ionization chamber, ε_{av} is the average energy lost by betas in the sensitive volume, equal to the average beta energy in the case of tritium (since the dimensions of the sensitive volume typically exceed the path length of tritium betas, the latter lose all their energy there), q is the electron charge, and W is the average energy required for one pair of ions to form in the working gas of the ionization chamber. For air, W is about 33 eV.[81]

The $n_\beta \varepsilon_{av}$ product represents the specific beta-flux power dP_β/dS, which can be estimated for $I_{IC} = (1.5 \text{ to } 3) \cdot 10^{-8}$ A/cm² by

$$\frac{dP_\beta}{dS} = n_\beta \cdot \varepsilon_{av} = I_{IC}\, W/q \qquad (2.46)$$

as 0.50 to 0.99 µW/cm².

Kherani and Shmayda[82] calculated specific beta-flux power by a procedure similar to that described in Section 2.2.3.1, except that integration with respect to the volume was simplified by multiplying by a factor of 0.5. This simplification probably yields less reliable results compared to these calculations; nevertheless, it is interesting to compare their results[82] with the present data. For example, for titanium tritide, Kherani and Shmayda[82] reported $D_{0.99}$ of 0.7 to 0.9 µm and the saturation-specific beta-flux power of 1.3 µW/cm².

Tompkins et al.[9] calculated the specific beta-flux power on the titanium tritide surface by the Monte Carlo method. Figure 2.14, curve 3, presents the corresponding dependence of the specific beta-flux power on the titanium tritide thickness; the calculated dependence is fairly close to that obtained in this work. The parameters of interest were estimated by Tompkins et al.[9] as maximal specific beta-flux power of about 0.85 µW/cm² and the $D_{0.99}$ layer thickness no greater than 1.2 µm.

All these calculations for titanium tritide are consistent in general with experimental data and data calculated by other methods. The most reliable results should be considered to be those calculated using the dose function of the point beta source (see Section 2.2.3.2). Correspondingly, the following data are calculated by this method (see Sections 2.2.3.3 to 2.2.3.6).

2.2.4.2 Characterization of Solid-State Beta Sources

Table 2.3 presents the calculated characteristics of beta sources based on tritium and other beta-emitting radionuclides. This table also characterizes the isotope ^{55}Fe, a source of x-ray photons (see below). The algorithm used in the table was described in Section 2.2.3.4.

In summary, at the layer thickness $D_{0.99}$ of titanium tritide–based beta sources, conversion efficiency of beta-particle energy to useful energy of beta flux coming to the surface, η_β, is 17%. This means that 1 Ci tritium in titanium tritide affords a usable power flux of 5.7 µW. When the tritium-containing substance layer has the thickness D_{opt}, the efficiency η_β is as large as 48%, and the corresponding specific (per unit activity) beta-flux power is 16 µW/Ci. At the same time, the specific beta-flux power per unit-specific activity at the optimal layer thickness $D_{opt} = 0.2$ μm does not exceed $6.4 \cdot 10^{-4}$ μW·g/(Ci·cm²). These values are fundamentally important when designing nuclear batteries or titanium tritide–based radioluminescent light sources.

Calculations show that lithium tritide–based beta source affords a relatively high specific beta-flux power. However, using lithium tritide poses some problems. For example, upon reacting with atmospheric air or water, lithium tritide degrades, releasing gaseous tritium. Also, even when stored in hermetically evacuated vessels, lithium tritide releases tritium at a rate of up to 0.14 Ci/(g·h).[5] Thus, lithium tritide is unsuitable as a beta source for devices converting beta energy to electrical energy, despite the high specific beta-flux power achievable on the lithium tritide surface. Zirconium tritide, similar to titanium tritide, is stable to air and moisture and does not exhibit pyrophoric properties.[5]

Table 2.3 suggests that titanium tritide has the greatest specific beta-flux power (except for lithium tritide) of the examined tritides. However, tritiated fullerenes or nanotubes with hydrogen stoichiometry already achieved using protium (see Section 2.1.4), specific activity, and, respectively, the specific beta-flux power for these tritiated species will be more than twice that of titanium tritide.

The specific activity of the substance A_m, Ci/g, can be calculated by the formula

$$A_m = \frac{\lambda \cdot N_A \cdot N \cdot \chi}{3.7 \cdot 10^{10} \cdot M_M} \qquad (2.47)$$

where λ is the isotope decay constant in sec⁻¹, N_A is the Avogadro constant, N is the total number of atoms of the given element in the molecule, χ is the ratio of radioactive isotope atoms to total number of atoms of the given element, and M_M is the gram-molecular mass in grams of the carrier substance comprising the element incorporating the radioactive isotope of interest. For example, titanium hydride can incorporate hydrogen with 100% tritium. Its specific activity will be equal to 1100 Ci/g. In other cases, the share of atoms of the radioactive isotope χ is less than unity. The practical achievable level of the specific activity of the carrier substance in many cases is governed by specific technology. Table 2.3 presents roughly estimated values of the specific activity of carrier substances suitable as thin layers in surface beta sources as well as corresponding values of the specific beta-flux power.

As mentioned earlier (see Table 2.1), the isotope ⁵⁵Fe emits Auger electrons with the energy of 5.2 keV and x-rays with the energy of about 5.9 keV. The mass absorption coefficient for electrons with such energies was estimated by Equation 2.38 as $2.6 \cdot 10^4$ cm²/g. The specific power of the electron flux on the surface of ⁵⁵Fe-activated iron, as calculated by Equation 2.36 and recalculated per unit-specific activity, is equal to 0.00017 μW·g/(Ci·cm²). At the same time, 5.9 keV x-rays have much greater penetrability. For x-rays, the flux power decreases exponentially.[66] The mass attenuation coefficient for x-rays with the energy indicated is about 150 cm²/g.

Similar to the case of beta particles, the theoretical equation for the specific power of the flux of x-rays escaping from the layer of the radioactive isotope–activated substance is analogous to Equation 2.36. The specific power of the flux of x-ray quanta on the surface of ^{55}Fe-activated iron, as calculated by this equation and recalculated per unit-specific activity, is equal to about 0.017 μW·g/(Ci·cm^2). Taking this into account, the power of the flux of ionizing radiation from the ^{55}Fe-activated substance is due only to that of the flux of x-ray quanta. Previously mentioned parameters for ^{55}Fe-activated iron were calculated on the basis of Equation 2.36 and presented in Table 2.3.

It is necessary to compare parameters of the real ionizing radiation sources with corresponding calculation results. (Parameters of some titanium tritide–based beta sources were presented earlier in this section.) For zirconium tritide–based source described in Mikhal'chenko,[80] the specific beta-flux power was estimated from the ionization current as 0.44 μW/cm^2 (against the calculated $(dP_\beta/dS)_{sat}$ of 0.54 μW/cm^2).

For a tritiated polystyrene-based source with A_m = 31 Ci/g and layer thickness of 3 μm, the ionization current density was estimated as $3.2 \cdot 10^{-10}$ amp/cm^2.[44] Hence, the specific beta-flux power can be estimated by Equation 2.46 as 0.011 μW/cm^2 against 0.028 μW/cm^2 estimated by Equation 2.35, using the dose function according to Equation 2.34 and the parameters indicated.

Catalogs for ionizing radiation sources typically contain information on the surface activity ($A_S = A_m d D_L$) and the ionization current of beta sources. ^{55}Fe-based sources are characterized by the x-ray photon flux. Basing on the ionization current I_{IC} to be generated by 1 cm^2 of the source (i.e., ionization current density, i_{IC}, amp/cm^2), one can estimate dP_β/dS by Equation 2.46. Knowing the photon flux Φ, quanta/sec, per unit surface area of 1-cm^2 source layer in the solid angle 2π, one can calculate dP/dS, μW/cm^2, as the photon flux Φ multiplied by the photon energy $h\nu$, μJ. By dividing the resulting dP/dS value by the $(dP/dS)_{sat} \cdot A_m^{-1}$ value from Table 2.3, one can determine the A_m value. By dividing A_S/d by A_m, one obtains the layer thickness, D_L. Similar calculations for tritium–based and ^{55}Fe-based beta sources yielded the values listed in Table 2.4. The specific activities and working layer thickness for ionizing radiation sources characterized in this table are generally consistent with those from the source manufacturing practice.

Table 2.4 Specific Activity and Thickness of the Working Layers of Tritium-Based Beta Sources and X-Ray Quantum Emitters

		Data from Catalog[a]		Calculation		
Isotope	Source Type	A_S, Ci/cm^2	I_{IC}, A/cm^2 (Φ, Quanta/(s·cm^2) in Solid Angle 2π)	$\dfrac{dP_\beta}{dS}$, μW/cm^2	A_m, Ci/g	D_L, μm
^{3}H	BITr-G3	0.234	$1.9 \cdot 10^{-8}$	0.62	700	0.85
	BITr-G1	0.212	$1.47 \cdot 10^{-8}$	0.48	540	1.0
^{55}Fe	IRIZh-1	$4.84 \cdot 10^{-3}$	$(4.8 \cdot 10^6)$	0.005	0.3	21
	IRIZh-2	$2.42 \cdot 10^{-2}$	$(2.4 \cdot 10^7)$	0.023	1.4	22
	IRIZh-3	$1.21 \cdot 10^{-1}$	$(1.2 \cdot 10^8)$	0.11	7.1	22

a From Kuznetsov, R.A. et al., in *Radionuclide-Based Sources and Preparations*, NIIAR Russian Federation State Research Center, Dimitrovgrad, 1998 [in Russian].

2.2.4.3 Characterization of Gaseous Tritium–Based Beta Sources

Results calculated for cylinders of different diameters, filled with gaseous tritium at different pressures, are shown in Figure 2.15 (curves). The plots in Figure 2.15 are normalized to their saturation levels, which for gaseous tritium are equal in all cases to 14 μW/cm². Figure 2.15 also presents experimentally measured specific power of the beta flux to the lateral surface of the cylinder as a function of the tritium pressure for different internal diameters. The experimental data were obtained by measuring the brightness of luminescence of the phosphor layer coating the lateral surface of the cylinder. The specific power of the beta flux is proportional to the phosphor layer brightness. The experimental data are normalized to their saturation levels. As seen from Figure 2.15, calculated and measured data are in fair agreement. However, for matching calculated and experimentally measured data, the gaseous tritium density d, when substituted into the upper integration limit with respect to dr in Equation 2.39, was calculated by the equation

$$d = \frac{1.6 \cdot 10^{-5} \cdot M_{3_{H_2}} \cdot 760 \cdot p \cdot k'}{T} \qquad (2.48)$$

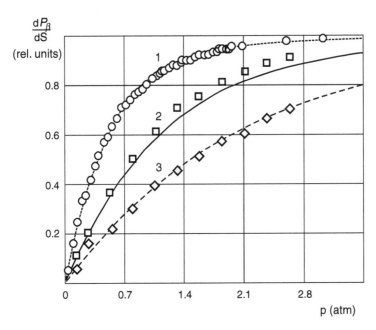

Figure 2.15 Calculated and experimental specific powers of the beta flux to the lateral surface of the cylinder filled with 100% gaseous tritium as functions of the tritium pressure at different diameters of the cylinder. (1) $D_{-} = 1.32$ cm; (2) $D_{-} = 0.60$ cm; (3) $D_{-} = 0.34$ cm, $\bigcirc$, $\square$, $\diamondsuit$ — experimental data ($\square$, $\diamondsuit$ according to Korin et al.[89]). The calculated dependences and experimental data are normalized to their saturation levels.

where $M_{3_{H_2}}$ is the molecular mass of tritium equal to 6, p is the gas pressure in atmospheres, T is the gas temperature in Kelvin, and k' is a correction factor of 0.6. This factor was introduced into the conventional equation for calculating the density of a gaseous substance to match the experimental and calculated data. The correction factor arises because the dose function in the form of Equation 2.34 is based on the absorption of tritium betas in air. For air, the Z/M_A ratio is equal to about 1/2, and for tritium, the Z/M_A ratio is about 1/3. The $(Z/M_A)_{tritium}/(Z/M_A)_{air}$ ratio is equal to approximately 0.66. With this factor introduced into the upper limit of integration (for the density d) with respect to dr in Equation 2.30, one can correct the dose function in the form of Equation 2.34 and apply it to the case of gaseous tritium. The results from Equation 2.35 (with the gaseous tritium density in the form of Equation 2.48), when substituted into the upper integration limit with respect to dr in Equation 2.35, are also consistent with those calculated by the Monte Carlo method (see Figure 2.16) for a gaseous tritium layer.

Similar calculations were carried out for the butt surface of the cylindrical gas cavity; the corresponding equations are presented in Section 2.2.3.5. Also, the specific power of the beta flux emerging on the cylinder butt surface as a function of the tritium pressure in the cylindrical cavity was obtained experimentally. Similar to the lateral surface case, the flux power was determined by measuring the luminescence brightness of the phosphor. Figure 2.17 presents the pressure dependence

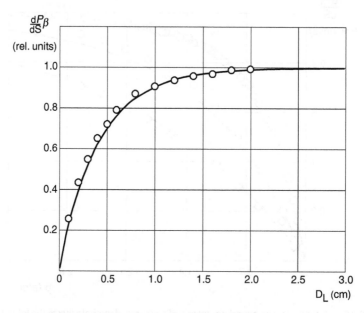

Figure 2.16   Specific power of the beta flux from gaseous tritium layer to its enclosing surface as a function of the layer thickness. Calculations by (curve) Equation 2.35 and (○) Monte Carlo method, (From Tompkins, J.A. et al., in *Radioluminescent Lighting Technology. Technology Transfer Conference Proceedings*, U.S. DOE, Annapolis, MD, 1990, 5-1.). The dependences are normalized to their saturation levels.

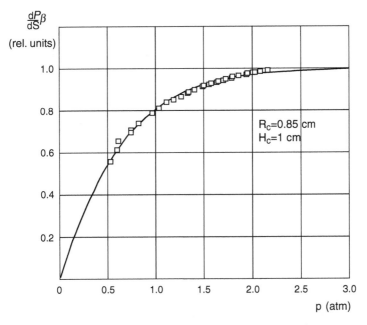

Figure 2.17 Calculated dependence and experimentally measured specific power of the beta flux from the cylindrical gas-filled cavity at the cylinder butt. (Curve) calculation by Equation 2.40, (□) experiment. The dependences are normalized to their saturation levels.

of dP_β/dS calculated by Equation 2.40 with the dose function described by Equation 2.34 and the gas density described by Equation 2.48, when substituted into the upper integration limit with respect to dr in Equation 2.40, as well as the experimentally measured data. The experimental data and calculated dependence with correction factor $k' = 0.6$ in Equation 2.48 are consistent in this case as well.

Calculation of the efficiency of the energy transfer from the volume of the gaseous tritium-filled cylinder to its lateral surface η_β will enable calculation of the efficiency of conversion of the beta decay energy to light for tritium-based gas-filled light sources. Typically, the latter are sealed, cylindrically shaped glass capsules with the inside lateral surface coated with a phosphor and the cavity filled with gaseous tritium. For technological reasons, the tritium pressure in such light sources is often held below 0.7-atm. The efficiency of the energy transfer from the gas volume to the phosphor coating can be calculated by the equation

$$\eta_\beta = \frac{\dfrac{dP_\beta}{dS} \cdot 4}{A_m \cdot d \cdot D_C \cdot P_{sp}} \tag{2.49}$$

Figure 2.18 shows the efficiency of the energy transfer as a function of the internal tube diameter at the tritium pressure of 0.7 atm. For example, this figure

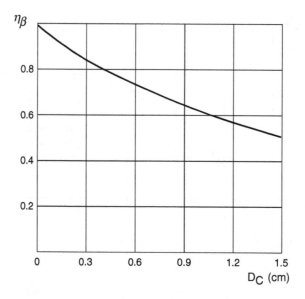

Figure 2.18   Efficiency of the energy transfer from the gaseous tritium volume to the lateral surface of the cylindrical cavity as a function of the cylinder diameter at the tritium pressure of 0.7 atm.

suggests that light sources with the internal diameter of about 3 mm (external diameter 5 mm) have 84% conversion efficiency, and the specific beta-flux power is equal to 3.6 μW/cm². Radioluminescent light sources with such efficiencies were utilized in the breadboard models of indirect-conversion batteries.

2.2.5   Doses in Materials under Tritium Beta Radiation

For estimating the dose to materials exposed to tritium beta radiation, cases of external and internal irradiation will be considered. In the former case, the material is exposed to a beta flux generated by an external beta source. Internal irradiation originates from tritium incorporated into the material.

Phosphors in RLSs-T and betavoltaics using titanium tritide–based beta sources are exposed to external tritium beta radiation. The values of the specific beta-flux power achievable with various beta sources were presented in Section 2.2.4. Now, it is necessary to calculate the dose rate and dose absorbed by materials exposed to the following beta sources:

- Titanium tritide–based beta source with the specific beta-flux power of 0.62 μW/cm² (see Table 2.4)
- Titanium tritide–based beta source with the maximal specific beta-flux power of 0.99 μW/cm² (see Figure 2.14)
- Cylindrically shaped gaseous tritium–filled beta source (tritium pressure 0.7 atm, $D_C = 3$ mm). In this case, the specific power of the beta flux on the lateral surface of the cylinder is equal to 3.6 μW/cm² (see Section 2.2.4.3)

- Gaseous tritium–based beta source with the maximal specific beta-flux power of 14 μW/cm^2 (see Section 2.2.4.3)

Assume betas are absorbed in a layer with thickness R'_m/d (R'_m is the maximal path length of tritium betas in the material equal to 700 μg/cm^2 and d is the material density in g/cm^3).[5] Correspondingly, the surface density of the radiation-exposed layer is equal to R'_m. The dose rate P_D, Gy/s, in the initial irradiation period can be calculated by the equation

$$P_D = \frac{dP_\beta/dS}{R'_m} \cdot 10^{-3} = 1.43 \cdot \frac{dP_\beta}{dS} \tag{2.50}$$

where dP_β/dS has the dimension of μW/cm^2 and R'_m is equal to 700 μg/cm^2. The dose rates for these cases calculated using Equation 2.50 are presented in Table 2.5.

The Monte Carlo calculations in Landsberg[90] showed that, for gaseous tritium, $(dP_\beta/dS)_{sat}$ decreases within a tritium half-life by a factor of about 2.8. This threefold change in the specific beta-flux power is evidently due not only to decreasing specific activity but also to enhanced absorption of electrons (betas) in a ^{3}He-^{3}H mixture compared to pure tritium. The latter fact is explained by the ^{3}He accumulation during tritium beta decay. Thus, the number of the electrons increases (one electron for tritium and two for helium atoms), but the number of atomic nuclei is unchanged. This is responsible for increased ionization loss of electron (beta) energy, proportional to Z/M_A in accordance with the Bethe–Born equation (see Equation 1.6). If one takes the exponential pattern of variation of the specific beta-flux power with the time, the exponent b is

$$b = \frac{1}{T_{1/2}} \cdot \ln 2.8 = 0.083 \text{ year}^{-1}$$

where $T_{1/2}$ = 12.34 years. If dP_β/dS varies with time only by the radioactive decay law, the exponent b is equal to 0.056 year^{-1}.

For titanium tritide–based beta source, one can assume that the specific power of the beta flux also varies with time by an exponential law, with the factor b in the exponent equal to the tritium decay constant $\lambda = \ln 2/T_{1/2} = 0.056$ year^{-1}.

Taking into account the change in the beta-flux power from radioactive decay, the dose D_{abs} in Gy, accumulated by the phosphor or betavoltaic by the time t in years, can be estimated using the equation

$$D_{abs}(t) = \int_0^t P_D(\tau)d\tau = 5.4 \cdot 10^8 \cdot \frac{dP_\beta}{dS} \cdot [1 - \exp(-b \cdot t)] \tag{2.51}$$

Table 2.5   Absorbed Doses in Materials Exposed to Tritium-Based Beta Sources

Beta Source	$\dfrac{dP_\beta}{dS}$, µW/cm²	P_D^o,[a] Gy/s	D_{abs} (1 Year), GGy	D_{abs} ($T_{1/2}$), GGy
Titanium tritide–based	0.62	0.9	0.018	0.17
Titanium tritide–based, with maximal specific beta-flux power	0.99	1.4	0.029	0.27
Cylindrical gaseous tritium–filled P_{3H_2} = 0.7 atm, D_C = 3 mm)	3.6	5.1	0.15	1.2
Gaseous tritium–filled with maximal specific beta-flux power	14	20	0.60	4.8

[a] P_D^o: the dose rate at $t = 0$.

The D_{abs} values calculated by Equation 2.51 for the period of 1 year and for the tritium half-life are presented in Table 2.5.

A significant share of beta energy released in radioactive decay is self-absorbed by the tritium-containing material (e.g., Section 2.1). This has a variety of consequences, such as radiation-chemical reactions and radiation-induced defect formation in the tritium-containing material. Also, gas formation occurs in the tritium-containing material owing to formation of ³He isotope as a result of the radioactive decay of tritium. These deleterious effects will be considered in detail in subsequent sections.

Now, one can estimate the integral dose absorbed by a tritium-containing material from self-irradiation. Let $A_{m,0}$ in Ci/g be the initial specific activity of the tritium-containing material, varying with time by the radioactive decay law for tritium. Then the dose rate $P_D(t)$, Gy/s, at the moment t can be calculated as

$$P_D(t) = A_{m,0} \cdot \exp(-0.056 \cdot t) \cdot 3.7 \cdot 10^{10} \cdot \varepsilon_{av} \cdot 1.6 \cdot 10^{-19} \cdot 10^3 =$$

$$0.033\, A_{m,0} \cdot \exp(-0.056t) \tag{2.52}$$

where t is the time in years and ε_{av} is the average energy of tritium betas in keV.

Correspondingly, the integral dose D_{abs}, Gy, accumulated by the time t can be calculated by the equation

$$D_{abs}(t) = \int_0^t P_D(\tau)d\tau = 1.86 \cdot 10^7 A_{m,0} \cdot [1 - \exp(-0.056 \cdot t)] \tag{2.53}$$

For the tritium-containing material with specific activity of 500 Ci tritium/g (see Table 2.3), doses equal to 0.5 GGy will be accumulated within 1 year and 4.7 GGy within one half-life of tritium. Major radiation loads experienced by materials exposed to both external and internal irradiation when in nuclear batteries operating under tritium beta radiation are responsible for major radiation-induced changes in the matter, most pronounced in the case of organic tritium–containing materials.

2.3 RADIOLUMINESCENT LIGHT SOURCES

Direct or indirect conversion of radioactive decay energy to electrical energy can occur. Indirect conversion of radioactive decay energy (initially to light and then to electrical energy) seems intuitively less advantageous than direct conversion (to electrical energy directly). However, the efficiency of direct conversion by semiconductor converters is fairly low. Low radiation stability of these converters reduces the conversion efficiency over time. Therefore, the optimal energy conversion scheme requires analysis and experimentation. Radioluminescent light sources (RLSs) are devices intended for generation of optical radiation whose operation depends on the radioluminescence phenomenon. RLSs comprise a radioactive substance, a phosphor, and design elements. The latter serve for arranging the phosphor and the radioactive substance as well as for their hermetic isolation and for environmental safety purposes.

The development of RLSs has a 100-year history.[80] Numerous designs of light sources have been developed and utilized; they differ in the radionuclide and phosphor employed, as well as in their design and other features. RLSs have a number of attractive, practically important, characteristics. Above all, they are autonomous devices that do not require a power supply. The service life of RLSs is typically long (10+ years). RLSs are operative at temperatures throughout the climatic range without maintenance and adjustment. The spectral composition of RLSs widely varies from UV to near-IR ranges.

All this makes RLSs competitive in a number of applications with conventional light sources, despite relatively low brightness levels. For example, gas-filled tritium-based radioluminescent light sources (RLSs-T) are employed for watch-dial lighting and instrument panel lights, as well as for lights for emergency exits, airfields, and runways.[20,80] Also, RLSs-T have been used in devices converting radioactive decay energy to electrical energy.

2.3.1 Energy-Conversion Efficiency

Every RLS — in particular, tritium-containing RLS-T — comprises a phosphor, radioactive material (in the latter case, tritium-containing), and design elements. In view of the low radiation power of tritium, every RLS-T must be designed in a way ensuring maximal possible efficiency of conversion of the radioactive decay energy to light energy. To compare efficiencies of variously designed RLSs-T, it is necessary to calculate the efficiency of conversion of the beta energy of decay to the light energy η_l as

$$\eta_l = \eta_\beta \eta_{\beta-l} = \frac{\Phi_e}{P} \qquad (2.54)$$

where Φ_e is the radiant flux emitted by an RLS-T as measured in energy units and P is the radiation power of the radioactive material in the RLS-T. The parameter P

can be calculated by multiplying the activity A of tritium in the RLS-T by the average energy of its beta particles ε_{av}.

The radiant flux of RLS-T can be calculated from the measured brightness B in cd/m² and the luminescence spectrum of the source. As a rule, when measuring the luminescence spectrum, one obtains the spectral density of the radiant flux in relative units per unit wavelength range $b(\lambda)$, relative units/nm. Taken as one relative unit is the intensity of the radiant flux at the maximum in the spectrum. Designate the radiant flux, as measured in energy units (for example, watts) and emitted by light sources in the wavelength range from λ_{max} to $\lambda_{max} + d\lambda$ (λ_{max} is the wavelength at the maximum in the spectrum), as f_m. The spectral density of the radiant flux $\Phi_e(\lambda)$, W/nm, can be represented as

$$\Phi_e(\lambda) = f_m \cdot b(\lambda) \tag{2.55}$$

Knowing the spectral density of the radiant flux $\Phi_e(\lambda)$, one can define the light flux Φ_V in photometric units (for example, lumens) as

$$\Phi_V = K_m \cdot \int_0^\infty U(\lambda) \cdot \Phi_e(\lambda) d\lambda = K_m \cdot f_m \cdot \int_0^\infty U(\lambda) \cdot b(\lambda) d\lambda \tag{2.56}$$

Here, $U(\lambda)$ is the spectral luminous efficiency function for photopic vision, which is a standard tabulated value[90]; K_m is the absolute maximum value of the luminous efficiency at the peak wavelength of 555 nm. $K_m = 683$ lm/W.[90]

The spectral luminous efficiency function for photopic vision differs from zero only in the wavelength region corresponding to visible light. Therefore, it is sufficient to calculate the integral in Equation 2.56 only for the visible wavelength region, which is within 380 to 780 nm.

Each surface unit of the gaseous tritium–based light source emits light uniformly into a solid angle 2π; that is, each surface unit is a Lambert light source. Therefore, the luminosity, S_V, in lm/m², of the surface unit is[90]

$$S_V = \pi \cdot B \tag{2.57}$$

Correspondingly, the light flux from the entire RLS-T will make

$$\Phi_V = \pi \cdot B \cdot S, \tag{2.58}$$

where S is the lighting surface area for RLS-T, m².

Comparison of Equations 2.56 and 2.58 yields the parameter f_m:

$$f_m = \frac{\pi \cdot B \cdot S}{K_m \cdot \int_{380}^{780} U(\lambda) \cdot b(\lambda) d\lambda} \tag{2.59}$$

Therefore, the spectral density of the radiant flux $\Phi_e(\lambda)$, W/nm, and the spectral density of the radiosity $S_e(\lambda)$, W/(m²·nm), of RLS-T will be defined by the equations

$$\Phi_e(\lambda) = \frac{\pi \cdot B \cdot S \cdot b(\lambda)}{K_m \cdot \int\limits_{380}^{780} U(\lambda) \cdot b(\lambda) d\lambda} \tag{2.60}$$

$$S_e(\lambda) = \frac{\pi \cdot B \cdot S \cdot b(\lambda)}{K_m \cdot \int\limits_{380}^{780} U(\lambda) \cdot b(\lambda) d\lambda} \tag{2.61}$$

The radiant flux and radiosity can be determined by integrating Equations 2.60 and 2.61 over the entire luminescence spectrum. The luminescence spectrum of RLS-T typically is in the visible region of the spectrum. Therefore, integration can be carried out in the wavelength range of 380 to 780 nm. Corresponding Equations 2.60 and 2.61 have the form

$$\Phi_e = \frac{\pi \cdot B \cdot S}{K_m \cdot Z_c} \tag{2.62}$$

$$S_e = \frac{\pi \cdot B}{K_m \cdot Z_c} \tag{2.63}$$

In Equations 2.62 and 2.63, the parameter Z_c was calculated by

$$Z_C = \frac{\int\limits_{380}^{780} U(\lambda) \cdot b(\lambda) d\lambda}{\int\limits_{0}^{\infty} b(\lambda) d\lambda} \tag{2.64}$$

One eventually obtains

$$\eta_l = \frac{\pi \cdot B \cdot s}{A \cdot K_m \cdot \varepsilon_{av} \cdot Z_c} = \frac{3.14 \cdot B \cdot S \cdot 10^{-4}}{3.7 \cdot 10^{10} \cdot 683 \cdot 5.7 \cdot 1.6 \cdot 10^{-16} \cdot A \cdot Z_c} = 0.0136 \frac{B \cdot S}{A \cdot Z_c} \tag{2.65}$$

where B is brightness in cd/m², S is the light-emitting surface area in cm², and A is the activity of the radionuclide in Ci.

Along with energy efficiency, each radioluminescent light source can be characterized by light output ξ, lm/Ci, i.e., the light flux, in lumens, gained from unit-loading activity, in curies. For RLSs-T, the corresponding equation has the form

$$\xi = \frac{\Phi_V}{A} = \eta_l \cdot K_m \cdot Z_c \cdot P_{sp} = 0.023 \eta_l \cdot Z_c \qquad (2.66)$$

2.3.2   Tritium Gas–Filled Radioluminescent Light Sources

2.3.2.1   Design and Characteristics

The most extensively used RLSs-T are sealed glass ampules of any desired shape coated on inside surfaces with a semitransparent layer of a phosphor and a tube cavity filled with gaseous tritium. Such light sources are manufactured by a number of firms in different countries — in particular, by MB-Microtec (Switzerland), SRB Technologies Ltd. (Great Britain), and Isotope Production Association (Russia). Fabrication of RLSs-T is regulated by standards (e.g., in Great Britain[91] and the U.S.[92]) or by technical specifications (in Russia[93,94]).

Gas-filled RLSs-T are commercially available in a variety of shapes and sizes. The diameter of RLSs-T designed as a straight tube, for example, can vary from 0.5 to 15 mm, and the length can vary from 2 to 200 mm. The brightness of such RLSs-T depends on their inside dimensions and on the tritium pressure. Figure 2.19 depicts the RLS-T brightness vs. tritium pressure plots for various internal tube diameters. Also, RLS-T brightness depends on phosphor type and phosphor coating density.[89,95] The plot[89] of the brightness vs. surface density for the phosphor coating has a maximum at the surface density of the phosphor coating of 3 to 5 mg/cm². The rise in brightness with increased phosphor thickness is due to better beta-phospher interactions while the drop is due to increased light absorption. The ascending branch in the dependence is due to increase in the light flux because increase in the amount of the phosphor per unit surface area exceeds the light absorption by the phosphor particle layer, and the descending branch, to enhancement of light absorption in the phosphor coating.

The brightness intensity of gas-filled RLSs-T varies with time. Figure 2.19 displays the relative change in brightness of these RLSs-T with time as reported by various manufacturers. This figure also shows the time dependence of the tritium activity relative to the initial one (in relative units) due to its radioactive decay. It is seen that the brightness intensity decreases in time at a rate almost twice that of the tritium activity decrease due to decay. For example, if the tritium activity decreases by half within 12 years, the luminescence intensity decreases about fourfold within the same period. This points to other significant processes in gas-filled RLSs-T, in addition to tritium decay, that reduce their luminance. These processes include:

1. Radiation-induced processes in the phosphor affecting the energy efficiency of luminescence

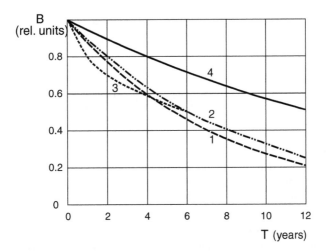

Figure 2.19 (1) to (3) Dependences of the relative brightness of gaseous tritium–filled RLSs-T on time according to References 93, 103, and 91, respectively, and (4) dependence of the tritium activity in RLS-T on time (in rel. units).

2. Radiation-induced surface reactions of the phosphor with tritium affecting the surface properties of the phosphor, leading to formation of a layer of nonluminescent or nontransparent compounds
3. Changes in the gas composition and pressure in RLS-T affecting the conditions of energy transfer from the gas volume to the phosphor surface

The decrease of tritium activity due to its diffusion through walls of the body of gas-filled RLSs-T is not significant. Radiation safety regulations stipulate that, when using RLSs-T, the leakage of tritium, e.g., from RLSs-T containing 10 Ci tritium, should not exceed 1800 Bq/h.[96] Within a 10-year period, this will produce no more than 4.3 mCi — significantly smaller than the loading activity of tritium. Evidently, this process does not affect the observed reductions in brightness.

The processes indicated in items 1 and 2 will be considered in Chapter 4. Here, discussion will be on how gas pressure and composition affect the energy transfer from the gas volume to the phosphor-coated surface. The gas composition in the RLSs-T varies with time. Radioactive decay of tritium (a diatomic gas) produces ^{3}He, a monatomic gas (one decaying tritium molecule gives two helium atoms). As a consequence, the pressure in the gas cavity of RLSs-T tends to increase with time. Variations of the partial pressures of helium p_{3He} and tritium p_{3H_2} and the total pressure of the gas mixture p_{tot} with time t can be described as

$$p_{3H_2} = p_0 \exp\left(-\frac{0.69t}{T_{1/2}}\right) \qquad (2.67)$$

$$p_{3_{H_e}} = 2p_0 \left[1 - \exp\left(-\frac{0.69t}{T_{1/2}} \right) \right] \qquad (2.68)$$

$$P_{tot} = p_0 \left[2 - \exp\left(-\frac{0.69t}{T_{1/2}} \right) \right] \qquad (2.69)$$

where p_o is the initial tritium pressure.

The gas mixture density remains constant in time. However, the Z/M_A ratio for the gas mixture of tritium and its decay product ^{3}He (for tritium $Z/M_A = 1/3$, and for ^{3}He $Z/M_A = 2/3$) changes. In accordance with the Bethe–Born equation (see Equation 1.6), the ionization loss of electrons (betas) propagating in the gas mixture with changed proportions increases in time. Thus, the description of beta propagation in the gas mixture and the corresponding calculations by the Monte Carlo method suggest[97] that the specific beta flux power declines with time at a rate exceeding that predicted by radioactive decay alone. The calculations show that within the half-life of tritium, the specific beta flux power decreases from 11.5 to 4 μW/cm^2, not to 5.75 μW/cm^2, as predicted by the radioactive decay law, i.e., by 65% rather than 50%.

One can assume that the Monte Carlo method gives fairly reliable results and the luminescence brightness of the phosphor coating is proportional (all other conditions being the same) to the specific power of the beta flux exciting the luminescence. This allows the following conclusions on how various processes contribute to the decline of the brightness of gas-filled RLSs-T. This 75% decrease in brightness relative to the initial brightness within one tritium half-life is due to a 65% decrease in the specific power of beta flux from the gas to the phosphor coating and a 29% decrease in the energy efficiency of radioluminescence of the phosphor coating.

2.3.2.2 Manufacturing Technology

The conventional technology of manufacturing gas-filled RLSs-T has the following elements. A borosilicate or quartz glass tube or capsule of any shape and size is coated on the inside surface with a phosphor layer using a binder (orthophosphoric acid, sodium silicate, potassium silicate, boron oxide, etc.[9,80]). Next, the glass tube is hermetically sealed on one end (using a gas-flame burner or CO_2 laser). The resulting half-finished product is hermetically connected to an apparatus for handling gaseous tritium (like that shown in Figure 2.2). Then it is evacuated under heating (for eliminating vapors of water and other volatile admixtures), whereupon it gets filled with gaseous tritium and hermetically sealed. The tritium pressure inside is typically 10 to 250 kPa. The half-finished product is sealed off using a gas-flame burner or CO_2 laser; the latter is suited for sealing off glass tubes less than 6 mm in diameter.[8]

At the same time, a CO_2 laser affords improved throughput of RLSs-T of a better quality. To this end one can use the conventional procedure above for manufacturing a long (30 to 60 cm) hermetically sealed glass tube filled with tritium and coated on the inside surface with a phosphor layer and then can slice and seal the tube with CO_2 laser cutting. This affords numerous hermetically sealed miniature gas-filled RLSs-T[98–102].

2.3.2.3 Energy Efficiency

One can calculate an efficiency of conversion of beta energy to light one $\eta_{\beta-l}$ for gaseous tritium–based light sources with external dimensions of diameter 5 mm and length 31 mm. Such light sources have been utilized in one of the breadboard models of an indirect-conversion betavoltaic battery (see Section 1.2.4). They contained about 0.34 Ci of gaseous tritium with pressure of 0.7 atm. Their brightness and the phosphors employed, as well as the radiosity calculated by Equation 2.63, are presented in Table 2.6. Brightness is usually measured with a photometer, and the relative spectral density of the radiant flux can be measured with various standard spectral-measuring instruments.

Figure 2.20 presents the wavelength dependencies of the spectral density of the radiosity for various RLSs-T as calculated from data measured by Equation 2.63. Because all the RLSs-T considered have identical dimensions and loading activity of tritium, in all the cases the specific power of the beta flux dP_β/dS from gas to phosphor is the same. The preceding calculations (see Section 2.2.3.5) showed that it is equal to 3.6 μW/cm². Then the efficiency of conversion of the beta particle flux to light will be

$$\eta_{\beta-l} = S_e \Big/ \frac{dP_\beta}{dS} \qquad (2.70)$$

The energy conversion efficiencies calculated by Equation 2.70 are presented in Table 2.6. The energy efficiency of luminescence for zinc-sulfide phosphors employed in blue-, green-, yellow-, and white-emitting sources is close to 20%. The energy efficiency of yttrium oxide–based phosphor utilized in red-emitting source is about 10% (see Table 3.1 in Chapter 3); at the same time, the parameter $\eta_{\beta-l}$ is lower, for reasons stated previously (9 to 10% and 5% for zinc sulfide– and yttrium oxide–based phosphors, correspondingly).

Table 2.6 Characteristics of RLSs-T

| Color of Emitted Light by RLS-T (Visual Inspection) | Phosphor | | Brightness of RLS-T, cd/m² | S_e, μW/cm² | $\eta_{\beta-l}$ |
	Trade-mark	Chemical Composition			
Blue	RST-450	ZnS:Cl	0.18	0.36	0.10
Green	FK-106z	ZnS:Cu	0.42	0.35	0.097
Yellow	RST-580	(Zn,Cd)S:Cu,Cl	0.51	0.33	0.092
Red	RST-612	Y_2O_3:Eu	0.17	0.18	0.05
White	RST-6500	Mixture ZnS:Cl+ (Zn,Cd)S:Cu,Cl	0.37	0.32	0.089

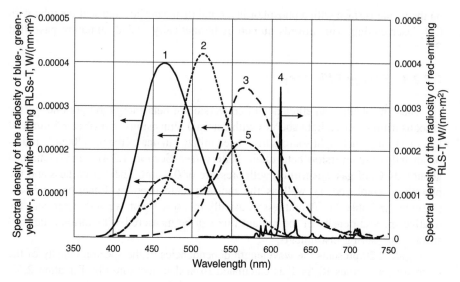

Figure 2.20   Spectral density of the radiosity of RLSs-T emitting light of different colors as a function of the wavelength. Light emitted by RLSs-T: (1) blue, (2) green, (3) yellow, (4) red, and (5) white. For characteristics of RLSs-T, see the text.

Section 2.2.4.3 calculated η_β for gas-filled RLSs-T with external diameter and length of 5 and 31 mm, respectively, and obtained $\eta_\beta = 0.84$. The energy efficiency of such RLSs-T as calculated by Equation 2.54 will be about 8% for blue-, green-, yellow-, and white-emitting sources and about 4% for red-emitting sources. The energy efficiency calculated by Equation 2.65 for green-emitting source will be equal to 8.4% (this RLS-T has $B = 0.42$ cd/m^2, $S = 2.8$ cm^2, $Z_c = 0.557$, and $A = 0.34$ Ci). The light output as calculated by Equation 2.66 is equal to 0.0011 lm/Ci.

Next, calculate energy conversion efficiency and light output for gas-filled RLSs-T based on characteristics available from manufacturers:

- In Reference 93 a cylindrical RLS-T with external diameter of 9 mm and length of 120 mm ($\varnothing$ 9 × 120 mm) had the following characteristics: brightness no less than 0.7 cd/m^2, loading activity no greater than 13.5 Ci, light-emitting surface area 21.5 cm^2, and green luminescence with $Z_c = 0.64$. This affords $\eta_l = 0.025$ and $\xi = 0.00037$ lm/Ci.
- In Reference 91, a $\varnothing$ 6.25 × 51 mm cylindrical RLS-T had the following characteristics: brightness greater than 1.88 cd/m^2, loading activity less than 3.8 Ci, light-emitting surface area 5.65 cm^2, and green luminescence with $Z_c = 0.64$. This affords $\eta_l = 0.059$ and $\xi = 0.00087$ lm/Ci. Also in this reference, a $\varnothing$ 3.5 × 51 mm cylindrical RLS-T had the following characteristics: brightness greater than 0.377 cd/m^2, loading activity less than 0.25 Ci, light-emitting surface area 2.86 cm^2, and green luminescence with $Z_c = 0.64$. This affords $\eta_l = 0.091$ and $\xi = 0.0013$ lm/Ci.
- In Reference 103, a $\varnothing$ 6.0 × 20 mm cylindrical RLS-T had the following characteristics: light flux 275 μcd, loading activity less than 1.48 Ci, and green luminescence with $Z_c = 0.64$. This affords $\eta_l = 0.039$ and $\xi = 0.00057$ lm/Ci.

The brightness of gas-filled RLSs-T cannot in principle be increased beyond a certain value because of self-absorption of tritium beta particles in gas and light self-absorption in semitransparent layers of the phosphor. To overcome these limitations, various designs of RLSs-T have been proposed. The primary aim is to ensure close proximity of the radiation source (tritium) and the phosphor and to reduce losses of energy by self-absorption. Options for such RLSs-T will be considered next.

2.3.3 Metal Tritide–Based Radioluminescent Light Sources

The suitability of metal tritides for RLSs-T has been discussed in a number of works.[9,80,82] Consider characteristics of the layers of various metal tritides (specific beta flux power, working layer thickness, etc.) that can be employed in RLSs-T (see Section 2.2.4.2). Tritium-based radioluminescent light sources with a metal tritide as beta source are typically represented by a combination of planar metal tritide film and planar semitransparent phosphor screen. Also, RLSs-T can be designed in a more original manner, as shown in Figure 2.21. Now we will consider the illumination-engineering characteristics achievable with metal tritide–based RLSs-T.

Experiments with metal tritides employed zinc sulfide–based phosphors. The light emitter was designed as a titanium tritide thin-film combined with a semitransparent phosphor layer; brightness varied[9,82] from 0.09 to 0.22 cd/m². The tritium surface activity content at the optimal titanium tritide layer thickness of 1 μm was 0.437 Ci/cm². Energy conversion efficiency and light output of this RLS-T were estimated from previous data by Equations 2.65 and 2.66 (with $Z_c = 0.64$ for green phosphor, $\lambda_{max} = 520$ nm) as $\eta_l = (0.0136 \cdot 0.22)/(0.437 \cdot 0.64) = 0.011$ and $\xi = 1.6 \cdot 10^{-4}$ lm/Ci, respectively.

2.3.4 Radioluminescent Light Sources Based on Tritiated Zeolites

To circumvent the major limitation of gas-filled RLSs-T, namely, significant energy loss by useless absorption in the gas, it was suggested to combine the beta source and the luminescence center in the same substance. A possible way of ensuring such a "close proximity" is the use of zeolites. On the one hand, rare-earth metal ions intended as luminescence centers can enter the zeolite lattice by ion exchange and, on the other hand, the molecules of water, in particular, tritiated water, can be sorbed in the pores of the material. When exposed to tritium betas, ions of the rare-earth element will luminesce, with minimum energy lost by absorption in the medium.[24,25]

Experiments used 4A zeolites ($Na_2O:Al_2O_3:2SiO_2$) and 13X zeolites ($Na_2O:Al_2O_3:2.5SiO_2$). The manufacturing procedure for tritiated zeolites was discussed in Section 2.1.3.3. According to Clough et al.[24] and Gill et al.,[25] 52 mg of the Eu:X zeolite sample containing 36 Ci of tritium emitted red light in a narrow spectral band at $\lambda_{max} = 620$ nm with brightness of 1.7 cd/m²; 53 mg of the Tb:X zeolite sample containing 36 Ci of tritium emitted yellow-green light in a narrow spectral band at $\lambda_{max} = 550$ nm with brightness of 2.6 cd/m². With the bulk density of the zeolites[99] d taken as 0.7 g/cm³, the light-emitting surface S for samples with the mass $m = 53$ mg was estimated as

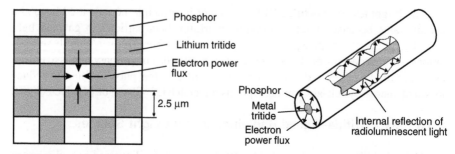

Figure 2.21   Designs of metal tritide–based RLSs-T. (From Kherani, N.P. and Shmayda, W.T., Radioluminescence using metal tritides, *Z. Phys. Chem.*, 183, 453, 1994.)

$$S = 4\pi \cdot \left(\sqrt[3]{\frac{3m}{4\pi d}} \right)^2 = 0.86 \text{ cm}^2$$

The parameter Z_c entering into Equations 2.65 and 2.66 for narrow (with a half-width of several nm) spectral bands that are the luminescence bands of rare-earth elements is approximately equal to $U(\lambda_{max})$. The energy conversion efficiency and the light output of such RLSs-T can be estimated as

$$\eta_l(\text{Eu:X}) = 0.0136 \cdot 1.7 \cdot 0.86/(36 \cdot 0.381) = 0.0014, \; \xi = 1.2 \cdot 10^{-5} \text{ lm/Ci}$$

$$\eta_l(\text{Tb:X}) = 0.0136 \cdot 2.6 \cdot 0.86/(36 \cdot 0.995) = 0.00085, \; \xi = 1.9 \cdot 10^{-5} \text{ lm/Ci}$$

The cited experiments do not characterize the stability of such RLSs-T in time. They mention only that thin layers of polymers [poly(methyl methacrylate), poly(styrenebutadiene)] incorporating the zeolite compositions of interest had a low initial brightness that rapidly dropped (within 2 and 7 days for poly(methyl methacrylate) and poly(styrenebutadiene), respectively), owing to tritium beta radiation–induced darkening of the polymers. It can be assumed that, despite a fairly high radiation resistance of zeolites (as oxide systems), the brightness of the RLSs-T rapidly declines owing to autoradiolysis of tritiated water and desorption of its degradation products (in particular, $^3H_2$).

2.3.5   Radioluminescent Light Sources Based on Tritiated Aerogels

Another conception for RLSs-T with phosphor particles uniformly dispersed throughout the volume and radioactive gas enveloping each phosphor particle is the use of tritium-saturated aerogel–phosphor mixtures.[20,22,23] The technology of preparing tritiated aerogel–phosphor mixtures was considered in Section 2.1.3.3. Tritium is loaded into the phosphor–aerogel matrix by three methods[20,21]: absorption of tritiated water vapor by the aerogel, incorporation of a tritiated organic substance

(TDEB, see Section 2.3.6) into pores of the aerogel, and isotopic exchange of gaseous tritium with hydrogen from the surface OH-groups.

The samples obtained by absorption of tritiated water had initial brightness of 15 cd/m^2. However, weak bonding of tritiated water to the aerogel causes activity to decrease rapidly and brightness to decline. Incorporation of an organic tritium carrier into the aerogel pores affords initial brightness of about 4.1 cd/m^2, which, however, decreases at the same rate (within 5 to 10 days), owing to darkening of the organic component with radiation-induced damage. The light source prepared by the isotopic exchange procedure (within several tens of days) will exhibit stable functioning with a brightness of 7 to 10 cd/m^2, according to theoretical calculations. In practice, the samples obtained in Ellefson et al.[23] had the following characteristics:

WA-1	MD-3B
Dimensions: diameter 1.27 cm, length 1.59 cm	Dimensions: diameter 1.27 cm, length 1.59 cm
Brightness: 1.1 cd/m^2	Brightness: 3.1 cd/m^2
Tritium content: 8.4 Ci/cm^3	Tritium content: 20 Ci/cm^3

A sample of tritiated aerogel was loaded with tritium by isotopic exchange within about 200 days. The initial brightness of this sample (after the surface was loaded with tritium) was 1.51 cd/m^2. After the first 1653 days of the experiment, or 1453 days after tritium was loaded into the aerogel surface, the brightness was 1.26 cd/m^2, i.e., about 83% of the initial value. Within the same period the activity of tritium in this sample of the aerogel decreased to 80% of the initial value. Evidently the brightness of this composition decreases mainly due to radioactive decay of tritium.

These characteristics of the WA-1 and MD-3B samples can be used for calculating energy conversion efficiency and light output by Equations 2.65 and 2.66 for light sources with surface area $S = 8.87$ cm^2, volume $v = 2.01$ cm^3, and $Z_c = 0.64$ (for green phosphor with $\lambda_{max} = 520$ nm) — sample WA-1: $\eta_l = 0.012$, $\xi = 1.7 \cdot 10^{-4}$ lm/Ci; sample MD-3B: $\eta_l = 0.014$, $\xi = 2.1 \cdot 10^{-4}$ lm/Ci.

2.3.6 Radioluminescent Light Sources Based on Tritiated Organic Compounds

Historically, the first RLSs-T based on organic compounds were radioluminescent paints (self-luminous paints). They have been developed since the 1960s as a substitute for radiotoxic paints based on radium compounds. Interestingly, production of all radium-containing self-luminous paints in the U.S. during World War II used a total of 190 g radium.[104]

The first self-luminous paints were mixtures of a tritium-containing organic compound and an inorganic luminophore, typically ZnS:Cu (in certain cases, added with a lacquer binder).[80] It was suggested that long-chain aliphatic alcohols or acids[105] and even such an exotic substance as tung oil[106] be used as tritium carriers.

However, this line has not received further development, probably because of insufficient radiation resistance and safety of low-molecular-weight compounds.

Much more promise was shown by tritiated polymers (see Section 2.1.4). A self-luminous paint based on polystyrene, ZnS, and lacquer binder studied in Reference 107 afforded the following technical characteristics of the coating:

- Luminophore particle size 5 to 20 μm
- Specific activity of the coating, less than 0.5 Ci/g
- Specific brightness 2.5 to 50 μcd/g
- Optimal thickness of the coating 0.2 mm
- Self-luminous paint consumption 0.05 g/cm^2

Equation 2.65 gave energy efficiency (the efficiency of conversion of the β-decay energy to light energy) of about 0.023 for this self-luminous paint.

The light output per unit activity varies from 8.5 to 340 mcd/(m^2·Ci), according to various authors,[27,47] with luminance 1 cd/m^2. As noted in References 27 and 107, the luminance of polystyrene-based self-luminous paints decreases at a rate of ca. 10% per year, which is twice that of tritium beta decay.[26] The luminance of self-luminous paints based on other polymer matrices decreased at a rate of 25 to 40% a year (nature of the matrices was not indicated).[27]

For self-luminous paints based on [^{3}H]polyethylene with specific activity of 600 Ci/g, the initial luminance, of 64 mcd/m^2 decreased by half within 100 days. The color of a polymer film with such initial characteristics changed within this time period from straw-yellow to dark brown.[40]

The desorption rate of tritium from tritium-based self-luminous paints is about 0.12 μCi per day per mCi of the loading activity.[26]

A new stage in the evolution of RLSs-T based on organic compounds began in the late 1980s. This was motivated by the search for new application fields for tritium as a nuclear power engineering by-product and increased interest in miniaturizing devices for conversion of radioactive decay energy to light energy. This stimulated development of original methods of synthesis of highly active and more stable tritium-containing polymer matrices.

The authors of References 20 and 107 to 112 suggested that 1,4-bis(phenylethynyl)benzene (DEB), whose molecule is able to accommodate eight tritium atoms, be catalytically hydrogenated with gaseous tritium:

$$C_6H_5-C{\equiv}C-C_6H_4-C{\equiv}C-C_6H_5 \rightarrow C_6H_5-(C^3H_2)_2-C_6H_4-(C^3H_2)_2-C_6H_5$$

DEB   TDEB

The hydrogenation product (TDEB) was dissolved in a polystyrene matrix containing a mixture of organic luminophores, which gave an all-organic optically clear light source — in other words, a self-luminous plastic scintillator. The solubility of TDEB in polystyrene is as low as 20 wt%, which enabled preparation of transparent matrices with an activity of about 160 Ci/g. The possibility of using m-isomer (m-TDEB) was also shown.

To shift the maximum in the emission spectrum of the light source to the region of the maximal wavelength sensitivity of the human eye, a mixture of three organic luminophores was used: 2-phenyl-5-(4'-biphenylyl)-1,3,4-oxadizole (PBD), 3-hydroxyflavone (3HF), and rubrene. The choice of these specific luminophores was dictated by the goal for maximal overlap of emission and absorption spectra in the chain of emitters PBD–3HF–rubrene (see Figure 2.22).[20,108]

The luminance of the source increases proportionally to the tritium content per unit volume and attains 3 cd/m^2, which is approximately as bright as a television screen. The achievable conversion efficiency of the beta decay to light energy was 1.8%.[20]

The conversion efficiency of the beta decay energy in organic, optically clear light sources tends to decrease over time because of radiation-induced degradation of the polymer base and radiation-induced degradation of organic phosphors. If one considers changes in the optical characteristics in the polymer base only, these changes can be attributed to an additional double-bond conjugation system arising in the polymer or to capture of electrons or radicals in traps of different natures (see Sections 5.2.4 and 5.3.3). As seen from Figure 2.23, in the tritium-containing polystyrene matrix an absorption "wall" arises, which gradually shifts from the UV to visible region of the spectrum with increasing internal irradiation dose.

The light emitted by the TDEB–polystyrene–PBD–3HF–rubrene system changes with time from orange to bluish, probably due to radiolytic degradation of the rubrene luminophore. Independent experiments with irradiating luminophores in toluene under inert atmosphere showed that a dose of 100 Mrad caused PBD, 3HF, and rubrene to degrade by 23, 52, and 74%, respectively. At the same time, p-terphenyl did not exhibit marked degradation under the same conditions.[20]

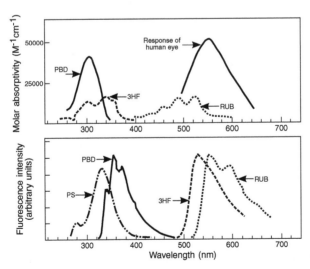

Figure 2.22 Absorption (top) and fluorescence (bottom) spectra of polystyrene, PBD, 3HF, and rubrene. (From Renschler, C.L. et al., Solid-state radioluminescent lighting, *Radiat. Phys. Chem.*, 44, 629, 1994.)

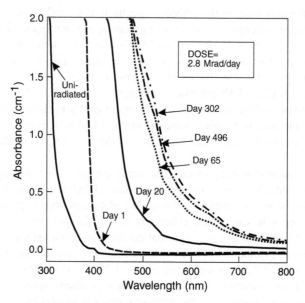

Figure 2.23   Absorption spectra for a mixture of 0.96 g of polystyrene with 0.016 g (9.4 Ci) of *m*-TDEB as a function of time. (From Renschler, C.L. et al., Solid-state radioluminescent lighting, *Radiat. Phys. Chem.*, 44, 629, 1994).

The possibility of sorption on silica gel is shown for the tritiation product of 1,3-bis(phenylethynyl)benzene (*m*-TDEB) as combined with inorganic lumino-phores to create a light source without employing polymer matrices.[20,23] Mullins et al.[40] employed 1,4-diphenylbutadiyne (DPB) in place of DEB as the initial unsaturated compound. DPB can add eight tritium atoms by catalytic hydrogenation, thus converting to tritiated diphenylbutadiyne (TDPB), i.e., to diphenyl[^{3}H$_8$]butane:

$$C_6H_5-(C\equiv C)_2-C_6H_5 \rightarrow C_6H_5-(C^3H_2)_4-C_6H_5$$

DPB   TDPB

TDPB was dissolved in a polystyrene matrix containing a mixture of the organic luminophores mentioned previously. Advantages of TDPB over TDEB were neither reported nor expected in view of a longer tritiated aliphatic chain in TDPB. The synthesis of some tritium-labeled organic luminophores (e.g., anthracene, stilbene, *p*-terphenyl, etc.) has been described.[26,113] However, given the low proportion of luminescent additives in plastic scintillators (see Section 3.3.1), this means of increasing tritium content per unit mass of the polymeric composition seems of little promise.

The coloring of polysiloxane matrices when irradiated is not as intensive due to the impossibility of forming double bonds (see Section 5.3.3). Therefore, Renschler et al.[20] used gaseous tritium to hydrogenate organosilicon-unsaturated compounds, namely, 1,3,5-trivinyl-1,1,3,5,5-pentamethyltrisiloxane and 1,3,5,7-tetravinyltetrame-thylcyclotetrasiloxane; the hydrogenation products were mixed with polydimethylsi-loxane and ZnS. The resulting tritium-containing matrices exhibit certain insignificant advantages over polystyrene matrices within the first 2 months of the operation, but they gradually release tritium-containing low-molecular-weight products.

Mullins et al.[40] concluded that high-activity tritiated organic polymer matrices undergo significant radiolytic degradation and exhibit significant light output losses within several months.

2.3.7 Other Radioluminescent Light Sources

One line in manufacturing gas-filled RLSs-T is creating so-called volumetric (in contrast to RLSs-T coated on the inside surface with a phosphor layer) light microsources (microspheres). Volumetric RLSs-T have in their interiors gaseous tritium and phosphor particles that are not fixed by a binder. Tritium surrounds these particles on all sides, causing them to emit light. The microspheres are prepared in the following manner.[114]

A tritium gas flow "blows" a glass microsphere through an annulus formed between the funnel at the bottom of the crucible with the molten glass and the capillary tube for transporting the mixture of tritium gas and phosphor particles. When transported to the capillary tube, tritium gas passes through the tube with a Ventura section and pulls the phosphor particles from the phosphor chamber forming the mixture. The microsphere with gaseous tritium and phosphor particles in its interior separates from the melt and solidifies. The size of the microsphere and its wall thickness, as well as the tritium pressure in its interior, are regulated by varying the pressure and flow rate ratios of tritium gas and gas above the molten glass layer. Thus, miniature gas-filled RLSs-T can be manufactured.

The density of crystalline zinc sulfide is 4 g/cm^3. At the same time, the bulk density of the zinc sulfide–based phosphor powder is about 2 g/cm^3. The void spaces among the particles are filled with gaseous tritium, so each phosphor particle is enveloped by tritium and the luminance is proportional to the gas pressure.[115,116] At the same time, the brightness of RLSs-T with the interior surface coated with the phosphor and gas cavity filled with gaseous tritium saturates with increasing pressure.

According to Ellefson et al.,[115] the luminance of the bulk phosphor RLS-T with an interior diameter of about 3 mm (1/8 in.; filled entirely with phosphor powder) at the tritium pressure of 25 atm will exceed that of the phosphor-coated surface RLS-T with an interior diameter of about 3 mm (1/8 in.) at the same pressure and will amount to about 6.8 cd/m^2 (ca. 2 fL). The phosphor powder will occupy half of the free volume; the loading activity of tritium in a tube with an internal diameter of 3 mm will be 2.3 Ci per unit length, and the lighting surface area per unit length will be 0.94 cm^2. The energy conversion efficiency of such light source (green phosphor, $Z_c = 0.64$) is 0.059, twice that of the phosphor–coated surface source; the activity of tritium is half that of the phosphor-coated RLS-T.

Such light sources (though with lower tritium pressure and smaller size) can be manufactured from glass tube segments filled with phosphor and tritium[115] or in the form of self-luminous microspheres. These miniature sources have been suggested for forming the desired light pattern. For example, according to calculations in Rivenburg et al.,[116] realization of a betavoltaic cell with indirect energy conversion to provide 50 μW of electrical power requires a 5 × 1.5 cm² panel comprising microspheres 250 μm in diameter.

Losses by absorption of the energy transferred from tritium to the phosphor can be reduced[80,117] by adding an inert gas such as xenon to tritium; then, almost all the tritium beta decay energy will be absorbed by the xenon atoms. Under beta excitation, xenon generates photons with $\lambda_{max} = 172$ nm (see Section 3.1.3 in Chapter 3). This radiation will reach the phosphor surface virtually without self-absorption in the gas and will excite luminescence of the phosphor.

To compare levels of the specific power released on the inside surface of a sphere filled with tritium and a tritium–xenon mixture, the radioluminescence efficiency of xenon, $\eta_{RL\text{-}Xe}$, containing 10 vol.% tritium, is about 0.1.[80] Assuming a tritium pressure of 1 atm in the pure tritium source and 10 atm for that in the tritium–xenon source, with 10 vol.% tritium and 90 vol.% xenon, the partial pressure of tritium in the tritium–xenon–based source will be equal to 1 atm and the volume activities will be identical in these cases. Under these assumptions, one can calculate the specific beta-flux power dP_β/dS released from the spherical tritium-based light source by Equation 2.42. The specific power of the vacuum UV photon flux dP_{VUV}/dS released on the surface in the tritium–xenon–based source will be determined using the equation

$$\frac{dP_{VUV}}{dS} = 3.7 \cdot 10^{10} \cdot A_o \cdot \varepsilon_{av} \cdot \frac{v}{S} \cdot \eta_{RL\text{-}Xe} = 3.7 \cdot 10^{10} \cdot A_o \cdot \varepsilon_{av} \cdot \frac{D_S}{6} \cdot \eta_{RL\text{-}Xe} \quad (2.71)$$

Figure 2.24 shows the dependence of the specific power released on the surface of tritium- and tritium–xenon–based sources. It is seen that the curve of dP_β/dS vs. the sphere diameter D_S described by Equation 2.42 attains saturation, and that described by Equation 2.71 exhibits a linear increase of dP_{VUV}/dS with increasing D_S. For $D_S > 13$ cm, the specific power of the flux of the excitation radiation for the tritium–xenon–filled sphere exceeds that for the tritium-filled one.

Assuming that luminescence efficiencies for beta radiation and vacuum UV irradiation are approximately the same, the luminance of the tritium–xenon–based source with $D_S > 13$ cm exceeds that of the tritium-based source with the same diameter and loading activity of tritium. Similarly, the energy efficiency of the tritium–xenon–based source with the parameters indicated will exceed that of the tritium-based source.

Testing crystalline hydrates with tritiated water for RLSs-T should also be mentioned.[80] These light sources lose their brightness very rapidly from autoradiolysis.

2.3.8 Comparison Characterization of Radioluminescent Light Sources

To compare the characteristics of various modifications of RLSs-T, energy conversion efficiencies and long-term stability are summarized in Table 2.7. The tabulated data show that the beta particle energy of radioactive decay is most efficiently

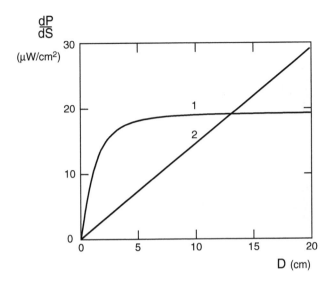

Figure 2.24 Specific power released on the surface of spherical (1) tritium and (2) tritium-xenone light sources as functions of their diameters.

converted to light energy in gas-filled RLSs-T. The RLSs-T utilizing tritiated aerogels are also fairly efficient and surpass gas-filled sources in stability. Bulk phosphor microsources exhibit high brightness and high efficiency and can be used for manufacturing illumination panels for lighting more economically, e.g., photovoltaic batteries. However, the stability of microRLSs-T still remains to be tested.

2.3.9 Radioluminescent Light Sources Suitable for Betavoltaic Battery Application

The efficiency of the electrical energy generation for any design of indirect-conversion battery will depend on the match of the emission spectra of the light source and the sensitivity of the photovoltaic. The efficiency of electrical energy generation by photovoltaics when illuminated by monochromatic light with a wavelength λ (denoted $\eta_{l\text{-}el}(\lambda)$) is defined by the equation (see Chapter 7)

$$\eta_{l\text{-}el}(\lambda) = \frac{P_m(\lambda)}{h\nu} \tag{2.72}$$

Here, $P_m(\lambda)$ is the power released to the load resistance with an optimally matched p–n junction of the photovoltaic and the external circuit per absorbed light quantum with the wavelength λ, and $h\nu$ is the energy of the absorbed light quantum with the wavelength $\lambda = c/\nu$. The $\eta_{l\text{-}el}(\lambda)$ dependence on the excitation light wavelength for betavoltaics is discussed in Chapters 7 and 8. The luminescence spectra of gaseous tritium–based radioluminescent light sources emitting light of different colors are shown in Figure 2.20.

Table 2.7 Initial Energy Conversion Efficiency and Long-Time Stability of RLSs-T

Radioluminescent Light Source	Characteristics of RLSs-T			Energy Conversion Efficiency	Light Output, ξ, mlm/Ci	Ref.
	Brightness, cd/m², or Light Flux	Loading Activity, Ci	Brightness Half-Life, Approx.			
Gas $^3H_2$ – ZnS, Ø9-120 mm, p_{3H_2} = 0.8 atm	No less than 0.7	No greater than 13.5	$0.5 \cdot i_{1/2}$ of tritium	0.025	0.37	93
Gas $^3H_2$ – ZnS, Ø6.25-51 mm, p_{3H_2} = 2.5 atm	1.88	3.8	$0.5 \cdot i_{1/2}$ of tritium	0.059	0.87	91
Gas $^3H_2$ – ZnS, Ø3.5-51 mm, $à_{tritium}$ = 0.8 atm	0.377	0.25	$0.5 \cdot i_{1/2}$ of tritium	0.091	1.3	91
Gas $^3H_2$ – ZnS, Ø6-20 mm, $p_{tritium}$ = 2.5 atm	(275)[a]	1.48	$0.5 \cdot i_{1/2}$ of tritium	0.039	0.57	103
Ti$^3H_2$-ZnS, Ti$^3H_2$ thickness 1 μm	0.22	(0.437)[b]	—	0.011	0.16	9
Eu:X:$^3H_2$O	1.7	36	—	0.0014	0.012	24, 25
Tb:X:$^3H_2$O	2.6	36	—	0.00085	0.019	24, 25
Aerogel-ZnS- absorb.$^3H_2$	1.1	16.9	$i_{1/2}$ of tritium	0.012	0.17	23
Aerogel-ZnS- absorb.$^3H_2$	3.1	40.2	$i_{1/2}$ of tritium	0.014	0.21	23
ZnS bulk-phosphor microsource - $^3H_2$ gas ($p_{tritium}$ = 25 atm)	6.8	(2.3)[c]	—	0.059	0.87	115
Self-luminescent paint polystyrene-ZnS	(up to 50)[d]	(0.5)[e]	$0.5 \cdot i_{1/2}$ of tritium	0.023	0.34	107

a Light flux, μcd.
b Surface activity, Ci/cm².
c Loading activity per unit length, Ci/cm.
d Specific light flux, μcd/g.
e Specific activity, Ci/g.

The power P_m released in the external circuit upon illumination of the photovoltaic by a light flux from RLS-T is given by

$$P_m = g \cdot S \int_\lambda \eta_{l-el}(\lambda) \cdot S_e(\lambda) d\lambda \tag{2.73}$$

S is the illuminated area of the photovoltaic (illuminance of the photovoltaic is assumed to be uniform) and g is the factor defining the illuminance of the photovoltaic by RLS-T and determined by design features of the illuminator. Integration is carried out over the entire wavelength range in which both $\eta_{l-el}(\lambda)$ and $S_e(\lambda)$ differ from zero. Estimating the overlap integral H of the $\eta_{l-el}(\lambda)$ and $S_e(\lambda)$ dependence,

$$H = \int_\lambda \eta_{l-el}(\lambda) \cdot S_e(\lambda) \cdot d\lambda \tag{2.74}$$

for various combinations of radioluminescent light source–photovoltaic. The maximum H value will correspond to the best option.

For example, H was calculated for gaseous tritium–based radioluminescent light sources emitting light of different colors (for characteristics, see Table 2.6) and one of the samples of the AlGaAs-based photovoltaics with known efficiency $\eta_{l-el}(\lambda)$ vs. wavelength plot. The results of calculations of the integral H in relative units are listed in Table 2.8. The highest calculated value of H is taken as unity. As can be seen from this table, the S309-1 sample with a yellow-emitting light source exhibits the greatest value of H.

The results of calculations were verified by measuring the current-voltage characteristics of the S309-1 sample when illuminated by light sources emitting light of different colors. The current-voltage characteristics measured, as well as the voltage dependence of the power released that was calculated on their basis, are presented in Figures 2.25a and 2.25b, respectively. The values of the maximal power P_m, in millivolts multiplied by nanoampere and in relative units (the maximal measured P_m value is taken for 1 relative unit), as well as the open-circuit voltage V_{oc} in millivolts, short-circuit current I_{sc} in nanoamperes, and the fill factor of current-voltage characteristic FF, are also listed in Table 2.8. The measured data suggest that the best performance characteristics are exhibited by the S309-1 sample–yellow-emitting light source pair. The measured P_m values (in relative units) are approximately equal to the calculated values of H for all photovoltaic–light source pairs studied.

Optimizing the performance of an indirect energy conversion battery primarily requires matching the response spectrum of the photovoltaic cell to the emission spectra of bright radiophosphors. This choice can be made using the procedure proposed in this chapter.

Table 2.8   **Calculated and Measured Performance Characteristics of S309-1 Sample Illuminated by RLSs-T Emitting Light of Different Colors**

Color of RLS-T (Visual Inspection)	S_e, W/m²	H, Rel. Units	Measured Parameters				
			V_{oc}, mV	I_{sc}, nA	FF	P_m, mV·nA	P_m, Rel. Units
Blue	0.0036	0.33	276	167	0.55	25,200	0.35
Green	0.0035	0.66	312	309	0.53	50,800	0.70
Yellow	0.0033	1.0	327	427	0.52	72,400	1.0
Red	0.0018	0.62	312	310	0.54	52,000	0.72
White	0.0032	0.73	312	315	0.53	51,800	0.71

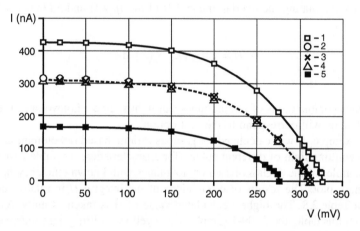

Figure 2.25a   Current-voltage characteristics of sample S309-1 when illuminated by sources emitting light of different colors: (1) yellow, (2) green, (3) red, (4) white, and (5) blue.

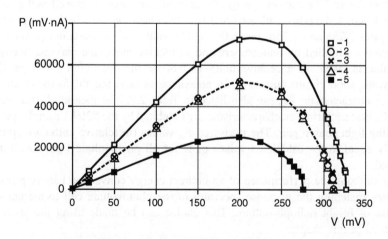

Figure 2.25b   Variation with voltage of the power generated by sample S309-1 when illuminated by sources emitting light of different colors: (1) yellow, (2) green, (3) red, (4) white, and (5) blue.

REFERENCES

1. Corliss, W.R. and Harvey, D.G., *Radioisotopic Power Generation,* Mir, Moscow, 1967 [Russian language].
2. Khol'nov, Yu.V. et al., *Characteristics of Radiation Emitted by Radioactive Nuclides Employed in the People's Economy. Estimated Data. Reference Book,* Atomizdat, Moscow, (1980) [Russian language].
3. Radiation Safety Norms (NRB-99), SP 2.6.1. 758-99, RF Public Health Ministry, St. Petersburg, 1999 [Russian language].
4. Principal Hygienic Rules of Ensuring Radiation Safety (OSPORB-99), SP 2.6.1. 799-99, RF Minzdrav, Moscow, 2000 [Russian language].
5. Belovodskii, L.F., Gaevoi, V.K., and Grishmanovskii, V.I., *Tritium,* Energoatomizdat, Moscow, 1985 [Russian language].
6. *Safe Handling of Tritium. Review of Data and Experience,* Technical Reports Series No. 324, IAEA, Vienna, 1991.
7. Gorlovoi, G.D. and Stepanenko, V.A., *Tritium-Based Emitters,* Atomizdat, Moscow, 1965 [Russian language].
8. Bokwa, S.R. et al., Handling high specific activity tritium, *Nucl. Instrum. Methods Phys. Res., Sect. A,* 257, 52, 1987.
9. Tompkins, J.A. et al., Tritide based radioluminescent light sources, in *Radioluminescent Lighting Technology. Technology Transfer Conference Proceedings,* U.S. DOE, Annapolis, MD, 1990.
10. Gallant, J.L. and Dmitrenko, P., The preparation of tritium, nickel-63 and carbon-14 large-area sources for document imaging, *Nucl. Instrum. Methods Phys. Res., Sect. A,* 257, 29, 1987.
11. Kherani, N.P. et al., Tritiated amorphous silicon for micropower applications, *Fusion Technol.,* 28, 1609, 1995.
12. Tam, S.-W., U.S. Patent 5,605,171, 1997.
13. Tam, S.-W., U.S. Patent 5,604,162, 1997.
14. Tam, S.-W., U.S. Patent 5,765,680, 1998.
15. Kherani, N.P., Shmayda, W.T., and Zukotinski, S., U.S. Patent 5,606,213, 1997.
16. Harry, M. and Zukotynski, S., The use of a direct current saddle-field plasma for the deposition of hydrogenated amorphous silicon, *J. Vac. Sci. Technol., A,* 9, 496, 1991.
17. Kruzelecky, R.V. and Zukotinski, S., The preparation of amorphous Si:H thin films for optoelectronic applications by glow discharge dissociation of SiH_4 using a direct-current saddle-field plasma chamber, *J. Vac. Sci. Technol., A,* 7, 2632, 1989.
18. Kostesky, T., et al., Tritiated amorphous silicon films and devices, *J. Vac. Sci. Technol., A,* 16, 893, 1998.
19. Kherani, N.P., Shmayda, W.T., and Zukotinski, S., U.S. Patent 5,118,951, 1992.
20. Renschler, C.L. et al., Solid state radioluminescent lighting, *Radiat. Phys. Chem.,* 44, 629, 1994.
21. Ashley, C.S. et al., U.S. Patent 5,240,647, 1993.
22. Ashley, C.S. et al., Aerogel matrices for volumetric light/power sources, in *Radioluminescent Lighting Technology. Technology Transfer Conference Proceedings,* U.S. DOE, Annapolis, MD, 1990.
23. Ellefson, R.E. et al., Tritiation of aerogel matrices: T_2O, tritiated organics and tritium exchange on aerogel surfaces, *Radioluminescent Lighting Technology. Technology Transfer Conference Proceedings,* U.S. DOE, Annapolis, MD, 1990.
24. Clough, R.L. et al., U.S. Patent 5,100,587, 1992.

25. Gill, J.T., Hawkins, D.B., and Renschler C.L., Solid state radioluminescent sources using zeolites, in *Radioluminescent Lighting Technology. Technology Transfer Conference Proceedings*, U.S. DOE, Annapolis, MD, 1990.

26. Evans, E., *Tritium and Its Compounds*, 2nd ed., Butterworths, London, 1974.

27. Rosenberg, J., U.S. Patent 3,342,743, 1967.

28. Zeller, A., U.K. Patent 1,002,426, 1962.

29. *Collection of Abstracts of the Scientific-Research Works on Isotopes in the Year 1964*, Gos. Inst. Prikl. Khim., Leningrad, 1965, 48 [Russian language].

30. Murray, A. and Williams, D.L., *Organic Synthesis with Isotopes*, Interscience Publishers, New York, 1958, part 2.

31. *Collection of Abstracts on Chemistry and Technology of Isotopes and Labeled Compounds in the Year 1968*, Gos. Inst. Prikl. Khim., Leningrad, 1970, 5 [Russian language].

32. *Collection of Abstracts of the Scientific-Research Works on Isotopes in the Year 1966*, Gos. Inst. Prikl. Khim., Leningrad, 1967, 12 [Russian language].

33. Evans, C.C. et al., U.S. Patent 3,210,288, 1965.

34. Evans, C.C. and Maynard, J.C., U.K. Patent 1,020,752, 1966.

35. Fisher, E. and Kaltenhauser, A., U.S. Patent 3,238,139, 1966.

36. *Collection of Abstracts of the Scientific-Research Works on Isotopes in the Year 1962*, Gos. Inst. Prikl. Khim., Leningrad, 1964, 9 [Russian language].

37. *Collection of Abstracts on Chemistry and Technology of Isotopes and Labeled Compounds in the Year 1969*, Gos. Inst. Prikl. Khim., Leningrad, 1971, 4 [Russian language].

38. *Collection of Abstracts on Chemistry and Technology of Isotopes and Labeled Compounds*, Gos. Inst. Prikl. Khim., Leningrad, 1(12), 1973, 3 [Russian language].

39. Futterknecht, R. and Neumann, W., U.S. Patent 3,325,420, 1967.

40. Mullins, D.F., Krasznai, J.P., and Mueller, D.A., Development of organic tritium light technology at Ontario Hydro, *Fusion Technol.*, 21, 312, 1992.

41. DeLeo, F.D. and Shapiro E., U.S. Patent 3,033,797, 1962.

42. Evans, E.A. et al., *Handbook of Tritium NMR Spectroscopy and Applications*, John Wiley & Sons, Chichester, 1985, 192.

43. *Compounds Labeled with Radioactive and Stable Isotopes. Catalogue*, V/O Izotop, Moscow, 1979, 15 [Russian language].

44. Gul'ko, V.M. et al., Beta radiation source based on tritiated polystyrene, *At. Energ.*, 48, 111, 1980 [Russian language].

45. Jensen, G.A., Nelson, D.A., and Molton, P.M., Tritiated polymers. Radioluminescent polymer light, in *Radioluminescent Lighting Technology. Technology Transfer Conference Proceedings*, U.S. DOE, Annapolis, MD, 1990.

46. Jensen, G.A., Nelson, D.A., and Molton, P.M., U.S. Patent 5,100,968, 1992.

47. Nelson, D.A., Molton, P.M., and Jensen, G.A., Radioluminescent polymers: preparation of deutero- and tritopolystyrene, *J. Appl. Polym. Sci.*, 42, 1801, 1991.

48. Sokolov, V.I., Chemistry of fullerenes: new allotropic modifications of carbon, *Izv. Akad. Nauk. Ser. Khim.*, 1211, 1999 [Russian language].

49. Beardmore, K. et al., The interaction of hydrogen with C_{60} fullerenes, *J. Phys. Condens. Matter*, 6, 7351, 1994.

50. Jimenez–Vazquez, H.A. et al., Hot-atom incorporation of tritium atoms into fullerenes, *Chem. Phys. Lett.*, 229, 111, 1994.

51. Khong, A., Cross, R.J., and Saunders, M., From $^3He@C_{60}$ to $^3H@C_{60}$: hot atom incorporation of tritium in C_{60}, *J. Phys. Chem.*, 104A, 3940, 2000.

52. Gol'dshleger, N.F. and Moravskii, A.P., Fullerene hydrides: preparation, properties, and structure, *Usp. Khim.*, 66, 353, 1997 [Russian language].

53. Tarasov, B.P., Gol'dshleger, N.F., and Moravskii, A.P., Hydrogen-containing carbon nanostructures: synthesis and properties, *Usp. Khim.*, 70, 149, 2001 [Russian language].

54. Tarasov, B.P. et al. "Synthesis and properties of crystalline fullerene hydrides," *Izv. Akad. Nauk. Ser. Khim.*, 2093, 1998 [Russian language].

55. Dunlap, B.I., Brenner, D.W., and Schriver, G.W., Symmetric isomers of $C_{60}H_{36}$, *J. Phys. Chem.*, 98, 1756, 1994.

56. Jijima, S., Helical microtubules of graphite carbon, *Nature*, 354, 56, 1991.

57. Smalley, R.E., Discovering fullerenes, *Usp. Fiz. Nauk,* 168, 323, 1998 [Russian language].

58. Dresselhaus, V.S., Dresselhaus, G., and Eklund, P.S., *Science of Fullerenes and Carbon Nanotubes*, Academic Press, New York, 1996.

59. Ivanovskii, A.L., Modeling nanotubular forms of matter, *Usp. Khim.*, 68, 119, 1999 [Russian language].

60. Dillon, A.C. et al., Storage of hydrogen in singlewalled carbon nanotubes, *Nature*, 386, 377, 1997.

61. Chen, P. et al., High H_2 uptake by alkali-doped carbon nanotubes under ambient pressure and moderate temperatures, *Science*, 285, 91, 1999.

62. Chambers, A., Hydrogen storage in graphite nanotubes, *J. Phys. Chem.*, B102, 4253, 1998.

63. Wang, Q. et al., Quantum sieving in carbon nanotubes and zeolites, *Phys. Rev. Lett.*, 82, 956, 1999.

64. Johnson, J.K., *http://www.engrng.pitt.edu/~chewww/johnson.html*.

65. Malyarov, V.V., *Basics of the Theory of Atomic Nuclear*, 2nd ed., Nauka, Moscow, 1967 [Russian language].

66. Babichev, A.P. et al., *Physical Values: Handbook,* Grigor'ev, I.S. and Meilikhov, E.Z., Eds., Energoatomizdat, Moscow, 1991 [Russian language].

67. Oliver, B.M. et al., Mass spectrometric determinations of the absolute tritium activities of NBS tritiated water standards, *Appl. Radiat. Isot.*, 40, 199, 1989.

68. Lewis, V.E., Beta decay of tritium, *Nucl. Phys.*, A, 151, 120, 1970.

69. Landau, L.D., On energy loss of fast particles by ionization, *Landau, L. D., Collection of Works*, 1, Lifshits, E.M., Ed., Nauka, Moscow, 1969, 482 [Russian language].

70. Fitting, H.-J., Transmission, energy distribution, and SE excitation of electrons in thin solid films, *Phys. Status Solidi, A*, 26, 525, 1974.

71. Gaber, M. and Fitting, H.-J., Energy-depth relation of electrons in bulk targets by Monte-Carlo calculations, *Phys. Status Solidi*, A, 85, 195, 1984.

72. Makhov, A.F., On electron penetration in solids. I. Electron beam intensity. Transverse path lengths of electrons, *Fiz. Tverd. Tela*, 11, 2161, 1960 [Russian language].

73. Makhov, A.F., On electron penetration in solids. II. Depth distribution of electrons, *Fiz. Tverd. Tela*, 11, 2172, 1960 [Russian language].

74. Pikaev, A.K., *Dosimetry in Radiation Chemistry*, Nauka, Moscow, 1975 [Russian language].

75. Everhart, T.E. and Hoff, P.H., Determination of kilovolt electron energy dissipation vs. penetration distance in solid materials, *J. Appl. Phys.*, 42, 5837, 1971.

76. Florinskii, V.Yu., Conversion of the beta radiation energy by structures based on hydrogenated layers of amorphous silicon, Ph. D. (Phys.-Math.) thesis, Ioffe Physico-Tecnical Institute, St. Petersburg, 1994 [Russian language].

77. Ivanov, V.I., *A Dosimetry Course*, Energoatomizdat, Moscow, 1988 [Russian language].

78. Bochkarev, V.V. et al., Distribution of absorbed energy from a point beta-source in a tissue-equivalent medium, *Int. J. Appl. Radiat. Isot.*, 23, 493, 1972.

79. Nemets, O.F. and Gofman, Yu.V., *Reference Book on Nuclear Physics*, Naukova Dumka, Kiev, 1975 [Russian language].

80. Mikhal'chenko, G.A., *Radioluminescent Emitters*, Energoatomizdat, Moscow, 1988 [Russian language].

81. Lomonosov, I.I. and Soshin, L.D, *Tritium Measurement*, Atomizdat, Moscow, 1968 [Russian language].

82. Kherani, N.P. and Shmayda, W.T., Radioluminescence using metal tritides, *Z. Phys. Chem.*, 183, 453, 1994.

83. *Radioisotopes and Radioisotope-Labeled Compounds. Catalogue*, Vneshtorgizdat, Moscow, 1978 [Russian language].

84. Gus'kova, V.I. and Merkina, T.S., *Radionuclide-Based Compounds and Articles. Catalogue*, TsNIIAtominform, Moscow, 1988 [Russian language].

85. Products and Services Catalogue, Bebig Trade Gmbh.

86. Kuznetsov, R.A. et al., *Radionuclide-Based Sources and Preparations*, Karelin, E.A., Ed., NIIAR Russian Federation State Research Center, Dimitrovgrad, 1998 [Russian language], *http:/www.niiar.simbirsk.su.*

87. Henley, E.J. and Johnson, E.R., *The Chemistry and Physics of High Energy Reactions*, University Press, Washington, D.C., 1969.

88. Gus'kova, V.I. et al., *Alpha, Beta, Gamma, and Neutron Radiation Sources*, Catalogue, 3rd ed., Izotop Publishing House, Moscow, 1980 [Russian language].

89. Korin, A., Givon, M., and Wolf, D., Parameters affecting the intensity of light sources powered by tritium, *Nucl. Instrum. Methods*, 130, 231, 1975.

90. Landsberg, G.S., *Optics*, 5th ed., Nauka, Moscow, 1976 [Russian language].

91. Nuclear Lamps Defence Standard 62-4/Issue 2, Ministry of Defence of G.B., 1972.

92. American National Standard N540; Classification of Radioactive Light Sources. (NBS Handbook 116), U.S. Government Printing Office, Washington, D.C., 1976.

93. SH3.R05 and SH3.R06 Type radioluminescent light sources. Technical specifications, TU 95 2639-97, 1997 [Russian language].

94. SH3.R07 Type radioluminescent light sources. Technical specifications, TU 95 2681-98, 1998 [Russian language].

95. Kavetsky, A.G., Dependences between luminous characteristics of tritium radioluminescent light sources and their design parameters. Preprint of Radium Institute. RI-253, Atominform, Moscow, 1998 [Russian language].

96. *Production and Application of Gaseous Tritium–Based Radioluminescent Light Sources and Articles Thereof: Sanitary Regulations*, SP 2.6.1. 543-96, Information & Publishing Center of Goskomsanepidnadzor of Russia, Moscow, 1996 [Russian language].

97. Walko, R.J. et al., Electronic and photonic power applications, in *Radioluminescent Lighting Technology. Technology Transfer Conference Proceedings*, U.S. DOE, Annapolis, MD, 1990.

98. Thuler, O., U.S. Patent 3,706,543, 1972.

99. Thuler, O., U.S. Patent 3,817,773, 1972.

100. Caffarella, T.E., Radda, G.J., and Watts, D.J., U.S. Patent 4,146,380, 1979.

101. Caffarella, T.E., Radda, G.J., and Watts, D.J., U.S. Patent 4,045,201, 1977.

102. Summers, T.L. and Yenawine D.L., U.S. Patent 4,273,398, 1981.

103. Traser, Self-Activated Light Sources. Mb-Microtec, *http://www.hsrd.ornl.gov/nrc/ssdr/ssdrindm.htm.*

104. Lubenau, J.O., Historical overview of radiation sources in United States, *IAEA Bull.*, 41, 49, 1999.

105. Shapiro, E., U.S. Patent 2,749,251, 1956.

106. McHutchin, J.D. et al., U.S. Patent 3,224,978, 1965.

107. *Autonomous Self-Luminous Coatings*, Prospect No 774-500, NIITEKhIM, Cherkassy, [Russian language].

108. Renschler, C.L., Clough, R.L., and Sheppodd, T.J., Demonstration of completely organic, optically clear radioluminescent light, *J. Appl. Phys.*, 66, 4542, 1989.

109. Sheppodd, T.J. and Smith, H.M., Hydrogen-tritium getters and their applications, in *Radioluminescent Lighting Technology. Technology Transfer Conference Proceedings*, U.S. DOE, Annapolis, MD, 1990.

110. Renschler, C.L. et al., All-organic, optically clear, radioluminescent lights, in *Radioluminescent Lighting Technology. Technology Transfer Conference Proceedings*, U.S. DOE, Annapolis, MD, 1990.

111. Gill, J.T. et al., Solid state radioluminescent sources: mixed organic/inorganic hybrids, in *Radioluminescent Lighting Technology. Technology Transfer Conference Proceedings*, U.S. DOE, Annapolis, MD, 1990.

112. Clough, R.L. et al., U.S. Patent 4,997,597, 1991.

113. Peng, C.T., The validity of the Perrin equation in solute quenching, in *Organic Scintillators*, Horrock, D.L., Ed., Gordon and Beach, 1968, 109.

114. Webb, R.D., U.S. Patent 4,677,008, 1987.

115. Ellefson, R.E., High-pressure bulk-phosphor tritium lamps, in *Radioluminescent Lighting Technology. Technology Transfer Conference Proceedings*, U.S. DOE, Annapolis, MD, 1990.

116. Rivenburg, H.C. et al., U.S. Patent 5,443,657, 1995.

117. Griror'ev, K.V. et al., Atomic sources of optical photons, *Izv. Akad. Nauk SSSR, Ser. Fiz.*, 50, 603, 1986 [Russian language].

CHAPTER 3

Nonradioactive Materials for Nuclear Batteries

A.G. Kavetsky, Y.L. Kaminski, G.P. Akulov, and S.P. Meleshkov

CONTENTS

3.1 INORGANIC LUMINESCENT MATERIALS

3.1.1 Parameters Describing Properties of Phosphors

3.1.1.1 Energy Efficiency of Radioluminescence of Phosphors

A critical characteristic of radiophosphors is their energy transfer as luminescence. For radioluminescence caused by soft tritium beta radiation, as well as for cathodoluminescence, the whole energy of the radiation incident on the phosphor is usually taken into account. Therefore, these cases utilize the concept of the energy efficiency of luminescence, i.e., the ratio of the emitted light energy to the energy of the incident radiation. Energy efficiency is less than energy output, i.e., the ratio of the energy of the light, emitted by the phosphor to the absorbed energy of the radiation exciting the luminescence,[1] taking into account reflection losses suffered by primary electrons and emission losses suffered by secondary electrons. Processes occurring in phosphors under tritium beta particles, which have a maximum energy of 18.6 keV,[2] are similar to those under cathode rays,[3] with energy between 5 and 20 keV. Therefore, studies of cathodoluminescence provide an excellent basis for describing processes of radioluminescence under soft beta radiation.

The energy efficiency η_{CL} of cathodoluminescence (as well as of radioluminescence η_{RL} under soft tritium beta radiation) can be represented as the product[3]

$$\eta_{CL}(\eta_{RL}) = (1 - r_{op}) \cdot \eta_T \cdot \eta_{st} \cdot \eta_q \qquad (3.1)$$

Here, r_{op} is the primary electron backscattering coefficient, η_T is the thermalization coefficient, η_{st} is the Stokes loss coefficient, and η_q is the internal and external quenching coefficient. It is now necessary to analyze the physical sense of each cofactor and its influence on $\eta_{CL}(\eta_{RL})$.

When phosphors are irradiated by cathode or beta rays (with the energy of primary electrons E_p), electrons lose their energy as a result of elastic and inelastic interactions with the crystal lattice ions. A share of the beam energy is carried away by backscatter electrons, which can be classified into three groups.[3,4] The first is secondary electrons with energy close to 3 to 5 eV that arise in a great number of ionization events due to electrons incident on the phosphor. The second group is primary electrons that underwent a number of inelastic collisions and lost a part of their energy by ionization. They escape the phosphor layer with an energy varying from practically zero to the value close to that of the incident primary electrons. The third group is the primary electrons that, after elastic collision without ionization, escape the phosphor surface at a rate equal to that of their incidence on the surface. The spectrum of the backscattered electrons is shown in Figure 3.1. The share of the backscattered electrons, γ, as dependent on the effective atomic number of the substance Z_{eff}, can be estimated using the equation[5]

$$\gamma = \frac{\ln Z_{eff}}{6} - 0.25 \tag{3.2}$$

The values calculated by Equation 3.2 are in agreement with those measured experimentally for single crystals. For example, zinc sulfide has $Z_{eff} = 23$ and, correspondingly, $\gamma = 0.25$. In the first approximation, the average energy of backscattered electrons can be taken as the half energy of the incident electrons.[3] Then, the primary electron backscattering coefficient r_{op} (that is, the share of energy carried away by backscattered electrons) will be approximately equal to $\gamma/2$. For powdered phosphors, the experiment shows that the parameter γ is less than that calculated by Equation 3.2.[5] Therefore, in estimation calculations the backscattering coefficient is taken as 0.1.[3] Most of the energy of the flux of electrons penetrating the phosphor is lost as a result of inelastic interactions with the lattice ions. These interactions generate x-rays, Auger electrons, secondary electrons, phonons, and electron–hole pairs.

The final stage of the relaxation of electronic excitations in phosphors is low-energy relaxed electron–hole pairs and excitons with an energy close or equal to E_g, where E_g is the bandgap energy of the crystal of phosphor. The relaxed electron–hole pairs migrate over the crystal and recombine near lattice defects. If such defects are radiative recombination centers, this will result in a light quantum emission. In this process, the energy released in recombination is only partly converted to electromagnetic radiation; a part of this energy goes into phonon excitation in restructuring the radiative recombination center during its excitation and relaxation. Thus, only part of the energy transferred to the phosphor goes into light emission. Now consider the values of the coefficients characterizing the share of the energy lost in various stages of electron beam energy conversion to light energy.

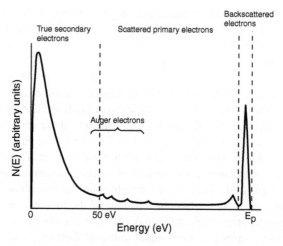

Figure 3.1 Spectral distribution of scattered electrons. (From Ozawa, L., *Cathodolumines-cence: Theory and Applications,* Codansha Ltd., Tokyo, VCH Publishers, New York, 1990.)

Generation and relaxation of the excited electronic states are responsible for origination of lattice phonons. As a result, the energy spent for electron–hole pair origination differs from E_g. The average energy E_{av} going into generation of the relaxed electron–hole pair can be estimated by the following empirical equation[5]:

$$E_{av} = 2.67\, E_g + 0.87 \qquad (3.3)$$

According to other data,[6-8] E_{av} can be represented as

$$E_{av} = \beta\, E_g \qquad (3.4)$$

β is a parameter depending on the material type. Under the assumption that the primary exciting particle loses its energy in creating longitudinal optical phonons and electron–hole pairs, the parameter β will be dependent on the so-called index of efficiency, $\Re,$[6] which can be calculated by the equation[6]

$$\Re = \left(\frac{1}{\varepsilon_{d,\infty}} - \frac{1}{\varepsilon_{d,S}} \right) \cdot \frac{\left(\hbar \omega_{LO} \right)^{\frac{3}{2}}}{1.5 \cdot E_g} \qquad (3.5)$$

$\varepsilon_{d,\infty}$, and $\varepsilon_{d,S}$ are the high-frequency and static dielectric permittivities of the phosphor base material, and $\hbar \omega_{LO}$ is the energy of longitudinal optical phonons of the phosphor base lattice. The β vs. $\Re$ dependence is presented in Figure 3.2 (solid line); this figure also demonstrates the positions of the indices of efficiency for various compounds constituting the phosphor base.

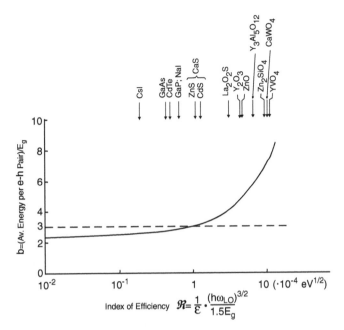

Figure 3.2   Ratio of the average energy going into generation of electron–hole pair to band gap as a function of the index of efficiency $\mathfrak{R}$. The top section demonstrates the positions of $\mathfrak{R}$ for selected materials constituting the phosphor base. (From Robbins, D.J., On predicting the maximum efficiency of phosphor systems excited by ionizing radiation, *J. Electrochem. Soc.*, 127, 2694, 1980.)

Using Equations 3.4 and 3.5 or Figure 3.2 and the bandgap energies for various phosphors, one can estimate the average energy going into generation of electron–hole pairs. For example, for ZnS-based phosphors with $\beta = 2.9$ and $E_g = 3.8$ eV, E_{av} can be estimated as 11 eV. The thermalization coefficient η_T, characterizing energy lost by phonon generation in formation of thermalized electron–hole pairs, can be estimated by the E_g/E_{av} ratio. For zinc sulfide–based phosphors, the bandgap energy is equal to about 3.8 eV,[6] and η_T is close to 0.35.

The energy lost by phonon origination in radiative recombination can be characterized by the ratio of the energy $h\nu$ of the quantum emitted (where $h\nu$ is the energy of the photon emitted at the spectrum maximum) to that of the relaxed electron–hole pair, which is about E_g. For example, for ZnS:Ag phosphor-emitting photons with $h\nu$ close to 2.75 eV, this ratio, termed the Stokes loss coefficient, η_{st}, is about 0.75.

The internal and external quenching coefficient, η_q, characterizes the probability of processes competing with luminescence, namely, nonradiative recombination at nonradiative recombination centers and nonradiative relaxation of the excited luminescence center. The activator concentration in the phosphor is designed so that the radiative recombination centers dominate, reducing probability of recombination at other centers to near zero. The probability of nonradiative relaxation of the excited luminescence center differs from zero only at elevated temperatures (for more detail, see Section 3.1.1.3). Therefore, in estimations, the coefficient η_q is taken as unity.

With these assumptions, the maximal energy efficiency of luminescence for phosphors can be estimated by the equation

$$\eta_{CL}\left(\eta_{RL}\right) = \left(1 - r_{op}\right) \cdot \frac{h\nu}{\beta \cdot E_g} \tag{3.6}$$

The results of calculations by this equation for selected phosphors are presented in Table 3.1. This table also presents $h\nu$, E_g, and β values according to Robbins[6] and the experimentally measured energy efficiencies. The tabulated data show approximate agreement between calculated and experimental energy efficiencies of luminescence.

Table 3.2 presents experimental values of the energy efficiency of cathodo- and radioluminescence of the phosphors not included in Table 3.1.

3.1.1.2 Spectral-Luminescent and Color Characteristics of Phosphors

Spectral composition of the light emitted by a phosphor is usually characterized by the spectral density of its radiant flux, $\Phi_e(\lambda)$. This is the ratio of the radiant flux in the infinitesimal wavelength range $d\lambda$, comprising λ, to $d\lambda$:

$$\Phi_e(\lambda) = \frac{d\Phi_e}{d\lambda} \tag{3.7}$$

The plot of the spectral density of the radiant flux $\Phi_e(\lambda)$ vs. wavelength is essentially the luminescence spectrum of the phosphor. $\Phi_e(\lambda)$ is measured in power per wavelength (e.g., watt/nanometer). In practice, the spectral density of the radiant flux is measured in relative units per nanometer, where one (or 100) relative unit is the maximal value of $\dfrac{d\Phi_e}{d\lambda}$ for that spectrum.

It is expedient to analyze the spectral composition of the light emitted by a phosphor in the following coordinate system: the ratio of the number of photons dN within the energy range dE, including photons with the energy E, to dE (i.e., the spectral density of the photon flux $N(E)$) vs. the photon energy E. The corresponding relations for converting the spectral density of the radiant flux per unit wavelength range $d\lambda$, $\Phi_e(\lambda)$, to that per unit frequency range $d\nu$, $\Phi_e(\nu)$, as well as relations between the spectral densities of the photon and radiant fluxes, have the forms[17,18]

$$\Phi_e(\nu) = \frac{\lambda^2}{c} \cdot \Phi_e(\lambda) \tag{3.8}$$

where $\nu = \dfrac{c}{\lambda}$, and

$$N(E) = \frac{\lambda^3}{c^2} \cdot \Phi_e(\lambda) \tag{3.9}$$

Table 3.1 Calculated and Experimental Values of the Energy Efficiency of Luminescence for Selected Phosphors

Phosphor (Chemical Composition)	$h\nu$, eV	E_g, eV	β	Energy Efficiency, % Calculation	Experiment Value	Ref.
ZnS:Ag	2.75	3.8	2.9	22	21	9, 10
					20.4	11
					19.6	12, 13
					20	1
ZnS:Cu	2.3	3.8	2.9	19	17.5	12
					15–20	1
$Y_3Al_5O_{12}:Tb^{3+}$	2.3	6.3	5.6	6	9	14
$Y_2O_3:Eu^{3+}$	2.0	5.6	4.6	6.5	7.5	9
					6.5	12, 13
					8	11
					10	1
$YVO_4:Eu^{3+}$	2.0	3.7	7.5	6.5	6.0	9
					6.0–6.7	13
					6.6	14, 15
					10	1
$La_2O_2S:Eu^{3+}$	2.0	4.4	3.9	10.5	11	6

where $E = \frac{h \cdot c}{\lambda}$. Here, c is the speed of light and h is Planck's constant. In practice, the relation $E(\text{eV}) = \frac{1240}{\lambda(\text{nm})}$ is often utilized.

The potential curve model[18] predicts that, when luminescence of the crystalline phosphor is due to single electronic transitions and the emission spectrum is represented by one band, the energy dependence of the spectral density of the photon flux can be described by the Gaussian

$$N(E) = N(E_{max}) \cdot \exp\left[-\alpha \cdot (E - E_{max})^2\right] \tag{3.10}$$

where E_{max} is the photon energy at which the spectral density of the photon flux has a maximum and α is the parameter related to the band half-width, δ, by Equation 3.9[18]:

$$\alpha = 4 \cdot \ln 2 / \delta^2 \tag{3.11}$$

In cases when the phosphor contains more than one luminescence center, the emission spectra are complex. They can be resolved into elementary components by a number of methods, such as the Alentsev–Fok method[19] and others.[20]

The emission spectrum of the phosphor, whose shape is close to Gaussian, is often characterized by the wavelength corresponding to the position of the maximum in the spectrum and the band half-width. It should be remembered, however, that the position of the maximum in the spectral distribution pattern plotted, e.g., in the $\Phi_e(\lambda) - \lambda$ coordinates, does not correspond to the energy at which the maximum in the spectral distribution pattern for the same radiation plotted in the $N(E) - E$ coordinates is situated.

Table 3.2 Energy Efficiency of Phosphor Luminescence Induced by Cathode Rays and Beta Particles

Phosphor (Chemical Composition)	Luminescence Color	η_{CL}, %	η_{RL}, %	Ref.
ZnS:Ag, Al	Blue	25		9, 10
		21.2		11
		24.1		14
ZnS:Cl		18		10
ZnS:Al		22		10
Zn_2SiO_4:Ti		8.5		10
$CaO\cdot MgO\cdot SiO_2$:Ti		7.5		10
$2CaO\cdot Al_2O_3\cdot SiO_2$:$Ce^{3+}$		4		10
			1.5–2	1
YBO_3:Ce^{3+}			4–6	1, 16
$Sr_3(PO_4)_2$:Eu^{2+}		5.5		11
			7–9	1
$MgWO_4$		2.9		12
		2		10
		2.5		9
$CaWO_4$:Pb		3.4		12–14
		3		9
LaOCl:Pr^{3+}			3	1
ZnS:Tm^{3+}			3.7	1
(Zn,Cd)S:Ag	Green	21		12
		15.4–19.8		11
			up to 20	1
(Zn,Cd)S:Cu,Al		16.8–18.4		11
ZnS:Cu,Al		25		10
ZnS:Au,Al		15.5		10
ZnO:Zn		11.8		12
		7.5		10
		5.1		12, 13
Zn_2SiO_4:Mn^{2+}		4.7–6.8		12, 13
		6–8		9
		7.4		11
			4 ± 1	1
La_2O_2S:Tb^{3+}		11.8		12
			12 ± 3	1
Y_2O_2S:Tb^{3+}		18.3		14
			12 ± 3	1
LaOCl:Ho^{3+}			3	1
(Zn,Cd)S:Ag	Yellow	18.7		12
		19.4		14
		19.5		10
		19		9
			up to 20	1
(Zn,Cd)S:Cl		17.5		10
$(ZnO,BeO)_2\cdot SiO_2$:Mn^{2+}		7		10
$(Zn,Mg)F_2$:Mn^{2+}			7	1
$ZnO\cdot SiO_2$:Mn^{2+}			2	1
Y_2O_2S:Eu^{3+}	Red	9.8		11
		13		12

Table 3.2 Energy Efficiency of Phosphor Luminescence Induced by Cathode Rays and Beta Particles (Continued)

Phosphor (Chemical Composition)	Luminescence Color	η_{CL}, %	η_{RL}, %	Ref.
$Gd_2O_3{:}Eu^{3+}$		9.6		12, 13
		8.7		14
			10	1
$YBO_3{:}Eu^{3+}$			7 ± 1	16
$Ca_2P_2O_7{:}Bi$		2		10
$Zn_3(PO_4)_2{:}Mn^{2+}$		3.1		12
		6		9
			5	1

Along with spectral characteristics, color characteristics are used for describing properties of phosphors. Colors of light emission can, in principle, be obtained for different spectral distribution patterns of radiation. Clearly, with gaseous tritium–based radioluminescent light sources (RLSs-T) employed in light signal facilities, the color of light emission of the phosphor utilized is important. When RLSs-T are employed as light sources for indirect-conversion nuclear batteries, the spectral distribution of photon radiation should be matched with the spectral sensitivity of the photovoltaic converter. Since phosphors are often classified according to the color of their light emission, one can take advantage of this convenient classification when describing properties of specific phosphors (see Section 3.1.2).

Without going into details of designing various colorimetric systems employed for color measurement, since these are described in sufficient detail in textbooks and user-oriented editions,[21-23] it can be noted that the color of light emitted by a phosphor can be characterized by the chromaticity coordinates, x and y. These can be calculated based on the spectral density of the radiant flux using the following equations[21-23]:

$$x = \frac{x'}{x' + y' + z'} ; y = \frac{y'}{x' + y' + z'} \qquad (3.12)$$

where

$$x' = \int_{380}^{780} \Phi_e(\lambda) \cdot \bar{x}(\lambda) d\lambda ; y' = \int_{380}^{780} \Phi_e(\lambda) \cdot \bar{y}(\lambda) d\lambda ; z' = \int_{380}^{780} \Phi_e(\lambda) \cdot \bar{z}(\lambda) d\lambda$$

In these equations $\bar{x}(\lambda)$, $\bar{y}(\lambda)$, and $\bar{z}(\lambda)$ are so-called ordinates of the addition curves (standard tabulated functions[23]) presented in Figure 3.3.

To determine the color of light emitted by phosphor based on chromaticity coordinates, it is sufficient to attribute the latter to the chromaticity coordinate areas corresponding to various colors. The so-called color plot shown in Figure 3.4 contains areas corresponding to different colors.[24] In what follows (see Section 3.1.2), the color of light emitted by phosphor will be determined, based on chromaticity coordinates with data presented in Figure 3.4.

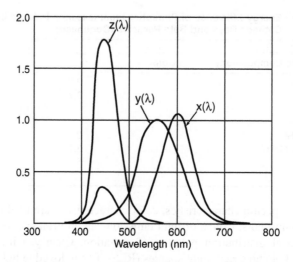

Figure 3.3 Color addition curves for XYZ system.

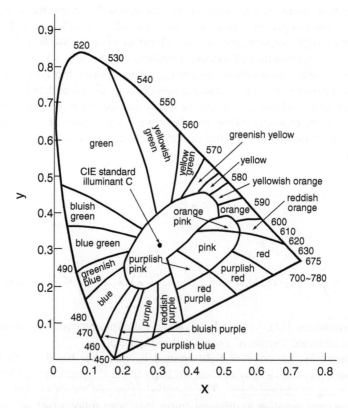

Figure 3.4 Color plot with different color areas indicated. (From Kelly, K.L., Color designations for lights, *J. Opt. Soc. Am.*, 33, 627, 1943.)

3.1.1.3 Physical Models Describing Properties of Phosphors and Temperature Quenching of Luminescence

Many regular features in the variation of spectral properties of crystalline phosphors can be explained by the band model and potential curve model. According to the band model,[18] when the phosphor base atoms form a crystal lattice, their electrons get collectivized, and the electron energy levels degenerate into energy bands. The upper energy band filled with electrons is termed the valence band. It is separated from the next band of energies allowed for electrons in crystals, the so-called conduction band, by the forbidden band. However, defects due to various disturbances of the crystal lattice — both intrinsic defects (vacant sites and interstitial atoms (ions)) and impurity defects (atoms (ions) of impurities at regular sites of the lattice) can present energy levels in the forbidden band. These local levels and the valence and conduction bands can be involved in electronic transitions, including those with light quantum emission. Impurities introduced into crystalline phosphors to create local energy levels to participate in electronic transitions with light quantum emission are termed activators.

On exposure to ionizing radiation, electron–hole pairs form in the crystalline phosphors. The electrons have energies close to the bottom of the conduction band. Holes, which are vacant sites among the valence electrons, move freely in the crystal with energies close to the valence band top. A hole can get to the local level of the activator. Recombination of such a hole with an electron from the conduction band can result in emission of a light quantum, termed recombination luminescence. The wavelength of the emission band maximum corresponds to the energy difference between the local level and the conduction band's bottom. This scheme of recombination luminescence is termed the Shoen–Klasens model.[4,25]

Electron–hole recombination can also follow another path. The center, which is the recombination site, can have an effective positive charge, which makes it an electron trap. A hole can recombine with an electron trapped by this center, resulting in light quantum emission. This scheme of recombination luminescence is termed the Lambe–Klick model.[4,26]

In addition to these recombination events is the situation when a hole and an electron are captured by different centers. Light emission results from electronic transition from the electron trap to the hole trap; this is the Prener–Williams donor–acceptor scheme of recombination.[4,27] Recombination luminescence is characteristic of many zinc sulfide–based phosphors.

If the activator forms energy levels that correspond to its ground and excited states in the forbidden band, electronic transitions between levels also become possible. This intracenter transition proceeds as follows. A hole from the valence band can be captured at the ground level of the activator, and an electron from the conduction band at an excited state. Recombination of the hole and electron localized at the ground and excited levels of the activator, respectively, gives rise to luminescence called intracenter luminescence.[18] It is characteristic of europium-activated yttrium oxide– or yttrium oxysulfide–based phosphors.

Along with radiative recombination centers, the crystal can contain other defects, which can be sites of nonradiative recombination of electron and hole. Thus, recombination can follow radiative and nonradiative pathways. With increasing numbers of nonradiative recombination centers, luminescence intensity tends to decline.

A hole trapped by the radiative recombination center is localized on for a limited time. The probability w_T for the hole to pass from the local center to the valence band is determined by the temperature; it can be represented by the expression[18]

$$w_T \sim \exp\left[-E_A/(kT)\right] \tag{3.13}$$

where E_A is the activation energy of the process corresponding to the energy difference separating a local level and the valence band top, k is the Boltzmann constant, and T is the temperature in Kelvin.

The probability of the radiative recombination w_r is temperature independent. Therefore, the relation between probabilities of the radiative recombination at the luminescence center to total probability of the hole transitions from the luminescence center, η_r, will determine (within a certain temperature range) the temperature dependence of the luminescence intensity. The corresponding equation has the form

$$\eta_r = \frac{w_r}{w_r + w_T} = \frac{1}{1 + q_v \cdot \exp\left(-\dfrac{E_A}{k \cdot T}\right)} \tag{3.14}$$

where q_v is the frequency factor. Equation 3.14 is a practical description of temperature dependence of the luminescence intensity for the phosphor in a number of cases. For example, temperature quenching of luminescence in ZnS:Ag, co-phosphor adheres to Equation 3.14.[28]

Shallow electron traps can also act as quenching centers. With lowering temperature, the time spent by electrons in these traps and, therefore, the probability of recombination of the latter with holes increases. This causes luminescence efficiency to decline with lowering temperature, which is responsible for the appearance of a maximum in the emission intensity vs. temperature plot.[18]

In many cases, emission spectra of phosphors are due to recombination processes at several recombination centers with varying energies in the forbidden band relative to the energy of conduction band bottom. As a result, the temperature dependence of the emission intensity of phosphors exhibiting recombination luminescence can deviate from Equation 3.14.

Experiment shows that, upon heating above a certain temperature, the luminescence intensity of the phosphor begins to decline sharply, even when this cannot be due to interception of excitation energy by the quenching centers.[18] In this case, nonradiative transitions can be explained in terms of the potential curve model.[18] This model allows for the oscillation motion of the luminescence center. Thus, its position in the energy level scheme of the crystalline phosphor is represented by a potential curve whose energy depends on the so-called configuration coordinate and not by a level characterized by a single energy value.

The configuration coordinate is the parameter related to the displacement of the luminescence center from its equilibrium position as a result of thermal oscillations. The potential curve corresponding to an excited state of the luminescence center is shifted relative to that of the ground state and is usually broader (see Figure 3.5). The electronic transitions from excited to ground state of the luminescence center, corresponding to a light quantum emission, are shown with vertical lines from the minimum in the curve of the excited state until intersection with the curve of the ground state.

The curves corresponding to the ground and excited states intersect at a point designated in Figure 3.5 as M. This means that, when in an excited state, the system can take a configuration identical to that of the ground state at a sufficiently high vibrational energy. In this case, nonradiative transition from an excited to the ground state takes place. The activation energy required for such a nonradiative transition is designated in Figure 3.5 as E_T. The probability that the vibrational energy fluctuation will reach the E_T value is proportional to $\exp[-E_T/(kT)]$. Taking the probability of the radiative transition as temperature independent, the ratio of probability of radiative transition to the total probability of the transition from excited to ground state can be expressed by an equation similar to Equation 3.14. Clearly, in this case the latter will describe the temperature dependence of the emission intensity as well. However, for internal quenching, the parameters it incorporates will have another meaning.

3.1.1.4 Models of Luminescence Centers and Energy Diagrams for Phosphors

Before discussing specific models of luminescence centers in phosphors employed in RLSs-T, it is necessary to note that the luminescent properties of phosphors are strongly affected by the methods of their synthesis. Zinc sulfide–based phosphors are synthesized in the presence of a flux, i.e., a substance promoting the growth of large crystals at the expense of small ones.[18] Commonly used fluxes are chlorides, in particular, sodium chloride.

Upon introduction into the phosphor base of an activator whose valence differs from that of the ions of the base (e.g., Ag^+ ions introduced into zinc sulfide), vacancies form in the anionic crystal sublattice for charge compensation, designated

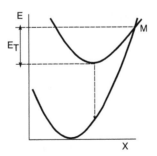

Figure 3.5 Potential curve model for the luminescence center.

as $V_S^{\bullet\bullet}$. Introduction of chloride ions into the anionic sublattice is energetically a favorable charge compensation approach. These defects are designated as $Cl_S^{\bullet}$. Also, with monovalent cations as activators for ZnS, trivalent cations (Al^{3+}, Ga^{3+}, In^{3+}) can be used for charge compensation. The impurities promoting dissolution of the activator in the phosphor base and creation of the luminescence centers are termed coactivators.[18,29] The latter can also be outside the luminescence center. The coactivator may alter the quenching center charge, thus decreasing the recombination cross section for them, which increases light output.

For zinc sulfide–based phosphors, copper and silver ions are frequently used as activators, while trivalent metal ions (Al^{3+}, Ga^{3+}, In^{3+}) or halide ions are used as coactivators. Depending on the concentration of the activator and coactivator ions in the zinc sulfide crystal, its luminescent properties are classified into five main types.[4,5,30] Figure 3.6 illustrates this classification. The activator and coactivator ion concentrations vary from 10^{-6} to 10^{-4} mole/mole of the basic substance.

The luminescence spectra for these five types of zinc sulfide–based phosphors are presented in Figure 3.7.[5,31] According to Shionoya,[5] the coactivator does not affect the luminescence spectrum, except for self-activated crystals. With copper replaced by silver as the activator, the luminescence spectra are blue-shifted by about 0.1 eV; gold is also used as an activator. The luminescence spectrum of ZnS:Au,Al, corresponding to G-Cu luminescence, is slightly shifted to longer wavelengths relative to phosphor with copper as activator.

The structure of the luminescence centers in crystals of different types of zinc sulfide–based phosphors is presented in Table 3.3 with the following designations:

- $Cu_{Zn'}$ for monovalent copper ion at the cationic sublattice site

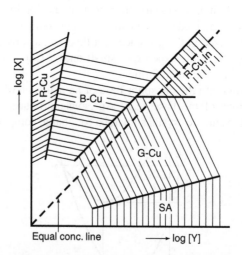

Figure 3.6 Relation between the color of light emission and concentrations of the activator (_) and coactivator (Y) in zinc sulfide–based phosphors. (G-Cu) green, (B-Cu) blue, (R-Cu) red, (SA) self-activated, blue, (R-Cu, In) with Al^{3+}, Ga^{3+}, In^{3+} as coactivators only. (From Van Gool, W., Fluorescence centers in ZnS, *Philips Res. Rept. Suppl.*, 3, 1, 1961.)

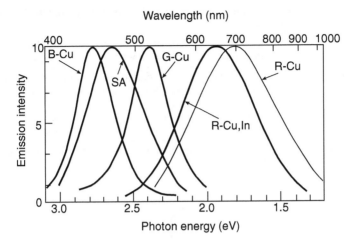

Figure 3.7   Luminescence spectra of different types of zinc sulfide–based phosphors at room temperature. The activators and coactivators in the phosphors whose spectra are presented in the figure are listed below.

Luminescence Type	Activator	Coactivator	Crystal Structure
G-Cu	Cu	Al	wurtzite
B–Cu	Cu	I	zinc blend
SA	—	Cl	zinc blend
R-Cu	Cu	—	wurtzite
R-Cu, In	Cu	In	zinc blend

(From Shionoya, S. et al., Nature of luminescence transitions in ZnS crystals, *Jap. J. Phys. Soc.*, 19, 1157, 1964; *Phosphor Handbook*, Shionoya, S. and Yen, W.M., Eds., CRC Press LLC, Boca Raton, FL, 1999.)

- $Al_{Zn'}$ - $Cl_S^{·}$ for aluminum and chloride ions at the neighbor sites of the cationic and anionic sublattices
- $V_{Zn''}$ - $Cl_S^{·}$ for doubly charged vacancy in the cationic sublattice and chloride ion at the neighbor site of the anionic sublattice
- $Al_{Zn'''}$ for trivalent aluminum ion at the cationic sublattice site
- $Cu_{Zn'}$ - $Cl_{int'}$ for monovalent copper ion at the cationic sublattice site and the chloride ion at the neighbor interstitial site
- $V_S^{·}$ - $Cu_{Zn'}$ for vacancy in the anionic sublattice and monovalent copper ion at the neighbor site of the cationic sublattice
- $Cu_{Zn'}$ - $In_{Zn'}$ for copper and indium ions at the neighbor sites of the cationic sublattice

The structure of one of the luminescence centers from Table 3.3 is shown in Figure 3.8. Models of luminescence centers pertain to the dominant band in the spectrum. Detailed analysis of the luminescence spectra of zinc sulfide–based phosphors using the Alentsev–Fok method shows[32] that the luminescence spectrum typically exhibits emission bands from several types of centers. Such centers include a single-charged anionic vacancy, $V_S^{·}$, single- and doubly charged cationic vacancies, $V_{Zn'}$ and $V_{Zn''}$, etc. The bands due to these centers overlap with the main band and have lower intensities.

Table 3.3 Structure of the Luminescence Centers in Zinc Sulfide–Based Phosphors

Luminescence Type	Phosphor	Type of Transition in Luminescence	Structure of the Center
G-Cu	ZnS:hu,Al(Cl)	Donor–acceptor pair	$Cu_{Zn'}$ and $Al_{Zn'}$ - $Cl_{S}^{\cdot}$
SA	ZnS:Cl(Al)	Donor–acceptor pair	$V_{Zn''}$ - $Cl_{S}^{\cdot}$ $(Al_{Zn''})$
B-Cu	ZnS:hu,I(Cl)	Intracenter	$Cu_{Zn'}$ - $Cl_{int'}$
R-Cu	ZnS:hu	Intracenter	$V_{S}^{\cdot\cdot}$ - $Cu_{Zn'}$
R-Cu, In	ZnS:hu,In	Intracenter	$Cu_{Zn'}$ - $In_{Zn'}$

Source: *Phosphor Handbook*, Shionoya, S. and Yen, W.M., Eds., CRC Press LLC, Boca Raton, FL, 1999.

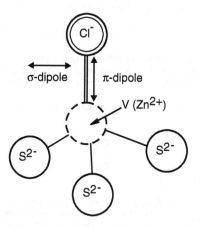

Figure 3.8 Structure of the $V_{Zn''}$ - $Cl_{S}^{\cdot}$ luminescence center in ZnS:Cl phosphor responsible for its blue luminescence band. (From *Phosphor Handbook*, Shionoya, S. and Yen, W.M., Eds., CRC Press LLC, Boca Raton, FL, 1999.)

In zinc sulfide phosphors, luminescence centers strongly interact with the lattice of the phosphor base. As a result, the luminescence spectra contain wide bands with half-widths of several tens of nanometers. In europium-activated yttrium oxide– and yttrium oxysulfide–based phosphors, the luminescence center is the Eu^{3+} ion. The electronic transitions responsible for luminescence occur within the $4f^6$ configuration.[33] The $4f^6$ shell of europium is screened from the external surrounding by the $5s^2$ and $5p^6$ shells, which considerably decreases the width of the electronic transition lines to 0.1 cm^{-1}. The energy level structure of the electronic state of europium in the lattice is determined by the Coulomb interaction of electrons with the nucleus and one another (tens of thousands cm^{-1}), spin-orbit coupling (several thousand cm^{-1}), and interaction with the crystal field of the lattice from incomplete screening of the 4f shell by the outer filled shells (hundreds cm^{-1}). Diverse electronic transitions between different electronic states of the Eu^{3+} ion are responsible for a great number of lines in the radioluminescence spectrum of europium-activated yttrium oxide– and yttrium oxysulfide–based phosphors.

It should also be noted that the europium ion substituting the yttrium ion in the Y_2O_3 lattice can occupy sites of different symmetries,[33] C_2 and S_6 (also denoted as C_{3i}). These two symmetry types of the cationic sites in the Y_2O_3 lattice are shown in Figure 3.9. With Eu^{3+} ions introduced into these cationic sites, the Eu^{3+} ($4f^6$) atomic J levels exhibit different patterns of crystal field splitting. This, in turn, increases the number of lines in the luminescence spectrum. The schemes of the electronic transitions for the Eu^{3+} center in the $Y_2O_3{:}Eu^{3+}$ phosphors[35,36] and $Y_2O_2S{:}Eu^{3+}$ phosphors[37] are presented in Figures 3.10 and 3.11, respectively.

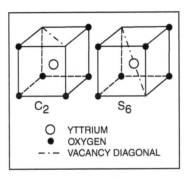

Figure 3.9 Two symmetry types for the Y^{3+} sites in the Y_2O_3 lattice. (From Forest, H. and Ban, G., Evidence for Eu^{+3} emission from two symmetry sites in $Y_2O_3{:}Eu^{+3}$, *J. Electrochem. Soc.*, 116, 474, 1969.)

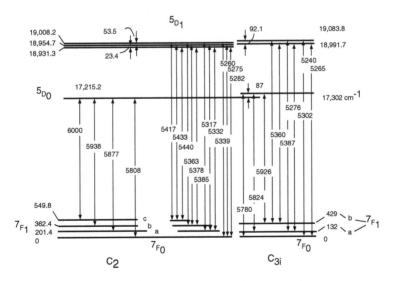

Figure 3.10 Scheme of the principal electronic transitions for the Eu^{3+} center in the $Y_2O_3{:}Eu^{3+}$ phosphor. (From Pappalardo, R.J. and Hunt, R.B., Dye-laser spectroscopy of commercial $Y_2O_3{:}Eu^{3+}$ phosphors, *J. Electrochem. Soc.*, 132, 721, 1985; Hunt, R.B. and Pappalardo, R.J., Fast excited-state relaxation of Eu–Eu pairs in commercial $Y_2O_3{:}Eu^{3+}$ phosphors, *J. Lumin.*, 34, 133, 1985.)

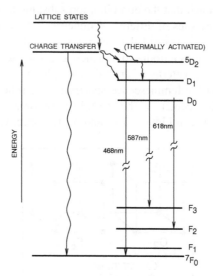

Figure 3.11 Scheme of the principal transitions for the Eu^{3+} center in the $Y_2O_2S:Eu^{3+}$ phosphor. (From Robbins, D.J., Cathodoluminescence saturation phenomena in $Y_2O_3S:Eu$, *J. Electrochem. Soc.*, 123, 1219, 1976.)

As seen from these figures, the spectra can contain many lines. The presence or absence of these lines in the spectrum will depend, along with specific crystal-field splitting patterns, on the transition probability of the sample excitation conditions, temperature, etc. The photoluminescence spectra of one of the $Y_2O_3:Eu^{3+}$ samples at 77 K at different excitation wavelengths are presented in Figure 3.12. This figure also shows the electronic transitions corresponding to each emission line. For example, the most intense line in the spectrum at $\lambda_{max} = 611$ nm is due to the $^5D_0 \rightarrow {}^7F_2$ electronic transition. This intense line is also present in the cathodo- and radioluminescence spectra.

Summarizing, the model views that concern the structure of the luminescence centers in phosphors suitable for RLS-T applications also explain the main regular features in formation of the luminescence spectra of these phosphors under tritium beta excitation.

3.1.1.5 Mechanisms of Radioluminescence of Crystalline Phosphors

The generally accepted conception[18,38] is that, in the radioluminescence of crystalline phosphors, exciting radiation is absorbed by the bulk material and redistributed but that light is emitted by individual submicroscopic luminescence centers. These centers originate primarily from impurities (activators), introduced in small concentrations (from thousandths of a percent to several percent) that form a solid solution with the base material. The latter is commonly termed the phosphor base. The luminescence spectrum is dependent on the nature of the activator (e.g., by that of silver ion in the case of the ZnS:Ag phosphor) though independent of nature and energy characteristics of the exciting radiation. Along with activators, impurities can

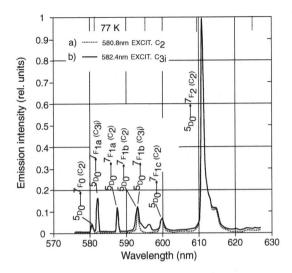

Figure 3.12   Luminescence spectra of Y_2O_3:Eu^{3+} at 77_ at different excitation wavelengths. (From Pappalardo, R.J. and Hunt, R.B., Dye-laser spectroscopy of commercial Y_2O_3:Eu^{3+} phosphors, *J. Electrochem. Soc.*, 132, 721, 1985.)

cause the luminescence to be strongly reduced. These impurities are termed quenchers and corresponding submicroscopic formations are quenching centers.

Energy efficiency of luminescence of the phosphor (the ratio of light flux power to that of excitation power) is determined by interaction of the radiation with the phosphor base. This suggests the existence of an efficient mechanism of energy transfer from the absorption site of an ionizing particle (e.g., electron) or a photon to the luminescence centers. This mechanism is related to the crystalline state of the substance. This generates the need in preparation of a phosphor under conditions favoring formation of fairly large crystal blocks (more than 0.1 μm in size). Ions constituting the phosphor base are close to one another in the regular crystal lattice sites; the valence electrons are collectivized and treated as belonging to the whole crystal. As previously mentioned, the energy levels of the phosphor base lattice ions are split and merge into one band (valence band). In a similar way, the excited states not occupied by electrons form a conduction band. In crystalline phosphors, similar to all semiconductors and dielectrics, these two bands are separated by an energy gap, namely, a forbidden energy band.

The position of the activator ions incorporated into the phosphor base lattice can be regular crystal lattice sites; this causes crystal lattice damages (defects). For ZnS:Ag phosphor, for example, these defects are the Ag^+ ions ($Ag_{Zn'}$). Such defects carry an effective negative charge. Since the crystal remains neutral, this suggests compensation by positively charged defects, which can be created by Cl^- ions supplied by the chloride flux. Chloride ion replaces S^{2-} in the lattice site and, owing to deficit of the negative charge compared to the S^{2-} ion, creates an efficient positively charged defect ($Cl_S^{\cdot}$). Crystal defects form upon incorporation of other activators in a similar way.

In the proximity of negatively charged defects such as $Ag_{Zn'}$ with their repulsion action, the minimal energy required for transition of an electron to the conduction band decreases. This is equivalent to rising of an energy level in the forbidden band on top of the valence band. If the activator atoms do not interact, the energy levels created by them preserve their discrete character. As a result, the electron occupying this level belongs to the defect (crystal space perturbed by the impurity) rather than the crystal as a whole.

In addition to defects formed by impurity atoms, the lattice of a crystalline phosphor can comprise sites not occupied with ions, namely, vacancies, and also atoms or ions of the base substance displaced into interstitial space, namely, interstitial atoms or ions. By contrast to impurity defects that are bound to atoms (or ions) of the impurities, these defects are termed intrinsic crystal defects. Similar to impurity defects, intrinsic defects create (without interacting with one another) local levels in the forbidden band of the crystal that can act as luminescence centers, quenching centers, or traps for electrons and holes. The role of the latter will be discussed next.

The radioluminescence process goes essentially by the following scheme.[38] The ionizing particle or quantum interacting with the crystal eventually generates low-energy electronic excitations, namely, electron–hole pairs and excitons. These migrate through the lattice and transfer their energy to luminescence centers, driving the latter to an excited state. The transition of the luminescence centers back to the ground state is accompanied by photon emission. Thus, the entire process in question can be divided into three stages: generation, migration, and intracenter stages.

In the generation stage, ionizing particles and hard photon radiation quanta are absorbed in the crystal and generate high-energy electrons and holes. Their relaxation gives rise to low-energy electronic excitations, electron–hole pairs, or excitons, that is, bound electron–hole pairs. The generation stage is responsible for a fundamental characteristic of the radiation processes such as the average energy of creation of one electron–hole pair. (See Section 3.1.1.1.)

Low-energy electronic excitations migrating over a crystalline phosphor lattice transfer their energy to luminescence centers, thus driving them into an excited state. Radioluminescence is excited by the same luminescence centers responsible for photoluminescence of the crystalline phosphor. This conclusion is based on the fact that the spectra of radio- and photoluminescence (that is, luminescence excited in the intracenter stage) are identical in the first approximation.[38] Also, the energy efficiency of radioluminescence is known to reach, e.g., for zinc sulfide–based phosphors, about 20% (see Tables 3.1 and 3.2). This exceeds by about two orders the analogous value calculated under the assumption that luminescence centers are excited directly by the electric field of particles exciting the crystal.[38]

This points to the existence of a rather effective mechanism of the excitation energy transfer from the base substance to the luminescence center. One considers two mechanisms, namely, electron–hole and exciton mechanisms. They imply that the excited state of a luminescence center can result either from interaction with an exciton or from the capture of a hole and then an electron (electron recombination luminescence) or of an electron and then a hole (hole recombination luminescence). The probability of annihilation of the exciton outside the luminescence center is much higher than for electron and hole. Moreover, the number of excitons generated is several times smaller than in the case of electron–hole pairs.[38] Thus, the exciton mechanism of energy transfer is several times less effective than the electron–hole one, which makes the latter the major mechanism of energy transfer to luminescence centers for radioluminescence.

Apart from recombination of electron–hole pairs on luminescence centers resulting in luminescence, they are also capable of nonradiative recombination on other lattice defects. Therefore, the energy transfer efficiency depends on the purity of a crystal, since the nonradiative recombination on defects will determine how much energy will be lost by migration. A characteristic of the migration stage is its inertia. The generation stage takes no more than 10^{-10} sec,[38] while the migration stage can last for a period many orders of magnitude longer. The reason is that migrating electrons and holes can localize for some time on various defects (electron or hole traps), for which the energy difference separating the energy levels from bottom of the conduction band or top of the valence band is comparable to the energy of the lattice thermal vibrations (phonons).

The final stage of radioluminescence is recombination of an electron and a hole on a luminescence center (intracenter stage). As a result of such recombination, the luminescence center transits into an excited state. The subsequent transition of a luminescence center from the excited to ground state is accompanied by emitting a light quantum. As mentioned earlier, luminescence centers can be both impurity and intrinsic lattice defects. Radioluminescence spectra depend on the energy characteristics of luminescence centers. The radioluminescence process described here is schematically shown in the band scheme of a crystalline phosphor presented in Figure 3.13.

In this figure, creation of high-energy electrons and deep core holes is shown with a vertical line (1) connecting the valence and conduction bands. Wavy lines show the relaxation of high-energy electronic excitations down to the bottom of the conduction band and up to the top of the valence band (thermalization). The thermalized holes can be captured by negatively charged defects (designated as A in

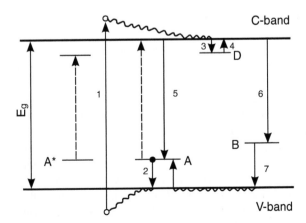

Figure 3.13 Energy band scheme for crystal-phosphor. (V-band) Valence band, (C-band) conduction band, and (E_g) bandgap energy; for other designations, see the text.

Figure 3.13), e.g., $Ag_{Zn'}$ (process 2 in the band scheme). At the same time, thermalized electrons in the conduction band can be entrapped by the site of the crystal containing an excessive positive charge (designated as D in Figure. 3.13), e.g., by a defect of the Cl_S^- type forming a shallow level under the bottom of the conduction band (electron trap) (transition 3 in the band scheme). Under the action of thermal vibrations (phonons), an electron can once again get in the conduction band (transition 4 in the band scheme) and recombine with a hole located on the $Ag_{Zn'}$ site (transition 5 in the band scheme).

Such a recombination is followed by emission of a quantum of light, i.e., by recombination luminescence. The location of the energy levels of the activator centers and their separation from the bottom of the conduction band correspond to a maximum in the luminescence spectrum. Along with the radiative recombination centers, the crystal can comprise nonradiative recombination centers (or quenching centers), designated as B (transitions 6 and 7 in Figure. 3.13). Notably, the most efficient luminescence quenchers for sulfide-based crystalline phosphors are iron, cobalt, and nickel ions that penetrate the zinc sulfide lattice as divalent species.

Along with recombination luminescence centers, the crystalline phosphor can comprise sites for which the ground and the excited levels are located inside the forbidden energy band (see Figure 3.13, A^* center). This is the case, for example, for the $Y_2O_3:Eu^{3+}$ phosphor, where recombination of an electron and a hole on this center involves its transition in an excited state. The subsequent transition to the ground state is accompanied by a light quantum emission (see Section 3.1.1.4).

Though approximate, this scheme of recombination luminescence mechanisms presents the major regular features of the temperature dependence of the luminescence intensity and the influence of ionizing radiation on crystalline phosphor luminescence.

3.1.2 Properties of Powered Phosphors for Radioluminescent Light Sources

3.1.2.1 Blue-Emitting Phosphors

Energy efficiencies of luminescence for blue-emitting phosphors when exposed to cathode or beta radiation suggest that these parameters take a maximal value in the case of ZnS:Ag,Cl (up to 20%) and ZnS:Cl (up to 22%) phosphors (see Tables 3.1 and 3.2). This makes these blue-emitting phosphors most suitable for RLS-T applications. Some commercial blue-emitting phosphors with chemical composition indicated are characterized in Table 3.4.

The luminescence spectra of ZnS:Cl and ZnS:Ag,Cl phosphors are presented in Figure 3.7, curve SA, and Figure 3.14, respectively. The cathodoluminescence spectrum is essentially a bell-shaped band with a half-width of several tens of nanometers. It should be noted that the position of the maximum in the cathodoluminescence spectrum of ZnS:Ag,Cl and ZnS:Cl phosphors varies with their synthesis conditions. Depending on the crystallization temperature, one can obtain a low-temperature modification of zinc sulfide with cubic crystal lattice (zinc blend type), and a high-temperature modification with hexagonal crystal structure (wurtzite type).

Table 3.4 Characteristics of Selected Blue-Emitting Phosphors Suitable for RLS-T Application

Trade-mark	Chemical Composition	Chromaticity Coordinates		λ_{max}, nm	Dispersity, μm	Manufacturer	Ref.
		x	y				
B-3s	ZnS:Ag	$0.148 \leq x \leq 0.165$	$0.105 \leq y \leq 0.125$	460	4.7–16 (>76%)	Luminophor Joint Stock Company, Russia	39, 40
KTC-450	ZnS:Ag,hl	$x \leq 0.16$	$y \leq 0.06$	450	≤14 (≥90%)	Luminophor Joint Stock Company, Russia	39, 40
KDC-450	ZnS:Ag	$x \leq 0.155$	$y \leq 0.055$	450	5–6[a]	Luminophor Joint Stock Company, Russia	41
FK-1	ZnS:Ag	—	—	460	—	Luminophor Joint Stock Company, Russia	40
RST-450	ZnS:hl	Inside a color quadrangle with indicated coordinates of vertices $x_1 = 0.08$, $y_1 = 0.15$; $x_2 = 0.15$, $y_2 = 0.21$; $x_3 = 0.17$, $y_3 = 0.13$; $x_4 = 0.14$, $y_4 = 0.03$		—	—	Luminophor Joint Stock Company, Russia	42
GL47/N-C1	ZnS:Ag	0.147	0.076	—	6.5[a]	Phosphor Technology Ltd., England	43
GL47/N-C2	ZnS:Ag	0.148	0.063	—	8.0[a]	Phosphor Technology Ltd., England	43
No data	ZnS:Ag	—	—	455	2.5–25	Applied Scintillation Technologies Inc, U.S.	44

[a] Average particle size.

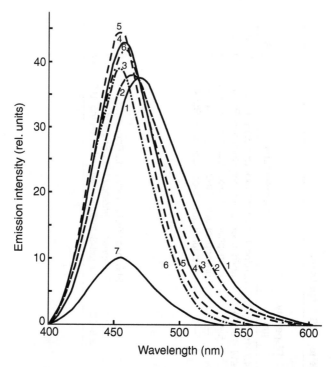

Figure 3.14 Cathodoluminescence spectra of the ZnS:Ag,Cl phosphor at various concentrations of the activator, %: (1) 0, (2) 0.002, (3) 0.005, (4) 0.008, (5) 0.032, (6) 0.128, and (7) 0.512. (From Leverenz, H.W., *An Introduction to Luminescence of Solids*, John Wiley & Sons, Inc., New York, 1950.)

The position of the maximum in the cathodoluminescence spectrum of the wurtzite-type phosphors is shifted to shorter wavelengths (by about10 nm) compared to the zinc blend–type phosphors of identical chemical composition. Increase in the activator concentration, silver ion, in the ZnS:Ag,Cl phosphor shifts the spectrum maximum to shorter wavelengths[45,46] (see Figure 3.14). The optimal concentration of the activator in this type of phosphor lies within 0.005 to 0.03 wt%.[5,46] The changes in the spectrum of ZnS:Ag,Cl with increasing activator concentration are responsible for changes in the color of emitted light from whitish blue to deep dark blue.

The spectral composition of the light emitted by the ZnS:Ag,Cl phosphor varies with the temperature. Levshin et al.[3] studied temperature dependence of the cathodoluminescence spectra of the ZnS:Ag,Cl phosphors over a wide temperature range (from 123 to 373 K). The position of the cathodoluminescence band maximum remains practically unchanged with rising temperature, and the band half-width changes from 60 nm at 123 K to 68 nm at 373 K.[3] Studies of ZnS:Cl phosphors in Shionoya et al.[31] showed that raising the temperature from liquid helium to ambient temperature not only increases the half-width of the emission band but also shifts their maxima insignificantly (by about 10 nm) to shorter wavelengths.

The temperature dependence of the relative luminescence intensity of ZnS:Ag,Cl phosphors (zinc blend type)[28] can be described by Equation 3.14. Loss of the relative luminescence intensity for these phosphors at 373 K reaches 50% of the maximal value measured at about 273 K (0°C). For ZnS:Ag,Cl phosphors (wurtzite type), reduced luminescence intensity within this temperature range is even more significant.[45]

Temperature-related instability of the energy efficiency of radioluminescence of phosphors, η_{RL}, can be characterized[1] by the coefficient K_T (in percent/degree K):

$$K_T = \frac{\left(\eta_{RL}\right)_{T_1} - \left(\eta_{RL}\right)_{T_2}}{\left(\eta_{RL}\right)_{T_1}\left(T_1 - T_2\right)} \cdot 100 \qquad (3.15)$$

Here, $(\eta_{RL})_{T_1}$ and $(\eta_{RL})_{T_2}$ are the energy efficiencies of radioluminescence at temperatures T_1 and T_2, respectively. For ZnS:Ag,Cl, the parameter K_T in the climatic temperature range was estimated[1] as minus (0.5 to 0.8) %/degree K, an estimate consistent with the data.[28,45]

It was shown[47] that, with minor amounts of Eu^{2+} (ca 0.002 wt%) introduced into ZnS:Cl phosphors, the luminescence brightness increases and stability of luminescence of these phosphors under electron beam excitation is enhanced. Rare-earth elements such as Nd, Gd, Dy, Ho, and Ce, when introduced into the composition of the ZnS:Ag,Cl phosphors in amounts up to 0.06 wt%, cause stability of cathodoluminescence to double.

3.1.2.2 Green-Emitting Phosphors

Extensive research has led to production of green- and yellowish-green-emitting phosphors based on oxysulfides of rare-earth metals (Y, La, Gd) activated with Tb^{3+}, as well as $Y_3Al_5O_{12}$ and $Y_3(Al,Ga)_5O_{12}$ activated with Tb^{3+} and Ce^{3+} and other matrices.[5,48] Despite this effort, the energy efficiency of cathodo- and radioluminescence of phosphors based on ZnS:Cu and (Zn,Cd)S:Ag remains highest (see Tables 3.1 and 3.2). Some of the commercial green-emitting zinc sulfide– and zinc-cadmium sulfide–based phosphors suitable for RLS-T applications are characterized in Table 3.5.

Luminescence spectra of ZnS:Cu and (Zn,Cd)S:Ag green-emitting phosphors are shown in Figures 3.7, curve G-Cu, and 3.15, respectively. As seen from these two figures, each spectrum is essentially a wide (with a half-width of several tens of nanometers) bell-shaped band. The position of the maxima in the cathodoluminescence spectra of these phosphors, similar to blue-emitting phosphors discussed previously, is only slightly dependent on their crystal structure and exhibits a blue shift on changing from the zinc blend to wurtzite structure.[4,45,46] Depending on the activator (copper ions) concentration in ZnS:Cu phosphors, the color of emitted light (and, correspondingly, the chromaticity coordinates) and their luminescence intensity vary under photo and electron beam excitation.[4,5,45,46]

Table 3.5 Characteristics of Selected Green-Emitting Phosphors Suitable for RLS-T Application

Trade-mark	Chemical Composition	Chromaticity Coordinates		λ_{max}, nm	Dispersity, µm	Manufacturer	Ref.
		x	y				
SPD-431B	ZnS:hu	0.244	0.497	—	6.3[a]	Toshiba Corp., Japan	49, 50
SPD-431E		0.269	0.562	—	5.8[a]	Toshiba Corp., Japan	49, 50
54131		0.260	0.540	—	10[a]	Riedel-de-Haën AG, Germany	49, 51
GL29/N-C1		On request		—	10[a]	Phosphor Technology Ltd, England	43
139		0.285	0.580	535	—	GTE Products Corp., U.S.	49, 52
RST-520		Inside a color quadrangle with indicated coordinates of vertices: $x_1 = 0.02$ $y_1 = 0.45$; $x_2 = 0.24$ $y_2 = 0.75$; $x_3 = 0.30$ $y_3 = 0.46$; $x_4 = 0.18$ $y_4 = 0.41$		—	—	Luminophor Joint Stock Company, Russia	42
FK-106z	ZnS:hu,Cl	—	—	520	—	Luminophor Joint Stock Company, Russia	40
P31-G1		0.265	0.558	530	—	Kasei Optonix Ltd, Japan	49, 53
KDC-525	ZnS: hu,Al	≥ 0.270	≥ 0.610	—	5–6[a]	Luminophor Joint Stock Company, Russia	41
KTC-525-1		≥ 0.28	≥ 0.610	—	≥ 7.5[a]	Luminophor Joint Stock Company, Russia	41
KTC-534-1	ZnS:Cu,Au,Al	≥ 0.305	≥ 0.615	—	≥ 7.5[a]	Luminophor Joint Stock Company, Russia	41
GL29A/N-C1		0.310	0.594	—	8.0[a]	Phosphor Technology Ltd, England	43
GJL47/N-C1	(Zn,Cd)S:Ag	0.297	0.571	—	8.0[a]	Phosphor Technology Ltd, England	43
ah-527		—	—	527 ± 5	—	Luminophor Joint Stock Company, Russia	41
JGL29/N-C1[b]	(Zn,Cd)S:Cu	0.358	0.524	—	12.0[a]	Phosphor Technology Ltd, England	43
KDC-530[b]	(Zn,Cd)S:Cu,Al	0.270–0.310	≥ 0.600	—	5–6[a]	Luminophor Joint Stock Company, Russia	41
JGL29/N-C1[b]	(Zn,Cd)S:Cu	0.358	0.524	—	12.0[a]	Phosphor Technology Ltd, England	43
KDC-530[b]	(Zn,Cd)S:Cu,Al	0.270–0.310	≥ 0.600	—	5–6[a]	Luminophor Joint Stock Company, Russia	41

[a] Average particle size.
[b] See section 3.1.2.3.

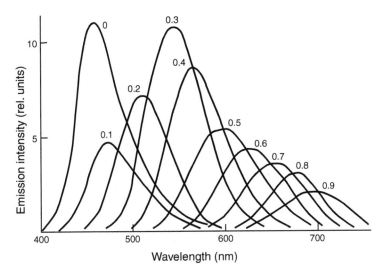

Figure 3.15   Cathodoluminescence spectra of (Zn,Cd)S:Ag phosphors at various concentrations of CdS (the numbers above the curves signify the mole fraction of CdS). (From Ozawa, L., *Cathodoluminescence: Theory and Applications*, VCH Publishers, New York, 1990.)

With increasing activator concentration in the ZnS:Cu phosphors from zero to about 10^{-2} wt%, the maximum in the emission spectrum shifts from the dark blue region to longer wavelengths to $\lambda_{max} \approx 540$ nm. At intermediate concentrations of the activator, the emission spectra exhibit two maxima. With further increase of the activator concentration (to ≈ 0.1 wt%), a dark blue luminescence band reappears in the emission spectra of the ZnS:Cu phosphors. This shifts the emission maximum to a shorter-wavelength (blue) region of the spectrum and concentration quenching becomes evident. At 0.3 wt% activator, light emitted by the phosphor is again green. The maximum relative intensity of cathodoluminescence of phosphors with the chemical composition of interest is observed at activator concentration of about 0.01 wt%; therefore, this is an optimum concentration for these phosphors. Commercial P31 (ZnS:Cu) phosphor is about 10^{-2} mol% activator. ZnS:Cu,Al phosphors are about 1.2×10^{-2} mol% activator and 2×10^{-2} mol% coactivator.[5]

The color of emitted light and intensity for (Zn,Cd)S:Ag phosphors depends on the ZnS to CdS ratio in the phosphor matrix and on the silver ion concentration in the phosphor. With increasing CdS in nonactivated (Zn,Cd)S phosphor base[54] and phosphor activated with silver ions,[4,5,18,46,55] the band maximum in the emission spectrum is red-shifted with replacement of Zn^{2+} ions by Cd^{2+} ions from the same group in the periodic system, differing in a polarizability of the electron shells.[3] Figure 3.15 presents the cathodoluminescence spectra of (Zn,Cd)S:Ag phosphors with different CdS concentrations. Cathodoluminescence intensity is largest for phosphors with composition $(Zn_{0.7},Cd_{0.3})S$:Ag.

Concentration behavior of the activator in (Zn,Cd)S:Ag phosphors[56] is similar to that in ZnS:Ag phosphors (see Section 3.1.2.1): with increasing concentration of the activator, the emission band maximum in the spectrum exhibits a blue shift.

The optimal concentration range is $5 \cdot 10^{-3}$ to $3 \cdot 10^{-2}$ wt% activator in the phosphor matrix.[5,46,56]

The temperature dependence of the cathodoluminescence spectra of the (Zn,Cd)S:Ag phosphor was studied from the liquid nitrogen temperature ($-196°C$) to $140°C$.[45,56] The relative photoluminescence intensity of (Zn,Cd)S:Ag phosphors increases with rising temperature from $-100°C$ to about $-10°C$ by 10 to 15%, then passes through a maximum and begins to decline. At $60°C$, intensity is 80% of the maximal value, and at $100°C$, 50%. Under electron beam excitation, luminescence intensity of the phosphors decreases by half at a higher temperature.[45] Impurities of the iron group metals (e.g., nickel in concentrations over $1 \cdot 10^{-5}$ wt%) reduce the temperature stability of (Zn, Cd)S:Ag phosphors.[45,57]

The photo- and cathodoluminescence intensity of ZnS:Cu phosphors tends to increase with rising temperature from $-100°C$ to about $200°C$, declining with further temperature rise.[45,58] The luminescence intensity decreases by half relative to the maximal value for ZnS:Cu phosphors at a temperature close to $347°C$.[45] Instability of energy efficiency of radioluminescence of ZnS:Cu phosphors caused by temperature variations within the climatic temperature range can be characterized by the coefficient $K_T =$ (0.04 to 0.06) %/degree K.[1]

The loss of the luminescence brightness of ZnS:Cu,Al phosphors at $100°C$ is no greater than 10% of that at $20°C$.[5] With rising temperature of ZnS:Cu,Al from liquid nitrogen temperature to $100°C$, the maxima of their luminescence bands red-shifts about 10 nm.[31] Thus, ZnS:Cu and ZnS:Cu,Al exhibit greater temperature stability than (Zn,Cd)S:Ag. Since toxic cadmium is preferentially excluded in the phosphor composition,[5] phosphors considered most suitable for RLSs-T application are ZnS:Cu and (Zn,Cd)S:Cu phosphors (see also Section 3.1.2.3), for which the content of cadmium is much lower than for (Zn,Cd)S:Ag phosphors as well as their additionally activated and coactivated modifications.

When analyzing suitability for RLS-T applications, one should not, however, exclude from consideration the previously mentioned green-emitting phosphors based on yttrium, lanthanum, and gadolinium oxysulfides, as well as phosphors based on other rare-earth matrices activated by terbium ions. These phosphors are highly resistant to electron beam irradiation.[59] Electron dose, $D_{el,50}$, describes the electron beam irradiation that produces a 50% loss of energy efficiency of phosphor luminescence and depends on the strength of the crystalline matrix chemical bonds of the phosphor. Phosphors based on rare-earth matrices with a crystal lattice energy E_{lat} of about 25 eV/bond had the greatest resistance (see Figure 3.16).

Resistance to electron beam irradiation for $A^{II}B^{VI}$ compounds (zinc sulfide phosphors) with E_{lat} of about 15 to 20 eV/bond is about half that exhibited by phosphors based on rare-earth matrices.[59] This makes expedient a study of the resistance of phosphors based on yttrium, lanthanum, and gadolinium oxysulfides, as well as on other rare-earth matrices activated by terbium ions, under tritium beta radiation. If phosphors based on rare-earth matrices are superior to zinc sulfide phosphors in the radiation resistance to tritium beta particles, the former may be preferred for RLS-T application. Despite comparing unfavorably with zinc sulfide phosphors in the initial energy efficiency of luminescence, these phosphors would be superior over the whole service life (10+ years).

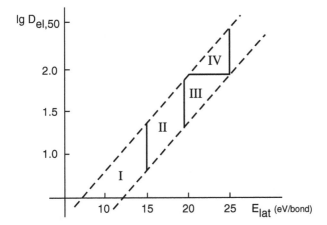

Figure 3.16   Electron dose, C/cm², producing a 50% loss of the luminescence brightness as a function of the crystal lattice energy of the inorganic matrix of various classes of phosphors. (I) Alkali and alkaline-earth metal halides; (II) $A_{II}B_{VI}$ compounds; (III) oxygen-containing matrices; and (IV) rare-earth matrices. (From Bol'shukhin, V.A. et al., Efficient cathodoluminophors for screens of oscillographic and indicator CRDs, *Electron. Tekhn., Ser. Elektrovak. Gazorazryad. Prib.*, (2)113, 28, 1986 [in Russian].)

Table 3.6 characterizes some commercially available phosphors based on various Tb^{3+}-activated rare-earth matrices. Table 3.7 presents physicochemical properties of the phosphors of interest, namely, their chemical composition, optimum activator concentration range, energy efficiencies of cathodoluminescence η_{CL} and radioluminescence η_{RL} under tritium beta radiation, and temperature instability coefficient K_T. It should be noted that the cathodoluminescence intensity of phosphors based on rare-earth matrices decreases with rising temperature from 20 to 100°C.[1,5,64,70,71] Table 3.7 contains, for comparison, the analogous data for the ZnS:Cu phosphor. The temperature instability coefficients K_T of the phosphors based on $Y_2O_2S:Tb^{3+}$, $La_2O_2S:Tb^{3+}$, and $Y_3Al_5O_{12}:Tb^{3+}$ (in the climatic temperature range) do not exceed that of the ZnS:Cu phosphor.

Tables 3.6 and 3.7 do not contain data for LaOBr:Tb^{3+} phosphors characterized by the greatest η_{CL} parameter among Tb^{3+}-activated (up to 20%[5]) phosphors but chemically unstable to ambient humidity and that undergo profound degradation under electron-beam irradiation.[69,72] Radiation resistance to electron-beam irradiation can be improved by partially replacing bromine in these phosphors by chlorine or iodine.[72] RLS-T manufacturing technology does not enable complete exclusion of moisture from RLS-T; this calls for special investigation of their suitability for RLS-T applications.

The cathodoluminescence spectra of green-emitting $Y_2O_2S:Tb^{3+}$, $La_2O_2S:Tb^{3+}$, $Gd_2O_2S:Tb^{3+}$, and $Y_3Al_5O_{12}:Tb^{3+}$ phosphors are shown in Figures 3.17 and 3.18, respectively. They contain several groups of narrow peaks, of which the most intense has a maximum at 544 nm. This corresponds to transition from the $^5D_4$ excited state of the Tb^{3+} ion to the $^7F_5$ ground state.[4,5] When aluminum ions are partially replaced by gallium ions in $Y_3Al_5O_{12}:Tb^{3+}$ phosphors, the luminescence intensity of

Table 3.6 Characteristics of Tb^{3+}-Activated Rare-Earth-Based Green-Emitting Phosphors Suitable for RLS-T Application

Trade-mark	Chemical Composition	Chromaticity Coordinates		λ_{max}, nm	Dispersity, µm	Manufacturer	Ref.
		x	y				
SPD-443	$Gd_2O_2S:Tb^3$	0.337	0.559	—	5.0[a]	Toshiba Corp., Japan	48, 50
P-43		0.334	0.561	544	7.2	Kasei Optonix Ltd., Japan	48, 49, 53
54144		0.331	0.557	—	8.8[a]	Riedel-de-Haën AG, Germany	48, 49, 51
UKL65D/N-C1		0.337	0.561	—	8.5[a]	Phosphor Technology Ltd, England	43
1830	$La_2O_2S:Tb^{3+}$	0.329	0.607	545	—	GTE Products Corp., U.S.	48, 52
P44		0.336	0.601	544	—	Kasei Optonix Ltd, Japan	48, 53
54145		0.305	0.590	—	7.5[a]	Riedel-de-Haën AG, Germany	48, 51
SKL65/N-C1[b]		0.346	0.590	—	7.0[a]	Phosphor Technology Ltd, England	43
SPD-445A	$Y_2O_2S:Tb^{3+},Dy^{3+}$	0.344	0.564	—	8.5[a]	Toshiba Corp., Japan	48, 50
KX-117A	$Y_2O_2S:Tb^{3+}$	0.339	0.570	544		Kasei Optonix Ltd, Japan	48, 53
1841		0.348	0.572	544		GTE Products Corp., U.S.	48, 52
KN-545-1		0.32–0.34	0.54–0.56	—	4–6[a]	Luminophor Joint Stock Company, Russia	41
150FG-53-01	$Y_3Al_5O_{12}:Tb^{3+}$	—	—	—	7–13[a]	USR Optonix Inc, U.S., Japan	48, 49
P53		0.342	0.572	545		Kasei Optonix Ltd, Japan	48, 49, 53
54208		0.368	0.539	—	4.5[a]	Riedel-de-Haën AG, Germany	48, 51
1271	$Y_3(Al,Ga)_5O_{12}:Tb^{3+}$	0.348	0.542	544	—	GTE Products Corp., U.S.	48, 52
QMPK65/N-C1		0.355	0.557	—	8.5[a]	Phosphor Technology Ltd, England	43

[a] Average particle size.

Table 3.7   Physicochemical Characteristics of Tb^{3+}-Activated Rare-Earth-Based Phosphors

Chemical Composition	Optimum Concentration of Activator, Mol%	η_{CL}	η_{RL}	K_T, %/Degree (Temperature Range)
$Y_2O_2S{:}Tb^{3+}$	3 – 5; 4, 60–62	from 13 to 19%; 1, 4, 14, 49, 59, 63	—	−0.045 – 0.02 (−50 – 50°h); 1
$La_2O_2S{:}Tb^{3+}$	0.5 – 2; 60, 61	from 10 to 12%; 12, 59, 63	(12 ± 3)%; 1	−0.045 – 0.02 (−50 – 50°h); 1
$Gd_2O_2S{:}Tb^{3+}$	0.1 – 1; 60, 61	up to 18%; 5, 49, 59, 63	(12 ± 3)%; 1	−0.4 – −0.5 (20 – 100°h); calculated by Equation 3.15 using data from 5, 64
$Y_3Al_5O_{12}{:}Tb^{3+}$	3 – 10; 65–68	up to 12%; 14, 49, 69	—	−0.04 – −0.05 (20 – 100°h); calculated by Equation 3.15 using data from 5, 64
$Y_3(Al,Ga)_5O_{12}{:}Tb^{3+}$ (Al/Ga) within 3/2–1/4	3 – 10; 65, 67, 68	exceeds $Y_3Al_5O_{12}{:}$ Tb^{3+}; 65, 69	—	−0.1 – −0.2 (20 – 200°h); calculated by Equation 3.15 using data from 65
ZnS:hu	about $1 \cdot 10^{-2}$; 5	up to 23%; 5	up to 20%; 1	0.04 – 0.06 (−50 – 50°h); 1

$Y_3(Al,Ga)_5O_{12}{:}Tb^{3+}$ phosphors increases relative to $Y_3Al_5O_{12}{:}Tb^{3+}$ phosphors, if activator concentrations are identical,[73,74] while the cathodoluminescence spectrum exhibits only minor changes.[65] As follows from the chromaticity coordinates for Tb^{3+}–activated rare-earth-based phosphors considered (see Table 3.6 and Figure 3.4), their cathodoluminescence can be characterized as yellowish-green.

3.1.2.3 Yellow-Emitting Phosphors

The cathodoluminescence of (Zn,Cd)S:Ag phosphors, depending on the ZnS-to-CdS ratio in the phosphor base, can generate light with a color varying from dark blue to red (see Figure 3.15). Although their relative emission intensity exhibits a maximum in the green spectral region[4,5,45,46] at the CdS content in the phosphor base of 30 to 45 wt%, phosphors with this chemical composition and a greater content of CdS in the base are suitable as yellow-emitting phosphors, as well. Replacement of silver ions with copper ones as activator in the (Zn,Cd)S matrix allows production of yellow-emitting phosphors with reduced CdS content. Aluminum is the most-used coactivator in such phosphors. Some commercial phosphors with the chemical composition indicated are characterized in Table 3.8.

A typical luminescence spectrum of (Zn,Cd)S:Cu yellow-emitting phosphors exhibits a broad (with a half-width of several tens of nanometers) bell-shaped band,

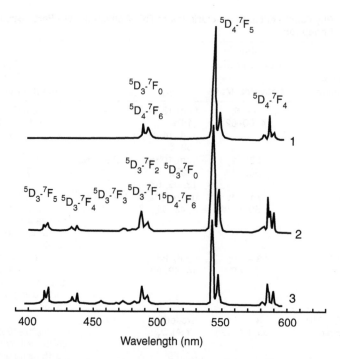

Figure 3.17 Cathodoluminescence spectra of $Ln_2O_2S:Tb^{3+}$ phosphors: (1) $La_2O_2S:Tb^{3+}$; (2) $Gd_2O_2S:Tb^{3+}$; and (3) $Y_2O_2S:Tb^{3+}$. (From Bol'shukhin, V.A. et al., Efficient cathodoluminophors for screens of oscillographic and indicator CRDs, *Electron. Tekhn., Ser. Elektrovak. Gazorazryad. Prib.*, (2)113, 28, 1986 [in Russian].)

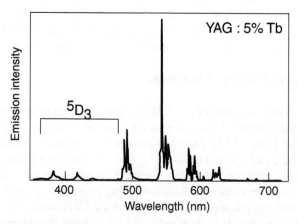

Figure 3.18 Cathodoluminescence spectrum of $Y_3Al_5O_{12}:Tb^{3+}$ phosphors. (From Ohno, K. and Abe, T., Effect of BaF_2 on the synthesis of the single-phase cubic $Y_3Al_5O_{12}:Tb$, *J. Electrochem. Soc.*, 133, 3, 638, 1986.)

Table 3.8 Characteristics of Selected Yellow-Emitting Phosphors Suitable for RLS-T Application

Trade-mark	Chemical Composition	Chromaticity Coordinates		λ_{max}, nm	Dispersity, μm	Manufacturer	Ref.
		x	y				
B-3zh	(Zn,Cd)S:Ag	$0.44 \leq x \leq 0.47$	$0.50 \leq y \leq 0.53$	560	4.7–16 (>70%)[a]	Luminophor Joint Stock Company, Russia	40
RST-580	(Zn,Cd)S:Cu	Inside a color quadrangle with indicated coordinates of vertices		—	—	Luminophor Joint Stock Company, Russia	42
		$x_1 = 0.43$	$y_1 = 0.48$				
		$x_2 = 0.42$	$y_2 = 0.57$				
		$x_3 = 0.61$	$y_3 = 0.39$				
		$x_4 = 0.53$	$y_4 = 0.38$				
L-15	(Zn,Cd)S:Cu	—	—	556 ± 3	5–15 (≥60%); 15–30 (≤40%)	Luminophor Joint Stock Company, Russia	40
FK-106g	(Zn,Cd)S:Cu,Al	—	—	540	—	Luminophor Joint Stock Company, Russia	40
No data	(Zn,Cd)S:Cu,Al	0.430	0.552	560	—	Kasei Optonix Ltd., Japan	49, 53

[a] Average particle size.

whose maximum position depends on the crystal structure of the (Zn,Cd)S:Cu phosphors; it exhibits a blue shift on changing from the zinc blend to wurtzite structure.[4,46] The color of emitted light and intensity for (Zn,Cd)S:Cu phosphors depend on the ZnS-to-CdS ratio in the matrix of the phosphor and on the copper ion concentration in it. Upon introduction of the optimum concentrations of copper activator into ZnS and (Zn,Cd)S, the emission band maxima for these phosphors exhibit a red shift (relative to nonactivated phosphors). It should be noted that activation of the same phosphor bases (matrices) with silver ions causes a blue shift of the emission spectra (see Section 3.1.2.1).

With increasing content of CdS in the (Zn,Cd)S:Cu phosphor base, the emission band maxima in the spectra exhibit a red shift.[5,46,55] Figure 3.19 shows the cathodoluminescence spectra for various concentrations of CdS in (Zn,Cd)S:Cu phosphors.[45,46] Analysis of the emission band maxima position as dependent on the CdS content in the phosphor base (see Figures 3.15 and 3.19) suggests that both (Zn,Cd)S:Cu and (Zn,Cd)S:Ag phosphors are suitable for generation of the light with identical positions of the band maximum. It should be noted, however, that the CdS content in (Zn,Cd)S:Cu phosphors is much lower. Changes in the cathodoluminescence spectra of (Zn,Cd)S:Cu phosphors with varying activator concentration are similar to those for ZnS:Cu phosphors (see Section 3.1.2.2). The optimal concentrations range from 2.10^{-3} to 3.10^{-2} wt% of the activator in the phosphor matrix.[5,46]

3.1.2.4 Red-Emitting Phosphors

The values of energy efficiencies of luminescence for red-emitting phosphors under cathode and beta excitation presented in Tables 3.1 and 3.2 suggest that these parameters take a maximal value in the case of europium-activated yttrium oxide Y_2O_3:Eu^{3+} (up to 10%) and yttrium oxysulfide Y_2O_2S:Eu^{3+} (up to 13%) phosphors. This makes these phosphors the most suitable for red RLS-T application. Table 3.9 characterizes phosphors with the chemical composition indicated, which are produced by various manufacturers.

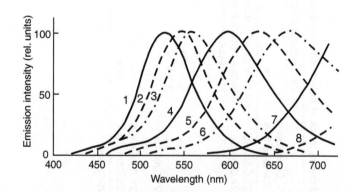

Figure 3.19 Cathodoluminescence spectra of (Zn,Cd)S:Cu phosphors at various concentrations of CdS, %: (1) 0; (2) 5; (3) 10; (4) 20; (5) 30; (6) 40; (7) 50; and (8) 60. (From Leverenz, H.W., *An Introduction to Luminescence of Solids*, John Wiley & Sons, Inc., New York, 1950.)

Table 3.9 Characteristics of Selected Red-Emitting Phosphors Suitable for RLS-T Application

Trade-mark	Chemical Composition	Chromaticity Coordinates		λ_{max}, nm	Dispersity, μm	Manufacturer	Ref.
		x	y				
RST-612	Y_2O_3:Eu^{3+}	Inside a color quadrangle with indicated coordinates of vertices: $x_1 = 0.60$, $x_2 = 0.64$, $x_3 = 0.73$, $x_4 = 0.60$	$y_1 = 0.36$, $y_2 = 0.37$, $y_3 = 0.26$, $y_4 = 0.29$	611	—	Luminophor Joint Stock Company, Russia	75
KDC-612		0.64–0.65	0.34–0.35	—	5–6[a]	Luminophor Joint Stock Company, Russia	41
QK63/N-C1		0.650	0.346	—	6.0[a]	Phosphor Technology Ltd., England	43
No data		—	—	611	4–10[a]	Applied Scintillation Technologies Inc, U.S./U.K.	41
1134		0.650	0.341	611	—	GTE Products Corp., U.S.	48, 52
1137		0.650	0.347	611	—	GTE Products Corp., U.S.	48, 52
RE610		—	—	—	—	USR Optonix Inc., U.S./Japan	48, 76
K-78	Y_2O_2S:Eu^{3+}	0.66 ± 0.01	0.34 ± 0.01	—	0–4 (<35%); >16 (<30%); 4–16 (the rest)	Luminophor Joint Stock Company, Russia	59, 77
QKL63/N-C1		0.647	0.343	—	7.5[a]	Phosphor Technology Ltd., England	43
1152		0.648	0.338	625	—	GTE Products Corp., U.S.	48, 52
RE555		—	—	—	—	USR Optonix Inc., U.S./Japan	48, 76

[a] Average particle size.

The luminescence spectra of Y_2O_3:Eu^{3+} and Y_2O_2S:Eu^{3+} phosphors consist of sets of narrow bands with a half-width of about 1 nm. In the spectrum of the Y_2O_3:Eu^{3+} phosphor, a band with $\lambda_{max} = 611$ nm dominates (see Figure 3.12), and the spectrum of the Y_2O_2S:Eu^{3+} phosphor contains a number of bands with comparable intensities, of which the most intense have maxima at 616 and 626 nm. Total luminescence intensity and intensity ratio for the bands in the luminescence spectra of these phosphors vary with the europium concentration in their matrices.

The concentration dependence of the cathodoluminescence intensity for Y_2O_3:Eu^{3+} and Y_2O_2S:Eu^{3+} phosphors was measured and calculated by various researchers.[4,63,78] With increasing Eu^{3+} concentration to 3 and 2 mol% for these phosphors, respectively, the relative intensity of cathodoluminescence tends to grow. With further increase of the activator ion concentration, cathodoluminescence decreases. For Y_2O_2S:Eu^{3+}, see Tseng et al.[79] For Y_2O_3:Eu^{3+} phosphors, the optimum Eu^{3+} concentration range is 2 to 6 mol%, and for Y_2O_2S:Eu^{3+} phosphors, it is 1 to 5 mol%.

Following changes in the spectral composition of light emitted by the Y_2O_2S:Eu^{3+} phosphor, the corresponding chromaticity coordinates vary with varying activator concentration in the phosphor matrix. The data on cathodoluminescence of Y_2O_2S:Eu^{3+} reported in Shionoya and Yen[5] and Tseng et al.[79] suggest that increasing Eu^{3+} concentration within 2 to 8 mol% changes the color of emitted light by the phosphor from orange to red.

It should be noted that a significant (up to several tens of percent) increase in the luminescence intensity of Y_2O_2S:Eu^{3+} phosphors can be achieved by introducing minor (10^{-4} to 10^{-2} at.%) amounts of Tb^{3+} or Pr^{3+} ions.[5,80,81]

The luminescence intensity of Y_2O_3:Eu^{3+} phosphor remains virtually unchanged with rising temperature from 20 to 100°C. Over the same temperature range, the luminescence intensity of the yttrium oxysulfide–based phosphor decreases over twofold relative to that at room temperature. This makes europium-activated, yttrium oxide–based, red-emitting phosphor more suitable for RLS-T application. Note that, according to Gaiduk et al.[33] and Johnson,[82] intensive luminescence of the Y_2O_3:Eu^{3+} phosphor is observed even at temperatures above 647°C. No other phosphor exhibits luminescent properties at temperatures so high.

3.1.2.5 White-Emitting Phosphors

White-emitting phosphors are conventionally produced by mechanically mixing blue- and yellow-emitting phosphors. Research on preparation and commercial production of white-emitting single-component phosphors, e.g., those based on Y_2O_2S:Tb^{3+}, is underway now.[43,51-53,76] However, the energy efficiency of cathodoluminescence of these phosphors ($\eta_{CL} = 18\%^5$) is lower compared to phosphors prepared by mixing. Performance characteristics of some brands of commercially available white-emitting phosphors are presented in Table 3.10. The typical luminescence spectrum of white-emitting phosphors is essentially a superposition of the spectra of the phosphors constituting the mechanical mixture forming the mixed phosphor.[4,5,40] The properties of mixed white-emitting phosphors are determined by the combination of the properties of their constituting phosphors.

Table 3.10 Characteristics of Selected Brands Of White-Emitting Phosphors Suitable for RLS-T Application

Trade-mark	Chemical Composition	Chromaticity Coordinates x	Chromaticity Coordinates y	λ_{max}, nm	Dispersity, μm	Manufacturer	Ref.
P4-K	ZnS:Ag + (Zn,Cd)S:Cu,Al	0.267	0.291	450, 560	—	Kasei Optonix Ltd, Japan	49, 53
1768		0.267	0.292	—	—	GTE Products Corp., U.S.	41, 52
KTB-1-1	ZnS:Ag + (Zn,Cd)S:Cu,Al,I	0.26 ± 0.01	0.27 ± 0.01	—	10 ± 1 [a]	Luminophor Joint Stock Company, Russia	40, 41
P40	ZnS:Ag + (Zn,Cd)S:Cu	0.216	0.208	440, 555	—	Kasei Optonix Ltd, Japan	49, 53
54141		0.250	0.266	—	8 [a]	Riedel-de-Haën AG, Germany	49, 51
150FG-40-01		—	—	—	8–12 [a]	USR Optonix Inc., U.S.	49, 76
NP-1004		—	—	460, 560	—	Nichia Chemical Industries Ltd, Japan	49, 83
183		0.203	0.173	—	—	GTE Products Corp., U.S.	49, 52
RST-6500		Inside a color quadrangle with indicated coordinates of vertices: $x_1 = 0.21$, $x_2 = 0.39$, $x_3 = 0.43$, $x_4 = 0.23$	$y_1 = 0.34$, $y_2 = 0.42$, $y_3 = 0.35$, $y_4 = 0.22$			Luminophor Joint Stock Company, Russia	75
W746S	ZnS:Ag + (Zn,Cd)S:Ag	—	—	—	8–10 [a]	USR Optonix Inc., U.S.	49, 76
1748		0.270	0.280	455	—	GTE Products Corp., U.S.	49, 52
KTB-3P		0.26 ± 0.01	0.26 ± 0.01	—	10 ± 1 [a]	Luminophor Joint Stock Company, Russia	40, 41
54146	$Y_2O_2S{:}Tb^{3+}$	0.250	0.312	—	7.5 [a]	Riedel-de-Haën AG, Germany	49, 51
WRE1055		—	—	—	2–10 [a]	USR Optonix Inc., U.S.	48, 76
1840		0.257	0.317	545	—	GTE Products Corp., U.S.	48, 52
P45		0.231	0.246	418, 544	—	Kasei Optonix Ltd, Japan	48, 55
QKL65/N-C1		0.263	0.342	—	7.5 [a]	Phosphor Technology Ltd, England	43
R14		—	—	—	4–10 [a]	Luminophor Joint Stock Company, Russia	41
KL65E/S-C1	$Y_2O_2S{:}Tb^{3+}$, Eu^{3+}	0.295	0.267	—	5.0 [a]	Phosphor Technology Ltd, England	43
54149	$Y_2O_2S{:}Tb^{3+}$, Eu^{3+}	0.321	0.328	—	7.5 [a]	Riedel-de-Haën AG, Germany	49, 51

[a] Average particle size.

The electron beam–excited $Y_2O_2S:Tb^{3+}$ phosphors emit bluish-white light at activator (Tb^{3+}) concentrations under 0.1 mol%.[4,5] The reason is that at such concentration of Tb^{3+}, the luminescence spectra of $Y_2O_2S:Tb^{3+}$ phosphors are dominated by bands located in the short-wave region of the spectrum from violet to blue. These are due to transitions from the $^5D_3$ excited state to the ^7F_j ground states of Tb^{3+}. Remember that at the terbium concentration of 3 to 5 mol%, bands in the yellow to green spectral region dominate in the luminescence spectrum of the $Y_2O_2S:Tb^{3+}$ phosphor.

The brightness of luminescence of $Y_2O_2S:Tb^{3+}$ (0.1 mol%) phosphor under electron-beam excitation is typically surpassed (slightly) by mixed phosphors based on ZnS and (Zn, Cd)S matrices. Upon prolonged exposure to ionizing radiation — in particular, beta radiation of tritium — the light emitted by mixed phosphors can change its color, owing to difference in degradation rates for yellow- and blue-emitting phosphors constituting these mixed phosphors. This drawback is not inherent in single-component $Y_2O_2S:Tb^{3+}$ phosphors that exhibit high radiation resistance. This makes $Y_2O_2S:Tb^{3+}$ phosphors suitable for a number of RLS-T applications.

3.1.3 Radioluminescence of Noble Gases

RLSs-T can employ powdered and also gaseous luminescent materials. Among gases of efficient luminescence under ionizing radiation are the noble gases: argon, krypton, and xenon. Interest in RLSs-T employing noble gases stems primarily from emission in the vacuum ultraviolet region. The energy efficiency of noble gas radioluminescence reaches 15 to 25%.[1] They are radiation-resistant substances because they are elemental. However, their high radiation resistance depends on keeping impurities from accumulating in the gas over time. A nonhermetic gas envelope and radiolysis of contaminants on the envelope walls introduce impurities that are radioluminescence quenchers.

The radioluminescence spectra of pure noble gases (argon, krypton, xenon) and their mixtures are shown in Figures 3.20a and 3.20b. The radioluminescence band of argon has a maximum at 127.5 nm, that of krypton at 146 nm, and that of xenon at 172 nm.

Radioluminescence of noble gases is due to formation of excimeric molecules upon gas irradiation.[1] Excimers are dimers stable only in the excited state; their ground state is a repulsive one. Excimers (A_2^*) form via three-body collisions of excited atoms (A^*), forming upon irradiation with noble gas atoms (A):

$$A^* + 2A \rightarrow A_2^* + A$$

or via formation of a noble gas molecular ion (A_2^+) in three-body collision of the noble gas ion (A^+), forming upon irradiation with noble gas atoms and subsequent recombination of A_2^+ with an electron e⁻:

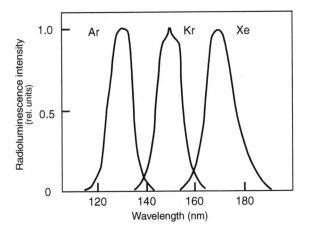

Figure 3.20a Radioluminescence spectra of noble gases. (From Mikhal'chenko, G.A., *Radioluminescent Emitters*, Energoatomizdat, Moscow, 1988 [in Russian].)

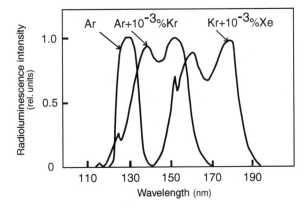

Figure 3.20b Radioluminescence spectra of noble gas mixtures. (From Mikhal'chenko, G.A., *Radioluminescent Emitters*, Energoatomizdat, Moscow, 1988 [in Russian].)

$$A^+ + 2A \rightarrow A_2^+ + A$$

$$A_2^+ + e^- \rightarrow A_2^*$$

Transition of the excimer from excited to ground state is accompanied by emission of a quantum of the vacuum ultraviolet radiation, and the molecule breaks down. This radiative transition proceeds as a first-order reaction:

$$A_2^* \rightarrow A + A + h\nu$$

In a noble gas mixture, heteroexcimeric molecules can form.[1] The position of the emission bands of heteroexcimeric molecules of noble gases is intermediate between those of the homoexcimers of the noble gases whose atoms constitute the heteroexcimeric molecule.

The maximum radioluminescence efficiency in a mixture with gaseous tritium is exhibited by xenon.[1] Figure 3.21 shows how the power of the vacuum ultraviolet radiation flux for a tritium–xenon mixture varies with the volume concentration of tritium, with the total pressure of the $^3H_2$ + Xe mixture equal to atmospheric pressure. Note that the power of the optical radiation is highest when the tritium concentration is several volume percent. For the mixture with the tritium volume concentration of 1%, energy efficiency of self-luminescence is about 10%. The temperature instability of luminescence of the tritium–xenon mixture in the climatic temperature range characterized by K_T is estimated as 0.5 to 1%/degree K, and the time instability is estimated as minus (10 ± 5)%/year.

Radioluminescent light sources employing noble gases as luminescent materials can find application as vacuum ultraviolet radiation sources. Furthermore, if the walls of the gas cavity filled with a tritium–xenon mixture are covered with a powdered phosphor, they will emit light under vacuum ultraviolet radiation. At certain pressures and gas cavity sizes, the tritium–xenon mixture will utilize the tritium activity at a higher efficiency than pure tritium. Relevant calculations are presented in Section 2.3 in the preceding chapter.

Interestingly, if a trace contaminant such as mercury vapor is introduced into a noble gas, the contaminant will efficiently accept the excitation energy in spite of the fact that the noble gas-to-mercury vapor concentration ratio is about 10^6. The radioluminescent spectrum of the noble gas–mercury vapor mixture will contain bands corresponding to electronic transitions for mercury atoms.[84] Also, with an electric field applied to the volume of the noble gas (or a noble gas–mercury mixture),

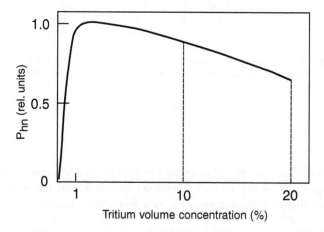

Figure 3.21 Vacuum ultraviolet radiation power for a tritium–xenon mixture as a function of the mixture composition. (From Mikhal'chenko, G.A., *Radioluminescent Emitters*, Energoatomizdat, Moscow, 1988 [in Russian].)

its radioluminescent intensity can be enhanced or attenuated, depending on field intensity.[85] Therefore, these sources can provide modulated light with modulation efficiency of about 95% for analytical instruments.[84]

3.2 ORGANIC LUMINOPHORES

3.2.1 Absorption and Emission Spectra

The number of known organic luminophores is large, but continuous search is underway for new luminescent materials because of an extremely broad range of applications and the constant desire for improved properties for specific applications. Most essential for luminophores employed in devices converting radioactive decay energy into light energy are spectral features, light output, radiation resistance, chemical inertness, and solubility in monomer and polymer matrices. According to their chemical structure, organic luminophores can be divided into the following groups[86]:

- Aromatic hydrocarbons and their derivatives
- Polyphenyl hydrocarbons
- Hydrocarbons with condensed aromatic rings
- Hydrocarbons with arylethylene and arylacetylene moieties
- Compounds with heterocycles
- Derivatives of five-membered cycles
- Derivatives of six-membered cycles
- Compounds with carbonyl group
- Compounds with two or several fluorophore moieties
- Complex compounds of metals with organic ligands

The most widely used for fabrication of plastic scintillators are polyphenyl hydrocarbons, 2,5-diaryl derivatives of oxazole, and 1,3,4-oxadiazole.

Among polyphenyl luminophores and luminophores with condensed aromatic rings, the most-used are naphthalene (I), p-terphenyl (pTP) (II), phenanthrene (III), p-quaterphenyl, tetraphenylbutadiene, 9,10-diphenylanthracene (IV), rubrene (V), and certain other compounds. The spectral properties of this group of luminophores are governed by the size of the π-conjugation system. The greater the delocalization of π electrons, the greater the bathochromic effect; i.e., the fluorescence spectrum is shifted from the UV region via the blue and green regions to the red region of the emission spectrum.

Efficient organic luminophores containing heteroatoms include 2,5-diphenyl-1,3,4-oxadiazole (PPD) (VI), 2,5-diphenyloxazole (PPO) (VII), 2-(4′-diphenyl)-6-phenylbenzoxazol (PBBO) (VIII), 2-phenyl-5-(4′-biphenyl)-1,3,4-oxadiazole (PBD) (IX), 2-(4′-$tert$-butylphenyl)-5-(4″-biphenyl)-1,3,4-ozadiazole ($butyl$-PBD) (X), 1,4-bis-(5-phenyloxazol-2-yl)benzene (POPOP) (XI), 2-(1′-naphthyl)-5-phenyloxazole, 3-hydroxyflavone (XII), etc. The majority of these luminophores emit light in the violet and blue regions of the spectrum. The structures are shown in Structures 3.1 and 3.2.

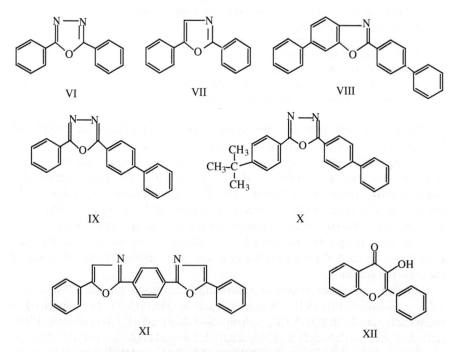

Structure 3.1 naphthalene (I), p-terphenyl (II), phenanthrene III, 9,10-diphenylanthracene (IV), and rubrene (V)

Structure 3.2 2,5-diphenyl-1,3,4-oxadiazole (VI), 2,5-diphenyloxazole (VII), 2-(4'-diphenyl)-6-phenylbenzoxazole (VIII), 2-phenyl-5-(4'-biphenyl)-1,3,4-oxadiazole (IX), 2-(4'-tert-butylphenyl-5-(4''-biphenylyl)-1,3,4-oxadiazole (X), 1,4-bis-4'-(5-phenylox-azolyl)benzene (XI), 3-hydroxyflavone (XII)

Research into the spectral-luminescent characteristics of organic luminophores was started long before they were adopted for manufacturing plastic scintillators. Initially, this research primarily involved photoexcitation of luminescence and, later on, excitation by ionizing radiation. It was found that the spectral-luminescent characteristics of luminophores varied little with the luminescence initiation procedure.

It is also interesting that luminophores' properties do not markedly differ on going from crystalline state to liquid or solid solutions. This is due to weak molecular interaction in solid organic compounds (in contrast to inorganic crystals), which slightly shifts electron energy levels and is responsible for the minor (ca. 20 nm) shift in absorption and emission spectra of solid organic luminophores relative to the analogous spectra of free molecules.[87] Upon transition of an organic luminophore from solid to liquid or solid solution, spectra in most cases do not change markedly, thus making the spectral-luminescent characteristics of crystalline luminophores suitable for roughly estimating the analogous parameters of their liquid and solid solutions.

The luminescence (emission) spectrum is determined by the energy of transition between an excited and unexcited (ground) state of the luminophore molecules. According to the Stokes law, the frequency of emitted light is always smaller compared to the absorbed light. Therefore, the emission spectrum typically lies in a longer-wavelength region relative to the absorption spectrum. For the majority of organic luminophores, the shift of the luminescence maximum relative to the absorption maximum is 50 to 70 nm, due to nonradiative energy loss by luminescence. In certain cases, when the molecule in an excited state takes a more-planar configuration or a proton transfer takes place (the latter characteristic of, e.g., 3-hydroxyflavone), the Stokes shift can be much greater, namely, 150 to 200 nm.[86] Figures 3.22 to 3.25 present the absorption and luminescence spectra of selected typical organic luminophores.[86]

The majority of organic luminophores have one peak in the spectra, but certain luminophores have several peaks of different intensity. With increasing numbers of peaks in the emission spectra, luminescence efficiency typically decreases; the best scintillators have clearly shaped spectra with symmetric narrow peaks.[88] The best spectral properties are exhibited by luminophores with similar absorption and emission spectra and with large Stokes shifts in them. These luminophores include 3-hydroxyflavone, 2-(4′-sulfofluoridophenyl)-5-(4″-dimethylaminophenyl)-1,3,4-oxadiazole, 4-methoxy-1,8-naphthoylene-1′,2′-benzimidazole, 9,10-dianilinoanthracene, and some other luminophores. Because organic luminophores have large Stokes shifts in the spectra, one can prepare two-component rather than three-component (base, primary luminophore, secondary additive) plastic scintillators.[89]

The absorption and emission spectra of organic luminophores (incidentally, similar to many other spectral characteristics) are determined by their molecular structure. Adding unsaturated groups to the conjugate system causes a further shift to longer wavelengths. Aromatic compounds with one ring absorb at 250 nm, naphthalenes derivatives absorb at 300 nm, and anthracene and phenanthrene derivatives absorb at 360 nm. Dependence of the spectral-luminescent characteristics on the molecular structure of organic luminophores will be analyzed in moe detail in Section 3.2.2. Here, Table 3.11 gives positions of the maxima in the absorption and emission spectra, along with the Stokes shifts, for the most important organic luminophores.

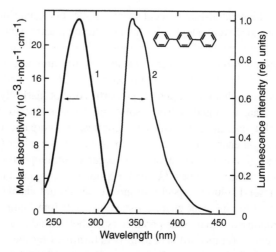

Figure 3.22   Absorption (1) and luminescence (2) spectrum of *p*-terphenyl in dioxane. (From Krasovitskii, B.M. and Bolotin, B.M., *Organic Luminophores*, Khimiya, Leningrad, 1976 [in Russian].)

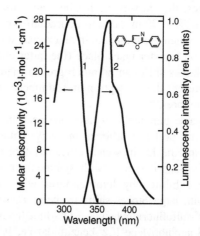

Figure 3.23   Absorption (1) and luminescence (2) spectrum of 2,5-diphenyloxazole (PPO) in toluene. (From Krasovitskii, B.M. and Bolotin, B.M., *Organic Luminophores*, Khimiya, Leningrad, 1976 [in Russian].)

Table 3.11 and the electronic emission spectra show that the majority of organic luminophores emit violet or dark blue light. However, in manufacturing plastic scintillators (especially those employed under high doses), preference is given to luminophores emitting in a longer-wavelength region (green, yellow, and red). The group of yellowish-green luminophores includes 3-methoxybenzanthrone, 9,10-dianilinoanthracene, 4-methoxy-1,8-naphthoylene-1′,2′-benzimidazole, 1,4-bis(1′,5′-diphenyl-Δ^2-pyrazolinyl-3′)benzene, and 1,8-naphthoylene-1′,2′-benzimidazole. The group of reddish-orange

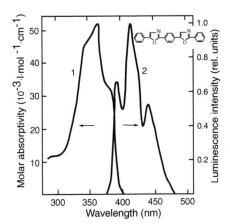

Figure 3.24 Absorption (1) and luminescence (2) spectrum of 1,4-bis-(2'-(5'-phenylox-azolyl)benzene (POPOP) in toluene. (From Krasovitskii, B.M. and Bolotin, B.M., *Organic Luminophores*, Khimiya, Leningrad, 1976 [in Russian].)

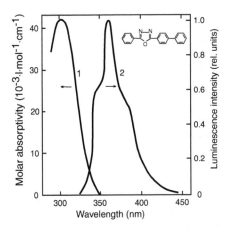

Figure 3.25 Absorption (1) and luminescence (2) spectrum of 2-phenyl-5-(4'-biphenylyl)-1,3,4-oxadiazole (PBD) in toluene. (From Krasovitskii, B.M. and Bolotin, B.M., *Organic Luminophores*, Khimiya, Leningrad, 1976 [in Russian].)

luminophores includes 4-diethylaminoanthraquininethiadiazole, 4-(1',5'-diphenyl-Δ^2-pyrazolinyl-3)-N-phenylnaphthalimide, and some other luminophores.

Because of self-absorption of the emitted light, the external (technical) luminescence spectrum is distinguished from the internal (molecular) spectrum by decreased intensity in the short-wavelength region. This decrease in the intensity is greater when the sample is thicker. Internal spectra are typically measured with thin sections of the samples. Technical spectra vary with the sample and, consequently, are most important from the practical viewpoint.

Table 3.11 Absorption (λ_{abs}) and Emission (λ_{em}) Maxima and Stokes Shifts ($\Delta\lambda$) for Selected Organic Luminophores

Luminophore	Abbr.	λ_{abs}, nm	λ_{em}, nm	$\Delta\lambda$, nm
2,5-Di-(5'-*tert*-butyl-2-benzoxazolyl)thiophene	BBOT	380	436	56
2,5-Bis-(4'-dimethylaminophenyl)oxadiazole	dmaPD	350	400	50
2-(4'-Dimethylaminophenyl)benzoxazole	BO	330	400	70
2-(4'-*tert*-Butylphenyl)-5-(4''-biphenylyl)-1,3,4-ozadiazole	bPBD	310	365	55
2-(4'-Biphenylyl)-5-phenyloxazole	BPO	330	390	60
2-(4'-Carbomethoxyphenyl)-5-(4''-dimethylaminophenyl)oxazole	cmPdaPO	353	460	107
7-Diethylamino-3-(benzimidazolyl-2')coumarin	Coum.7	460	510	50
7-Diethylamino-3-(3'-methylbenzimidazolyl-2')coumarin	Coum. 30	420	480	60
4-Dimethylamino-1,2-(1',8'-naphthoylene)benzimidazole	dma-NBI	—	600	—
5-(4'-Dimethylaminophenyl)-2-(1''', 8'''-naphthoylene-1'',2''-benzimidazolyl)oxadiazole	dmaPD-NBI	440	565	125
1,3-Diphenyl-2-pyrazoline	DP	–	450	–
1,5-Diphenyl-3-styryl-2-pyrazoline	DSP	390	460	70
3-Hydroxyflavone	3HF	350	530	180
2-(4'-Methoxyphenyl)-3-hydroxyflavone	M3HF	350	430	80
1-Methylnaphthalene	1MN	320	400	80
2-(4'-Methylphenyl)-5-(4''-biphenylyl)-1,3,4-oxadiazole	mPBD	—	365	—
1-(*o*-Methoxyphenyl)-3,5-diphenyl-2-pyrazoline	mPDP	360	460	100
Naphthalene	Nph	310	330	22
2-Phenyl-5-(4'-biphenylyl)-1,3,4-oxadiazole	PBD	305	365	60
4-(5''-Phenyloxadiazolyl)-4'-(3''',5'''-diphenylpyrazolinyl-1''')stilbene	PDdPPS	450	560	110
4,4'-Bis-(2''-(5''-phenyloxazolyl))stilbene	PEP	385	450	65
1,4-Bis-(2'-(5'-phenyloxazolyl))benzene	POPOP	365	420	55
2,5-Diphenyloxadiazole	PPD	280	350	70
2,5-Diphenyloxazole	PPO	310	365	55
p-Terphenyl	pTP	290	360	70
2-(4'-Sulfofluoridophenyl)-5-(4''-dimethylaminophenyl)-1,3,4-oxadiazole	sfPDdmaP	370	495	125
2-(4'-Sulfofluoridophenyl)-5-(4''-dimethylaminophenyl)-1,3-oxazole	sfPOdmaP	400	513	113
4-(4',5'-Benzothiazolyl-2)naphthalic acid anhydride	—	370	450	80
4-(5'-Phenyloxazolyl-2)naphthalic acid anhydride	—	400	480	80
1,2-Diphenylethylene	dPE	295	345	50
1-Phenyl-2-naphthylethylene	PNE	380	385	5
1,2-Dinaphthylethylene	dNE	330	400	70
1,2-Distyrylbenzene	1,2dSB	320	405	85
1,3-Distyrylbenzene	1,3dSB	300	375	75
1,4-Distyrylbenzene	1,4dSB	350	405	55
Stilbene	St	308	355	47
Anthracene	An	360	400	40
9,10-Diphenylanthracene	dPA	380	420	40
9,10-Dianilinoanthracene	dAA	420	530	110
1,4-Diphenylbutadiene	—	340	390	50
2-(2'-Hydroxyphenyl)benzimidazole	HPBI	360	420	60
1,3,5-Triphenylpyrazoline-Δ^2	tPP	360	40	80

Table 3.11 Absorption (λ_{abs}) and Emission (λ_{em}) Maxima and Stokes Shifts ($\Delta\lambda$) for Selected Organic Luminophores (Continued)

Luminophore	Abbr.	λ_{abs}, nm	λ_{em}, nm	$\Delta\lambda$, nm
1,4-Bis-(1',5'-diphenyl-Δ^2-pyrazolinyl-3')benzene	—	420	470	50
1,8-Naphthoylene-1',2'-benzimidazole	NBI	390	480	90
4-Methoxy-1,8-naphthoylene-1',2'-benzimidazole	mNBI	380	530	150

Sources: Krasovitskii, B.M. and Bolotin, B.M., *Organic Luminophores*, 2nd ed., Khimiya, Leningrad, 1984 [in Russian]; Britvich, G.I. et al., *Prib. Tekhn. Eksper.*, 1, 75, 1994 [in Russian]; Britvich, G.I. et al., *Prib. Tekhn. Eksper.*, 1, 109, 1993 [in Russian]; Vasil'chenko, V.G. et al., *Prib. Tekhn. Eksper.*, 5, 85, 1995 [in Russian]; Ambrosio, C.D., *Nucl. Instr. Methods Phys. Res.*, A307, 430, 1991.

Luminescence intensity of luminophores is determined to a significant extent by absorption intensity. The greater the number of the quanta absorbed by the lumino-phore molecules, the greater will be the number of the quanta emitted.

For comparative assessment of the emission intensity of luminophores, a relative parameter is used, showing by which factor the emission intensity of the sample studied exceeds that of the reference. The latter is typically anthracene, exhibiting high light output when in the crystalline state. The majority of organic crystalline scintillators have much lower light outputs than anthracene. For example, the relative light output of *p*-terphenyl is estimated at 0.65, *p*-quaterphenyl and 1,4-diphenyl-butadiene at 0.85, *trans*-stilbene at 0.6 to 0.7, and naphthalene at 0.2.[86] According to Sangster and Irvine,[88] some crystalline scintillators (e.g., 1,2-benzanthracene, 1,2, 5,6-dibenzanthracene, coronene) have light outputs exceeding that of anthracene.

It should be noted that published values of the relative light outputs of organic luminophores exhibit considerable scatter, due, above all, to the different degrees of purity of the luminophores utilized. Even for anthracene, which usually serves as reference, luminescence intensity depends on the purification methods and, after storage for several months, its light output decreases by 15 to 25% because of resulting impurities.[87]

Apart from purity, the luminescence intensity of organic luminophores is affected by magnetic and induction temperature. These factors will be discussed in greater detail in Sections 3.3.5 and 3.3.6. Figure 3.26 compares relative luminescence intensities of a great number of materials (organic luminophores, solvents, NaI (Tl), etc.) at two different temperatures (+30°C and –70°C).

3.2.2 Luminescent Property Relations to Molecular Structure of Organic Luminophores

A certain correlation exists between the spectral-luminescent characteristics of luminophores and their structural features, determined primarily by the mobility of the π electrons of the conjugate molecular system and its extension. The basic moieties in the majority of organic luminophores are benzene rings, either condensed (naphthalene, anthracene, chrysene, etc.) or forming a linear structure (*p*-terphenyl, quaterphenyl, etc.). The mobility of the aromatic electronic systems can be improved

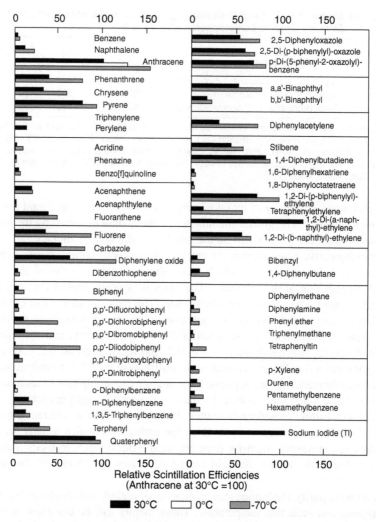

Figure 3.26 Luminescence intensity of selected substances relative to anthracene at +30°C and −70°C. (From Sangster, R.C. and Irvine, J.W., Study of organic scintillators, *J. Chem. Phys.*, 24, 670, 1956.)

by introducing nitrogen or oxygen atoms into the aromatic rings , e.g., in oxadiazole luminophores.

An extended conjugate system of bonds favors luminescence and emission shift to longer-wavelength regions of the spectrum. For example, by replacing the methyl group in 2-methyl-5-phenyloxazole, which exhibits a fairly weak luminescence, with a benzene ring, one obtains an efficient and popular lumiphore, diphenyloxazole (PPO, VII).

With lengthening of the chain of conjugate bonds, luminescence intensity tends to increase. However, excess linear conjugation favors nonradiative energy loss by intramolecular rotation and vibration, which can adversely affect luminescence intensity. This luminescence intensity and wavelength of the emitted light are strongly affected by electron-donating and electron-accepting substituents in aromatic and heterocyclic moieties. Electron-donating substituents in most cases increase the light output of the luminophore and shift the luminescence maximum to longer wavelengths. Electron-accepting substituents (especially nitro group) are luminescence suppressors. Luminescence efficiency is higher for higher-symmetry molecules and those lacking steric hindrance.

The spectral-luminescent characteristics of luminophores are also affected by the structural isomerism of the molecules. For example, 1,2-di-1′-naphthylethylene (XIII) and 1,2-di-2′-naphthylethylene (XIV), differing only in affixions for the naphthyl radicals to the ethylene group, differ significantly in the emission spectra (λ_{em}) and the Stokes shifts ($\Delta\lambda$). For luminophore (XIII), λ_{em} = 502 nm and $\Delta\lambda$ = 112 nm, and for luminophore (XIV), λ_{em} = 476 nm and $\Delta\lambda$ = 86 nm.[93]

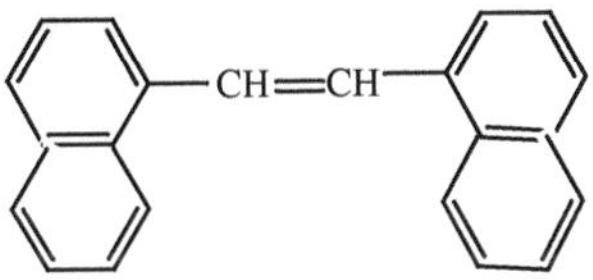

XIII

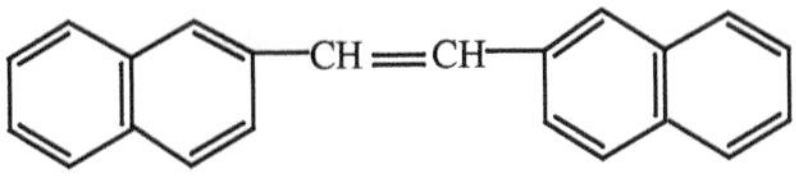

XIV

Structure 3.3 1,2-di-1′-naphthylethylene (XIII); 1,2-di-2′naphthylethylene (XIV)

The difference in the spectral characteristics of these luminophores is explained by the fact that the molecules of compound (XIII) in the excited state have a more planar configuration than in the ground state; upon transition to the excited state, the configuration of the molecules (XIV) changes insignificantly.[93]

The influence of isomerism on the spectral-luminescent characteristics of organic luminophores is especially clear in the case of isomers of distyrylbenzene. Upon transition from 1,4-isomer (XV) to 1,3-isomer (XVI) and 1,2-isomer (XVII), the hypsochromic effect in the absorption spectra is enhanced, the Stokes shift increases, and the luminescence intensity decreases. This is explained by shortening of the conjugation chain as well as by the probable enhancement of the steric hindrance in this series of isomers.[86]

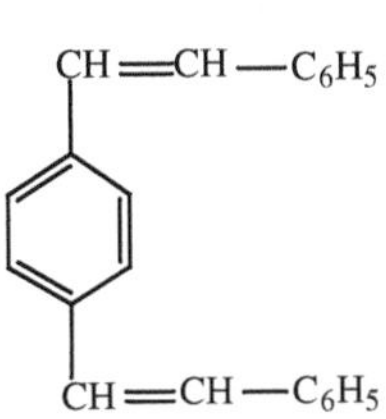

XV

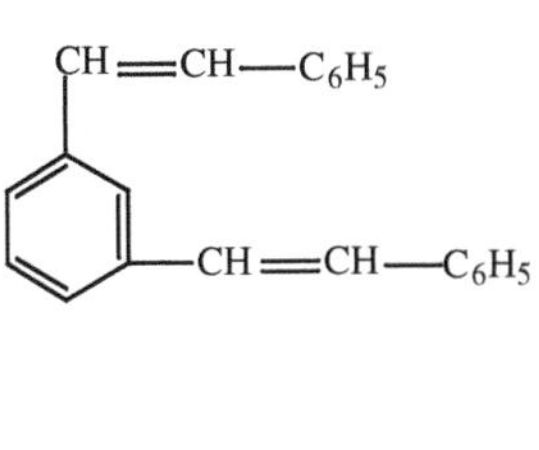

XVI

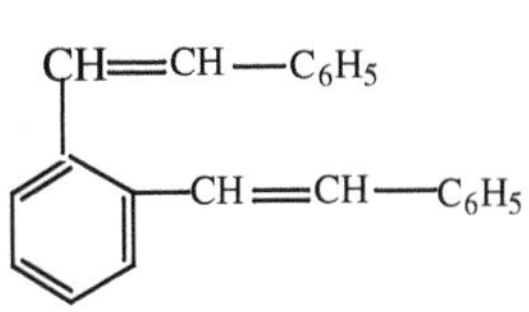

XVII

Structure 3.4 1,4-distyrylbenzene (XV); 1,3-distyrylbenzene (XVI); 1,2-distyrylbenzene (XVII)

The spectral-luminescent characteristics of organic luminophores are also affected by the nature and number of the heteroatoms involved in the conjugation system. When the oxygen atom in a fairly efficient luminophore such as 2,5-diphenyl-1,3,4-oxadiazole (VI) is replaced by the sulfur atom (2,5-diphenyl-1,3,4-thiadiazole), the substance does not luminesce.

Theoretically, the correlation between the spectral-luminescent characteristics and molecular structure of organic luminophores was most comprehensively formulated by Sangster and Irvine[88] and Nurmukhametov[94] and further developed in the quantum-chemical context by Maier and Danilova[95] and Maier.[96] Those investigations begin from the relative position of the electronically excited states differing in the nature of the orbital and multiplicity. On this basis, a classification of luminophore molecules was suggested according to the relative position of the energy levels of excited states.

According to this classification, there are five basic types of molecules differing in spectral-luminescent characteristics (electronic states are listed according to growing energy):

1. S_0, $T_{n\pi*}$, $S_{n\pi*}$, $T_{\pi\pi*}$, $S_{\pi\pi*}$
2. S_0, $T_{n\pi*}$, $T_{\pi\pi*}$, $S_{n\pi*}$, $S_{\pi\pi*}$
3. S_0, $T_{\pi\pi*}$, $T_{n\pi*}$, $S_{n\pi*}$, $S_{\pi\pi*}$
4. S_0, $T_{\pi\pi*}$, $T_{n\pi*}$, $S_{\pi\pi*}$, $S_{n\pi*}$
5. S_0, $T_{\pi\pi*}$, $S_{\pi\pi*}$, $T_{n\pi*}$, $S_{n\pi*}$

The majority of organic luminophores belong to type 5; i.e., their lowest states are represented by both levels of the same type, $S_{\pi\pi*}$ and $T_{\pi\pi*}$.

Analysis of a large body of relevant quantum-chemical calculations shows that with extending π-system, the energy of the $S_{\pi\pi*}$ and $T_{\pi\pi*}$ states decreases, thereby shifting the spectra to a longer-wavelength region.[96] The energy of $n\pi*$ transitions typically varies weakly with the extent of the π-system.

The semiquantitative method of estimating the light output of organic luminophores[88] implies calculating the number of the quinoid structures with the single-shot charge distribution possible for the molecules of interest. Figure 3.27 shows the simplest quinoid structures with single-shot charge distribution for the example benzene molecule. As energy transfer and scintillation probabilities are determined by the dipole moment of the transition, any correlation with the electronic structure of the molecule should take into account charge distribution in the molecule.

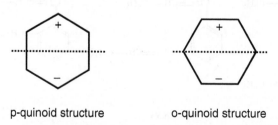

p-quinoid structure   o-quinoid structure

Figure 3.27   Quinoid structures for benzene with single-shot charge distribution.

For calculating the mobility of π electrons for each possible quinoid structure for a given molecule, one identifies the squared projections of the charge distribution vector onto two mutually perpendicular axes. These squared components are summed over all possible structures. The light output L_L for a large number of compounds is found to increase monotonically with increasing sum of the squared components.

Taking into account the structural features of molecules, one can select the reduction factor (structural parameters) f such that all the plots of L_L vs. the sums derived will coincide. These reduction factors can be calculated for each molecule by the formula

$$f = \left[n'\sqrt{2} + 2m' + 2l' + \left(\sqrt{2}\right)^{k'} + S' \right]^{-1} \tag{3.16}$$

where m' is the number of ordinary bonds in the continuous chain of conjugate benzene rings in the molecules of polyphenyls; l' is the number of five-membered aromatic rings in the molecule (e.g., for fluoranthene and acenaphthene, $l' = 1$); k' is the number of olefin bonds linking the aromatic systems; S' is the structural parameter ascribed to various pairs of aromatic systems linked by ordinary or olefin bonds ($S' = 0$ for phenyl, $S' = 4$ for biphenyl, $S' = \sqrt{2}$ for α-naphthyl, and $S' = 5$ for β-naphthyl); and n' is the number of angles formed by the straight lines passing through the centers of the neighbor aromatic rings. For example, for phenanthrene (III), $n' = 1$; chrysene (XVIII), $n' = 2$; and triphenylene (XIX), $n' = 3$.

XVIII   XIX

Structure 3.5   chrysene (XVIII), triphenylene (XIX)

The reduction factor accounts for various barriers to the movement of π electrons reducing the contribution from structures with charge distribution to the first excited state. For example, the "angularity" characterized by the parameter n' is a hindrance to the movement of electrons. "Smooth" molecules such as anthracene do not present this hindrance.

A great number of luminophores exhibit correlation between the structural parameters calculated and the light output, as seen in Figure 3.28. High spectral-luminescent characteristics of luminophores with five-membered aromatic rings (derivatives of oxazole and oxadiazole) are explained by the fact that, compared to luminophores with six-membered aromatic rings, they have a greater number of

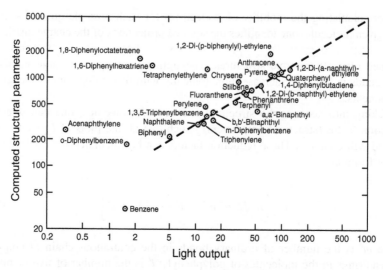

Figure 3.28   Correlation between the calculated structural parameters and the light output of organic luminophores. (From Sangster, R.C. and Irvine, J.W., Study of organic scintillators, *J. Chem. Phys.*, 24, 670, 1956.)

possible quinoid structures with single-shot charge distribution. In six-membered rings, the charge distribution centers can be only the atoms in *ortho* and *para* positions to each other, while in five-membered rings any two atoms can act as distribution centers.[88]

3.2.3   Temperature Effect on Emission Intensity of Organic Luminophores

Most organic luminophores exhibit a decrease in luminescence efficiency with rising temperature; at the melting point, efficiency declines to practically zero. Temperature quenching of luminescence of organic luminophores is due to intramolecular conversion of the electronic excitation energy into vibrational energy of the molecule. With rising temperature, the stored vibrational energy tends to increase, thereby increasing the probability of intramolecular deactivation.

The rate of decrease of luminescence efficiency with rising temperature varies with the luminophore and can be characterized by the temperature coefficient of stability as expressed in %/°C. For example, naphthalene is more sensitive to temperature changes than anthracene, and stilbene is less sensitive than naphthalene and anthracene.[87] Table 3.12 presents the temperature coefficients of stability for selected organic luminophores as determined from the light outputs at +30°C and −70°C.

The temperature coefficients of light output stability reported by different authors do not always coincide (e.g., −0.4, −0.3, and −0.25%/°C for anthracene and −0.6 and −0.8%/°C for naphthalene[97]). Nevertheless, these coefficients can be helpful for rough extrapolation of light output to the temperature at which the luminophore will be exploited. Light outputs of single-crystalline and powdered organic luminophores vary with the temperature in a similar way.[87] In view of temperature dependence of

Table 3.12 Change in Light Output (%) Caused by Temperature Change by 1°h for Some Organic Luminophores

Luminophore	Temperature Coefficient, %/°h
Anthracene	−0.54
Naphthalene	−1.06
Phenanthrene	−0.96
Diphenyl	−1.10
p-Terphenyl	−0.40
Quaterphenyl	−0.05
2,5-Diphenyloxazole	−0.42
2,5-Di-(4′-biphenylyl) oxazole	−0.20
1,4-Di-(2′-(5′-phenyloxazolyl))benzene	−0.19
Diphenylacetylene	−1.41
Stilbene	−0.31
Diphenylbutadiene	−0.08
Pyrene	−0.20
Triphenylene	−0.39
Fluorene	−1.49
1,6-Diphenylhexatriene	−0.18
α,α′-Dinaphthyl	−0.49
β,β′-Dinaphthyl	−0.39
Chrysene	−0.85
Fluoranthene	−0.26

Source: Sangster, R.C. and Irvine, J.W., Study of organic scintillators, *J. Chem. Phys.*, 24, 670, 1956.

the light output exhibited by the organic luminophores considered, it is expedient to cool the scintillation materials under operating conditions.

One can take advantage of this dependence in optimization of the luminophore application conditions and in measuring temperature. Temperature quenching of luminescence of organic luminophores found application in manufacturing temperature indicators.[98,99] Luminophores differ by the temperatures above which they do not luminesce. With calibrated standards, temperatures ranging from room temperature to 300°C can be monitored by the disappearance of luminous spots. These temperature indicators are reusable, since upon cooling they regain their luminescing ability. The temperature-related changes in light outputs from organic and inorganic luminophores are about the same.[97]

3.3 PLASTIC SCINTILLATORS

3.3.1 Brief Description of Energy-Transfer Processes in Plastic Scintillators

Major characteristics of the efficiency of energy conversion in plastic scintillators are the light output and scintillation efficiency. The light output L_L (absolute technical light output) is equal to the ratio of light energy emitted by the scintillator to ionizing radiation energy taken up. (Section 2.3.1 in Chapter 2 discussed a similar characteristic of luminophores, the energy efficiency of luminescence η.) The scintillation efficiency η' (physical light output, conversion efficiency) is equal to the ratio of the energy converted to the light to the energy of ionizing radiation taken up. This suggests that the light output and scintillation efficiency are interrelated as

$$L_L = k_L \eta' \tag{3.17}$$

where k_L is the light concentration coefficient ($k_L < 1$) characterizing the transparency of the scintillator to the light of its intrinsic fluorescence and the light loss by reflection and refraction at the scintillator–medium interface.

For small samples (up to 10 mm thick), the concentration coefficient, in the opinion of some authors,[100] can approach unity, though a universal limiting thickness independent of the spectral region and the structure of the polymer matrix can hardly be identified. In view of difficulties with measuring absolute light outputs, one often utilizes relative light outputs as measured relative to a sample taken as reference.[100]

There is extensive literature on the photophysical processes in organic molecules such as polymers and plastic scintillators (e.g., Vekshin[101] and Birks[102]). As applied to energy transfer from radioactive material to the photomultiplier, these processes can be described by the following simplified sequences[103]:

1. Excitation of the monomer by collision with the nuclear particle ejected or other processes of the excitation energy transfer in polymers (see Section 5.2.1 in Chapter 5):

   $M \leadsto M^*$

2. Nonradiative deactivation of the monomer:

   $M^* \rightarrow M$

3. Luminescence of the monomer:

   $M^* \rightarrow M + h\nu_0$

4. Formation of an excimer (unstable dimer that is essentially an associate of the excited and nonexcited states):

   $M^* + M \rightarrow (M^*M)$

5. Nonradiative energy transfer from the excited monomer to the impurity molecule:

   $M^* + Q \rightarrow M + Q^*$

6. Nonradiative deactivation of the impurity:

   $Q^* \rightarrow Q$

7. Energy transfer from the excited monomer to the luminescent additive:

$$M^* + A \rightarrow M + A^*$$

8. Energy transfer from the excimer to the luminescent additive:

$$(M^*M) + A \rightarrow 2M + A^*$$

9. Energy transfer from the impurity to the luminescent additive:

$$Q^* + A \rightarrow Q + A^*$$

10. Nonradiative deactivation of the luminescent additive:

$$A^* \rightarrow A$$

11. Fluorescence of the luminescent additive:

$$A^* \rightarrow A + h\nu$$

The contribution from each process depends on structure of the elementary unit of the polymer, presence of impurities, nature and concentration of the luminescent additive, position of the absorption band of the luminescent additive, etc.[100,103] The simplified scheme presented needs to be complemented with consideration of the role played by secondary solvents and secondary luminescent additives (the latter are often termed spectrum shifters).[86,100,103]

If the polymer matrix [e.g., poly(methyl methacrylate)] has no aromatic fragments, its repeating unit cannot be transferred to an excited state with a lifetime sufficient for further energy transfer. To ensure effective uptake of the nuclear particle energy, such matrices are loaded with up to 15 wt% of low-molecular-weight aromatic compounds (most often, naphthalene).[103,104] By analogy with liquid scintillation counting, such compounds are sometimes termed secondary solvents.

As known, the polymer bases are nontransparent to their intrinsic luminescence light. The luminescence spectrum of plastic scintillators corresponds to that of the luminescent additives introduced. Theoretically, the primary luminescent additive can receive energy from the primary (or secondary) solvent by nonradiative and by radiative pathways. The path to be taken is largely determined by the concentration of the primary luminescent additive. The majority of the primary additives employed afford the maximal light output in 1 to 4 wt% concentrations. At these concentrations, the transfer of energy of the primary luminescent additive is predominantly by nonradiative pathway.[89,105-107] Figure 3.29[108,109] shows how the concentration of the primary luminescent additive affects the efficiency of radiative and nonradiative processes of energy transfer in plastic scintillators.

Primary luminescent additives typically emit light at a shorter wavelength than the maximum of spectral sensitivity of the photomultiplier detector or human eye. This problem can be solved by introduction into the polymer base of a secondary luminescent additive (spectrum shifter), which absorbs light emitted by the primary

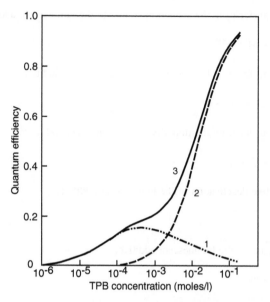

Figure 3.29 Quantum efficiency of energy transfer from the matrix to the primary luminescent additive - in the polystyrene - 1,1,4,4-tetraphenylbutadiene system. (1) Radiative transfer, (2) nonradiative transfer, and (3) overall transfer. (From Birks, J.B., *The Theory and Practice of Scintillation Counting*, Pergamon Press, Oxford, 1964, chap. 9; Birks, J.B. and Kuchela, K.N., Energy transfer in fluorescent plastic solutions, *Disc. Faraday Soc.*, 27, 57, 1959.)

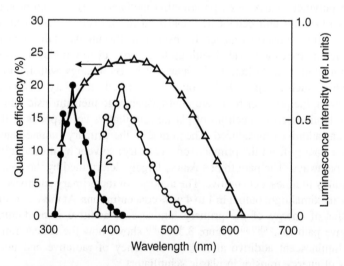

Figure 3.30 Quantum efficiency of the photomultiplier FEU-84-3 with a multialkaline photo-cathode and the luminescence spectra of (1) primary luminescent additive, *p*-terphenyl and (2) the secondary luminescent additive, 1,4-bis-(2′-(5′-phenylox-azolyl))benzene (POPOP). (From Vasil'chenko, V.G. and Solov'ev, A.S., Plastic scintillators with antirad additions and enhanced light output, *Prib. Tekhn. Eksper.*, 4, 46, 1998 [in Russian].)

additive and re-emits it to a longer-wavelength region (Figure 3.30[112]). The spectrum shifter concentration is typically 0.01 to 0.02% and, more rarely, up to 0.05%. At higher concentrations of luminescent additives, self-absorption of light takes place.[87,89,105-107]

If one spectrum shifter is not sufficient, a second shifter is introduced into the polymer matrix that shifts the light emission maximum to even longer wavelengths.[110] Also known are cases when a third spectrum shifter is introduced (e.g., Yuan et al.[111]), but with increasing numbers of shifters their efficiency typically decreases because of re-emission losses.

The effect of the luminescent additives includes minimal loss in the excitation energy transfer from the polymer base to the additive, high quantum output of luminescence, minimal concentration quenching, and a maximally close match between the luminescence spectrum and the spectral response of the photomultiplier.[100] Luminescent additives are typically introduced into the polymer matrix as solid solution. Exceptions are anthracene, fluorene, and tetraphenylbutadiene, which get chemically bonded to the polymer chain during polymerization.[87,104] In some cases, the luminescent additives are copolymerized, using their vinylated derivatives as comonomers.[86,113]

Unreacted monomer is a scintillation quencher that decreases light output and changes the scintillator to blue over time.[104,114]

As the molecular weight of the polymer base of the scintillator increases to a certain limiting value (e.g., 10^5 for polyvinyltoluene), the light output increases[115] (Figure 3.31). This is explained by the increased quantum output of fluorescence for macromolecules with a larger molecular weight.[100] Other conditions being the same,

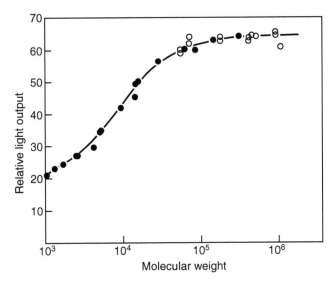

Figure 3.31   Relative light output of polyvinyltoluene scintillator obtained in the absence (o) and in the presence (◊) of the polymerization chain transfer agent as a function of the molecular weight of polymer. (From Funt, B.L. and Hetherington, A., The influence of chain length on the luminescent out of plastic scintillators, *Int. J. Appl. Rad. Isot.*, 4, 189, 1959.)

a higher light output is characteristic of scintillators with narrower molecular-mass-distribution pattern of the polymer base.[89] The luminescent additives introduced into the monomer (before polymerization) do not affect the molecular weight of the resulting polymer, except for previously mentioned cases of their chemical bonding.[87,108,115]

Plastic scintillators should be uniform and isotropic, and for this reason only amorphous polymers are employed as their base.[104] Noted that the efficiency of conversion of radioactive decay energy into light (the light output) for plastic scintillators as expressed in absolute units is no more than 4%,[1,107] 4 to 6%,[110] and 4 to 7%.[116] This is 1.5 to 2 times lower than the corresponding values for organic crystals (e.g., anthracene),[87,107] but the radiation resistance of small organics is an order of magnitude lower compared to plastics.[103,107] At the same time, the quantum output of luminescence for organic compounds can approach 100%, which accounts for the desirability of liquid scintillation for beta assay.[117]

The light output of plastic scintillators in recording heavy particles (protons, α-radiation, etc.) is noticeably less than in recording β- or γ-radiation. The α/β ratio of light output is only slightly dependent on the nature of the scintillator, namely, 0.07 to 0.15.[103]

3.3.2 Matrix Influence on Plastic Scintillator Transparency

Transparency is one of the most important characteristics of plastic scintillators, determining their suitability as large-sized blocks or extended fibers.[118,119] Quantitative characteristics of the transparency are 1) the transmission spectrum of the sample with a known thickness, 2) the integrated linear light absorption coefficient, and 3) the light attenuation length. The light attenuation length is the length at which the light intensity is attenuated by e times. By definition, the product of the light attenuation length and the linear light absorption coefficient is equal to unity. High-quality optical materials should exhibit an integrated light absorption coefficient under 0.004 cm^{-1},[119] though for smaller articles this coefficient[118] can be under 0.01 cm^{-1}. The transparency of the scintillator can be increased by selecting the appropriate polymer base exhibiting (other things being equal) a minimum absorption throughout the spectral region of emission by all the system components.

Intrinsic light absorption by a polymer is due to the presence in its molecule of chromophore groups such as carbonyl $>C=O$, nitrile $-C\equiv N$, phenyl $-C_6H_5$, etc. For example, light absorption by polyacrylates and polymethacrylates in the 190 to 210 nm region is due to the carbonyl group. At the layer thickness of 10 μm, poly(methyl methacrylate) transmits practically all light with wavelengths above 200 nm; for samples with layer thickness of 2 mm, the transmission edge is close to 260 nm.[119]

Materials whose molecules contain aromatic rings have markedly greater light absorption coefficients. For example, plastic scintillators based on polystyrene and polyvinyltoluene[118] have the light absorption coefficient of 0.02 to 0.05 cm^{-1}. However, many applications of scintillation materials do not require extended samples, and polymers containing aromatic rings noticeably surpass poly(methyl methacrylate) in radiation stability (see Section 5.3.7 in Chapter 5).

Figure 3.32 shows the transmission spectra of various polymer scintillator bases of practical importance. It is seen that polystyrene is less transparent than mixtures

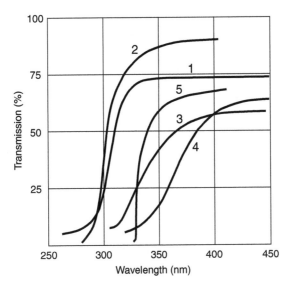

Figure 3.32   Transmission spectra of various polymer bases: (1) poly(methyl methacrylate), (3) polystyrene, and (5) poly(methyl methacrylate) added with 15% of naphthalene (all samples with thickness of 10 mm); (2) poly(methyl methacrylate) and (4) polystyrene (all with thickness of 2 mm, as reported by other authors). (From Gunder, O.A. et al., Plastic scintillators with enhanced transmittance, *Prib. Tekhn. Eksper.*, 3, 66, 1969 [in Russian].)

of poly(methyl methacrylate) with 15% naphthalene. Based on poly(methyl methacrylate), naphthalene, and luminescent additives, one can manufacture scintillators whose transparency exceeds by two to three times that of polystyrene scintillators,[118] but such poly(methyl methacrylate)-based scintillators are slightly less efficient (see Section 3.3.3).

The transmission spectra of polysiloxanes (Figure 3.33) show that, at wavelengths above 290 nm and layer thickness of 50 μm, the majority of polysiloxanes exhibit 90 to 98% light transmission, the least transparent being the phenyl-containing polymer.[120]

3.3.3   Matrix Influence on Plastic Scintillator Light Output

The plastic scintillators known by the middle 1980s were classified into four groups,[100,103] based on the chemical nature of the polymer and the features of the energy transfer:

1. Scintillators based on polystyrene and its derivatives, containing a substituent in the benzene ring or in the alpha position (group I)
2. Scintillators based on vinyl monomers with polyphenyl or condensed aromatic and heterocyclic rings (group II)
3. Scintillators with an inactive polymer base whose elementary unit lacks π conjugation, though with low-molecular-weight aromatic compounds as the secondary solvent (group III)
4. Scintillators based on copolymers of various composition (group IV)

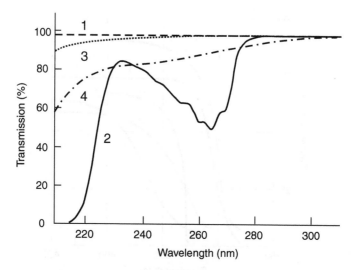

Figure 3.33   Transmission spectra of polysiloxanes at the layer thickness of 50 ± 10 μm: (1) polydimethylsiloxane, (2) polymethylphenylsiloxane, (3) polymethyltrifluoropropylsiloxane, and (4) polymethylnitrilesiloxane. (From Andryushchenko, L.A. and Grinev, B.V., Organosilicon materials for scintillation detectors, *Prib. Tekhn. Eksper.*, 4, 5, 1998 [in Russian].)

These four groups can now be supplemented with organosilicon scintillators containing one or two phenyl groups in the elementary unit.[120] Light outputs achievable with group I plastic scintillators are given in Table 3.13.

Scintillation activity does not markedly increase with increasing size of the alkyl group; the methyl group in the alpha position somewhat decreases the efficiency. Other electron-donating groups in the ring (e.g., methoxy group) increase the light output to a lesser degree than the methyl group. Halogen atoms in the benzene ring sharply reduce the scintillation efficiency. Thus, the optimal substituents are methyl groups in positions 2 or 4.[103] It should be noted that plastic scintillators, whose light output is optimal, do not necessarily exhibit optimal radiation resistance. For example, the rate constants of the radiation degradation of polystyrene, polyvinyltoluene, and polyvinylxylene (under β radiation) are in the ratio of 1:3.5:5.[105,121] Specific operation conditions and requirements for the scintillator dictate optimization criteria.

For group II plastic scintillators (see Table 3.14), increase in the number of phenyl groups in the elementary unit (both substituting and main-chain) somewhat increases scintillation efficiency but deteriorates performance characteristics of plastic scintillators. Block polymerization yields fragile and less transparent materials; also, purification of the monomers intended for synthesis of such scintillators presents considerable difficulties. On the whole, this group of scintillators does not find extensive application,[103] though it is superior to polystyrene in radiation resistance.[113]

Group III plastic scintillators (see Table 3.15) are characterized by fairly low light outputs, although for sufficiently large scintillators, a higher transparency to the emitted light can be significant. The efficiency of such poly(methyl methacrylate)

Table 3.13 Relative Light Output of Some
Group I Plastic Scintillators
(Light Output of Polystyrene
Scintillator Is Taken as 100%)[a]

Substituent in the Benzene Ring or in the Side Chain of Styrene	Light Output, %
d	100
2-Methyl	117
3-Methyl	105
4-Methyl	125
2,4-Dimethyl	160
2,5-Dimethyl	150
3,4-Dimethyl	140
2,4,5-Trimethyl	170
α-Methyl	86

[a] Luminescent additives: pTP 2%, POPOP 0.1%.

Sources: Gunder, O.A., *Polymeric Systems and Their Scintillation Properties*, All-Union Research Institute of Single Crystals, Khar'kov, 1975 (in Russian); Barashkov, N.N. and Gunder, O.A., *Florescent Polymers*, Khimiya, Moscow, 1987 (in Russian; English translation: Ellis Horwood, Chichester, 1994.).

Table 3.14 Relative Light Output of Some
Group II Plastic Scintillators (Light
Output of Polystyrene Scintillator
Is Taken as 100%)[a]

Initial Monomer	Light Output, %
4-Vinyldiphenyl	120
3,3'-Dimethyl-4-vinyldiphenyl	180
1-Vinylnaphthalene	130
2- Vinylnaphthalene	120
N- Vinylcarbazole	100
2-Vinylfluorene	60

[a] Luminescent additives: PPO 1%, POPOP 0.1%.

Sources: Gunder, O.A., *Polymeric Systems and Their Scintillation Properties*, All-Union Research Institute of Single Crystals, Khar'kov, 1975 (in Russian); Barashkov, N.N. and Gunder, O.A., *Florescent Polymers*, Khimiya, Moscow, 1987 (in Russian; English translation: Ellis Horwood, Chichester, 1994.).

Table 3.15 Relative Light Output of Some Group III Scintillators Based on Poly(methyl methacrylate) (Light Output of Polystyrene Scintillator Is Taken as 100%)

Secondary Solvent and Its Content, %	Luminescent Additives and Their Content, %	Light Output, %
Naphthalene (15)	POPOP (0.4)	40
Phenanthrene (20)	POPOP (0.4)	40
Naphthalene (8)	POPOP (0.08)	30
Naphthalene (10)	PPO (1.5); POPOP (0.08)	40
Naphthalene (15)	PPO (1.5); POPOP (0.08)	60

Sources: Gunder, O.A., *Polymeric Systems and Their Scintillation Properties*, All-Union Research Institute of Single Crystals, Khar'kov, 1975 (in Russian); Barashkov, N.N. and Gunder, O.A., *Florescent Polymers*, Khimiya, Moscow, 1987 (in Russian; English translation: Ellis Horwood, Chichester, 1994.).

scintillators containing naphthalene is 60 to 80% of that of polystyrene.[86,103] Some authors report even 100% efficiency relative to polystyrene scintillators.[118]

In this context, it will be appropriate to touch on how naphthalene additives affect the light output of polystyrene scintillators. Samples containing 5 to 15% naphthalene proved to be unstable in storage.[122] Even within 1 to 2 months, the evaporation of naphthalene caused cracking of the surface — not surprising in view of the fact that the solubility of naphthalene in polystyrene does not exceed 4%.[103] Such effects were not observed for naphthalene in amounts of 0.5 to 2%,[122] but this low content is characteristic of the primary scintillation additives, of which naphthalene is not the best (see Section 3.3.4), or for radioprotectors (see Chapter 5, Section 5.5.2). With an eye to improving the quality of scintillator base and, correspondingly, the light output, this line does not seem promising.

In this connection, group IV scintillators seem of interest (see Table 3.16). Moreover, copolymerization of styrene with polynuclear vinylaromatic monomers increases radiation resistance of the scintillator.[113]

The organosilicon scintillators surpass all other groups of scintillators in thermostability, but they are worse than other scintillators as solvents for luminescent additives and exhibit slightly poorer mechanical characteristics.[89] When in air, they exhibit light output and radiation resistance close to those of polystyrene, but in absence of air these characteristics surpass those of polystyrene scintillators.[89,120]

3.3.4 Luminescent Additive Influence on Plastic Scintillator Spectral Characteristics

The spectral properties of organic luminophores discussed in Section 3.2 concerned individual compounds. In the case of plastic scintillators, one takes into account the mutual influence of the components (matrix, primary additive, and spectrum shifter) determining the spectral characteristics of the scintillator as a whole. Figure 3.34 presents the light output of polystyrene scintillators as influenced by the nature of the primary luminescent additive and its concentration.

Table 3.16 Relative Light Output of Some Group IV Copolymeric Plastic Scintillators (Light Output of Polystyrene Scintillator Is Taken as 100%)

Monomer 1 and Its Content, %	Monomer 2 and Its Content, %	Light Output, %
Methyl methacrylate (100)	—	10[a]
Methyl methacrylate (60)	Styrene (40)	80[a]
Methyl methacrylate (30)	N-Vinylcarbazole (70)	120[a]
4-Vinyldiphenyl (30)	Styrene (70)	140[a]
4-Vinylterphenyl (20)	Styrene (80)	138[a]
1-Vinylnaphthalene (30)	Styrene (70)	136[b]
2-Vinylnaphthalene (30)	Styrene (70)	156[b]

[a] Luminescent additives: pTP 2%, POPOP 0.1%.
[b] Luminescent additives: PPO 1%, POPOP 0.1%.

Sources: Gunder, O.A., *Polymeric Systems and Their Scintillation Properties*, All-Union Research Institute of Single Crystals, Khar'kov, 1975 (in Russian); Barashkov, N.N. and Gunder, O.A., *Florescent Polymers*, Khimiya, Moscow, 1987 (in Russian; English translation: Ellis Horwood, Chichester, 1994.).

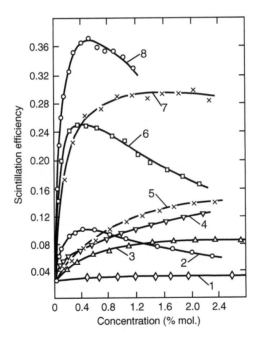

Figure 3.34 Light output of plastic scintillators (relative to anthracene) as influenced by the nature and concentration of the primary luminescent additive: (1) diphenyl, (2) stilbene, (3) naphthalene, (4) diphenylbutadiene, (5) anthracene, (6) PPO, (7) *p*-terphenyl; and (8) tetraphenylbutadiene. (From Swank, R.K. and Buck, W.L., *Phys. Rev.*, The scintillation process in plastic solid solutions, 91, 927, 1953.)

The figure shows that in all cases, the light output increases with increasing concentration of the primary additive. At a certain concentration specific for each activator, light output attains a maximum. Decrease in light output at increased concentrations of the additives is due to the concentration-quenching phenomenon.

The emission spectra of the majority of commercial plastic scintillators have their maxima at 420 to 490 nm with band half-width of 40 to 60 nm.[124] Typical scintillation time for plastic scintillators is small (2 to 5 ns). The luminescence light attenuation length for the majority of plastic scintillators is 100 to 400 cm.[124-126]

Unfortunately, the use of different references for the spectral-luminescent characteristics of various plastic scintillators complicates comparison of data reported by different authors. Sometimes, the spectral-luminescent characteristics of plastic scintillators with various additives are compared using a particular scintillator as the reference — most often, the widely used polystyrene scintillator added with 2% pTP and 0.025% POPOP.[91] Table 3.17 lists the light outputs for some polystyrene scintillators with various primary and secondary additives relative to this scintillator.

Design requirements for activating additives and spectrum shifters intended for producing plastic scintillators with good spectral-luminescent characteristics are as follows:

1. The luminescent additives have high light outputs.
2. The overlap region of the absorption spectra of the primary additive and the emission spectra of the polymer base are matched.
3. The absorption spectrum of the secondary additive effectively overlaps with the emission spectrum of the primary additive.
4. The luminescent additives exhibit good solubility in the polymer base.
5. The additives do not inhibit polymerization of the monomers.

The spectral-luminescent characteristics of plastic scintillators are also affected by how luminescent additives are introduced into the plastic base. The light output can

Table 3.17 Light Outputs of Polystyrene Scintillators with Various Luminophore Additives (Relative to PS + 2% pTP + 0.025% POPOP Scintillator)

Scintillator Composition	Light Output, %
2% pTP + 0.025% POPOP	100
2% 1,2-Bis-(5'-phenyloxazolyl-2')benzene	62
0.7% 2,5-Bis-(styryl-4)oxadiazole + 0.025% pTP	103
2% pTP + 0.025% 2-(hydroxyphenyl-2)quinolinyloxazole	42
1.5% pTP + 0.025% 2-(2'-ethoxynaphthyl)-6-diethylamino-benzoxazole	86
1.5% pTP + 0.025% (1'-cyano-2',4'-diphenylpyrido-1,2)benzimidazole	65
1.5% pTP + 0.025% 9,10-bis-(phenylethylene)anthracene	55
1.5% pTP + 0.025% 1,4-di-2-(4'-methyl-5'-phenyloxazolyl)benzene	102
2% pTP + 0.02% 1,3-diphenyl-2-pyrazoline	105
1% 2-(Dimethylaminophenyl)-5-methylbenzoxazole	95
1% 2-(4'-Dimethylaminophenyl)-5,7-di(tert-butyl)benzoxazole	82
1% 1,4-Bis-(5',7'-di(*tert*-butyl)benzoxazolyl-2)benzene	89

Source: Vasil'chenko, V.G. et al., Study of the radiation stability of plastic scintillators, *Prib. Tekhn. Eksper.*, 5, 85, 1995 [in Russian].

be significantly increased by copolymerization of vinylated luminophores with the appropriate monomer, i.e., by introducing the activator molecules into the polymer chain. This increases the efficiency of migration of the excitation energy from the polymer to the activating additive. For example, the light output of the copolymer of styrene with 2-vinyl-9,10-diphenylanthracene is twice that of the solid solution of 9,10-diphenylanthracene in polystyrene, with the same additive concentration.[86]

3.3.5 Temperature Effects on Plastic Scintillators

Rising temperature results in deformation of plastic scintillators. Deformation temperatures of the most widely used plastic scintillators are 82°C for polystyrene, 95°C for cross-linked polystyrene, 73°C for polyvinyltoluene, 100°C for poly(methyl methacrylate), and 200°C for polyphenylsiloxanes.[120] Thus, virtually all polymers, except for organosilicon compounds, have working temperatures below 100°C. The thermal stability of polymers is characterized by various parameters such as glass transition temperature, softening temperature, and deformation onset temperature with and without different loads. These parameters are not identical in absolute value, but their relative order for various polymers is often preserved (see, for example, Speranskaya and Tarutin,[127] and Lipatov et al.[128]).

It was suggested that the thermostability of scintillators be improved by using cross-linked polymers, such as poly-2,4-dimethylstyrene cross-linked by diisopropylbenzene. This increases the softening point by 10 to 15°C (compare to polystyrene and cross-linked polystyrene) and does not significantly decrease the light output.[103] Note that for plastic scintillators of identical composition, the temperature dependence of the mechanical properties is, to a certain degree, determined by the polymer matrix-preparation procedure.[129]

It is believed that with rising temperature, the initial light output of plastic scintillators decreases, the temperature coefficient being –0.1 to –0.4%/°C.[130] However, the actual behavior of plastic scintillators can be varied. For example, for some scintillators, the light output remains virtually unchanged between –173 and –33°C and varies at the –0.1 to –0.3%/°C gradient in the –33 to +67°C range.[108] For other plastic scintillators, the light output in the temperature range from –10 to +20°C is stable, and in the range from +20 to +60°C it exhibits a temperature coefficient of –0.125%/°C. For certain samples, this coefficient is as high as –0.3%/°C.[91] In some cases, the plot of the light output–working temperature dependence has a weakly pronounced maximum[107,129] (Figure 3.35), and with rising temperature from 20 to 80°C, the light output decreases by 14%.[129] This corresponds to an averaged temperature coefficient of –0.25%/°C. On average, the light output of plastic scintillators is less temperature dependent than organic crystalline luminophores (see Table 3.12).

The stability of the light output of plastic scintillators exploited at elevated temperatures deserves special consideration. On prolonged heating of polystyrene scintillator (110 to 120°C, 1000 h) the light output of the scintillation block declines first at its surface and then to a depth of up to 5 mm. The decreased transparency of the scintillators is manifested, above all, in the blue wavelength region of the

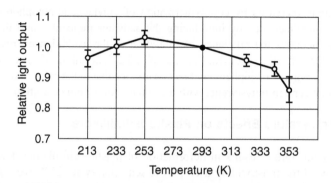

Figure 3.35   Relative light output of the polystyrene scintillator prepared from melt as a function of the working temperature (the light output at 293 K is taken as unity). (From Gen, N.A. et al., Performance characteristics of plastic scintillators obtained from melt, *Prib. Tekhn. Eksper.*, 4, 35, 1988 [in Russian].)

spectrum. Therefore, with all other conditions the same, the light output of scintillators with longer-wavelength absorption is less affected by temperature.[130]

At high temperatures, plastic scintillators exhibit a fast and irreversible decrease in the light output.[130] Their serviceability deteriorates with long storage at medium temperatures, as well (Figure 3.36). For example, it was found[131] that the service life of polystyrene scintillators is reduced by about half with rising storage temperature by every 6 to 12°C.

In prolonged storage of plastic scintillators under ambient conditions, the light output decreases exponentially with time:

$$(L_L)_t = (L_L)_0 \exp(-\lambda' t) \tag{3.18}$$

where $(L_L)_0$ and $(L_L)_t$ are the initial light output and the light output at time t, respectively; λ' is the aging constant. The time in which the light output of plastic scintillators decreases by half is 55 years for PS + 4% pTP + 0.04% POPOP, 114 years for PS + 4% pTP + 0.08% POPOP, and 106 years for PS + 4% pTP + 0.2% POPOP.[131]

Evidently, the aging of plastic scintillators in prolonged operation or storage at medium and high temperatures is due to thermal-oxidative reactions of the polymer base. The products yielded by these reactions can accept the excitation energy and absorb the light emitted by the luminescent additives. Handling plastic scintillators in a nitrogen or argon atmosphere improves the temperature stability of their optical characteristics.[132]

Light output tends to increase with lowering temperature, but at very low temperatures (e.g., liquid nitrogen temperature), it sharply decreases. Of certain interest is how the samples, irradiated at low temperatures, behave upon subsequent heating. This is important in the context of the radiothermoluminescence phenomenon (RTL), which is light emission by samples preirradiated at low temperature.[133] Many organic compounds, including polymer compounds and plastic scintillators, exhibit RTL.[133-137]

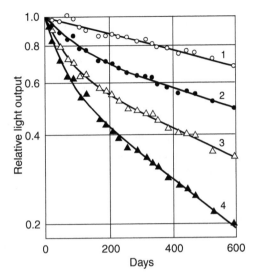

Figure 3.36 Relative light output of SPS-B10 polystyrene scintillator as a function of the storage time at different temperatures: (1) 36°C, (2) 44°C, (3) 52°C, and (4) 60°C. (From Zyablin, L.V. et al., On durability of plastic scintillators, *Prib. Tekhn. Eksper.*, 2, 73, 1986 [in Russian].)

This phenomenon is due to long-lived (many days) metastable luminescence centers under low-temperature irradiation and the subsequent light emission upon heating. In the early stages of RTL investigations, radicals were regarded as potential luminescence centers. More recent investigation pointed to the primary importance of charged particles. The ionic character of the emission centers for RTL is corroborated by the following arguments[138]:

1. RTL is often observed at temperatures much lower than those characteristic of the reactions involving radicals.
2. No luminescence occurred upon heating of the samples that accumulated radicals yielded by mechanical degradation at low temperatures.
3. RTL intensity typically attains a maximum at doses of several megarad, while radicals exhibit linear accumulation up to those of 50 Mrad.

Different compounds are characterized by different temperature regions of accumulation of light energy under irradiation and of its release under subsequent heating. Some compounds accumulate light energy at very low temperatures, while others accumulate at moderately low, and in certain cases, room, temperatures. The reasons for this temperature dependence are the structural features of the substance of interest and different efficiencies of the traps retaining the reactive intermediates formed from radiolysis. The temperature regions of radiothermoluminescence correspond to those of structural transitions.

Thermoluminescence curves (luminescence intensity vs. temperature plots) for samples of low- and high-density polyethylene were presented in an early work devoted to RTL.[136] The irradiation dose was 10^6 rad at −173°C; the luminescence was measured upon subsequent defrosting of the sample at a rate of 15°C/min. For low-density polyethylene (LDPE), two radiothermoluminescence scintillations peaked at −120 and −40°C; for high-density polyethylene (HDPE), there was a maximum near −120°C as well as a poorly resolved maximum at −59°C. These scintillation curves are adequately explained in view of the different structures of the samples and the temperatures of the structural transitions.

LDPE is characterized by a higher degree of branching than HDPE and exhibits two distinct structural transitions at −105 to −125°C and −20 to −70°C. The temperature of the latter transition is termed the glass transition temperature. HDPE does not exhibit the second maximum at all, due to the low degree of branching of the molecules. The observed radiothermoluminescence may be due to defrosting of the hindered molecular motion in the polymer.[136] Since radiothermoluminescence maxima are typically observed in the glass transition and melting temperature regions of samples, radiothermoluminescence appears to be a useful probe for studying structural transitions due to temperature-related changes in a substance.[133]

The majority of polymers and plastic scintillators exhibit radiothermoluminescence at rather low temperatures. There are examples, however, of light emission in RTL at room temperature and higher temperatures. Poly(tetrafluoroethylene), for example, has an RTL peak at +22°C,[136] SKB rubber has an RTL peak at +60°C,[136] and polycarbonate, irradiated at room temperature, has a scintillation peak at ca. +130°C.[137] Introduction into the latter of luminophore additives increases the RTL intensity several times.[139] Certain polystyrene and polyvinyltoluene scintillators exhibit blue luminescence under beta irradiation at the dry ice temperature (ca. −70°C) and subsequent heating at room temperature.[134]

A high-temperature RTL peak (+90°C) was observed for polyethylene with an ultra-high molecular weight, as well as for high-density polyethylene (+72°C) when irradiated by low-energy (5 to 30 keV) electrons in the absence of oxygen.[135]

For luminescence be at a maximum, low-temperature irradiation should be carried out in the absence of oxygen (in a vacuum or in an inert atmosphere) and in the dark. Different substances show different patterns of variation of the maxima in the scintillation curves with the dose. Some exhibit almost linear increase in the RTL intensity up to doses of 100 Mrad, and some exhibit this increase only up to 1 to 5 Mrad, whereupon the RTL intensity remains unchanged or slowly decreases.

Polymers and plastic scintillators exhibiting high-temperature radiothermoluminescence may have application in dosimetry.[135] As expected, these materials can be utilized for measuring doses up to several kilorad. Importantly, organic RTL dosimeters experience dose absorption equivalent to that absorbed by human tissues.

The luminescence time in RTL depends on the material nature, irradiation dose, and heating rate. RTL can be manifested as short-time scintillation or for several hours. Numerous publications concern the RTL of samples irradiated by external sources of ionizing radiation; however, there are no data on RTL of labeled compounds under their intrinsic radiation. This kind of RTL could be termed self-radiothermoluminescence.

In the context of conversion of decay energy into electrical power, it can be assumed that radiothermoluminescence of tritium-containing compositions could be utilized to create betavoltaic devices to accumulate tritium decay energy as "frozen" luminescence centers and to convert this into light and then into electrical power upon defrosting the article. One can expect much larger currents than with other betavoltaic sources, since decay energy is accumulated during low-temperature exposure rather than consumed uniformly.

Calculations show that for tritium-containing compounds for which the specific activity of, for example, 1 Ci/g can easily be achieved, luminescence can grow in intensity over a 1-year period for those compounds for which RTL intensity grows up to doses of 100 Mrad. For compounds for which the luminescence intensity grows up to doses of 1 to 5 Mrad, the luminescence maxima will be attained within 1 to 2 weeks. The period required for attaining the RTL intensity maximum can be widely varied by varying the specific activity of the radiation source. Self-radiothermoluminescence of tritium-containing compounds could possibly find application in pulse devices (light and electrical).

3.3.6 Magnetic Field Effects on Plastic Scintillator Light Output

Another factor influencing luminescent characteristics of plastic scintillators is magnetic induction such as that experienced in modern colliders. Most studies concerned with the influence of the magnetic field on the light output of plastic scintillators were carried out under magnetic induction to 0.5 T.[140-142] In all cases, the light output of the plastic scintillators increased in a magnetic field. For example, the light output of polystyrene scintillator SCSN-38 increased 0.3%, 0.9%, 1.1%, and 3.3% at magnetic induction of 1, 10, 100, and 450 mT, respectively.[142] The response to the magnetic field from polystyrene and polyvinyltoluene scintillators is about 4 to 5 times weaker compared to polyacrylate scintillators.[141]

To get a more comprehensive idea of the influence of magnetic field on the light output of plastic scintillators, one can use a superconducting-coil electromagnet with an induction of up to 5 T.[143] In this case, the light output increased with enhancement of the magnetic field (to 15 to 17% at the magnetic field induction of 2 to 2.5 T; with further increase of the induction, the light output slowly decreased to 12% at 5 T). Interestingly, patterns of variation of the light output with the magnetic field induction were virtually identical in those experiments where the field induction was increased and in those where it was decreased.

Independent of the type of ionizing radiation, the light output of all plastic scintillators in the magnetic field initially increases. In magnetic fields with induction up to 1.5 T, 1) the light output grows rapidly and virtually linearly, 2) whereupon the growth flattens, and at induction above 2.5 T the light output decreases; 3) the spectral characteristics of the light emission are independent of the magnetic field; 4) the influence of the magnetic field on the light output is independent of its sense; and 5) with growing thickness of the plastic scintillators, the influence of the field is enhanced.

The mechanism by which the magnetic field affects the light output of plastic scintillators still remains unclear. It can be assumed that this effect is related to orientation of the polar molecules of light-emitting additives in the magnetic field. The light output may be influenced by splitting the levels emitting light quanta.[143]

3.3.7 Manufacturing Optimization for Plastic Scintillator Efficiency

Production of plastic scintillators requires extra-pure-grade substances. Here, the deciding role is played by the purification processes of all the initial substances, development of purity-checking methods, and maintenance of all stages of the process. This should be even more relevant to manufacture of radioluminescent light sources based on tritium-labeled organic compounds. However, the problems of microsynthesis, specific features of handling organic substances with high specific activity, and product sensitivity to radiation degradation complicate extension of the methods developed for synthesis of nonradioactive plastic scintillators to manufacturing RLSs-T.

Plastic scintillators are manufactured mainly by block polymerization of the monomer with the luminescent additive dissolved in it or dissolution of the luminescent additive in the polymer melt. The resulting homogeneous mixture can be processed by injection molding, extrusion, lacquer method, etc.[103,108]

The process conditions of manufacturing plastic scintillators with desired chemical structure should be such that their light output is at a maximum. This requires distribution of luminescent additives as uniform as possible in the polymer, optimal molecular weight, optical isotropy and homogeneity of the block, lack of residual stresses, and minimal residual content of the monomer.[104]

Self-luminous paints are manufactured predominantly with mixtures of a tritium-labeled polymer, a scintillator (typically ZnS), and a binder of a suitable solvent. Application of the resulting mixture to the surface of the article is finished with evaporation of the solvent.[144]

All-organic optically clear light sources were manufactured by radiation block polymerization induced by radiation from the organic tritium carrier. The latter, along with luminescent additives, was dissolved in the liquid styrene mass. At room temperature, radiation polymerization was complete within several days.[114,145]

The next section briefly outlines some technological factors whose influence on light output was unambiguously established for ordinary plastic scintillators. The irreproducibility of these parameters is specifically responsible for variation between performance characteristics reported by different authors for scintillators of the same composition, namely, transparency, light output, and radiation resistance.

3.3.7.1 High Chemical Purity of Monomers

This is one of the most important requirements, as the initial monomer makes primary contribution to the mass of the scintillator. The monomer should be free from impurities such as polymerization inhibitors and luminescence quenchers as well as any substances exhibiting appreciable additional absorption in the UV or visible regions of the spectrum. Various batches of unpurified monomer can differ substantially in their impurity content, manifested in the difference of the absorption spectra in the region of 320 to 420 nm.[104] The content of the basic substance in the monomer should exceed 99.8[104] or 99.9%.[146]

The first stage in treatment of the monomer to remove impurities consists in removal of the inhibitor (especially important for fairly volatile inhibitors such as hydroquinone). To this end, one can wash the monomer with an alkali solution to remove phenolic inhibitors such as hydroquinone or *tert*-butylpyrocatechol,[104] followed by flushing the monomer with water and drying with sodium sulfate or calcium chloride.[119] A widely used alternative is to pass the monomer through a basic aluminum oxide chromatographic column.[104,147,148] The adsorbent can be used in dynamic (passing through a column) and in batch conditions (mixing in the bottle). Low-volatility stabilizers, e.g, *p*-quinonedioxime, are not necessarily removed before distillation.[104,119]

The second stage consists of efficient vacuum distillation at residual pressure of about 10 mm Hg in a nitrogen atmosphere. At such reduced pressure, styrene boils at ~31°C, and vinyltoluene boils at ~50°C.[87] Occasionally, repeated distillation runs (up to four runs involving thorough fractionation[104] and taking the middle fraction[147]) are needed.[87,108] It has been shown that glass has advantage over metal in distillation equipment.[104,119,148] Styrene purified to the degree suitable for producing optically transparent polymers should exhibit a transmission not less than 70% at 350 nm and not less than 20% at 320 nm; methyl methacrylate should exhibit a transmission not less than 100% at 350 nm and not less than 80% at 320 nm. The refractive index of styrene should lie within 1.5465 to 1.5470, and that of methyl methacrylate should lie within 1.4140 to 1.4160.[104,119]

In storage in air, the transparency of styrene is rapidly deteriorated, especially in the short-wavelength region.[104]

3.3.7.2 High Degree of Chemical Purity of Luminescent Additives and Secondary Solvents

The necessary but insufficient indicators of luminescent additive purity are melting point, intensity and position of luminescence maximum of the additives in the crystalline state, and light output of their toluene solutions.[104]

The first stage in the treatment of luminescent additives consists in their recrystallization. However, in order to achieve high and reproducible light outputs for plastic scintillators, in certain cases one must utilize additional methods of purification of the luminescent additives and secondary solvents, e.g., zone melting or column chromatography.[104,148] Although these methods do not significantly change the indicators of the additives, they can substantially improve the quality of plastic scintillators. For example, chromatographic purification of *p*-terphenyl increases light attenuation length of plastic scintillators by a factor of 1.5 to 2. The positive influence of these additional methods of purification is immediately evident from increase in the optical transmission of concentrated toluene solutions of the purified luminescent additives or secondary solvents.[104]

Luminescent additives, except for those chemically bonded to the polymer chain, do not affect polymerization kinetics.[104]

3.3.7.3 High Quality of Polymerization Initiator

With polymerization initiators, the molecular weight of polymers typically decreases.[103,115] It is desirable to use active initiators, which have low degradation temperature and do not form UV-absorbing degradation products — e.g., 2,2-azo-diisobutyronitrile. UV-absorbing degradation products or impurities in the initiator decrease scintillator efficiency[104,108]; the same effect is produced by an increase in initiator concentration. Therefore, it is advisable to do without initiators or to utilize them in minimal amounts.[108] Free-radical initiation by the tritium itself is a unique advantage of RLS-T. At a temperature as low as room temperature, styrene and methyl methacrylate even undergo radiation-induced polymerization by external sources of γ radiation, with methyl methacrylate an order of magnitude more reactive than styrene.[153]

3.3.7.4 No Contacts with Air during Polymerization and Heat Treatment

Before polymerization is started, one must remove dissolved oxygen from the monomer by prolonged nitrogen blowing or by thorough evacuation.[104,107,108] The latter process is carried out as several successive freezing–evacuation–defrosting runs or as a single prolonged evacuation from the frozen sample[108,112,115] (see Figure 3.37).

In the presence of oxygen in the polymer, fluorescence is quenched (Figure 3.38) and oxidation products are formed that reduce transmission of the polymer in the near-UV region.[104] Polymerization in air reduces the light output of the scintillator by 20 to 30%.[103,104,149]

Oxygen affects polymerization kinetics as well. Methyl methacrylate reacts more slowly and has a longer induction period than styrene in oxygen.[104] Polyvinyltoluene samples produced in vacuum and in nitrogen atmosphere have identical molecular weights.[115]

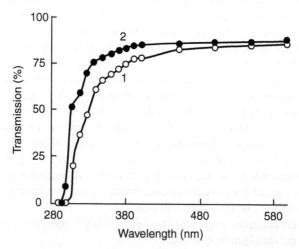

Figure 3.37   Transmission spectra of 0.5-cm-thick polystyrene samples. Sample 1 was prepared by the conventional procedure (10- to 15-min evacuation for 100 ml of styrene), and sample 2 was prepared by the prolonged-evacuation (600 min) procedure. (From Vasil'chenko, V.G. and Solov'ev, A.S., Plastic scintillators with antirad additions and enhanced light output, *Prib. Tekhn. Eksper.*, 4, 46, 1998 [in Russian].)

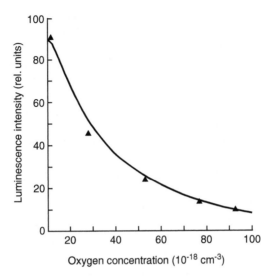

Figure 3.38 Luminescence intensity of the polystyrene scintillator as a function of the dissolved oxygen concentration. (From Bol'bit, N.M. et al., Novel plastic scintillator with excellent radiation stability, *Prib. Tekhn. Eksper.*, 4, 31, 2000 [in Russian].)

3.3.7.5 Optimal Polymerization Temperature

Polymerization is exothermic: the heat of formation of polystyrene is estimated at 17 kcal/mol and that of poly(methyl methacrylate) at 13.6 kcal/mol. Therefore, at high temperature of polymerization, especially in the case of large blocks, the mass can boil, the internal pressure increases, and defects form inside the block.[104]

Several polymerization regimes exist for vinylaromatic monomers.[103,107,108,114] Their temperature and temporal characteristics, reported by different investigators, are fairly close though slightly dependent on the nature of the monomer and the amount and nature of initiator:

Regime 1: low-temperature (~50 to 60°C) polymerization with initiator
Regime 2: medium-temperature (~120 to 140°C) polymerization with initiator
Regime 3: medium-temperature (~140 to 170°C) polymerization without initiator
Regime 4: high-temperature (~170 to 200°C) polymerization without initiator
Regime 5: two-step polymerization, first at a low or a medium temperature and then at a high temperature

Regime 2 polymerization is complete within 60 to 90 h, regime 3 within 7 to 8 days, and regime 4 within 12 to 38 h.[87,91,108,114] In regime 5, depending on the amount and nature of the initiator utilized, styrene is polymerized at 50 to 90°C, 80 to 90°C, or 120 to 130°C, and then at 160°C, 150 to 200°C, or 170 to 200°C.[103,104,114] Tentative polymerization time, is first, 10 h at 120 to 140°C and then 48 h at 160°C.[150] Vinyltoluene was first polymerized at 110°C for 3 days and then at 170°C for 1 day.[108]

Regime 1 receives very little use because the residual amounts of the initiator considerably depress the light output.[108,122] Regime 5 (two-step) has received wide application. Polymerization by this regime (especially in the case of large blocks) yields polymers with optimal molecular weight, uniform distribution of the luminescent additives, lack of strain (in the block polymer), and optical homogeneity.[87,104,114,119] In the case of low-temperature polymerization, the content of the residual monomer can reach 3%, and in the case of high-temperature or two-step polymerization, it can be below 0.8 to 1%.[103,104,150]

The lower the polymerization temperature, the higher the molecular weight of the polymer, all other factors being the same. For example, the molecular weight of polyvinyltoluene prepared[115] at 80°C was $1.1 \cdot 10^6$; at 100°C, $9.3 \cdot 10^5$; at 125°C, $4.5 \cdot 10^5$; at 150°C, $1.8 \cdot 10^5$; at 175°C, $7.1 \cdot 10^4$; and at 200°C, $5.6 \cdot 10^4$. Similar trends are exhibited by polystyrene and polyvinylxylene.[103,104]

Low-temperature polymerization of methyl methacrylate (at ~20°C) using an optimal concentration of the luminescent additives is difficult, since this low temperature reduces the additive solubility.[118]

Block polymerization of large-sized objects creates problems in keeping the temperature regime in the bulk of the sample uniform. In such cases, it is expedient to polymerize smaller objects with efficient heat removal from all sides and then to assemble a large block from the smaller pieces.[151]

Solidification of molten polystyrene results in up to 15% shrinking.[123] To prevent internal strain leading to reduced transparency or even cracking, the temperature after polymerization is lowered slowly 2 to 7°C/h; extra annealing at a temperature close to glass transition temperature can also be helpful.[87,104,114,152]

To prevent pollution and a decrease in the light output of the plastic scintillators manufactured by injection molding, molding additives should be avoided.[108]

REFERENCES

1. Mikhal'chenko, G.A., *Radioluminescent Emitters*, Energoatomizdat, Moscow, 1988 [Russian language].
2. Khol'nov, Yu.V. et al., *Characteristics of Radiation Emitted by Radioactive Nuclides Employed in the People's Economy: Reference Book*, Atomizdat, Moscow, 1980 [Russian language].
3. Levshin, V.L. et al., Studies of cathodoluminescence of zinc-sulfide and some other cathodoluminophors, in *Researches on Luminescence, Proc. P.N. Lebedev Physical Inst., Academy of Sciences*, 23, 64, 1963 [Russian language].
4. Ozawa, L., *Cathodoluminescence: Theory and Applications*, Codansha Ltd., Tokyo, VCH Publishers, New York, 1990.
5. Shionoya, S. and Yen, W.M., Eds., *Phosphor Handbook*, CRC Press LLC, Boca Raton, FL, 1999.
6. Robbins, D.J., On predicting the maximum efficiency of phosphor systems excited by ionizing radiation, *J. Electrochem. Soc.*, 127, 2694, 1980.
7. Stoffers, C. et al., Activator recycling in low voltage cathodoluminescent phosphors, *Appl. Phys. Lett.*, 71, 1759, 1997.

8. Shea, L.E., McKittrick, J., and Philips, M.L.F., Predicting and modeling the low-voltage cathodoluminescent efficiency of oxide phosphors, *J. Electrochem. Soc.*, 145, 3165, 1998.

9. Bril, A., Absolute efficiencies of phosphors with ultra-violet and cathode-ray excitation, in *Luminescence of Organic and Inorganic Materials*, Kallmann, H.P. and Spruch, G.M., Eds., John Wiley & Sons Inc., New York, 479, 1962.

10. Bril, A. and Klasens, H.A, Intrinsic efficiencies of phosphors under cathode-ray excitation, *Philips Res. Rep.*, 7, 401, 1952.

11. Meyer, V.D., Measurements of the absolute radiative efficiency of cathodoluminescent phosphors, *J. Electrochem. Soc.*, 119, 920, 1972.

12. Ludwig, G.W. and Kingsley, J.D., The efficiency of cathode-ray phosphors. I. Measurement, *J. Electrochem. Soc.*, 117, 348, 1970.

13. Kingsley, J.D. and Ludwig, G.W., The efficiency of cathode-ray phosphors. II. Correlation with other properties, *J. Electrochem. Soc.*, 117, 353, 1970.

14. Alexander, G., Ramakrishnan, P., and Mukherjee, T.K., Measurement of quantum and cathodoluminescent efficiencies of inorganic powder phosphors using silicon photodiode detector, *Indian J. Pure Appl. Phys.*, 31, 531, 1993.

15. Alexander, G. et al., Development of a demountable electron beam excitation system for the study of cathodoluminescence of the rare-earth phosphors, *Indian J. Pure Appl. Phys.*, 31, 23, 1993.

16. Griror'ev, K.V. et al., Atomic sources of optical photons, *Izv. Akad. Nauk SSSR, Ser. Fiz.*, 50, 603, 1986 [Russian language].

17. Epshtein, M.I., *Measurements of Optical Radiation in Electronics*, Energoatomizdat, Moscow, 1990 [Russian language].

18. Gurvich, A.M., *Introduction to Physical Chemistry of Crystalline Phosphors*, Vysshaya Shkola, Moscow, 1982 [Russian language].

19. Fok, M.V., Resolution of complex spectra into individual bands using the generalized Alentsev method, *Luminescence and Nonlinear Optics, Proc. P.N. Lebedev Physical Inst.*, 59, 3, 1972 [Russian language].

20. Kavetsky, A.G., Photo-, Thermo-, and Radiation-Stimulated Electronic and Ionic Processes in Silver Bromide Single Crystals, Cand. Sci. (Chem.) dissertation, Leningrad, 1983 [Russian language].

21. Krivosheev, M.I. and Kustarev A.K., *Color Measurements*, Energoatomizdat, Moscow, 1990 [Russian language].

22. Luizov, A.V., *Color and Light*, Energoatomizdat, Leningrad, 1989 [Russian language].

23. *Reference Book on Illumination Engineering*, Aizenberg, Yu.B., Ed., Energoatomizdat, Moscow, 1995 [Russian language].

24. Kelly, K.L., Color designations for lights, *J. Opt. Soc. Am.*, 33, 627, 1943.

25. Klasens, H.A., On the nature of fluorescent centers and traps in zinc sulfide, *J. Electrochem. Soc.*, 100, 72, 1953.

26. Lambe, J. and Klick, C.C., Model for luminescence and photoconductivity in the sulfides, *Phys. Rev.*, 98, 909, 1955.

27. Prener, J.S. and Williams, F.E., Activator systems in zinc sulfide phosphors, *J. Electrochem. Soc.*, 103, 342, 1956.

28. Hill, C.G. and Klasens, H.A., The influence of temperature on the efficiencies of zinc sulfide phosphors containing silver and cobalt, *J. Electrochem. Soc.*, 96, 275, 1949.

29. Fok, M.V., Photoluminescence issues, *Recombination Luminescence and Laser Spectroscopy, Proc. P. N. Lebedev Physical Inst.*, 117, 80, 1980 [Russian language].

30. Van Gool, W., Fluorescence centers in ZnS, *Philips Res. Rept. Suppl.*, 3, 1, 1961.
31. Shionoya, S. et al., Nature of luminescence transitions in ZnS crystals, *Jap. J. Phys. Soc.*, 19, 1157, 1964.
32. Miloslavskii, A.G. and Suntsov, N.V., Defect structure and luminescence centers of zinc-sulfide phosphors, *Fiz. Tekh. High Press.*, 7, 94, 1997 [Russian language].
33. Gaiduk, M.I., Zolin, V.F., and Gaigerova, L.S., *Luminescence Spectra of Europium*, Nauka, Moscow, 1974 [Russian language].
34. Forest, H. and Ban, G., Evidence for Eu^{+3} emission from two symmetry sites in Y_2O_3:Eu^{+3}, *J. Electrochem. Soc.*, 116, 474, 1969.
35. Heber J. et al., Energy levels and interaction between Eu^{3+}-ions at lattice sites of symmetry C_2 and symmetry C_{3i} in Y_2O_3, *Z. Phys.*, 237, 189, 1970.
36. Pappalardo, R.J. and Hunt, R.B., Dye-laser spectroscopy of commercial Y_2O_3:Eu^{3+} phosphors, *J. Electrochem. Soc.*, 132, 721, 1985.
37. Robbins, D.J., Cathodoluminescence saturation phenomena in Y_2O_2S:Eu, *J. Electrochem. Soc.*, 123, 1219, 1976.
38. Aluker, E.D., Lusis, D.Yu., and Chernov, S.A., *Electronic Excitations and Radioluminescence of Alkali-Halide Crystals*, Zinatne, Riga, 1979 [Russian language].
39. Gol'dina, O.A. et al., *Chemical Reagents and High-Purity Chemical Substances: Catalogue*, Khimiya, Moscow, 1990 [Russian language].
40. Averbuh, V.M. et al., *Luminophors and Chemicals. Technical Information Bulletin*, Part 1, NIITEKhIM, Stavropol, 1990 [Russian language].
41. LUMINOPHOR Joint Stock Company. *http://www.luminophor.ru/*.
42. RST-450, RST-520, RST-580, RSK-450, RSK-520, and RSK-580 phosphors, Technical Specifications TU 6-09-31-361-92 [Russian language].
43. Phosphor Technology Ltd., England, *http://www.phosphor.demon.co.uk/*.
44. Applied Scintillation Technologies Inc., U.S., *http://www.appscintech.com/*.
45. Leverenz, H.W., *An Introduction to Luminescence of Solids*, John Wiley & Sons, Inc., New York, 1950.
46. Markovskii, L.Ya., Pekerman, F.M., and Petoshina, L.N., *Phosphors*, Khimiya, Moscow–Leningrad, 1966 [Russian language].
47. Vygonyailo, O.M., Guretskaya, Z.I., and Galaktionov, S.S., Ways to increase the efficiency of blue-emitting zinc-sulfide phosphors, *Neorg. Mater.*, 29, 1356, 1993 [Russian language].
48. Bunin, A.M., Rare-earth elements: market situation and world practice of application in phosphors, in *Investigations, Production of Phosphors and Articles Thereof on the World Market*, v. 40, All-Russia Research Inst. of Luminophors, Stavropol, 24, 1991 [Russian language].
49. Bunin, A.M. et al., Phosphor coatings for screens of monochrome display tubes manufactured abroad, *Investigations of Luminescent Materials*, v. 41, All-Russia Research Inst. of Luminophors, Stavropol, 3, 1991 [Russian language].
50. *Phosphors of the Toshiba Firm: Catalogue*, Japan, 1985.
51. *LUMILUX Luminescent Pigments. Catalogue*, Riedel-de Haën AG, BRD, 1988.
52. *Inorganic Phosphor and Related Chemicals. Catalogue*, GTE Products Corp., U.S., 1988.
53. *KYOKKO Phosphors. Catalogue*, Kasey Optonix Ltd., Japan, 1986.
54. Drozd, L., On the influence of the composition of the base of nonactivator ZnS·CdS phosphors on their luminescence, *Izv. Akad. Nauk SSSR, Ser. Fiz.*, 23, 1300, 1959 [Russian language].
55. Antonov–Romanovskii, V.V., *Kinetics of Photoluminescence of Crystalline Phosphors*, Nauka, Moscow, 1986 [Russian language].

56. Gugel', B.M. and Krupenina, A.Ya., On the nature of short-wavelength luminescence bands in ZnS·CdS:Ag,Cl phosphors, in *Luminescent Materials and Extra-Pure Grade Substances*, v. 3, All-Russia Research Inst. of Luminophors, Stavropol, 9, 1970 [Russian language].

57. Zhang, W. et al., The preparation and the characteristic of temperature-sensitive phosphors, *J. Lumin.*, 40–41, 850, 1989.

58. Shionoya, S., Kallmann, H.P., and Kramer B., Behavior of excited electrons and holes in zinc sulfide phosphors, *Phys. Rev.*, 121, 1607, 1961.

59. Bol'shukhin, V.A. et al., Efficient cathodoluminophors for screens of oscillographic and indicator CRDs, *Elektron. Tekhn., Ser. Elektrovak. Gazorazryad. Prib.*, 2(113), 28, 1986 [Russian language].

60. Klaassen, D.B.M., Mulder, H., and Ronda, C.R., Excitation mechanism of cathodoluminescence of oxisulfides, *Phys. Rev., B*, 39, 42, 1989.

61. Ozawa, L., Nonlinear dependence of cathodoluminescence on activator concentrations, *Appl. Phys. Lett.*, 33, 586, 1978.

62. Ozawa, L. et al., The effect of exciting wavelength on optimum activator concentration, *J. Electrochem. Soc.*, 118, 482, 1971.

63. Klaassen, D.B.M., Mulder, H., and Ronda, C.R., Excitation mechanism of cathodoluminescence of solid solutions of oxisulfides, *J. Electrochem. Soc.*, 136, 2754, 1989.

64. Ohno, K. and Abe, T., Effect of BaF_2 on the synthesis of the single-phase cubic $Y_3Al_5O_{12}$:Tb, *J. Electrochem. Soc.*, 133, 638, 1986.

65. Ohno, K. and Abe, T., Bright green phosphor, $Y_3Al_{5-x}Ga_xO_{12}$:Tb, for projection CRT, *J. Electrochem. Soc.*, 134, 2072, 1987.

66. Scholl, M.S. and Trimmier, J.R., Luminescence of YAG:Tm, Tb, *J. Electrochem. Soc.*, 133, 643, 1986.

67. Robbins, D.J. et al., Investigation of competitive recombination processes in rare-earth activated garnet phosphors, *Phys. Rev., B*, 19, 1254, 1979.

68. Robbins, D.J. et al., The relationship between concentration and efficiency in rare-earth activated phosphors, *J. Electrochem. Soc.*, 126, 1556, 1979.

69. Welker, T., Recent developments on phosphors for fluorescent lamps and cathode-ray tubes, *J. Lumin.*, 48–49, 49, 1991.

70. Blasse, G., Grabmaier, B.C., *Luminescent Materials*, Springer–Verlag, Berlin, 1994.

71. Robbins, D.J. et al., The temperature dependence of rare-earth activated garnet phosphors. II. A comparative study of Ce^{3+}, Eu^{3+}, Tb^{3+}, and Gd^{3+} in $Y_3Al_5O_{12}$, *J. Electrochem. Soc.*, 126, 1221, 1979.

72. Ronda, C.R. et al., The degradation behavior of LaOBr:Tb under cathode-ray excitation, *J. Appl. Phys.*, 75, 4636, 1994.

73. Ohno, K., Abe, T., and Kusunoki, T., Reduction of the deterioration of luminescence efficiency of P-53 phosphor due to electron bombardment, *Jap. J. Appl. Phys.*, 29, 103, 1990.

74. Matsukiyo, H. et al., Reduction of electron-beam induced degradation by rare-earth doping in $Y_3(Al,Ga)_5O_{12}$:Tb phosphors, *J. Electrochem. Soc.*, 145, 270, 1998.

75. RST-612, RST-6500, RSK-612, and RSK-6500 Phosphors, Technical Specifications TU 6-09-31-373-93 [Russian language].

76. *Cathode Ray Phosphors. Catalogue*, USR Optonix Inc., U.S., 1989.

77. Phosphor K–78, Technical Specifications TU 48-4-328-87 [Russian language].

78. Buchanan, R.A. et al., Cathodoluminescent properties of the rare-earths in yttrium oxide, *J. Appl. Phys.*, 39, 4342, 1968.

79. Tseng, Y.-H. et al., Spectral properties of Eu^{3+}-activated yttrium oxysulfide red phosphor, *Thin Solid Films*, 330, 173, 1998.

80. Yamamoto, H. and Kano T., Enhancement of cathodoluminescence efficiency of rare-earth activated Y_2O_2S by Tb^{3+} or Pr^{3+}, *J. Electrochem. Soc.*, 126, 305, 1979.

81. Royce, M.R., U.S. Patent 3,418,246, 1968.

82. Johnson, P.D., Oxygen-dominated lattices, in *Luminescence of Inorganic Solids*, Goldberg, P., Ed., Academic Press, New York, 1966, chap. 5.

83. *Phosphor Index (Inorganic Luminescent Materials). Catalogue*, Nichia Chemical Industries Ltd., Japan, 1989.

84. Mikhal'chenko, A.G., Zhurikh, M.Yu., and Chumak, N.V., Radioluminescnt emitters and radiation-chemical processes in them, in *Radiation-Chemical Transformations in Inorganic and Organic Materials*, Vasil'ev, I.A., Ed., St. Petersburg State Technol. Inst. (Techn. Univ.), St. Petersburg, 18, 2000 [Russian language].

85. Kholodov, A.A., Terekhov, S.A., and Mikhal'chenko, A.G., Effect of Band Quenching of Recombination Luminescence, in *Radiation-Chemical Transformations in Inorganic and Organic Materials*, Vasil'ev, I.A., Ed., St. Petersburg State Technol. Inst. (Techn. Univ.), St. Petersburg, 8, 2000 [Russian language].

86. Krasovitskii, B.M. and Bolotin, B.M., *Organic Luminophores*, Khimiya, Leningrad, 1976 [Russian language].

87. Schram, E. and Lombaert, R., *Organic Scintillation Detectors*, Elsevier, Amsterdam, 1963.

88. Sangster, R.C. and Irvine, J.W., Study of organic scintillators, *J. Chem. Phys.*, 24, 670, 1956.

89. Britvich, G.I., Vasil'chenko, V.G., and Lapshin, V.G., Radiation stability of plastic scintillators, *Prib. Tekhn. Eksper.*, 1, 75, 1994 [Russian language].

90. Britvich, G.I. et al., Study of the radiation stability of polystyrene-based scintillators, *Prib. Tekhn. Eksper.*, 1, 109, 1993 [Russian language].

91. Vasil'chenko, V.G. et al., Study of the radiation stability of plastic scintillators, *Prib. Tekhn. Eksper.*, 5, 85, 1995 [Russian language].

92. Ambrosio, C.D., Organic scintillators with large Stokes shifts dissolved in polystyrene, *Nucl. Instr. Methods Phys. Res.*, A307, 430, 1991.

93. Heller, A.J., Organic liquid scintillators. VI. Substituted distyrylbenzenes: scintillation properties and spectra of absorption and fluorescence, *J. Chem. Phys.*, 40, 2839, 1964.

94. Nurmukhametov, R.N., *Absorption and Luminescence of Organic Compounds*, Khimiya, Moscow, 1971 [Russian language].

95. Maier, G.V. and Danilova, V.I., *Quantum Chemistry, Structure, and Photonics of Molecules*, Tomsk Gos. Univ., Tomsk, 1984 [Russian language].

96. Maier, G.V., Orbital nature of electronically excited $\pi\pi$ states and spectral-luminescent properties of organic molecules, *Izv. Akad. Nauk SSSR, Ser. Fiz.*, 54, 413, 1990 [Russian language].

97. Rabin, N.V., Electron–photon calorimeters. Properties of the materials of calorimeter detectors, *Prib. Tekhn. Eksper.*, 6, 8, 1992 [Russian language].

98. Abramovich, B.G., *Thermal Indicators and Their Application*, Energiya, Moscow, 1972 [Russian language].

99. Bolotin, B.M. et al., USSR Inventor Certificate 368288; Invent. Bull., 9, 1973.

100. Gunder, O.A., *Polymeric Systems and Their Scintillation Properties*, All-Union Research Inst. of Single Crystals, Khar'kov, 1975 [Russian language].

101. Vekshin, N.A., *Excitation Transfer in Macromolecules*, VINITI, Moscow, 1989 [Russian language].

102. Birks, J.B., Scintillation in organic solids, in *Physics and Chemistry of Organic Solid State*, Fox, D., Labes, M.M., and Weissberger, A., Eds., Interscience, New York, 1965, chap. 5.

103. Barashkov, N.N. and Gunder, O.A., *Fluorescent Polymers*, Khimiya, Moscow, 1987 [Russian language]. (Engl. transl.: Barashkov, N.N. and Gunder, O.A., *Fluorescent Polymers*, Ellis Horwood, Chichester, 1994).

104. Gunder, O.A., *Physicochemical Basics of Production of Plastic Scintillators*, NIITEKhIM, Moscow, 1979 [Russian language].

105. Majewski, S. and Zorn, C., Designing radiation-hard plastic scintillators, ACS Symposium Ser., 475, 569, 1991.

106. Gao, F. et al., New fluors for radiation tolerant scintillators, in *Scintillator and Phosphor Materials, Materials Res. Soc. Symp. Proc.*, 348, 173, 1994.

107. Rozman, I.M. and Kilin, S.F., Luminescence of plastic scintillators, *Usp. Fiz. Nauk*, 69, 459, 1959 [Russian language].

108. Birks, J.B., *The Theory and Practice of Scintillation Counting*, Pergamon Press, Oxford, 1964, chap. 9.

109. Birks, J.B. and Kuchela, K.N., Energy transfer in fluorescent plastic solutions, *Disc. Faraday Soc.*, 27, 57, 1959.

110. Renschler, C.L. et al., Solid state radioluminescent lighting, *Radiat. Phys. Chem.*, 44, 629, 1994.

111. Yuan, H. et al., Energy transfer in multicomponent plastic scintillators, *J. Lumin.*, 31–32, 833, 1984.

112. Vasil'chenko, V.G. and Solov'ev, A.S., Plastic scintillators with antirad additions and enhanced light output, *Prib. Tekhn. Eksper.*, 4, 46, 1998 [Russian language].

113. Gunder, O.A. and Koba, V.S., IR-spectroscopic study of the radiation stability of a series of vinylaromatic polymers and copolymers, *Radiokhimiya*, 11, 119, 1969 [Russian language].

114. Vlasov, V.G. and Titskaya, V.D., Influence of the polymerization parameters on the quality of polystyrene scintillators, *Plast. Massy*, 8, 12, 1989 [Russian language].

115. Funt, B.L. and Hetherington, A., The influence of chain length on the luminescent output of plastic scintillators, *Int. J. Appl. Rad. Isot.*, 4, 189, 1959.

116. Bol'bit, N.M. et al., Novel plastic scintillator with excellent radiation stability, *Prib. Tekhn. Eksper.*, 4, 31, 2000 [Russian language].

117. Rozman, I.M., Andreeshchev, E.E., and Kilin S.S., On luminescence mechanism of plastic scintillators, *Izv. Akad. Nauk SSSR, Ser. Fiz.*, 23, 102, 1959 [Russian language].

118. Gunder, O.A. et al., Plastic scintillators with enhanced transmittance, *Prib. Tekhn. Eksper.*, 3, 66, 1969 [Russian language].

119. Barashkov, N.N. and Sakhno, T.V., *Optically Clear Polymers and Materials Thereof*, Khimiya, Moscow, 1992 [Russian language].

120. Andryushchenko, L.A. and Grinev, B.V., Organosilicon materials for scintillation detectors, *Prib. Tekhn. Eksper.*, 4, 5, 1998 [Russian language].

121. Gunder, O.A. and Koba, V.S., Stability of methyl-substituted polystyrene derivatives under beta irradiation, *Khim. Vys. Energ.*, 8, 83, 1974 [Russian language].

122. Gorbachev, V.M. et al., Scintillating plastics with enhanced radiation stability, *Atom. Energ.*, 38, 427, 1975 [Russian language].

123. Swank, R.K. and Buck, W.L., The scintillation process in plastic solid solutions, *Phys. Rev.*, 91, 927, 1953.

124. Vyazemskii, V.O. et al., *Scintillation Method in Radiometry*, Gosatomizdat, Moscow, 1961.

125. Hurlbut, C.R., Plastic scintillators: a survey, *Trans. Amer. Nucl. Soc.*, 50, 20, 1985.

126. Bross, A.D., Workshop on Scintillation Fider Detector Development for the SSC, Fermilab, Batavia, 1988.

127. Speranskaya, T.A. and Tarutina, L.I., *Optical Properties of Polymers*, Khimiya, Leningrad, 1976 [Russian language].

128. Lipatov, Yu.S. et al., *Reference Book on Polymer Chemistry*, Naukova Dumka, Kiev, 1971 [Russian language].

129. Gen, N.A. et al. Performance characteristics of plastic scintillators obtained from melt, *Prib. Tekhn. Eksper.*, 4, 35, 1988 [Russian language].

130. Akimov, Yu.K., Nuclear-radiation detectors based on plastic scintillators, *Fiz. Elem. Chast. Atom. Yadra*, 25, 496, 1994 [Russian language].

131. Zyablin, L.V. et al., On durability of plastic scintillators, *Prib. Tekhn. Eksper.*, 2, 73, 1986 [Russian language].

132. Berezin, I.A. et al., Thermal stabilization of optical characteristics of plastic scintillators, *Prib. Tekhn. Eksper.*, 1, 53, 1981.

133. Zlatkevich, L., *Radiothermoluminescence and Transitions in Polymers*, Springer–Verlag, New York, 1987.

134. Pannel, J.H. and Manning, B., Production of a blue color in irradiated plastic scintillators, *J. Chem. Phys.*, 23, 1368, 1955.

135. Crist, B., Radiothermoluminescence with *in situ* electron irradiation, *J. Polym. Sci. Part B — Polym. Phys.*, 28, 1641, 1990.

136. Nikol'skii, V.G. and Buben, N.Ya., Radiothermoluminescence of organic compounds, *Dokl. Akad. Nauk SSSR*, 134, 134, 1960 [Russian language].

137. Maharil, S.V. and Deshmukh, B.T., On the possibility of using organic phosphors in thermoluminescence dosimetry, *J. Polym. Sci. Part B — Polym. Phys.*, 27, 2577, 1989.

138. Dole, M., *The Radiation Chemistry of Macromolecules*, v. 1, Academic Press, New York, 1972.

139. Vanderschueren, J., Linkens, A., and Niezette, J., Effect of doping on thermoluminescence in polymers: I. 9,10-Phenantrenequinone-doped poly(dian carbonate), *J. Polym. Sci. Part B — Polym. Phys.*, 24, 697, 1986.

140. Bertolucci, S. et al., The CDF central and endwall hadron calorimeter, *Nucl. Instr. Meth. Phys. Res.*, A267, 301, 1988.

141. Cumalat, J.P. et al., Effects of magnetic fields on the light yield of scintillators, *Nucl. Instr. Meth. Phys. Res.*, A293, 606, 1990.

142. Blomker, D. et al., Response of plastic scintillators in magnetic fields, *IEEE Trans. Nucl. Sci.*, 37, 220, 1990.

143. Turchanovich, L.K. and Korneev, Yu.P., Study of the light output of a plastic scintillator in magnetic field, *Prib. Tekhn. Eksper.*, 2, 36, 1996 [Russian language].

144. Autonomous Self-Luminous Coatings [Prospect], NIITEKhIM, Cherkassy, 774–500 [Russian language].

145. Mullins, D.F., Krasnai, J.P., and Mueller, D.A., Development of organic tritium light technology at Ontario Hydro, *Fusion Technol.*, 21, 312, 1992.

146. Sychova, M.V., Korneeva, O.G., and Revinskii, V.S., Study of the influence of the vacuum rectification conditions on the quality of 2,4-dimethylstyrene and light output of plastic scintillators thereof, in *Optical and Scintillating Materials*, Coll. Scient. Works All-Union Inst. of Single Crystals, Khar'kov, 9, 159, 1982 [Russian language].

147. Lapshin, V.G. et al., *Styrene Purification Methods for Scintillator Manufacturing*, Preprint of Inst. of High-Energy Physics 94-24, Protvino, 1994 [Russian language].

148. Bross, A.D. and Pla–Dalmau, A., Radiation damage of plastic scintillators, *IEEE Trans. Nucl. Sci.*, 39, 1199, 1992.

149. Brekhovskikh, V.V. et al., Radiation stability of polystyrene scintillators and spectrum-shifting fibers, *Prib. Tekhn. Eksper.*, 6, 95, 1992 [Russian language].

150. Senchishin, V.G. et al., A new radiation stable plastic scintillator, *Nucl. Instr. Methods Phys. Res.*, A364, 253, 1995.

151. Golutvina, I.G. et al., On polymerization of blocks of large-sized plastic scintillators, *Prib. Tekhn. Eksper.*, 4, 91, 1976 [Russian language].

152. Kopina, I.V. and Korneeva O.G., Influence of the polymerization regime on the physicomechanical characteristics of polyvinylxylene scintillators, in *Physics and Chemistry of Solids,* Coll. Scient. Works of All-Union Inst. of Single Crystals, Khar'kov, 10, 70, 1983 [Russian language].

153. Ivanov, V.S., *Radiation Chemistry of Polymers*, Khimiya, Leningrad, 1988 [Russian language].

CHAPTER 4

Radiation-Induced Processes in Phosphors

A.G. Kavetsky and S.P. Meleshkov

CONTENTS

Phosphors most extensively employed in RLSs are zinc sulfide– and yttrium oxide–based phosphors. Therefore, this chapter primarily considers radiation-induced processes occurring in these systems.

4.1 INFLUENCE OF IRRADIATION ON LUMINESCENCE OF PHOSPHORS

4.1.1 Ionizing Radiation–Induced Degradation of Phosphors

Along with excitation luminescence, exposure of crystalline phosphors to ionizing radiation causes radiation damage. Radiation damage to crystalline solids (radiation defects) has been usefully classified[1] into point and extended (space) defects. Point defects include vacancies, interstitial atoms, impurity atoms, and combinations of several elementary point defects. Extended defects include dislocations, displacement wedges, etc.

Point defects arise in irradiated crystals from transition of atoms (ions) from regular lattice sites to irregular sites. The displacement of an atom requires displacement energy, the magnitude of which depends on nature of the material and mass of the atom displaced; this parameter varies between 5 and 80 eV.[2,3] Energy transfer to the lattice site causing transition of the atom to an interstitial site under crystal irradiation can follow a number of paths. (Mechanisms of defect creation will be discussed in Section 4.2.) These depend on the nature of the crystal and the ionizing radiation as well as on characteristics such as energy and linear energy transfer.

Radiation defects (and their combinations) can create energy levels of various depths in the forbidden band of the crystalline phosphor. These levels will have diverse functions in recombination luminescence kinetics, namely, those of electrons and hole traps, as well as of the quenching sites (see Sections 3.1.1.4 and 3.1.1.5 in Chapter 3). Point defects can interact with radiative recombination centers, thus changing their energy and, correspondingly, the luminescence characteristics. Also, point defects can act as color centers in crystals and have absorption bands in the luminescence region of the radiative recombination centers. Radiation defects generally lead to luminescence intensity degradation.

Various designs of radioluminescent light sources typically include a binder for fixation of the phosphor. The binder is also subject to the ionizing radiation field and can undergo radiation-induced transformations. In a number of cases, degradation of the phosphor with a binder proceeds at a higher rate than phosphor alone due to binder interaction with the phosphor surface in the ionizing radiation field.

Table 4.1 shows how luminescence intensity of phosphors is influenced by the irradiation.[4,5] Energy-conversion efficiency of phosphors typically decreases upon irradiation, depending on a number of factors:

- Ionizing radiation type and energy
- Absorbed dose and dose rate
- Phosphor type
- Presence and type of binder

As shown in Mikhal'chenko,[4] the extent of decomposition vs. absorbed dose dependence is typically not linear. The initial stage of irradiation (up to absorbed doses of about 10^4 Gy) in many cases involves variation of the radiation change coefficient K_{RC}, depending on the absorbed dose D_{abs}. K_{RC} can be determined as

$$K_{RC} = \frac{1}{\eta_{RL}} \cdot \frac{d\eta_{RL}}{dD_{abs}} \tag{4.1}$$

The coefficient K_{RC} can be negative and positive. In subsequent stages, the negative K_{RC} value can remain constant with increasing D_{abs}.

4.1.2 Cathode Ray–Induced Degradation of Phosphors

The energy of tritium betas lies in the energy range of electrons in electron-beam devices and field emission displays. Therefore, the operation conditions for phosphors in tritium-based radioluminescent light sources correspond to those in the devices already mentioned. It is reasonable to consider aging, i.e., deterioration of the luminescent properties (degradation) of phosphors under cathode excitation. However, when applying data on phosphor degradation in cathode-ray tubes to phosphor aging in RLSs-T, one should take into account that, in such devices, phosphors are subject to fairly high-vacuum conditions and also to more severe radiation impacts than in RLSs-T.

Careful analysis of degradation processes in various phosphors highlights many reasons for degradation in each material.[6] In $Sr_2Al_6O_{11}$:Eu^{3+} phosphors, color centers form, thus decreasing luminescence efficiency due to self-absorption of optical radiation in the phosphor. Nonluminescent layers form on the Zn_2SiO_4:Mn^{2+} phosphor surface that decrease excitation efficiency of the base material due to absorption of a part of the electron beam and also partially attenuate the optical radiation emitted. In Y_2SiO_5:Ce^{3+} phosphor, the concentration of the Ce^{3+} luminescence centers is decreased, leading to degradation of this phosphor. The luminescent properties of zinc sulfide phosphors deteriorate, owing to formation on their surface of nonluminescent and nontransparent layers.[7-14]

Degradation mechanisms for zinc sulfide phosphors were studied when exposed to electron beam by simultaneously measuring the luminescence intensity and determining the chemical species on the surface (by Auger electron and x-ray photoelectron spectroscopy[7-12] as well as by Raman spectroscopy.[13,14]) Table 4.2 summarizes the experimental conditions, approximate electron doses decreasing the luminescence intensity almost twice, and also compositions of the surface layers formed.

Formation of such layers on the phosphor surface is explained in terms of the ESSCR (electron-stimulated surface chemical reactions) model advanced in references 7 to 12. This model is based on the postulate that electron beam stimulates dissociation of various molecules of the surrounding gaseous phase that are adsorbed on the surface of zinc sulfide phosphor (such as H_2O, H_2, O_2, etc.) The reactive atomic species thus formed (e.g., hydrogen or oxygen radicals) rapidly react with S^{2-}, forming volatile products such as SO_x or H_2S.

In oxidizing conditions, zinc remaining on the surface forms nonluminescent ZnO. In a reducing (hydrogen) environment, sulfur is removed as H_2S from the surface retaining metallic zinc. The zinc atoms diffuse into the bulk of phosphor and create traps in a layer about 1 μm thick. These traps participate in recombination processes and cause luminescence decay. Also, zinc can volatilize from the surface under electron beam. Figure 4.1a illustrates the scheme of the processes described in terms of the ESSCR model. This model explains the observed acceleration of degradation for silicon dioxide–coated phosphors as due to facile adsorption of hydrogen on SiO_2 surface compared to zinc sulfide surface. This accelerates adsorption, the electron-stimulated dissociation of hydrogen, and subsequent processes involving hydrogen radicals, illustrated in Figure 4.1b.

Table 4.1 Influence of Irradiation on the Luminescence Intensity of Phosphors

Characteristics of Phosphor		Characteristics of Irradiation		Power Flux		Dose,	Ratio of the Luminescence Intensities of Irradiated and Nonirradiated Phosphors
Chemical Composition	Binder Type	Radiation Type	Radiation Source and Energy	$\mu W/cm^2$	Gy/h	Gy	
$LaPO_4:Pr^{3+}$	K_2SiO_3	β	Zr^3H_2, ε_{av} = 5.69 keV	0.44	$3.3 \cdot 10^3$	$6.4 \cdot 10^7$	0.83
	K_2SiO_3	γ	^{60}Co, 1.25 MeV	—	$8.28 \cdot 10^3$	$7 \cdot 10^6$	0.94
	Na_2SiO_3	e^-	Accelerator, 2.5 MeV	—	10^7	10^7	0.82
	B_2O_3	e^-	Accelerator, 2.5 MeV	—	10^7	10^7	0.31
$YPO_4:Nd^{3+}$	K_2SiO_3	γ	^{60}Co, 1.25 MeV	—	$8.28 \cdot 10^3$	$7 \cdot 10^6$	1.33
	K_2SiO_3	β	Zr^3H_2, ε_{av} = 5.69 keV	0.44	$3.3 \cdot 10^3$	$6.4 \cdot 10^7$	1
	Na_2SiO_3	e^-	Accelerator, 2.5 MeV	—	10^7	10^7	0.54
	B_2O_3	e^-	Accelerator, 2.5 MeV	—	10^7	10^7	0.41
ZnS:Ag	Na_2SiO_3	e^-	Accelerator, 2.5 MeV	—	10^7	10^7	0.44
	K_2SiO_3	γ	^{60}Co, 1.25 MeV	—	$8.28 \cdot 10^3$	$0.7 \cdot 10^7$	0.41
	Absent	α	$^{238}PuO_2$, 5.5 MeV	37.6	$1.17 \cdot 10^4$	$0.56 \cdot 10^7$	0.50
	Absent	β	Zr^3H_2, ε_{av} = 5.69 keV	0.44	$3.3 \cdot 10^3$	$7.5 \cdot 10^7$	1
ZnS:Cu	Absent	γ	^{60}Co, 1.25 MeV	—	$7.2 \cdot 10^2$	$1.8 \cdot 10^7$	0.98
	K_2SiO_3	γ	^{60}Co, 1.25 MeV	—	$7.2 \cdot 10^2$	$0.2 \cdot 10^7$	0.75
	$AlH_3(PO_4)_2$	γ	^{60}Co, 1.25 MeV	—	$7.2 \cdot 10^2$	$0.2 \cdot 10^7$	0.84
$ZnS:Tm^{3+}$	Absent	β	Zr^3H_2, ε_{av} = 5.69 keV	0.44	$3.3 \cdot 10^3$	$7.5 \cdot 10^7$	0.95
	K_2SiO_3	γ	^{60}Co, 1.25 MeV	—	$8.28 \cdot 10^3$	$7 \cdot 10^6$	0.81
$Sr_3(PO_4)_2:Eu^{2+}$	Absent	β	Zr^3H_2, ε_{av} = 5.69 keV	0.44	$3.3 \cdot 10^3$	$7.5 \cdot 10^7$	0.43
	K_2SiO_3	γ	^{60}Co, 1.25 MeV	—	$8.28 \cdot 10^3$	$7 \cdot 10^6$	0.83
$YPO_4:In$	B_2O_3	e^-	Accelerator, 2.5 MeV	—	10^7	10^7	0.77
$CaSO_4:Eu^{2+}$	K_2SiO_3	γ	^{60}Co, 1.25 MeV	—	$8.28 \cdot 10^3$	$7 \cdot 10^6$	0.53
$CaWO_4$	Absent	α	$^{238}PuO_2$, 5.5 MeV	37.6	$1.17 \cdot 10^4$	$0.14 \cdot 10^7$	0.50

$Y_2O_3{:}Eu^{3+}$	K_2SiO_3	γ	^{60}Co, 1.25 MeV	—	$7.2 \cdot 10^2$	$0.2 \cdot 10^7$	0.75
	$AlH_3(PO_4)_2$	γ	^{60}Co, 1.25 MeV	—	$7.2 \cdot 10^2$	$0.2 \cdot 10^7$	0.76
	Absent	γ	^{60}Co, 1.25 MeV	—	$7.2 \cdot 10^2$	$0.2 \cdot 10^7$	0.78
	Absent	α	$^{238}PuO_2$, 5.5 MeV	37.6	$1.17 \cdot 10^4$	$0.14 \cdot 10^7$	0.50
$Y_2O_2S{:}Tb^{3+}$	Absent	β	Zr^3H_2, $\varepsilon_{av} = 5.69$ keV	1.09	$3.3 \cdot 10^3$	$7.5 \cdot 10^7$	0.83
	Absent	γ	^{60}Co, 1.25 MeV	—	$7.2 \cdot 10^2$	$1.8 \cdot 10^7$	0.76
	K_2SiO_3	γ	^{60}Co, 1.25 MeV	—	$7.2 \cdot 10^2$	$0.2 \cdot 10^7$	0.65
	$AlH_3(PO_4)_2$	γ	^{60}Co, 1.25 MeV	—	$7.2 \cdot 10^2$	$0.2 \cdot 10^7$	0.72

Sources: Mikhal'chenko, G.A., *Radioluminescent Emitters*, Energoatomizdat, Moscow, 1988 [Russian language]; Okaminov, V.M. et al., in *Problems of Production and Application of Isotopes and Nuclear Radiation Sources in the People's Economy of the USSR*, TsNIIAtominform, Moscow, 1988 [Russian language].

Table 4.2 Irradiation Conditions and Degradation Behavior for Selected Phosphors

Phosphor	Electron-Beam Parameters		Gas Pressure and Composition	Electron Dose Corresponding to Lowering Luminescence Intensity by Half, C/cm²	Degradation Behavior	Ref.
	Energy, keV	Current Density, mA/cm²				
ZnS:Cu,Al,Au	2	2	$1.2 \cdot 10^{-8}$ Torr, d_2O, d_2, hd_4, traces he, he_2, e_2	50	Formation of the ZnO and $ZnSO_4$ layer	7
ZnS:Cu,Al,Au	2	64	$1 \cdot 10^{-6}$ Torr, e_2	1000	Formation of the ZnO layer	8
ZnS:Cu,Al,Au	2	52	$5 \cdot 10^{-7}$ Torr, e_2	4000	Formation of the ZnO layer	9
ZnS:Ag,Cl	2	52	$5 \cdot 10^{-7}$ Torr, e_2	1000	Formation of the ZnO layer	9
ZnS:Ag,Cl	2–5	0.272	10^{-6} Torr, H_2O	3–5	Formation of the ZnO layer	11
			10^{-8} Torr, H_2	3–5	Formation of the Zn layer	
ZnS:Ag ZnO:Zn Y_2O_3:Eu³⁺	1.5	0.02	$(0.7–1.1) \cdot 10^{-6}$ Torr, residual gas composition not indicated	0.2	Formation of a graphite layer from hydrocarbon impurities contained in the gas	13
ZnS:Ag+(Zn,Cd)S: Ag in RLS-T	5.7	$1.7 \cdot 10^{-6}$	760 Torr, ³H_2	0.5	—	—

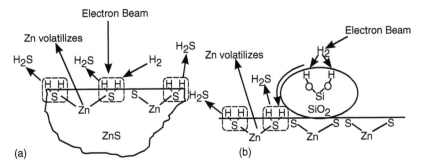

Figure 4.1 Scheme of degradation of zinc sulfide phosphors in the presence of hydrogen according to the ESSCR model (for details, see the text). (From Abrams, B.L. et al., Electron beam–induced degradation of zinc sulfide–based phosphors, *Surf. Sci.*, 451, 174, 2000.)

To a first approximation, volatilization of zinc under electron beam explains why degradation of the zinc sulfide phosphor is slower at higher electron beam current densities (see Table 4.2). Indeed, when exposed to a more intense electron beam, zinc volatilizes from the surface faster and the nontransparent oxide layer forms more slowly. At small current densities of the electron beam, practically no volatilization occurs, and virtually all zinc atoms remain on a surface. When electron beam is characterized by current densities smaller than those indicated in Table 4.2, a dark imprint forms on the phosphor surface.[15] Upon irradiation, the metals contained in the phosphor appear in the subsurface layer, e.g., in the case of (Zn,Cd)S:Cu,Al phosphor, metallic zinc and cadmium. At the same time, the dark imprint disappears on heating in a vacuum at 300 to 400°C, and a condensate (several rings) forms on cold parts of a test tube. As shown by chemical analysis for (Zn,Cd)S:Cu,Al phosphor, the first ring contains primarily condensed cadmium; the second contains zinc, sulfur, and minor quantities of other elements.

Quantitative data on the decline of luminescence efficiency and those on absorption of the energy of electrons exciting the luminescence and light emitted in luminescence in the forming layers cannot be fully matched.[9,11,13] Therefore, it has been suggested[9,11,13] that defects form in the surface layers of the phosphor and participate in recombination luminescence, contributing to lowering of its energy efficiency. The defects can arise either by the impurity-ionization mechanism or via implantation into the crystal lattice of the products that form from radiation-induced reactions between adsorbed gases and the phosphor base (e.g, oxygen[7] or zinc[11] atoms).

4.1.3 Phosphors under Beta and Cathode Irradiation

Now phosphor irradiation conditions in gas-filled RLSs-T with those in the previously described experiments will be compared. Table 4.2 presents the corresponding energies of electrons, electron current densities on the phosphor surface, and the electron dose corresponding to lowering luminescence intensity by half $(D_{el,1/2})$. The electron dose for the phosphor employed in a gas-filled RLS-T was estimated. According to Figure 2.19 in Chapter 2, brightness decreases about fourfold after a half-life of tritium; brightness decreases by half with lowering of tritium activity by half. (See also Section 2.3.2.1 in Chapter 2.) Therefore, within a half-life of tritium, the brightness of RLS-T will decrease via degradation of phosphor by about half. Take the initial specific beta-flux power dP_β^0/dS as about 10 µW/cm² (see Section 2.2.4.3 in Chapter 2). The electron dose responsible for lowering the brightness by half will be

$$D_{el,1/2} = \frac{dP_\beta^0}{dS} \cdot \frac{q}{\varepsilon_{av}\lambda} \cdot 3.15\cdot 10^7 \left[1 - \exp\left(-\lambda T_{1/2}\right)\right] \qquad (4.2)$$

and is equal to 0.5 C/cm². In Equation 4.2, λ is the tritium decay constant equal to 0.056 year^{-1} (3.15·10⁷ is the number of seconds in a year).

As seen from Table 4.2, with low electron beam densities and with hydrogen in the surrounding atmosphere, the sample will degrade because of formation of zinc metal on the phosphor surface. Apparently, one reason for degradation in gas-filled RLSs-T operating under the described irradiation conditions is formation of metallic zinc on the phosphor surface. The darkening of the phosphor coating in RLSs-T with long service life does not contradict this conclusion. Evidently, analyses of the phosphor surface and gas composition for long-service-life RLSs-T provide additional information on the degradation mechanism in the phosphors under electron beam, in particular, under tritium beta radiation.

Apparently, interaction between the contacting molecules of gaseous tritium and phosphor is not confined to irradiation of the surface with beta particles; doubtless, other processes take place, as well.

Self-absorption of beta particles in gaseous tritium is accompanied by its autoradiolysis[16] with formation of ³H• radicals. It can be assumed that under permanent self-irradiation, a steady-state concentration of ³H• radicals exists. Both tritium molecules and radicals are sorbed on the phosphor surface. As shown previously, reactions between hydrogen radicals and the phosphor base result in degradation of the latter; this is evidently true for ³H• as well.

Radical recombination can also proceed on the phosphor surface. The energy emitted in recombination of hydrogen radicals is equal to 4.48 eV.[17] Recombination of hydrogen (tritium) radicals on the phosphor surface can follow either the knock-on mechanism (involving recombination of radicals from the gas phase with radicals on the surface) or the diffusion mechanism (involving recombination following diffusion of the radicals adsorbed on the surface).[17] In recombination on the surface, the crystal lattice acts as a third body to which the emitted energy is transferred. This causes luminescence of the phosphor (radical-recombination luminescence), emission of electrons and ions from the surface, or generation of electron–hole pairs.[17-19]

In the case of radical-recombination luminescence, the phosphor luminescence spectrum can contain bands missing from the spectra measured under different excitation conditions.[20] These bands arise from chemiluminescent reactions accompanying adsorption of hydrogen atoms on the phosphor surface.[17] In this context, it would be interesting to compare the luminescence spectra of phosphors when exposed to gaseous tritium (i.e., when radical-recombination luminescence is possible) and other tritium-based (e.g., titanium tritide–based) beta sources. A study of the possible differences in these spectra may provide additional information on mechanisms of electron-stimulated surface reactions.

In Grankin,[18] examples of (Zn,Cd)S:Ag and ZnS:Tm phosphors showed that recombination of hydrogen radicals primarily involves emission of activator ions from the surface to the gas phase. Their surface concentration is restored via diffusion from the volume of phosphor to its surface. In the case of prolonged selective emission of the activator under hydrogen radical recombination on the phosphor surface, concentration of the activator in the phosphor decreases irreversibly with consequent loss of luminescence intensity. Evidently, such a process can occur on the phosphor surface when subjected to tritium. Owing to recombination of ^{3}H$^•$ radicals on the surface, the activator concentration in the phosphor can irreversibly decrease, producing loss of radioluminescence intensity under the action of beta particles.

4.2 MECHANISMS OF RADIATION-INDUCED DEGRADATION OF PHOSPHORS

Point defects can be created in crystalline solids — in particular, in semiconductors (including most phosphors) — under radiation by the "knock-on" and subthreshold mechanisms.[21]

4.2.1 "Knock-on" Mechanisms

"Knock-on" mechanisms[22] relate defect creation to elastic collisions of incident particles (e.g., electrons) with nuclei of atoms or ions of crystals. The primary interaction event is a pair impact collision. For the primary particle with kinetic energy E_o and mass m_0 (the mass of the nucleus of atom or ion in the crystal being M_A), the atom (ion), displaced from the regular lattice site at an angle φ with respect to the direction of incidence of the primary particle, acquires energy, E:

$$E = E_o \cdot \frac{4 \cdot M_A \cdot m_o}{\left(M_A + m_o\right)^2} \cdot \cos^2 \varphi \qquad (4.3)$$

When the energy transferred exceeds a threshold value E_d, a pair of defects is created. The displaced atom or ion moves from the new vacancy, via interstices (diffusion or chanelling) and via the chain of focusing collisions, along the densely packed atom (ion) rows. With $E >> E_d$, the primary displaced atoms (ions) can generate a cascade of secondary displacements, giving rise to several defects. For high energies of incident particles, the elastic collision model is inapplicable, so one considers interaction of the primary particle with a certain space in the crystal lattice rather than with one atom (ion). Such interaction results in disordering of regions in the crystal ("displacement wedges").

Threshold displacement energy of elastic collision is the kinetic energy imparted to the atom or ion at rest in the crystal lattice site as a result of collision with the particle bombarding the crystal. The atom or ion is displaced by a distance precluding, at low temperatures, its recombination with the empty site (vacancy) formed. When the threshold defect creation energy in crystal solids is taken as 25 eV,[21,22] the energy of the incident electron with mass m_e required for defect creation in zinc sulfide ($M_{Zn} = 1.2 \cdot 10^5\ m_e$, $M_S = 5.9 \cdot 10^4\ m_e$, where m_e is the mass of electron) is about 750 and 370 keV for zinc and sulfur ions, respectively. When phosphors are subjected to tritium beta particles whose energy does not exceed 18.6 keV, radiation defect creation by knock-on mechanisms therefore seems unlikely.

4.2.2 "Subthreshold" Mechanisms

It is now time to discuss possible mechanisms of creation of radiation defects when the energy of the incident particle is insufficient for direct ejection of an ion from the crystal lattice site.[1,3,21,22]

One possible mechanism is that the energy emitted in degradation of the electronic excitation (exciton, electron–hole pair, etc.) can, in certain cases (along with processes leading to heat liberation or light emanation), be transferred to an atom (ion) in the crystal lattice site, thus forming a pair of defects: a vacancy and an interstitial atom or ion. This requires a number of conditions, of which the most important is that electronic excitation energy must be greater than the defect creation energy; also, the lifetime of the electronic excitation on the fixed site of the crystal lattice must be longer than the period of an effective lattice vibration.

These conditions are realized in alkali halide crystals in which defects are created; the process is efficient even under ultraviolet radiation. In zinc sulfide phosphors, where the electronic excitation energy ($E_g = 3.7$ eV) is lower than the defect creation energy (more than 5 eV), creation of defects by this mechanism is unlikely.

The second, so-called electrostatic or impurity-ionization, mechanism (the generalized Varley mechanism) postulates formation of a number of multiply ionized atoms at the lattice sites. Ionization of inner (deep) shells under excitation and the consequent Auger process create multiply charged positive ion. When such an ion is formed in proximity of a positively charged impurity ion, either the multiply charged ion of the base or the impurity ion can be knocked out to the interstice owing to Coulombic repulsion. Consequently, a pair of Frenkel defects (interstitial ion-vacancy) or a combination of the vacancy and the interstitial impurity atom is formed. This mechanism requires that the Coulomb repulsion energy exceed the energy of defect creation. Defect creation follows the impurity-ionization mechanism in semiconductor crystals.[21,22] The impurity-ionization mechanism of radiation-induced defect creation probably applies to zinc sulfide phosphors and to europium-activated yttrium oxide–based phosphors when subjected to tritium beta particles.

4.2.3 Tritium-Based Radioluminescent Light Source Phosphor Degradation

Radiolytic degradation mechanisms for phosphors have been discussed. Now, available data on mechanisms of radiation degradation of phosphors when exposed to tritium irradiation will be summarized.

Phosphors employed in RLSs-T are exposed to tritium beta radiation. Differently designed RLSs-T are characterized by different radiation dose rates for the phosphors. Table 2.5 in Chapter 2 lists dose rates and the doses absorbed by a phosphor exposed to various tritium-based beta sources. Thus, due to the exposed dose, energy efficiency of the luminescence of phosphors in gas-filled RLSs-T decreases by no greater than half of the initial value (see Section 2.3.2 in Chapter 2). Zinc sulfide phosphors employed in RLSs-T with a zirconium tritide beta source show luminescence energy efficiency change at a dose of $0.7 \cdot 10^8$ Gy of less than 5% (see Table 4.1).

Factors responsible for radiation-induced degradation of phosphors include

- Radiation-induced formation of defects in the crystal lattice of the phosphor base by the impurity-ionization mechanism (see Section 4.2.2) and their involvement in the kinetics of recombination luminescence of crystalline phosphors (see Section 3.1.1.5 in Chapter 3)
- Formation on the phosphor surface of nontransparent or nonluminescent layers due to electron-stimulated reactions of molecules adsorbed on the phosphor surface with the phosphor base (see Section 4.1.2)

4.2.4 Phosphor Degradation Suppression

Overcoming radiation-induced degradation of phosphors improves the long-term stability of gas-filled RLSs-T. For RLSs-T employing other types of tritium-based beta sources, lowering of their luminescence intensity in time is due to other factors (primarily, radioactive decay of tritium; see Section 2.3 in Chapter 2 and Table 4.2). Phosphor luminescence degradation in gas-filled RLSs-T is due mainly to formation of nontransparent or nonluminescent layers on the phosphor surface.

As shown by Abrams et al.,[11] the presence of a thin (approximately 20 nm thick) silicon oxide layer coating accelerates degradation of zinc sulfide–based phosphors under cathode irradiation. However, one can formulate this coating to inhibit phosphor degradation. For example, In_2O_3 can be deposited on the phosphor by the sol-gel method. This coating decelerates the aging of zinc sulfide phosphor under low-voltage cathode irradiation.[23] Figure 4.2 presents the time dependence of the luminescence intensity for ZnS:Ag,Cl phosphor mixed with In_2O_3 powder and for ZnS:Ag,Cl phosphor with In_2O_3 coating applied by the sol-gel method when exposed to electron beam. As seen from this figure, a uniform In_2O_3 coating essentially decelerates aging of the phosphor. The electron beam energy in these experiments was 0.4 keV, and the current density was 100 $\mu A/cm^2$. It is undoubtedly of interest to investigations of prototype gas-filled RLSs-T with In_2O_3-coated phosphor and its long-time performance stability.

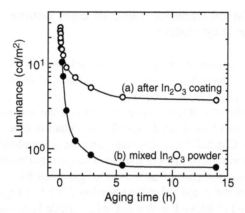

Figure 4.2 Time dependences of the cathodoluminescence intensity of the ZnS:Ag,Cl phosphor when exposed to electron beam with the electron energy of 0.4 kV and the current density of 100 μA/cm^2: (a) with In$_2$O$_3$ coating applied by the sol-gel method and (b) in a mixture with powdered In$_2$O$_3$. (From Kominami, H. et al., Low-voltage cathodoluminescent properties of phosphors coated with In$_2$O$_3$ by sol-gel method, *Appl. Surf. Sci.*, 113–114, 519, 1997.)

Another factor that can be responsible for radiation-induced degradation of phosphors is subthreshold formation of radiation-induced crystal defects in the crystal lattice of the phosphor base and their involvement in recombination luminescence kinetics. An approach to overcoming phosphor degradation for this reason can be based on the following.

The optical characteristics of irradiated crystalline phosphors can be restored in some cases by either thermal or photoannealing of defects.[24,25] Thermal annealing of defects is effective at temperatures close to phase transition; photo annealing is effective with exposure of the crystalline phosphor to high-intensity visible light (100 to 1000 W/cm^2). Under combined action of heating and lighting, the parameters for the first and second factors can be much lower.[25] Furthermore, when exposed to thermo-optical treatment immediately during irradiation, the phosphor can preserve its transparency and luminescent properties. For example,[25] the CsI:Tl crystal under electron bombardment at (0.3 to 3)·10^{13} cm^{-2}·s^{-1} preserves its transparency and luminescent properties indefinitely at a temperature of 100°C and under exposure to a light flux with a power of 0.3 to 0.5 W/cm^2.

Long-time stability of RLSs-T can be improved by thermo-optical treatment of the phosphor to reduce its radiation-induced degradation. Trykov et al.[26] reported on manufacturing a radioluminescent light source whose operation is underlain by the preceding principle. These authors used the ^{35}S isotope, a pure beta emitter with the average energy of 167 keV and half-life of 87 days.[27] The RLS in question was essentially an assembly of alternating layers of the beta source and CsI:Tl phosphor plates. The overall initial activity of the ^{35}S isotope was 16.6 Ci. Heating to about 100°C was carried out with an electrical heater, and self-radiation of the RLS provided the optical impact as well. Monitoring for 20 days showed that the light flux changed due to ^{35}S radioactive decay only.[26]

Although thermo-optical treatment to prevent radiation-induced degradation of phosphors appears fairly efficient, the need for external heat reduces practical application. On the other hand, radiation-damaged RLS-T may be restored by thermo-optical treatment. Heat treatment probably anneals radiation defects and eliminates nontransparent or nonluminescent layers from the surface (see Section 4.1.2). However, assessment of these suggestions requires additional study. Deposition of thin-layer inhibiting coatings and thermo-optical treatment are the most promising approaches to overcoming radiation-induced degradation of phosphors.

REFERENCES

1. Rodnyi, P.A., *Physical Processes in Inorganic Scintillators*, CRC Press, NewYork, 1997.
2. Pikaev, A.K., *Modern Radiation Chemistry. Solids and Polymers: Applied Aspects*, Nauka, Moscow, 1987 [Russian language].
3. Klinger, M.I. et al., Defect creation in solids under electronic excitation break-down, *Usp. Fiz. Nauk*, 147, 523, 1985 [Russian language].
4. Mikhal'chenko, G.A., *Radioluminescent Emitters*, Energoatomizdat, Moscow, 1988 [Russian language].
5. Okaminov, V.M. et al., Phosphors for radioluminescent light sources (RLS), in *Problems of Production and Application of Isotopes and Nuclear Radiation Sources in the People's Economy of the USSR, Abstracts of Papers*, TsNIIAtominform, Moscow, 1988, 125 [Russian language].
6. Klaassen, D.B.M., de Leeuw, D.M., and Welker, T., Degradation of phosphors under cathode-ray excitation, *J. Lumin.*, 37, 21, 1987.
7. Swart, H.C. et al., Degradation of zinc sulfide phosphors under electron bombardment, *J. Vac. Sci. Technol.*, A, 14, 1697, 1996.
8. Oosthuizen, L. et al., ZnS:Cu,Al,Au phosphor degradation under electron excitation, *Appl. Surf. Sci.*, 120, 9, 1997.
9. Swart, H.C. et al., The difference in degradation behaviour of ZnS:Cu,Al,Au and ZnS:Ag,Cl phosphor powders, *Appl. Surf. Sci.*, 140, 63, 1999.
10. Darici, Y. et al., Electron beam dissociation of CO and CO_2 on ZnS thin films, *J. Vac. Sci. Technol.*, A, 17, 692, 1999.
11. Abrams, B.L. et al., Electron beam-induced degradation of zinc sulfide–based phosphors, *Surf. Sci.*, 451, 174, 2000.
12. Itoh, S., Kimizuka, T., and Tonegava, T., Degradation mechanism for low voltage cathodoluminescence of sulfide phosphors, *J. Electrochem. Soc.*, 136, 1819, 1989.
13. Seager, C.H., Tallant, D.R., and Warren W.L., Cathodoluminescence, reflectivity changers, and accumulation of graphitic carbon during electron beam aging of phosphors, *J. Appl. Phys.*, 82, 4515, 1997.
14. Seager, C.H., Warren, W.L., and Tallant, D.R., Electron-beam-induced charging of phosphors for low voltage display applications, *J. Appl. Phys.*, 81, 7994, 1997.
15. Zarembo, V., Krongauz, V.G., and Podluzhnyi V.V., Physicochemical modeling of transformation in phosphors based on $A^{II}B^{VI}$ compounds, *Neorg. Mater.*, 29, 1350, 1993 [Russian language].
16. Pikaev, A.K., *Modern Radiation Chemistry. Radiolysis of Gases and Liquids*, Nauka, Moscow, 1986 [Russian language].

17. Vol'kenshtein, F.F., Gorban', A.N., and Sokolov, V.A., *Radical-Recombination Luminescence of Semiconductors*, Nauka, Moscow, 1976 [Russian language].
18. Grankin V.P., Chemiluminescence of ZnS–Tm; ZnS, CdS–Ag surface in atomic hydrogen and oxygen atmosphere, *Zh. Prikl. Spektrosk.*, 63, 444, 1996 [Russian language].
19. Styrov, V.V., Emission of charged particles from solid surface when chemical reaction occurs on it, *Pis'ma Zh. Ehksp. Teor. Fiz.*, 15, 242, 1972 [Russian language].
20. Sokolov, V.A. and Gorban', A.N., in *Luminescence and Adsorption*, Vol'kenshtein, F.F., Ed., Nauka, Moscow, 1969 [Russian language].
21. Lushchik, Ch.B. and Lushchik, A.Ch., *Decay of Electronic Excitations with Defect Formation in Solids*, Nauka, Moscow, 1989 [Russian language].
22. Vavilov, V.S., Kekelidze, N.P., and Smirnov, L.S., *Influence of Radiation on Semiconductors*, Nauka, Moscow, 1988 [Russian language].
23. Kominami, H. et al., Low voltage cathodoluminescent properties of phosphors coated with In_2O_3 by sol-gel method, *Appl. Surf. Sci.*, 113–114, 519, 1997.
24. Kolontsova, E.V., Radiation-induced states in crystals with ionic-covalent bonds, *Usp. Fiz. Nauk*, 151, 149, 1987 [Russian language].
25. Trykov, O.A., Increasing the transparence resource of solid scintillation media, Preprint of Physical-Energy Institute, Obninsk, 2482, 1995 [Russian language].
26. Trykov, O.A. et al., High-intensity autonomous light sources based on radioluminescent principle of conversion of the nuclear particle energy to electromagnetic optical radiation in the visible range, Preprint of Physical-Energy Institute, Obninsk, 2696, 1998 [Russian language].
27. Babichev, A.P. et al., in *Physical Values: Handbook*, Grigor'ev, I.S. and Meilikhov, E.Z., Eds., Energoatomizdat, Moscow, 1991 [Russian language].

CHAPTER 5

Radiation Stability of Organic Materials for Nuclear Batteries

Y.L. Kaminski and G.P. Akulov

CONTENTS

5.1 RADIOLYTIC DEGRADATION OF TRITIUM-CONTAINING ORGANIC COMPOUNDS

Primary internal radiation degradation is transformation of the molecule containing the nucleus that decays.[1-3] Primary external radiation degradation is due to interaction of the surrounding molecules with the nuclear particle ejected. Secondary radiation effects arise from interaction of the molecules with electrons, free radicals, ions, and excited particles formed in the irradiated material. As applied to tritium-containing polymeric compositions, where tritium may be in the polymer structure and in low-molecular-mass additive (see Chapter 2, Sections 2.1.4 and 2.3.6), these degradation mechanisms have identifiable features.

Primary internal radiation entails degradation of the molecule incorporating the tritium. Since the half-life of tritium is about 12.3 years, radioactive decay reduces the amount of tritium in a tritium-containing matrix by approximately 5% a year. This would be the limit of activity reduction if no other effects were also relevant.

The primary external radiation effect is due to interaction of beta particles with the polymer base of the matrix and incorporated substances. The maximal energy of the beta particles of tritium is estimated as 18.6 keV, and the average energy as 5.7 keV.[1] The maximal path length of tritium beta particles in a material with density of 1 g/cm^3 is estimated as 6 μm[4] when surrounding atoms have a size on the order of 1 Å (10^{-4} μm). The energy of the chemical bonds amounts to several electron volts.[5] Therefore, one beta particle ejected in decay generates several hundred diverse reactive species, irrespective of origin, suggesting that the site of tritium incorporation in the polymer matrix does not make a significant difference to the total damage from this effect.

The secondary radiation effect can involve both polymer and low-molecular-weight additives, leading to chemical modification and changes in physical characteristics of the system such as transparency and embrittlement. The theoretically weak dependence of the primary effect on the tritium location site in the matrix applies to the secondary radiation effect as well.

It can be assumed that the combined action of the primary external and secondary radiation effects can be modeled to a certain extent by external radiation exposure of nonradioactive polymers. Such modeling is more reliable, the closer the nature and energy of external irradiation are to those of internal irradiation and the more uniform irradiation is throughout the material. Such a comparison is difficult in practice, however, so one extrapolates from numerous results of external irradiation of nonradioactive polymer when addressing the stability of their tritiated analogs, although the approximate character of such analogies is recognized.[1]

The sequence of the processes in degradation of tritium contained in an organic molecule R–^{3}H is as follows (the primary internal radiation effect)[6-8]:

$$R - {}^3H \rightarrow \left[R - He^+\right] + \beta^-$$

$$\left[R - He^+\right] \rightarrow \left[R^+ - He\right] \rightarrow R^+ + He$$

The first chemical consequence of beta transformation of tritium is that a single-charged helium ion appears in the composition of the initial molecule. Because the ionization potential of the helium atom (24.6 eV) significantly exceeds ionization potentials of organic radicals (under 10 eV), the primary molecular ion undergoes rapid charge redistribution, thus forming an intermediate ion [R+-He]. This is followed by breakage of the very weak C+-He bond (energy 0.18 eV) within a time period very close to the oscillation period of atoms (ca. 10^{-13} sec). The R+ ion formed and the helium atom become separated in space, owing to the pulse that the daughter atom receives as a result of radioactive recoil. Thus, every event of radioactive decay in tritium-containing compounds involves formation of a primary R+ ion, a helium atom, and a beta particle.

The resulting molecular ion R+ contains a trivalent carbon bearing a formal positive charge; these structures belong to carbocations traditionally termed carbonium ions.[9] Since the carbon atom in the cation has a vacant orbital, it should be treated as an electrophilic particle. All subsequent reactions of these carbocations (and those formed by other pathways) will consist essentially in the cation attacking accessible electrons of the surrounding molecules, including unshared electron pairs of heteroatoms of π bonds (including those of aromatic rings) and electrons of σ bonds.

Helium atoms originated from beta decay of tritium escape the initial molecule with a recoil energy determined by Equation 5.1[6]:

$$E_{max} = \frac{548}{M} \varepsilon_{max} + \frac{536}{M} \varepsilon_{max}^2 \qquad (5.1)$$

where E_{max} is the maximal recoil energy in eV; M is the mass of the recoil atom; and ε_{max} is the maximal energy of beta particles in MeV.

Calculation yields E_{max} of 3.35 eV. Cases in which the recoil energy of helium atoms is at a maximum are extremely rare; in ~90% of cases, the recoil energy in decay will be under 2 eV, and in 50% it will be under 1 eV. With this low recoil energy, helium atoms will have path lengths insufficient for immediate escape from the matrix of the initial compound.

Helium can accumulate on storage of solid-state tritium-containing preparations. NMR studies show helium initially retained as microscopical bubbles 1 to 10 nm in radius.[10] As helium content attains a certain limiting value, the helium diffuses out. This limiting content depends on the structure of the material as well as on temperature and other storage conditions. The amount of helium accumulated can be calculated by the radioactive decay law: 1 Ci of a labeled compound gives $1.2 \cdot 10^{18}$ helium atoms, or only 0.045 ml per year.

Note that the behavior of stabilized beta particles (nucleus-originated electrons) resembles electrons from other processes. Such free electrons are unstable in the condensed phase and within a short time are stabilized as captured electrons. Time and temperature dependence of electron capture by traps has been theoretically analyzed.[11] Along with stabilization of electrons when physically captured by various traps, they can be chemically captured by electron acceptors, forming anions or radical anions. Chemical capture of electrons is typical for molecules with low-lying orbitals. This applies to many aromatic compounds as well as compounds with carbonyl or nitrile functional groups that have antibonding π orbitals.

Because of the short path length of tritium beta particles, virtually all beta decay energy is absorbed in the bulk of the material unless in a thin film whose thickness is comparable to path length of the beta particles of tritium. The absorbed dose can be calculated by Equation 5.2[1,12]:

$$D_{abs} = \frac{1.602 \cdot 10^{-12}}{100m} N_\beta \varepsilon_{av} \tag{5.2}$$

where D_{abs} is the absorbed dose, rad; N_β is the number of decays; ε_{av} is the average energy of the beta particles ejected in eV, and m are the mass of the compound, in grams.

For a tritium-containing compound with specific activity of 1 Ci/g, the dose absorbed within a year will be equal to $1.1 \cdot 10^8$ rad $\approx 10^2$ Mrad ≈ 1 Mgy.[1] By definition, 1 Gy = 1 J/kg = 100 rad = $6.24 \cdot 10^{15}$ eV = 10^4 erg/g. With actually attainable specific activity of tritium-containing matrices of $\approx 10^2$ Ci/g (see Sections 2.1.4 and 2.3.6 in Chapter 2) and the desirable durability of energy-emitting tritium-containing articles of ≈ 10 years, the dose absorbed in this case will be $\approx 10^5$ Mrad $\approx 10^3$ MGy ≈ 1 Ggy. (Since the purpose is estimation of the order of magnitude only, there is no correction for reduction in the number of the decays per unit mass with time owing to decay of tritium; more precise calculations can be carried out using Equation 2.53 in Chapter 2).

It is interesting to compare this D_{abs} value with the rated dose for polymers employed in other fields of science and technology. Efficient modern plastic scintillators intended for operation with external radiation sources for 10 years require a dose of 10^1 to 10^3 Mrad.[13-15] The rated dose for polymer coatings of cables intended for 10-year operation at nuclear power plants is around 10^2 Mrad.[16]

The radiation-chemical yield, G, is the number of molecules or reactive species participating in the reaction (entering into reaction or forming in it) or the number of certain interaction events per 100 eV of the energy absorbed. Radiation-chemical yield makes no sense if the specific process is not indicated. In formation of an ion pair in air, $G \approx 3$; in radiation-related transformations of macromolecules under external radiation, G is 10^{-2} to 10^2, and in chain reactions of radiation polymerization,[17] G is as high as 10^6. For autoradiolysis of labeled compounds (i.e., for decrease in their content relative to the initial state), the radiation-chemical yield is designated as $G(-M)$; in the case of low-molecular-mass compounds,[1] $G(-M)$ is 10^2.

The radiation-chemical yield of products is independent of the dose to a first approximation.[16] This independence implies direct proportionality between chemical yield of the products and the absorbed dose (at a constant dose rate, direct proportionality between the yield and irradiation time). This is not true for all complex totality of radiation-induced consecutive-parallel reactions. Indeed, in most cases chemical yields are directly proportional to the dose only in the initial sections of the "yield–dose" kinetic curves, i.e., at conversion degrees that are not very high. Chemical yield curves can "attain saturation" or descend upon passing through a maximum.[16-18] The assumed independence of radiation-chemical yield of the dose and the dose rate and nature of radiation is equally approximate.[19]

The "extent of decomposition of the substance" is a term taken from analysis of consequences of the primary external radiation effect and secondary radiation effect as applied to labeled compounds whose radiolysis products were radioactive. The extent of decomposition can be calculated by the empirical formula[1,2,12]

$$\alpha_d = \left[1 - \exp\left(-6.14 \cdot 10^{-10} \psi \varepsilon_{av} A_{mol} GT\right)\right] \cdot 100\% \qquad (5.3)$$

where α_d is percent decomposition, ψ is the inherent radioactive decay energy absorbed by the material of interest, ε_{av} is the average energy of beta radiation of tritium in eV, W_{mol} is the molar activity of the labeled compound in Ci/mol, $G = G(\text{-M})$ is the radiation-chemical yield for irreversible degradation, and t is time in seconds. $(100 - \alpha_d)$ is the maximal "radiochemical purity" that can be preserved by the labeled compound within time t. The parameter ψ depends on the geometry of the system and for real tritium-containing matrices is taken as unity. An essential consequence of the preceding formula is the exponential dependence of the extent of decomposition on the molar activity of the tritium-labeled substance. The formula is presented in approximate form, taking no account of the change in the amount of tritium in the material due to its decay. More precise calculations require changing from t to $(T_{1/2}/0.693)[1 - \exp(-0.693t/T_{1/2})]$, where $T_{1/2}$ is the half-life.

The extent of decomposition of a hypothetical substance with molecular mass 125 was calculated as a function of the absorbed dose at different $G(\text{-M})$ values.[12] As follows from Table 5.1, at a dose of 10^3 Mrad, the extent of decomposition is considerable.

As applied to labeled macromolecules, terms such as "radiochemical purity" and "extent of decomposition" look ambiguous, since this specific case involves the problem of taking into account the change of the molecular-mass distribution due to radiolysis. Large α_d values suggest transformation of the initial material to other compounds with loss of intrinsic physicochemical properties.

An essential feature distinguishing polymers from low-molecular-mass compounds is that comparatively small radiation doses cause changes to a significantly larger fraction of molecules. For example, let radiolytic degradation in a polymer with density of 1 g/cm^3 and molecular mass of $3 \cdot 10^5$ be characterized by a yield G

Table 5.1 Extent of Decomposition of a Substance with Molecular Mass 125 as a Function of the Absorbed Dose and $G(\text{-M})$

Dose, Mrad	Extent of Decomposition, %, at Different $G(\text{-M})$ Values		
	$G(\text{-M}) = 5$	$G(\text{-M}) = 10$	$G(\text{-M}) = 20$
1	0.06	0.12	0.24
10	0.59	1.19	2.38
10^2	5.5	11.3	21.4
10^3	45.0	70.2	90.9

Source: Tolbert, B.M., Radiation decomposition of labeled compounds, *Adv. Tracer Methodol.*, 1, 64, 1963.

= 1. At a dose of 10 kGy, 1 cm^3 of the polymer undergoes $6.24 \cdot 10^{17}$ changes. Provided that one macromolecule undergoes only one transformation, this number of changes corresponds to transformation of 30% of the molecules. A hydrocarbon with density of 0.66 g/cm^3 and molecular mass of 86, at the same dose and radiation-chemical yield, will affect only 0.006% of the molecules.[20]

5.2 FORMATION OF REACTIVE INTERMEDIATES IN POLYMER RADIOLYSIS

Under ionizing radiations, polymers form highly reactive intermediate species such as free radicals, ions, and excited states. These species initiate further chemical transformations strongly affecting chemical structure and physical properties of polymers.

5.2.1 Excited States

When energy transferred in the interaction of radiation with the material is insufficient to cause ionization of the molecule, the latter converts to an electronically excited state[21]:

$$RH \rightsquigarrow RH^*$$

An electronically excited state can also result from reaction of the carbocation with an electron having sufficient energy[21]:

$$RH^+ + e^- \rightarrow RH^*$$

The excited-state molecule can return to the ground state via radiationless decay (e.g., by dissipating the excitation energy in collisions) or via light emission (i.e., by phosphorescence or fluorescence)[21]:

$$RH^* \rightarrow RH \quad [+h\nu]$$

Excited states can also undergo homolytic or heterolytic degradation reactions, yielding free radicals or an ion pair[20,21]:

$$RX^* \rightarrow R^\bullet + X^\bullet$$

$$RX^* \rightarrow R^+ + X^-$$

More typical for organic polymers is homolytic degradation.

Organic compounds are classified into three groups, based on the type of orbitals of their first excited state, which is the difference in energy of the ground and first excited state. Molecules with C–H and C–C bonds have electronic levels of the first excited states ($\sigma\sigma^*$ and $n\sigma^*$) separated from the ground state by 7 to 8 eV. Molecules with chemical bonds containing n and π electrons have $n\pi^*$ as the first excited state, approximately 6 eV higher than the ground state. Aromatic systems with the first excited state, $\pi\pi^*$, are separated from the ground state by 2.5 to 5 eV. The first group exhibits highest yields of the radiolysis products, and the third exhibits the lowest.[22]

Excited states can be further classified on their multiplicity into singlet (S) and triplet (T) states. The S and T states differ by over 1 eV for $\pi\pi^*$ states and by 0.3 to 0.5 eV for $n\pi^*$ states. The $\pi\pi^*$ and $n\pi^*$ states of different multiplicity can be located relative to each other in five ways, depending on the structure of the molecule and the nature of the surrounding.[23] Irradiation can give rise to various excited states of macromolecules: singlet, triplet, highest singlet, highest triplet, superexcited with energies exceeding the first ionization potential, as well as collective excited states of molecule ensembles.[16]

Excited states from exposure to ionizing radiation should not, in principle, differ from those formed in a different way (e.g., by light exposure). The yield of specific excited states depends, however, on irradiation characteristics. Therefore, under ionizing radiation, molecules can transition to states not yielded by conventional (other than laser) photolysis, e.g., superexcited or excited states of ions. The primary role in radiation-chemical reactions belongs to the lowest triplet states. Radiolysis, in contrast to photolysis, can involve exciting triplet states and bypassing singlet states.[24] Apart from higher energies, the excited states in radiation chemistry differ from those in photochemistry by the lower degree of localization in fragments of the molecules ("photochromic groups").[21,25]

Each electronically excited state can have several vibrational and rotational states. The neighboring rotational levels differ by about 0.01 eV, the vibrational levels by about 0.1 eV, and the electronically excited states by several electron volts.[26]

Theoretically, electronically excited states can decay into radicals by two different pathways: directly from electronically excited states of the molecules or from high-excited vibrational states of the ground electronic level that resulted from nonradiative conversion. However, experimental data show that a high rate of vibrational relaxation makes the latter pathway of minor importance.[16]

Electronically excited molecules can decay into radicals by two mechanisms: predissociative and nonadiabatic. The former involves transition from delocalized excited singlet or triplet state to localized excitation state via spin-orbit and electron–electron exchange interaction. Each chemical bond is characterized by a localized excitation with repulsion term. For C–H and C–C bonds, the latter can be only a triplet term. Bonds involving N, O, and Si atoms with unshared electron pair can have both singlet and triplet terms. This suggests that predissociative mechanisms dissociate C–H and C–C bonds when in the triplet state only, with the radical pair formed in the triplet state as well. Its recombination requires conversion to the singlet state.[16]

The nonadiabatic mechanism of bond cleavage consists in transition of the molecular system from discrete states of the electronically excited bonding term to the continuous spectrum of the lower term lying above the dissociation threshold of the bond. This mechanism is operative only for transitions from high-excited states, when bond energy significantly decreases (e.g., from 4 to 4.5 to 1 to 1.5 eV for the C–H bond).[16]

Specific features of the polymer structure (long chains and periodic structure) favor radiationless migration of excitation energy over comparatively long ranges. The intramolecular energy transfer is confined to the same macromolecule, while the intermolecular transfer involves different molecules (e.g., the excited macromolecule and the low-molecular-mass luminophore). Intramolecular energy transfer follows the internal conversion (the multiplicity does not change) or intercombination conversion mechanism (the multiplicity changes).[20]

The intermolecular energy transfer follows the exchange-resonance or induction-resonance mechanism.[16,20,24,27] The first implies overlap of the electron shells of the donor and acceptor molecules; the excited donor molecule and the acceptor molecule exchange energy. This mechanism applies to cases where interacting molecules are separated by small distances (ca. 1 nm). The second mechanism emphasizes by the dipole–dipole interaction of molecules in which excitation energy can be transferred over ca. 10 nm.[20,27] For polystyrene, for example, the transfer of singlet excitation energy extends to 7 to 8 phenyl groups, with energy transfer between two adjacent groups taking ca. 30 psec.[20]

The exchange-resonance mechanism applies to the singlet–singlet and the triplet–triplet energy transfer, and the induction-resonance to the singlet–singlet transfer only,[20,25] although this is not a commonly held opinion.[16,24,27]

5.2.2 Free Radicals

External irradiation and autoradiolysis of polymers give rise to diverse free radicals, i.e., molecules or molecular fragments with unpaired electrons.[20,21,22,28,29] They can originate from decay of excited molecules or other radicals:

$$RX^* \rightarrow R^{\bullet} + X^{\bullet}$$

$$R^{\bullet} \rightarrow R'^{\bullet} + R'' - R'''$$

or interaction of electron-accepting groups of macromolecules (e.g., halogen atoms) with electrons:

$$RY + e^- + R^{\bullet} + Y^-$$

or reactions of hydrogen atoms (e.g., detachment of another hydrogen atom to form gaseous hydrogen or addition of hydrogen atoms at the double bond):

$$RH + H \rightarrow R^{\bullet} + H_2$$

$$\sim CH{=}CH\sim + H \rightarrow \sim CH_2 - C^{\bullet}H\sim$$

etc. The radiation-chemical yield of free radicals in organic materials varies from 0.01 to 10, depending on nature of the polymer and irradiation conditions.[16]

Free radicals resulting from radiolysis of polymer matrices can be classed into macroradicals, low-molecular-mass radicals, and radicals originated from polymer additives. Low-molecular-mass radicals arise from detachment of side and terminal groups from the macromolecule. Hydrogen atoms are regarded as free radicals.

Two types of macroradicals are recognized: terminal macroradicals resulting from abstraction of a terminal group in the polymer chain, and middle macroradicals resulting from abstraction of a group from the polymer midchain.

Irradiated polyethylene forms predominantly alkyl radicals, $\sim CH_2C^\bullet HCH_2\sim$, with the radiation-chemical yield of approximately 6 at 77 K. Alkenyl, $\sim CH_2-C^\bullet H-CH=CH-CH_2\sim$, and dienyl and polyenyl, $\sim CH_2-C^\bullet H-(CH=CH)_n-CH_2\sim$, radicals form, though in smaller amounts. At room temperature, the number of repeating units, n, approaches five.[20,29] Low-temperature irradiation of polypropylene results in formation of several types of alkyl radicals: $\sim CH_2-C^\bullet H-CH_2\sim$, $\sim CH_2-C^\bullet(CH_3)-CH_2\sim$, $\sim CH_2-CH(C^\bullet H_2)-CH_2\sim$ and $\sim CH(CH_3)-C^\bullet H-CH(CH_3)\sim$.[29]

For poly(methyl methacrylate), radicals such as $h^\bullet de$, $h^\bullet H_3$, $C^\bullet OOCH_3$, $\sim CH_2C(CH_3)(COOC^\bullet H_2)CH_2\sim$, and $\sim CH_2C^\bullet(CH_3)(COOCH_3)\sim$ have been identified among the radiolysis products. The overall yield of radicals is $G \approx 5$ at 77 K.

In the case of polystyrene, free radicals form in very low yield ($G < 0.2$), with benzyl and cyclohexadienyl macroradicals dominating.[20] The relative content of the latter tends to increase with the dose.[30]

Structure 5.1   Free radical in polystyrene backbone vs stabilized radical on phenyl substituent.

For polydimethylsiloxane at the liquid nitrogen temperature, $\equiv Si-C^\bullet H_2$, $\equiv Si^\bullet$, and $C^\bullet H_3$ radicals are observed; their total radiation-chemical yield is equal to 5.8.[18]

Radiolysis of polymers can result in formation, along with radicals, of radical ions, i.e., free radicals with excess positive or negative charge.[22,31] A radical pair yielded by decay of an excited molecule can recombine; the probability of this reaction in condensed phases is fairly high and is called the "cage effect."[29]

Radicals formed in crystalline regions have longer lifetimes than those formed in amorphous segments. The lifetime of radicals tends to decrease with rising temperature, but even at room temperature it can be as high as several weeks.[21,29] Free radicals can be stabilized on various structural defects of the polymer (impurities, crystal surfaces, etc.) and their local concentration can exceed the average value by 100 times.[29]

Macroradicals and low-molecular-mass radicals can dimerize, disproportionate, isomerize, etc. The close proximity required for macroradicals to enter into bimolecular reactions can be achieved by physical diffusion and chemical relay. Physical diffusion implies migration of the macroradicals, and chemical relay is formal migration of the free valence throughout the molecule, which essentially consists of a sequence of uniform reactions, such as $\sim CH_2C{\cdot}HCH_2\sim$ + $\sim CH_2CH_2CH_2\sim$ → $\sim CH_2CH_2CH_2\sim$ + $\sim CH_2C{\cdot}HCH_2\sim$. Physical diffusion dominates in the amorphous regions of crystalline polymers and in elastomers, while chemical relay is primary in rigid polymers.[20]

There is a limiting concentration of free radicals for specific polymers under specific conditions, due to a first approximation to the fact that, in high concentrations, the radicals interact with one another. For example, for polymethacrylates the maximal concentration is attained at a dose of 10 to 30 Mrad.[29]

5.2.3 Carbocations and Carbanions

Regardless of the mechanism, the primary outcome of interaction between radiation and the polymer is formation of a carbocation and a high-energy electron.[21]

$$RH \rightsquigarrow RH^+ + e^-$$

Strictly speaking, the carbocation formed by this reaction is the radical cation, but for simplicity's sake, this will not be discussed in greater detail.

For tritium-containing compounds, the primary event is beta decay (see Section 5.1), yielding a carbocation (carbonium ion), beta particle (electron), and helium atom:

$$R-{}^3H \rightarrow R^+ + e^- + He$$

High-energy primary electrons can cause ionization of other molecules by knocking out secondary, or so-called δ, electrons[21]:

$$RH + e^- \rightarrow RH^+ + 2\,e^-$$

Carbocations originated from beta decay of tritium carry a positive charge on the carbon atom previously bonded to tritium[6] and represent only one kind of carbocation that can form upon exposure of a material to beta particles or gamma quanta. No grounds exist for treating them as dominating among carbocations; on the contrary, the primary radiation-chemical yield upon tritium decay is only (100 eV)/(5700 eV) = 0.018.

The resulting carbocations can enter into reactions of electrophilic substitution, addition at double bonds, recombination with electrons, etc. For the polymer matrix containing the molecules of the substance, M, to be able to accept electrons or positive charges, reactions

$$c + R^+ \rightarrow M^+ + R$$

$$M + e^- \rightarrow M^-$$

are possible. Which of the reactions will dominate depends on the relationship between the ionization potential and the electron affinity of the polymer R and additive M.[29] Fragments of the polymer macromolecule can also act as additives.

5.2.4 Captured Electrons

Electrons can originate from external irradiation, beta decay of tritium-labeled molecules, or abstraction of a secondary electron from a neutral molecule with a primary electron. The behavior of electrons depends on their energy, dielectric characteristics of the medium, and number of traps.[29]

Having caused a series of ionization and excitation events, electrons of sufficiently high energy finally become thermal and are taken up by a neutral molecule far from their origination to form anions or radical anions. For example, in polyvinyl acetate, radical anions $\sim CH_2CH(OC\cdot O^-CH_3)\sim$ are formed.[20] In the solid phase, excess electrons are also captured by various traps such as crystalline regions, boundaries between crystalline and amorphous phases, cavities between macromolecules, and intercrystalline boundaries. At 77 K the radiation-chemical yield of captured electrons typically ranges from 0.06 for polystyrene to 0.45 for high-density polyethylene.[22]

Immediately after origination, the electron is again captured by the parent ion to form an excited molecule that subsequently decomposes into radicals. Since the radiation-chemical yields of free radicals (G = 3 to 5 on average) usually exceed those of ions ($G \leq 0.3$), electron capture by the parent ion prevails.[29]

The critical distance beyond which the electron is not subject to noticeable effect of the parent ion field is about 5 nm in hydrocarbons.[29] Electrons whose distance from the parent ion is so large that Coulomb interaction energy is lower than thermal energy are termed quasi-free electrons.[22]

5.3 RADIATION-INDUCED CHANGES IN POLYMER PROPERTIES

5.3.1 Chemical Changes

As mentioned in Section 5.1, the average energy of tritium beta radiation of 5.7 keV significantly exceeds that of chemical bonds in molecules of several electron volts. This difference is even more prominent in the case of external beta irradiation with an energy of 1 to 2 MeV. The 1- to 2-eV energy differences between different chemical bonds — subtle conformational differences (about 0.1 eV and less) — seem insignificant relative to these high-energy sources. These speculations underlie the views on nonselectivity of chemical bond cleavage in macromolecules under irradiation. Referred to as the "big truncheon" concept,[31] this implies statistical (random) distribution of radiation damages.

Indeed, primary absorption of ionizing radiation energy, unlike absorption of light energy, is nonselective. However, the intra- and intermolecular redistribution of electron excitation energy is responsible for localization of radiation damages. Among other factors, so-called "cage effects" need to be mentioned. These exert an action opposite to primary radiation damage and introduce some differentiation among damaged areas. Radiation-chemical processes are fairly selective, in fact, though not to the degree typical of ordinary chemical reactions.[29,31,32-35]

Typical radiation chemistry of polymers involves cross-linking and scission reactions, as well as unsaturation changes, gas evolution, and oxidation. A number of authors even introduce "radiation-sensitive" groups in polymers (–COOH, –X, >C=C<, etc.).[33]

Cross-linking is an irreversible radiation-chemical process, yielding new bonds between macromolecules and thus increasing molecular weight of the polymer. Cross-linking involves gradual transformation of the linear polymer structure to a three-dimensional network, reducing swelling and solubility in organic solvents, increasing its strength characteristics, etc. Cross-links are classed into those formed from 1) adding an active terminal group of one macromolecule to the middle section of another, 2) linking two middle segments of two macromolecules, and 3) neighboring segments of the same macromolecule.[17,29]

Cross-linking occurs predominantly in amorphous domains of polymers, where it is favored by the higher mobility and interlying segments of different macromolecules. In crystalline domains (lamellas), formation of intramolecular cross-links dominates. The ratio of the contents of the amorphous and crystalline domains depends not only on the chemical structure but also on the morphology of the polymer.[20,29] The fraction of intramolecular cross-links is negligible if irradiation temperature exceeds melting temperature of the crystalline phase of the polymer.

The cross-linking process can be quantitatively characterized by the radiation-chemical yield of cross-links $G(X)$, which is the number of cross-links formed per 100 eV of energy absorbed.[20,27]

At a certain dose, cross-linking yields a polymer fraction insoluble in all solvents, termed the gel fraction, in contrast to the soluble sol fraction. The dose corresponding to gelation onset is termed gel dose or gelation dose. At the gel point, there is one intermolecular bond per polymer molecule. The radiation-chemical yield of cross-links and the gelation dose are interrelated by the Charlesby equation[17,20,36]:

$$G(X) = 0.48 \cdot 10^5 \frac{1}{M_w D_g} \qquad (5.4)$$

where D_g is the gelation dose in kGy and M_w is the mass-average molecular weight of the polymer.

Gelation doses typically amount to hundreds of kilograys. The preceding equation holds if one assumes the independence of $G(X)$ to dose, random distribution of cross-links in the network, invariance of the molecular-mass distribution, and lack of radiation degradation.

Practically, the gelation dose D_g is estimated by extrapolating the plot of the content of gel fraction vs. absorbed dose D_{abs} where it intersects the abscissa. The content of the gel fraction g_c and that of the sol fraction s_c are dimensionless parameters related as $g_c + s_c = 1$. The gelation dose can be determined from the $(1/M_w) = f(D_{abs})$ dependence.[17,20] $G(X)$ can also be determined by measuring elasticity modulus, swelling, and light scattering of the irradiated polymer. ^{13}C NMR spectroscopy is one of the best methods for determining $G(X)$, by directly measuring the concentration of tertiary cross-link carbon atoms, and is suitable for doses under D_g.[20] Table 5.2 includes radiation-chemical cross-link and scission yields for key polymers.

A number of mechanisms have been proposed for polymer cross-linking,[20,29] the most popular (e.g., for polyethylene and polystyrene) of which is the free-radical mechanism. This mechanism amounts to dimerization of the alkyl macroradicals of adjacent chains[20,24]:

$$\sim CH_2{}^\bullet CHCH_2\sim \qquad\qquad \sim CH_2 CHCH_2\sim$$

$$+ \qquad\qquad \rightarrow \qquad\qquad |$$

$$\sim CH_2{}^\bullet CHCH_2\sim \qquad\qquad \sim CH_2 CHCH_2\sim$$

That pairs of radicals in close proximity could arise by independent primary processes is unlikely because of the statistical distribution of radical centers along the macromolecule. Therefore, one hypothesis is that an escaping hydrogen atom from the first polymer molecule splits off another hydrogen atom from the neighboring molecule, forming a hydrogen molecule and the second macroradical. Alternatively, neighboring radicals can be due to intra- and intermolecular migration of free valency, resulting from detachment of the hydrogen atom from the other macromolecule, to the reactive center by the "chemical relay" pathway (see Section

Table 5.2   Radiation-Chemical Yields of Cross-Linking and Scission
for Polymers in a Vacuum at 300 K

Polymer	$G(X)$, Cross-Links/100 eV	$G(S)$, Scissions/100 eV
High-density polyethylene	0.5–2.3	0.25
Polypropylene	0.5	0.6
Polystyrene	0.02–0.1	0.003–0.02
Polyvinyltoluene	0.1	0.01
Poly(vinyl acetate)	0.1	0.004
Poly(methyl methacrylate)	—	1.8–2.3
Polycarbonate	—	0.5
Poly(ethylene terephthalate)	—	0.4
Polydimethylsiloxane	2	—
Polyisobutylene	—	3–7.6
Polytetrafluoroethylene	—	5.4

Source: Radiation Stability of Organic Materials. Handbook, Milinchuk, V.K. and Tupikov, V.I., Eds., Energoatomizdat, Moscow, 1986.

5.2.2). A third option is physical diffusion of macroradicals. These three mechanisms help explain the partial, not complete, suppression of cross-linking with addition of trace free-radical inhibitors.[29,37]

Deuterated polystyrene cross-links can involve main chains and benzene rings.[38] In polystyrenes with a substituted alkyl group in the benzene ring, cross-linking can involve two macroradicals, of which one carries an unpaired electron in the main chain and the other carries an unpaired electron on the alkyl group.

Cross-linking can involve the vinylene bonds formed, e.g.:

$$\sim CH_2{}^\bullet CHCH_2\sim + \sim CH{=}CHCH_2\sim \rightarrow \sim CH_2CH\big(CH_2{}^\bullet CHCH_2\sim\big)CH_2\sim$$

In a number of cases (such as irradiation of natural rubber), cross-linking involves double bonds already occurring in the polymer.[29]

Upon irradiation of siloxanes with only main-chain silicon and oxygen atoms, macroradicals can arise through detachment of the hydrogen atoms from methyl groups and detachment of the latter from the silicon atom. Recombination of such radicals gives the Si-Si, Si-CH$_2$-Si and Si-CH$_2$-CH$_2$-Si cross-links.[18,19] In certain cases, new bonds can form in polymers that differ from cross-link bonds. For example, in polydimethylsiloxane at room temperature, an Si-H bond forms ($G = 0.42$).[18]

Broadly, polymer degradation is an irreversible radiation-chemical process involving rupture of the main chain, failure of the three-dimensional network at the cross-links, and detachment of the side groups and macromolecular fragments.[17,20] As a result, the molecular mass of the polymer decreases. In a narrower sense, scission is rupture of the main polymer chains only, which simplifies the mathematical modeling of the process.

Radiation scission differs in principle from thermal depolymerization (pyrolysis). The latter results in splitting out of the monomer, with average molecular mass of the polymer varying insignificantly and total mass of the sample decreasing. By contrast, radiation scission gives randomly distributed ruptures of the main chain, with each polymer molecule giving, on average, two molecules of a smaller size; virtually no monomers form and molecular mass of the polymer decreases.[29]

Scission can be characterized by radiation-chemical yield $G(S)$, i.e., the number of ruptures in the main chain per 100 eV of energy absorbed. The radiation-chemical yield of scission $G(S)$ is determined by the Alexander–Charlesby–Ross equation[17,20]:

$$G(S) = \frac{0.965 \cdot 10^5}{D_{abs}}\left(\frac{1}{M_n^D} = \frac{1}{M_n^0}\right) \tag{5.5}$$

where D_{abs} is the absorbed dose, kGy, and M_n^0 and M_n^D are the average molecular masses of the polymer before and after irradiation, respectively.

In actual irradiation conditions, cross-linking and scission of polymers are concomitant reactions. Mathematical modeling can be performed assuming the following conditions: 1) the molecular-mass distribution of the polymer is described by statistical law; 2) cross-links and ruptures are randomly distributed along the main chain; 3) radiation-chemical yields of the cross-linking and scission are dose independent; and 4) the absorbed dose exceeds the gelation dose. If these four conditions are met, $G(X)$ and $G(S)$ can be determined from the Charlesby–Pinner equations[17,20]

$$s_c + s_c^{0.5} = \frac{G(S)}{G(X)} + \frac{1.92 \cdot 10^5}{M_w^0 G(X) D_{abs}} \tag{5.6}$$

$$s_c + s_c^{0.5} = \frac{G(S)}{G(X)} + \frac{9.6 \cdot 10^4}{M_n^0 G(X) D_{abs}} \tag{5.7}$$

where D_{abs} is the absorbed dose, kGy; s_c is the share of the sol fraction; and M_w^0 and M_n^0 are the mass-average and average molecular masses, correspondingly [under statistical distribution, $(M_w/M_n) = 2$]. The Charlesby–Pinner equation can be modified by replacing the molecular masses by the gelation dose[17]:

$$s_c + s_c^{0.5} = \frac{G(S)}{G(X)} + \left[2 - \frac{G(S)}{G(X)}\right] \cdot \frac{D_g}{D_{abs}} \tag{5.8}$$

According to these equations, the $s_c + s_c^{0.5}$ parameter depends on the inverse dose $1/D_{abs}$, the linear fit of which enables calculation of $G(X)$ and $G(S)$ (see Figure 5.1). Table 5.2 presents radiation-chemical yields of degradation for selected polymers.[22]

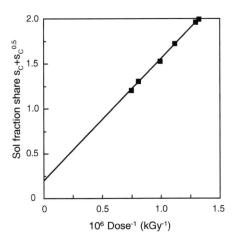

Figure 5.1 Charlesby–Pinner plot for polystyrene irradiated in a vacuum at 303 K. (From O'Donnel, J. H., Chemistry of radiation degradation of polymers, *ACS Symp. Ser.*, 475, 402, 1991.)

Although in reality cross-linking and scission occur simultaneously, polymers are grouped according to their predominant tendency to cross-link or to scission. Cross-linkers include polystyrene, polyethylene, polyacrylamide, polyamides, polysiloxanes, etc. Scission polymers include polytetrafluoroethylene, poly(methyl methacrylate), polyisobutylene, poly(α-methylstyrene), poly(ethylene terephthalate), etc.[20,24,38] Polymers with the formula ~CH_2–$CH(R)$~ primarily cross-link, while those with the formula ~CH_2–$C(CH_3)(R)$~ primarily scission. This is commonly attributed to steric hindrance in the latter polymers. Pyrolysis of polymers that predominantly undergo scission upon irradiation typically gives monomers in a good yield.[38]

There is no correlation between chemical and radiation resistances of polymers. Polytetrafluoroethylene is one of the most chemically resistant, though one of the least radiation-resistant, polymers.

Changes in the degree of unsaturation — the formation and loss of double carbon–carbon bonds in the polymer — affect mechanical, optical, and electrical properties of the material. Double bonds form from detachment of two substituents at neighboring carbon atoms of the main chain or from radical chain termination in disproportionation. The resulting double bonds can participate in chain cross-linking; polyenes can readily cross-link as well.[20] One can distinguish several kinds of carbon double bonds and, correspondingly, several unsaturated species forming in polymers upon irradiation.[20,29]

trans-vinylene unsaturation is due to formation of the R–CH=CH–R′ type moieties. They can arise by the reactions

$$\text{~}CH_2\text{–}C^+H_2\text{~} \rightarrow \text{~}CH{=}C^+H\text{~} + H_2$$

$$\text{~}CH{=}C^+H\text{~} + e^- \rightarrow \text{~}CH{=}CH\text{~}$$

and

$$\text{~}CH_2\text{–}C^{\bullet}H\text{~} + H \rightarrow \text{~}CH{=}CH\text{~} + d_2$$

Vinyl unsaturation is due to formation of R–CH=CH_2 moieties and also can be due to detachment of two hydrogen atoms from neighboring carbon atoms terminating the polymer chain.

Vinylidene unsaturation is due to formation of R′–$C(R'')$=CH_2 moieties. This originates in polyethylene, polypropylene, poly(methyl methacrylate), and other polymers upon irradiation. In polyisobutylene, vinylidene bonds can form by the reaction

$$\text{~}CH_2\text{–}C(CH_3)_2\text{–}CH_2\text{–}C(CH_3)_2\text{~}$$

$$\rightarrow \text{~}CH_2\text{–}C^{\bullet}(CH_3)_2 + C^{\bullet}H_2\text{–}C(CH_3)_2\text{~}$$

$$\rightarrow \text{~}CH_2\text{–}C(CH_3){=}CH_2 + (CH_3)_3C\text{~}$$

Polymer irradiation gives rise to gaseous products, including hydrogen, methane, alkanes, alkenes, carbon monoxide, carbon dioxide, formaldehyde, methyl formate, sulfur dioxide, and, more rarely, monomers and other products. The set of gaseous products forming in each specific case is governed by chemical composition of the polymer and its structure.[17,20,22]

For most polymers, hydrogen dominates among the gaseous products formed under irradiation (except for polycarbonate).[22,39] For tritiated polymers, tritium will be in this hydrogen. In the case of polyethylene and similar polymers,[20] hydrogen can form by two mechanisms. The first involves interaction between the hydrogen atoms and the polymer chains, giving a hydrogen molecule and an alkyl radical; the latter product enters into further reactions (cross-linking, etc.):

$$H+ \sim CH_2CH_2CH_2 \sim \; \rightarrow H_2 + \sim CH_2C^{\bullet}HCH_2 \sim$$

The second mechanism involves monomolecular detachment of hydrogen from intermediate radiolysis products (carbocations, excited molecules, etc.), forming a *trans*-vinylene bond:

$$\sim CH_2CH_2C^+H_2 \sim \; \rightarrow \sim CH_2CH{=}C^+H \sim + H_2$$

For polyethylene, a dose of 1 MGy produces about 10 cm^3 of gases per gram of the polymer.[21] For a number of polyethylene samples with molecular weight ranging from 1500 to 114,000, the radiation-chemical yield of hydrogen $G(H_2)$ increases with molecular mass of the polymer:

$$\frac{d[H_2]}{dt} = 1.03 \cdot 10^{-6} M_n^{0.1} P_D \qquad (5.9)$$

where $(d[H_2]/dt)$ is the hydrogen evolution rate in mol/h, P_D is the dose rate in Mrad/h, and M_n is the average molecular mass. Increase in the molecular mass by two orders of magnitude causes the hydrogen yield to increase by 35%.[39]

There is a correlation between the increase in $G(H_2)$ and decrease in the radiation-chemical yield of vinyl unsaturation. The terminal vinyl groups act as acceptors for some hydrogen atoms produced by irradiation. The capacity of this "trap" accounts for increased $G(H_2)$ with increased molecular mass.[39]

Upon going from polyethylene to polypropylene and polyisobutylene, the methane yield increases; the same is observed when going from polystyrene to poly-α-methylstyrene (Table 5.3[22]). Methane yield is extremely high for polydimethylsiloxane. Methane also forms from methyl groups attached to the tertiary or quaternary carbon atom of the main chain or, more rarely, terminal methyl groups.[19]

Gaseous products form upon radiolysis of poly(methyl methacrylate) due mainly to detachment of side ester groups, such as $G(C^{\bullet}OOCH_3) \approx 0.7$ to 0.8.[20] The resulting radicals dissociate:

Table 5.3 Radiation-Chemical Yield G(G) and Composition of the Gas Mixture Formed Upon Polymer Radiolysis in a Vacuum

Polymer	G(G), Molecules/100 eV	Composition (vol%)
High-density polyethylene	2.9–3.5	Hydrogen (99), hydrocarbons (1)
Polypropylene	2.3–3.8	Hydrogen (85–98), methane (2–12), other hydrocarbons (0.2–0.3)
Polyisobutylene	2.3–2.5	Hydrogen (79), methane (20), isobutylene (1)
Polystyrene	0.02–0.05	Hydrogen (99.7), hydrocarbons (0.3)
Poly(α-methylstyrene)	0.02	Hydrogen (85), methane (15)
Polyoxymethylene	1.7	Hydrogen (99.4), carbon monoxide (0.6)
Poly(vinyl acetate)	4.7	Hydrogen (6), methane (8), carbon monoxide (12), carbon dioxide (15), acetic acid (59)
Poly(methyl methacrylate)	1.6–3	Hydrogen (15), carbon monoxide (45), carbon dioxide (36), monomer, etc. (4)
Polycarbonate	1.3	Hydrogen (2), carbon monoxide (54), carbon dioxide (43.5), methane (0.3), other (0.2)
Nylon-6	1.8–3.7	Hydrogen (51), carbon monoxide (49)
Polyimides	0.02–0.06	Hydrogen (20–30), methane (1), carbon monoxide (15–40), carbon dioxide (40–55)
Polyacrylonitrile	0.2–0.3	Hydrogen (85), hydrogen cyanide (15)
Polydimethylsiloxane	2.5–3.9	Hydrogen (36–42), methane (33–57), ethane (3–29)
Polyvinylchloride	20	Hydrogen chloride (95), hydrogen, chloroalkanes, etc. (5)

Source: Radiation Stability of Organic Materials Handbook, Milinchuk, V.K. and Tupikov, V.I., Eds., Energoatomizdat, Moscow, 1986.

$$C^{\bullet}OOCH_3 \rightarrow CO_2 + C^{\bullet}H_3$$

$$C^{\bullet}OOCH_3 \rightarrow O^{\bullet}CH_3 + CO$$

or interact with other radicals:

$$C^{\bullet}OOCH_3 + h^{\bullet}H_3 \rightarrow CH_3COOCH_3$$

$$C^{\bullet}OOCH_3 + H \rightarrow HCOOCH_3$$

Hydrogen chloride is formed upon radiolysis of chlorine-containing polymers, and sulfur dioxide is formed from sulfur-containing polymers.

A specific feature of autoradiolysis of tritium-containing polymer matrices is escape of significant amounts of helium, namely, 0.77 ml per Ci of tritium decayed. In a matrix with 100 Ci/g initial tritium, about 5 Ci/g will decay within a year, producing 4.5 ml He/g matrix.

An important role in gas evolution is played not only by the chemical nature of polymers but also by their aggregation states, microstructure, and microdefects.[19] Evolution of heavy gaseous products is hindered by a low diffusion coefficient. Irradiated polyethylene not heated prior to analysis retained 1, 18, 72, and 100% of the methane, propane, pentane, and octane produced, respectively.[39] Upon heating polyethylene for one hour at 423 K, 98% of all volatile products is removed.

Even hydrogen can be retained in polymers. For example, immediately after poly(methyl methacrylate) irradiation at a dose of 100 Mrad and 293 K, virtually all gas formed is retained in the polymer. Within several months the gas diffuses out of it. Polyethylene irradiated at a dose of 6250 Mrad at 163 K retains 95% of the resulting hydrogen.[39]

In rigid plastics with low gas permeation constants, gases cannot escape in large amounts from the polymer matrix via diffusion; they gather into bubbles. As a result, the matrix increases in volume, frothes, and degrades, and the polymer may undergo cracking or frothing.[21] Mechanical degradation is more pronounced as rigidity, thickness of the matrix, and dose rate increase.

The previous discussion of chemical reactions accompanying the radiolytic degradation of polymers assumed use of degassed polymers in a vacuum or inert gas atmosphere. In practice, polymers contain dissolved oxygen and their radiolysis proceeds in the presence of oxygen from air diffusing into the polymer. Both these factors are responsible for reactions not typical for irradiated or radioactive polymer when in an inert atmosphere.[21] The key role in oxidation of polymers is played by peroxide macroradicals yielded by radiolysis.[20,29,40,41]

The simplest mechanism of polymer oxidation implies the following chain reactions of macroradicals:

chain propagation $R^\bullet + O_2 \rightarrow RO_2^\bullet$

$RO_2^\bullet + RH \rightarrow ROOH + R^\bullet$

chain branching $ROOH + ROOH \rightarrow RO^\bullet + RO_2^\bullet + H_2O$

$ROOH \rightarrow RO^\bullet + HO^\bullet$

participation of $RO_2^\bullet$, $RO^\bullet$, and $HO^\bullet$ radicals in polymer cross-linking, scission, and chemical modifications, e.g.,

$$RO_2^\bullet \rightarrow R'-CO-R'' + HO^\bullet$$

$$RO_2^\bullet \rightarrow R'-CHO + HO^\bullet$$

$$RO_2^\bullet \rightarrow R'H + R''H + CO_2$$

$$RO^\bullet + RH \rightarrow ROH + R^\bullet$$

$$HO^\bullet + RH \rightarrow R^\bullet + H_2O$$

chain termination $\quad RO_2^\bullet + RO_2^\bullet \rightarrow R\text{–OO–}R + O_2$

$$RO_2^\bullet + RO_2^\bullet \rightarrow R'\text{–CO–}R'' + R'\text{–CH(OH)–}R'' + O_2$$

$$RO_2^\bullet + R^\bullet \rightarrow R\text{–OO–}R$$

$$R^\bullet + R^\bullet \rightarrow R - R$$

Polymer oxidation can also follow other mechanisms.[21] Formation of ozone from atmospheric oxygen, when exposed to radiation, enters into reactions with polymers. Ozone yield on exposure of atmospheric air to radiation is about 0.4 molecules/100 eV.[16] Excited-state singlet molecular oxygen can form from molecular oxygen in the presence of sensitizers. Aromatic (typically polycyclic) sensitizers with long-lived, excited-state triplet states are populated under high-energy irradiation. The energy of the excited triplets exceeds that of the first excited state of oxygen and so can excite molecular oxygen. Chemical oxidation of irradiated polymers gives alcohols, aldehydes, ketones, peroxides, carbon dioxide, and water.

Upon irradiation in the presence of oxygen, the ratio of the polymer cross-linking and scission rates can change significantly.[29] The addition of oxygen to radicals and double bonds prevents polymer cross-linking, though the formed C–O–O–C bonds themselves act as a "weak" bridge at low temperatures.[29] The scission rates either remain unchanged or increase slightly. As a result, in the presence of oxygen, cross-linking polymers may predominantly scission.

The changeover from nonoxidative to oxidative irradiation conditions affects polymers of different natures differently. For example, the properties of polysiloxanes change insignificantly, but the radiation resistance of polystyrene decreases by a factor of almost 40.[21]

Oxygen is responsible for a number of phenomena complicating investigation of polymer radiolysis. The outcomes of the process become more heavily dependent on the irradiation time; dose rate; mass, geometry, and density of the sample; temperature; oxygen pressure; and post-irradiation effects (as well as on any factors affecting oxygen diffusion into polymers).[20,21]

When irradiated in air, polymers undergoing cross-linking acquire a double-layer structure resulted from oxidation of the external layer and cross-linking of the internal volume inaccessible for oxygen. This heterogeneous material experiences the growth of mechanical stresses, resulting in microcracking. The microcracks act as channels for oxygen coming from outside to the bulk of the material, which favors intergrowth of microcracks deep inside the material and even more profoundly modifies the physicochemical properties of the material.[16]

Radiation processes initially involve the dissolved oxygen and, as the consumption of the latter progresses, the contribution to oxidation from atmospheric oxygen tends to increase. This is supported by identical initial rates of radiation degradation of the polymer in air, nitrogen, and vacuum.[29]

Given oxygen diffusion coefficients for polymers at 298 K on the order of (10^{-9} to 10^{-6}) cm^2/s, there are $(0.3$ to $3.0) \cdot 10^{-3}$ mol of oxygen per kg of the polymer exposed to air.[16] For many polymers, the irradiation dose at which the whole of the oxygen dissolved in the polymer is consumed in oxidation reactions is on the order of 10 kGy.[21]

As applied to autoradiolysis of tritium-containing polymer matrices, oxidation will lead to release of a certain amount of tritiated water, formed both directly in the oxidation reaction and in the isotopic exchange reaction between the moisture contained in air and the tritiated hydroxy and carboxy groups arising in the polymer.

5.3.2 Mechanical Changes

Ionizing radiation can bring about changes in the mechanical properties of polymer matrix such as hardness, elasticity, formability, and creep. These changes are, for the most part, irreversible due to irreversible changes in the chemical structure (cross-linking, scission, etc.) and supramolecular structure (degree of crystallinity).[22] The mechanical properties of polymers undergoing scission are typically deteriorated, as tensile strength decreases and fragility and crackability increase. These changes enhance monotonically with increasing absorbed dose. Mechanical property changes of polymers undergoing cross-linking are ambiguous and governed by their structure. For example, elasticity and hardness of crystalline polymers decrease under radiation, owing to a decrease in their crystallinity; elasticity and hardness of amorphous polymers are typically improved. Dependence of the mechanical changes on the absorbed dose is also different in these cases. The ultimate strength typically increases (or remains unchanged) up to a certain absorbed dose and then monotonically decreases.

The extent to which radiation affects the mechanical properties of different polymers can vary by factors of orders of magnitude. These changes can be quantitatively characterized by the dose whose absorption causes ultimate strength or deformability to decrease twofold. This dose is estimated at 100, 5, and 0.3 MGy for polyimides, polystyrene, and poly(methyl methacrylate), respectively.[22]

5.3.3 Optical Changes

Ionizing radiation can markedly change polymeric materials' optical properties such as light transmission, reflection and absorption, and color changes. Radiation exposure can initially cause yellowing of previously colorless transparent polymer. Both reversible and irreversible changes can occur, but their relative contributions vary with the polymer and the irradiation conditions.[22] Reversible changes are due to accumulation in the polymer of intermediate reactive particles such as radicals, charged and excited species, and free and captured electrons. They can diffuse out or chemically react stoichiometrically within a certain period after irradiation, thereby eliminating radiation-induced changes in the optical properties. Irreversible changes in the optical properties are due to radiation-induced formation or degradation of various chromophores (double bonds, conjugated polyenes, oxygen-containing groups, etc.) as well as to formation of intermediate long-lived species, e.g., free radicals, in certain polymers.

Change in optical properties is due to longer wavelength absorption by intermediate reactive species or their transformation products and higher molar absorptivities ($\sim 10^3$ to 10^5 l·mol^{-1}·cm^{-1}). Reduced transparency, that is, higher optical absorption, arises from new low-lying electronic levels. Optical absorption of the irradiated sample can be characterized qualitatively by λ_{max} and quantitatively by the optical density (absorption coefficient) at this wavelength.

Almost all polymers color under irradiation. The dose at which the color becomes noticeable is strongly dependent on the polymer nature. Poly(methyl methacrylate) turns reddish-brown at doses of about 10 Mrad; the same dose gives a slight yellow color to polyethylene. Poly(vinyl chloride) turns dark green under gamma irradiation for 1 h at a dose of 10 Mrad.[42] Poly(vinyl chloride) can turn red, brown, or even black at doses of 5 to 15 Mrad.[21] Polystyrene and polysiloxanes (especially) exhibit poor coloration. In the latter case this is due to the fact that double bonds cannot form in the main chains of these polymers.[38] Radiation-induced color intensity is poorly reproduced on going from sample to sample, due to microimpurities, the presence of stabilizers, monomer traces, etc.[21,38]

Studies of a polystyrene matrix with TDEB as tritium carrier revealed a short-wave "absorption wall" which shifts with time from the UV to the visible range of the spectrum (see Section 2.3.6 in Chapter 2).[43]

[^{3}H]poly(methyl methacrylate) with a specific activity of 10 mCi/g received a self-radiation dose of about 1 Mrad over 15 months, which did not cause marked optical changes in the sample. Another [^{3}H]poly(methyl methacrylate) sample with a specific activity of 316 Ci/g turned brown after self-radiation of 45 Mrad and collapsed readily into powder when touched with a pair of forceps.[1] [^{3}H]polyethylene with a specific activity of 600 Ci/g turned dark brown within a several-month period.[44]

Any increase in the optical density of the polymer matrix, especially near the absorption or emission maxima of the primary or secondary luminophore additives, adversely affects the efficiency of visible light emission. This has been an important limiting factor in the use of tritiated scintillation polymers as light sources for indirect conversion.

5.3.4 Influence of Dose and Dose Rate on Radiolysis of Polymers

Important factors influencing radiation-induced changes in irradiated materials are the absorbed dose and the absorbed dose rate. The physicochemical properties of polymers vary upon irradiation in a complex manner, depending on the specific material being irradiated and the service parameter concerned. At low irradiation doses, most properties of polymeric materials either do not vary with the dose or vary linearly. At large doses, deviations from linearity are commonly observed, due to either disappearance of some components (e.g., dissolved oxygen) from the material or accumulation of radiolysis products and their subsequent participation in reactions with primary active species.

Upon irradiation of polyethylene to a dose of 0.15 MGy, the optical density at $v = 888$ cm^{-1} (absorption of vinylidene groups) linearly decreases; upon further irradiation it remains constant. This effect is due to radiation-induced elimination by cross-linking of vinylidene bonds initially present in polyethylene.[20]

On irradiation of plastic scintillators in an inert atmosphere or in a vacuum, the light output initially decreases with increasing dose and then, after reaching a certain limit, remains constant because of consumption of dissolved oxygen with radicals formed by radiolysis. The products of such reactions contain hydroxy and carbonyl groups absorbing light of the larger wavelength and reducing emission yield. As the dose increases and oxygen is consumed, these processes cease. At the same time, irradiation in air with the light output continues to decrease because of dissolution and diffusion of oxygen in the sample. Light output is also reduced for liquid scintillators in the presence of air.[45]

The dependence of mechanical properties of polymers on the absorbed dose is complex. For the majority of polymers, the ultimate strength does not change up to a certain dose level and then monotonically decreases. For some polymers, such as polyethylene, strength increases initially with dose and then decreases. The ultimate strain of polymeric materials decreases proportionally to increasing absorbed dose.

Radiation-induced changes in polymers can strongly depend on the dose rate, owing primarily to changes in the concentration of intermediate reactive species. As their concentration increases, the relative contribution of reactions between these species increases, and the contribution of their reactions with the surrounding molecules decreases. As a result, at a high dose rate radiation-induced degradation of samples is lower, sometimes by several orders of magnitude, than at a low dose rate.

With 600-keV pulse irradiation of polyvinyltoluene containing p-terphenyl and diphenylstilbene, light output linearly decreases with increasing dose and is independent of the dose rate.[46] At a dose rate higher than $5 \cdot 10^{10}$ Gy/sec, the luminescence intensity is linear with the square root of the dose rate. This trend is explained by formation at high dose rates of p-terphenyl negative ions that absorb light emission from diphenylstilbene.

Further studies have shown that the decisive role in this case is played by the polymer matrix rather than by additives.[47] Plastic scintillators prepared with introduction of irradiated additives had light output similar to unexposed commercial samples. The effect of the dose rate on degradation of polymers is due to participation of oxygen in radiation-chemical processes and radiation-induced heat-up.

With irradiation in a vacuum or in an inert atmosphere in the absence of radiation-induced sample heating, the absorbed dose rate has practically no effect on the physicochemical properties of polymers. Depending on the relative rates of oxygen supply into the reaction area and its consumption in radical reactions, radiation-induced oxidation can be diffusion- or kinetically controlled. With increasing dose rate, the rate of radiation-induced oxygen uptake starts to exceed the rate of oxygen supply to the reaction area, and the reaction mode changes from kinetic control to diffusion control. In this case, radiation-induced degradation of samples at the same irradiation dose noticeably decreases. For example, irradiation at room temperature of isotactic polypropylene to a dose of 0.5 MGy with increasing dose rate from 0.28 to 200 Gy/sec gives decreasing radiation-chemical yields of hydroperoxides, carbonyl compounds, and alcohols from 21 to 0.85, 22 to 14, and 26 to 17, respectively.[48]

With increasing temperature, this difference becomes less significant because of increase in the solubility and diffusion rate of oxygen in the sample.

For polypropylene irradiated in air at low dose rates, the molecular mass depends on the dose rate; at higher dose rates, the dependence flattens out. At very high dose rates, the molecular mass decreases somewhat, probably due to radiation-induced heat-up. At a dose rate of 0.0064 MGy/sec, the temperature of the radiation-induced heat-up of polymers reaches 383 to 403 K (110 to 130°C) after 45 sec.[29]

The effect of the dose rate on radiolysis of polymers is studied due to the need for testing the radiation resistance of plastic scintillators in modern accelerators. Such scintillators usually serve for many years and are subject to doses of up to 10 kGy annually. Development of accelerated testing procedures is very important in the search for new components and formulations for plastic scintillators. The possibility of accelerated testing is also important for development of isotope current sources, since their designed service life is too long to perform tests under natural conditions.

In discussions of the effect of the dose rate on the properties of plastic scintillators, it was noted that the plastic matrix is of decisive importance, rather than the scintillating additives whose radiation resistance is very high. In accelerated tests, a sample is irradiated at a high dose rate to take up within a short period the dose accumulated in the course of prolonged service under real conditions. Such tests usually give overestimated radiation resistance of polymers.

The following considerations may make results of accelerated tests more adequate to real conditions[49,50]:

- Use of elevated oxygen pressure to better simulate the total available oxygen over time
- Series of experiments with varied dose rates and extrapolation of the results to real conditions
- Use of thin samples
- Empirical correlation between degradation and dose rate
- Study of the chemical kinetics that reveal stages responsible for variation of the given property, and determination of their rate constants

5.3.5 Interaction of Low-Energy Electrons with Polymer Matrices

Available papers on radiolysis of polymers concern mainly the impact of high-energy radiation, insofar as polymers are widely used in various nuclear power installations as radiation detectors, electric insulators, coatings, lubricants, moderators, heat carriers, etc.[51-55] At the same time, the effect on organic materials of low-energy electrons, especially of tritium beta particles, has been studied to a lesser extent. To what extent are the data on polymer radiolysis with high-energy electrons applicable to conditions of low-energy irradiation in RLSs-T? Above a certain level, electron energy does not appreciably change its interaction with a substance. This concerns the main pathways of energy transfer, the nature of rising primary active species, physicochemical changes in the irradiated sample, methods for radiation protection of materials, etc. However, there are some differences.

First, the probability of interaction with atomic nuclei in an irradiated material is different for high- and low-energy electrons. In the general case, passing through a substance, an electron loses its energy in nonelastic collisions with orbital electrons and by braking in the Coulomb field of nuclei, termed ionization loss and radiation loss, respectively. The ratio of energy losses in these two kinds of interactions is given by Equation 1.8 in Chapter 1.

The electron energy loss for bremsstrahlung is significant only for high-energy electrons and for absorbers with high effective atomic numbers. This loss can be as high as 20 percent. In irradiation of organic materials with soft tritium beta particles, this kind of loss is insignificant.

Also, it is necessary to take into account the influence of ionizing radiation energy on the concentration of active species generated within a track, which may result in different rates of their interaction with each other and with surrounding molecules. Therefore, it is appropriate to discuss the structural features of tracks produced in a substance by electrons of various energies.

The spatial distribution of the absorbed energy in a substance is characterized by linear energy transfer (LET), defined as the ratio of the energy transferred to a substance by a charged particle in collisions along the length of its path. This concept is similar to specific energy loss defined in Chapter 1 but does not include bremsstrahlung, which can be deposited some distance from the path. According to the Bethe formula (see Chapter 1, Section 1.1.2), LET is a function of the rate of the ionizing particle. Figure 5.2 shows how LET varies with the energy of ionizing electrons.[56]

Electron energy loss due to interaction with the medium results in electron deceleration and increase in LET along the track. Therefore, one should distinguish the instant and path-averaged values of LET. Because of increase in LET with

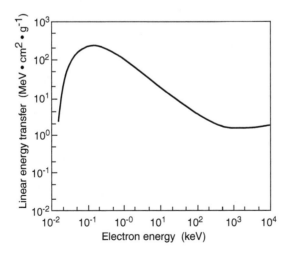

Figure 5.2   LET as a function of ionizing electron energy. (From Savinskii, A.K., Popov, V.I., and Kulyamin, V.A., *LET Spectra and Quality Coefficients of Incorporated Radionuclides: Handbook*, Energoatomizdat, Moscow, 1986 [in Russian].)

decreasing electron energy, the distance between the successive events of energy transfer along the main track decreases from several micrometers at an energy of about 1 MeV to hundredth of fractions of a micrometer at an average energy of tritium beta particles (5.7 keV).

At high LET, the concentration of the intermediate reactive species in tracks can be high and the distance between them can be low, so that interaction of these reactive species with each other, including recombination of radicals and excited species, neutralization of charges, etc., is likely. High concentrations of intermediate species can affect the luminescence efficiency of organic scintillators by increasing the fraction of absorbed energy emitted as light pulses.[51]

Low-energy electrons produce areas of high activation density in which the probabilities of generation of the triplet (T) and singlet (S) states are in an approximately 3:1 ratio; beyond these areas the ratio is about 10^{-6}:1.[57] It is known that the probability of deactivation of triplet states by phosphorescence is negligibly low, since the characteristic phosphorescence time ($\geq 10^{-2}$ sec) is long compared to radiationless excitation dissipation, which therefore occurs with greater probability. At a high concentration of triplet states, triplet annihilation becomes possible[58]:

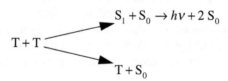

$$S_1 + S_0 \rightarrow h\nu + 2\,S_0$$
$$T + T$$
$$T + S_0$$

The singlet channel in such interaction results in generation of a luminescence quantum. The time delay of this quantum depends mainly on the concentration of T states and their diffusion in the irradiated material. Thus, for low-energy electrons producing a high concentration of triplet states, delayed radioluminescence in the course of annihilation of triplet states becomes possible.

Similar to LET, stopping power with respect to radiation characterizes the rate at which a particle loses its energy on passing a path segment, irrespective of site of energy absorption by the medium. LET is less than or equal to the stopping power with respect to radiation. The stopping powers of some polymers and one plastic scintillator are listed in Table 5.4 for electrons with energy close to that of tritium beta particles.[59] As seen from this table, stopping power depends on nature of the polymer and electron energy (similarly to LET). For mixtures of organic compounds, the stopping power can be calculated as an additive quantity with accuracy sufficient for practical purposes.

The maximal electron path length (see Chapter 1, Section 1.1.2) is approximated by the empirical formula[60]

$$R_{e,max} = aE_{max} - b \tag{5.10}$$

where a and b are empirical constants depending on the electron energy and effective atomic number of the absorber. For electrons with energy ≥ 0.5 MeV, coefficients

Table 5.4   Stopping Power for Electrons with Energy Ranging from 1 to 20 keV, MeV·cm²/g

Energy E, keV	Poly- ethylene	Poly-(methyl- methacrylate)	Poly- propylene	Poly- styrene	Poly-(vinyl- chloride)	Polyvinyl- toluene Scintillator
20	14.3	—	—	13.1	11.3	—
10	24.4	22.0	23.8	22.2	19.3	22.7
8	29.1	26.1	28.4	26.4	22.9	27.0
6	36.4	32.6	35.5	33.0	28.4	33.7
5	—	—	—	—	—	38.7
4	49.7	44.3	48.5	44.9	38.2	45.9
3	61.8	54.8	60.2	55.7	46.9	56.9
2	83.3	73.4	81.1	74.7	62.0	76.5
1	135	117	131	120	95.7	123

Source: Stopping Powers for Electrons and Positrons, ICRU Report 37, Woodmont, 1984.

b are practically equal for all the polymers (0.10 to 0.11), and coefficients a for polystyrene, polyethylene, polytetrafluoroethylene, and poly(vinyl chloride) are 0.55, 0.52, 0.61, and 0.60, respectively. The values of $R_{e,max}$ for electrons of various energies in some polymers are listed in Table 5.5.[22] As the table shows, the maximal path length of tritium beta particles in polymers does not exceed 7 to 9 μm.

The geometry and structure of tracks of tritium beta particles have some specific features and merit brief consideration of the description of steric and energetic characteristics of fast electron tracks.[61] Along an electron track, relatively small portions of energy (about 40 eV) are transferred to the medium with formation of small areas of ionization and excitation called spurs (see Figure 5.3[62]). The initial

Table 5.5   Maximal Path Length of Electrons $R_{e,max}$ in Some Polymers, g/cm²

Energy E, MeV	Polyethylene	Polystyrene	Poly(vinyl- chloride)	Polytetrafluoro- ethylene
0.01	$2.25 \cdot 10^{-4}$	$2.47 \cdot 10^{-4}$	$2.58 \cdot 10^{-4}$	$2.60 \cdot 10^{-4}$
0.02	$7.75 \cdot 10^{-4}$	$8.45 \cdot 10^{-4}$	$9.60 \cdot 10^{-4}$	$9.80 \cdot 10^{-4}$
0.03	$1.61 \cdot 10^{-3}$	$1.76 \cdot 10^{-3}$	$2.00 \cdot 10^{-3}$	$2.05 \cdot 10^{-3}$
0.05	$3.99 \cdot 10^{-3}$	$4.34 \cdot 10^{-3}$	$4.95 \cdot 10^{-3}$	$5.09 \cdot 10^{-3}$
0.07	$7.20 \cdot 10^{-3}$	$7.81 \cdot 10^{-3}$	$8.94 \cdot 10^{-3}$	$9.19 \cdot 10^{-3}$
0.10	$1.33 \cdot 10^{-2}$	$1.44 \cdot 10^{-2}$	$1.65 \cdot 10^{-2}$	$1.69 \cdot 10^{-2}$
0.2	$4.19 \cdot 10^{-2}$	$4.53 \cdot 10^{-2}$	$5.19 \cdot 10^{-2}$	$5.32 \cdot 10^{-2}$
0.3	$7.88 \cdot 10^{-2}$	$8.52 \cdot 10^{-2}$	$9.71 \cdot 10^{-2}$	$9.93 \cdot 10^{-2}$
0.5	$1.66 \cdot 10^{-1}$	$1.79 \cdot 10^{-1}$	$2.05 \cdot 10^{-1}$	$2.09 \cdot 10^{-1}$
0.7	$2.63 \cdot 10^{-1}$	$2.83 \cdot 10^{-1}$	$3.22 \cdot 10^{-1}$	$3.27 \cdot 10^{-1}$
1.0	$4.15 \cdot 10^{-1}$	$4.47 \cdot 10^{-1}$	$5.04 \cdot 10^{-1}$	$5.14 \cdot 10^{-1}$
2	$9.38 \cdot 10^{-1}$	1.01	1.12	1.15
3	1.46	1.56	1.73	1.76
5	2.48	2.65	2.90	2.96
8	3.97	4.24	4.59	4.66
10	4.42	—	—	—

Source: Radiation Stability of Organic Materials. Handbook, Milinchuk, V.K. and Tupikov, V.I., Eds., Energoatomizdat, Moscow, 1986.

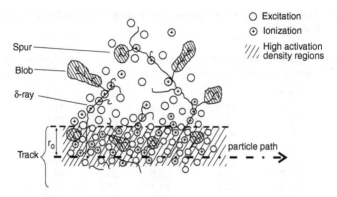

Figure 5.3 Schematic of the track structure for ionizing particle passing through the substance. (From Laustriat, G., The luminescence decay of organic scintillators, *Mol. Crystals*, 4, 127, 1968.)

spur radius is usually 0.008 to 0.01 µm, and each spur contains 2 to 3 ion pairs. As follows from the preceding discussion, the distance between spurs depends on LET; for tritium beta particles the spurs are located fairly close to each other (0.01 µm or less). Spurs are produced by secondary electrons with energy of 10 to 100 eV.

Along with low-energy secondary electrons, a certain number of electrons with energy above 100 eV forms δ electrons. These electrons form large spurs resembling pear-shaped drops (so-called blobs). At a still higher energy of secondary electrons (> 500 eV), short δ rays are formed in which spurs are located so close to each other that they form a continuous cylindrical column (short track). Short tracks are cylindrical areas with increased concentration of ionized and excited species; their maximal energy is about 5 keV, i.e., close to the average energy of tritium beta particles.

The ratio of the number of spurs, blobs, and short tracks is determined by energy distribution of secondary electrons, which, in turn, depends on primary electron energy. Impact of electrons with energy close to that of tritium beta particles on polymers increases the fraction of high-energy secondary electrons. As a result, the number of blobs and short tracks increases. Thus, ionizing electrons with energies of 1 and 5 keV produce no spurs and blobs, and all the energy is transferred in short tracks; at energies of 10 and 100 keV, the ratio of the numbers of spurs, blobs, and short tracks is 25:10:65 and 50:15:35, respectively.[61]

Some features of radiation-chemical processes induced by low-energy (kilovolts) electrons in thin polymer films have been reported. The path lengths of electrons in a solid are usually considered constant for a given electron energy and a given material. However, Bubnov and Frankevich indicate that, for low-energy electrons, the character of the energy exchange with the medium and the electron path lengths depends on the geometry of the irradiated sample and the state of the boundaries.[63,64] Thus, in thin films of thickness about 10 nm, the mechanism of volume electron energy loss differs from that in bulk samples. The electron path lengths in such films appear to be anomalously large. Without dwelling upon detailed theoretical explanation of the experimental fact, Bubnov and Frankevich consider this to be due to reflection of longitudinal polarization waves excited by low-energy electrons from the boundaries of the media. A specific feature of a film is that the dielectric properties of the media surrounding the film affect the spectrum of excitations produced by an electron.

Considering whether data obtained at irradiation with high-energy electrons are applicable to conditions of low-energy irradiation in RLSs-T leads to the following conclusions. The majority of data on polymer radiolysis with high-energy electrons are applicable to low-energy irradiation with tritium beta particles, including data on relative radiation resistance of polymers, role of luminophore additives, effect of the environment, and efficiency of radiation-protecting additives and other protection means. However, increased density of radiation damage from low-energy electrons should be taken into account.

5.3.6 Temperature Effect on Radiation-Chemical Processes in Polymer Matrices

Temperature affects reaction kinetics and mechanism in polymer matrices in four ways: 1) as a result of radiation alone; 2) in a radiation and thermal zone of reaction; 3) where thermal processes are beginning to dominate ionization; and 4) at high temperatures where reactive radicals are formed by thermal dissociation faster than by radiation ionization. This classification was first proposed for the radiative-oxidation reactions of hydrocarbons but is suitable for all organic compounds, especially those capable of reacting with atmospheric oxygen[65]:

- The radiation area covers fairly low temperatures. Here, free radicals are generated by ionizing radiation only, the reactions proceeding are nonchain processes, radiation-chemical yields do not exceed 10, and activation energy is no greater than 2 kcal/mol.
- In the radiation-thermal area, chain propagation reactions can proceed, the activation energy increases to 7 to 12 kcal/mol, and the radiation-chemical yields can display certain dependence on the dose rate.
- In the thermal-radiation area, chain process becomes branched, and activation energy increases to 20 to 25 kcal/mol.
- In the thermal area, the rate of free radical formation by thermal dissociation exceeds that of formation by the ionizing radiation energy uptake.

The quantitative temperature effect on nonoxidative radiation-chemical processes in polymers with different structures when the temperature is varied from 70 to 470 K is typically no greater than an order of magnitude of G. An important characteristic of radiation-chemical processes in polymers is the ratio between the radiation-chemical yields of scission and cross-linking: when these processes occur simultaneously, the ratio of their rates affects many of the physicochemical properties of polymeric materials. For polystyrene, this ratio is 0.2 at 303 K, 1.22 at 373 K, and 2.75 at 423 K.[16]

Polymers in nonoxidative conditions usually exhibit one or several inflection points in physical properties with increasing temperature. With the degradation of polyisobutylene, for example, the Arrhenius plot exhibits three sections. At temperatures above 200 K (the glass transition point of the polymer), the motion of the polymer chain segments is free of hindrance, which depresses recombination of the terminal radicals in the cage. Above 83 K, a barrier of hindered rotation for the methyl groups is overcome, favoring their detachment. Thus, the rate of radiation-chemical conversion is determined to a considerable extent by the motion of large fragments of macromolecules as well as by local mobilities of molecular groups.[16] Even soft beta radiation has high energy relative to the chemical bonds in an organic molecule and is much more than sufficient for local instantaneous heat-ups.[19]

5.3.7 Relative Radiation Stability of Polymers Differing in Chemical Structure

The radiation stability of polymers is governed by the electronic structure of macromolecules, supramolecular structure, and irradiation conditions, as well as by uncontrolled impurities and additives with a special purpose of imparting desired performance characteristics.[16] Depending on these factors, the resistance of organic polymers can vary by more than five orders of magnitude.[66]

Based on the decay mechanism of excited states (see Section 5.2.1), organic molecules are classified into three radiation stability groups according to the energy difference between the lowest triplet level, E_T, and the C–H bond being disrupted, E_{C-H}.[16,24] Incorporating aliphatic compounds, the first group is characterized by E_T of 5 to 7 eV, which significantly exceeds E_{C-H} of 3.9 to 4.1 eV. This group is characterized by high radiation-chemical yield ($G > 1$). The second group, incorporating aromatic compounds, has E_T of 2.8 to 4 eV, which is smaller than E_{C-H} of approximately 3.7 to 4 eV and $G < 1$. Finally, for the third group, namely, compounds incorporating aromatic and aliphatic moieties, $G \approx 1$. In the case of polymers containing heteroatoms such as N, O, S, and other atoms with unshared electron pair, decay for the triplet and also for the singlet state is possible. Therefore, tentative estimation of radiation stability of such molecules requires comparison of the E_{C-H} and the energy of the lowest singlet excited state.[16]

Polymer materials with identical chemical composition can nevertheless differ in their supramolecular structure. They can have different degrees of crystallinity, spatial sitting, and arrangement of lateral groups. Polymers contain physical defects such as cavities, vacancies, intercrystallite domains, and interfaces. Defects and other morphological features can markedly affect radiation-chemical processes occurring in polymers since energy transfer is localized close to defects.[16]

Radiation stability of polymers can be characterized by the threshold absorbed dose corresponding to change of a characteristic by a preset value. The choice of such a characteristic is governed by the polymer application field: mechanical characteristics are typically investigated for structural materials, dielectric properties for materials intended for electrical engineering purposes, plastic properties for lubricants, light yield for plastic scintillators, etc.[15,26] The size of the relative change is chosen arbitrarily; the available literature contains those relative changes ranging from 10 to 100% (most often, from 10 to 25%).[16,26,37]

The threshold doses can be found in numerous tables and diagrams that often do not indicate the polymer property investigated and do not quantitatively characterize what is meant by "weak," "notable," and "very strong, preventing further utilization of the polymer" relative changes.

Very rough assessment of the radiation stability of organic polymeric materials is that the onset of degradation occurs at 0.1 Mrad, notable degradation at 1 Mrad, significant degradation at 10 Mrad, and critical degradation at 10^2 Mrad.[42] The expected dose for a tritium-labeled polymer useful as a beta source is estimated at 10^5 Mrad (see Section 5.1); therefore, difficulties in using polymers in this manner appear overwhelming.

A more differentiated assessment requires individual analysis of the data for each polymer, taking into account all the factors affecting radiation stability. It should be noted that threshold doses for the same polymer estimated from different properties can differ by several orders of magnitude[16,29]; additional scatter is due to dissimilar irradiation conditions. Therefore, only the nature of the structural moieties imparting radiation stability and instability to the polymer, identification of classes of polymers treated as radiation stable, and tentative estimates of dose levels at which the most radiation stable polymers retain their serviceability will be addressed.

The relationship between the chemical structure of the polymer and its relative radiation stability can be illustrated by the sequence of macromolecules arranged in order of decreasing radiation stability[20] found in Structure 5.2.

Structure 5.2

In this sequence, Compound I, polystyrene, is one of the most stable polymers, and Compound XIII, containing quaternary carbons in the main chain, is one of the least stable. Nearly identical data has been reported,[17] with emphasis on polymers with radiation stability that is very low [polytetrafluoroethylene, poly(methyl methacrylate), butyl rubber], low [poly(vinyl chloride), cellulose acetate, phenol-formaldehyde resins, polyamides], medium [polyethylene, poly(ethylene terephthalate)], and high (polystyrene).

However, polystyrene is not the only radiation-stable polymer. According to other data,[20]10 Mrad does not severely damage polystyrene and its copolymers, as well as poly(ethylene terephthalate), phenol-formaldehyde resins, and poly-N-vinyl-carbazole. All these polymers contain aromatic rings. Also, it is worthwhile to mention the high radiation stability of polycondensation products of pyromellitic acid anhydride and diaminodiphenyl ether that withstand doses of 10^3 Mrad[17,21] and even 10^4 Mrad[16] (See Structure 5.3).

Structure 5.3 High radiation resistant polymer.

Polymeric lubricants were arranged according to their degradation under radiation: polyphenyls > polyphenyl ethers > polyvinyl aromatic species; these polymers are serviceable at dose levels of 10^3 to 10^5 Mrad.[26]

In Clough,[21] approximately the same radiation stability was reported for polyphenylsulfides, aromatic polyesters, aromatic sulfones, and aromatic amides (See Structure 5.4).

Structure 5.4

Polymers with aromatic groups in the main chain (polyimides, polyaryl ethers, polyarylketones, etc.) are widely viewed as the most radiation-stable polymers.[16] The stability of these materials was found to be ordered as seen in Structure 5.5:

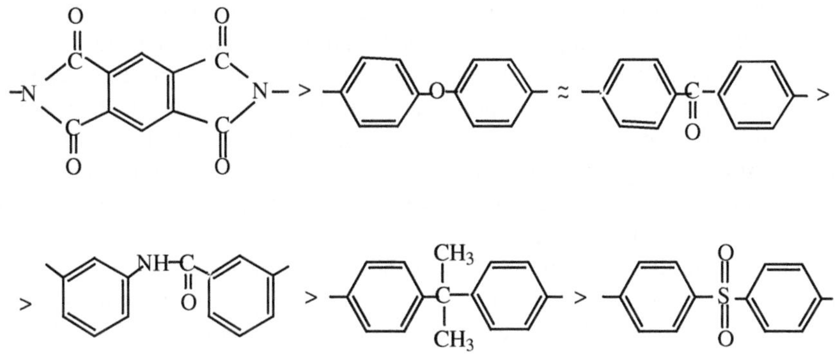

Structure 5.5

To summarize, radiation-stable polymers contain aromatic systems, the most radiation-stable spreading the conjugated system over the entire macromolecule. However, even these polymers are not very radiation-resistant materials relative to inorganic materials (see Table 5.6).

The expected radiation stability of fullerenes and their hydrides (tritides), as well as of nanotubes, deserves special attention. As mentioned in Section 2.1.4 in Chapter 2, hydrogenation of fullerene yields tertiary carbon atoms with one C–H bond each.

Table 5.6   Radiation-Chemical Yields (G) of Radiolysis Products for Various Materials

Material	G
Metals, ceramics	10^{-6}–10^{-4}
Glasses	10^{-4}–10^{-1}
Halides, silicates, and sulfates of alkali metals	10^{-4}–1
Water	0.02–1.2
Polymers with conjugated double bonds*	10^{-4}–1
Unsaturated polymers	0.1–1
Saturated polymers	0.1–10
Monomers	10^{-2}–10^{6}

* Note: This group evidently includes aromatic polymers as well.

Source: Milinchuk, V.K., Klinshpont, E.R., and Tupikov, V.I., Basics of Radiation Resistance of Organic Materials, Energoatomizdat, Moscow, 1994.

This is accompanied by gradual disappearance of conjugated double bonds that would enhance the radiation resistance of fullerene hydrides. The all-hydrogenated fullerene $C_{60}H_{60}$ can be likened (to a certain extent) to polyisobutylene, one of the least radiation-resistant polymers (see Section 5.3.1). The hydride $C_{60}H_{36}$, which supposedly contains four benzenoid rings, might be a "stability island." However, this hydride contains 36 aliphatic C–H bonds and 4 benzenoid rings, i.e., 9 bonds per ring. For comparison, in polystyrene there are three ordinary aliphatic C–H bonds per benzene ring. Though very approximate, this approach is fairly realistic. Thus, one should not expect that even the most resistant fullerene hydrides will be more radiation stable than polystyrene. Indirect evidence in favor of low radiation resistance of fullerene structures is the low yield for $^{3}H@C_{60}$ synthesized by the "hot atoms" method (see Chapter 2, Section 2.1.4).

The graphite-like structure and C–H bond absence in nanotubes improve their radiation stability over that of fullerene hydrides. Graphite is known to exhibit stability even in neutron flux. Direct verification of the radiation stability and safety of tritium-filled nanotubes is desirable.

5.4 RADIOLYSIS OF ORGANIC LUMINOPHORES

In contrast to inorganic phosphors whose luminescent properties are largely determined by structure of the crystal lattice, molecular structure is of primary importance for organic luminophores where light emission is due mainly due to electronic transitions in isolated molecules.

Organic luminophores are characterized by the presence of a developed conjugated system and aromatic and heterocyclic fragments (see Section 3.2 in Chapter 3). Owing to the presence of delocalized ™ electrons, energy absorbed by luminophore molecules is redistributed over the whole molecule rather than localized on particular bonds. In such compounds the first excited state is 2.5 to 5 eV from the ground state, which is considerably smaller than in other classes of compounds.[67] Organic luminophores, as a rule, exhibit high radiation resistance.

Ionizing radiation generates various intermediate active species with organic luminophores. In comparison to other classes of organic compounds such as alkanes, the fraction of excited molecules formed is larger, and the fraction of charged species is smaller.

The yields of excited states of organic luminophore molecules were mainly estimated from fluorescence (for singlet states) and phosphorescence (for triplet states) intensities. The relative distribution of singlet and triplet states can be illustrated by the example of electron irradiation of naphthalene solutions in cyclohexane. The radiation-chemical yield for singlet excited states ranges from 0.05 to 1.8, depending on the naphthalene concentration in solution. The yield of triplet excited states is 1.5 to 2 times that of singlet states.[68]

The yield of excited states of organic luminophores is influenced by the nature of both luminophore and solvent (Table 5.7[69]). The radiation-chemical yield of free ions from naphthalene and anthracene under the action of ionizing radiation is as low as

Table 5.7   Radiation-Chemical Yields of Triplet Excited States ($G_{(t)}$) of Some Luminophores in Solutions at 300 K

Compound	Solvent	Concentration, 10^{-2} M	$G_{(t)}$
Diphenyl	Acetone	1.7	1.1
Diphenyl	Benzene	5	1.5
Diphenyl	Polystyrene matrix	3	0.5
Naphthalene	Acetone	5	3
Naphthalene	Benzene	10	2.3
Naphthalene	Polystyrene matrix	5	0.8
Anthracene	Acetone	—	1.1
Anthracene	Benzene	5	1.5
Anthracene	Polystyrene matrix	4	0.4

Source: Alfimov, M.V., Triplet states in radiolysis of solid solutions, *Int. J. Phys. Chem.*, 8, 43, 1976.

0.094 and 0.1, respectively.[70] Radiation-chemical yields of free radicals with irradiation of diphenyl, diphenylethane, and terphenyl are 0.045, 0.1, and 0.04, respectively.[29]

Qualitative and quantitative composition of the radiolysis products of organic luminophores is determined by the nature and concentration of intermediate active species and also by pathways and rate constants of their reactions with molecules of the medium. The radiolysis products of diphenyl, its alkyl derivatives, and isomeric terphenyls have been studied in detail. Organic luminophores with annelated rings and heterocyclic structures are less studied.

The main radiolysis products of diphenyl are gases and polymers.[71] The gases contain 90% hydrogen and acetylene, with the remainder saturated and unsaturated hydrocarbons. The radiation-chemical yields of hydrogen and acetylene are independent of the irradiation dose, but they depend on the LET. The yield of hydrogen from the solid phase is approximately ten times lower space than in the melt. The majority of polymeric products are polyphenyls and hydrogenated compounds. The average molecular mass of the polymeric fraction corresponds to dimers, which means that the probability of cleavage of the bond between the rings is low. The yield of polymers decreases with increasing dose.

The radiolysis products of terphenyl and its derivatives also consist mainly of hydrogen ($G = 0.01$ to 0.02) and polymers ($G = 0.2$ to 0.4). Also, small amounts of methane, ethane, and ethylene were detected in the gaseous products. On the whole, terphenyls are more radiation resistant than diphenyl.[22] In radiolysis of *p*- and *m*-terphenyls, the hydrogen yield is lower than in radiolysis of *o*-terphenyl. The polymeric fraction consists of 50% insoluble tars. The yields of gaseous and polymeric radiolysis products increase with increasing LET.

Irradiation of luminophores with annelated aromatic rings also results in formation of small amounts of gaseous products (mainly hydrogen) and polymers (Table 5.8).

With an increasing number of conjugated aromatic rings, the yields of gaseous radiolysis products and of polymers decrease. In all cases, introduction of methyl groups increases the yield of gaseous products and decreases the yield of polymers.

Table 5.8 Radiation-Chemical Yields of Radiolysis Products of Some Aromatic Hydrocarbons with Annelated Rings at 618 K

Hydrocarbon	G(gas)	G(polymer)
Naphthalene	0.086	0.080
1-Methylnaphthalene	0.153	0.43
2-Methylnaphthalene	0.111	0.45
Acenaphthene	0.111	0.37
Fluorene	0.092	0.24
Anthracene	0.025	0.50
Phenanthrene	0.034	0.42
2-Methylphenanthrene	0.068	0.33
3-Methylphenanthrene	0.060	0.41
Fluoranthene	0.023	0.42
Pyrene	0.027	0.38
Chrysene	0.020	0.16
2,2'-Dinaphthyl	0.031	0.28
Coronene	0.029	0.08

Source: Foldiak, G., *Radiation Chemistry of Hydrocarbons*, Akademiai Kiado, Budapest, 1981.

It is interesting to compare radiation resistance of luminescent additives containing aromatic and heterocyclic rings. Comparative analysis uses IR and UV spectroscopy of the resistance to electron- and 3-radiation of luminophore additives introduced at equal concentrations into polymer matrices.[72] Compounds with polyphenyl rings are usually the most radiation resistant, and those with heterocyclic structures are the least radiation resistant. With increasing numbers of conjugated bonds in 2,5-diphenyl-1,3-oxazole and 2,5-diphenyl-1,3,4-oxadiazole with replacement of the phenyl substituent by the biphenyl group, radiation resistance increases. Methyl substitution on the benzene ring reduces radiation resistance.

5.5 RADIATION PROTECTION OF POLYMERS

Radiation protection of polymers is aimed at enhancing polymer resistance to ionizing radiation. Development of the main principles of radiation protection of polymers was initiated in the late 1950s.[26,36]

External radiation protection is distinguished from internal,[16,24,36] which consists in structural modification of a polymer macromolecule by introducing additional fragments (elementary units) enhancing the radiation resistance. External protection involves addition to a polymer of radiation degradation inhibitors, radioprotectors, or antirads. In many cases certain radiation-protecting effects can be exerted by various components of polymeric formulations having other primary purposes, such as luminophores, fillers, plasticizers, and antioxidants.[29]

Radiation protection is semiquantitatively evaluated by the protection coefficient[16,17,22,24] introduced by Charlesby[36]:

$$p' = \frac{D_{abs} - (D_{abs})_0}{D_{abs}} \qquad (5.11)$$

where $(D_{abs})_0$ is the absorbed dose causing definite changes in the initial polymer and D_{abs} is the absorbed dose causing the same changes in the protected polymer. In the physical sense, the protection coefficient should be equal to the fraction of the radiation energy drawn off by the additive from the polymer[17]; i.e., the protection coefficient can be calculated on the basis of comparison of the radiation-chemical yields or specific physicochemical characteristics of the polymer rather than on the basis of comparison of doses. There are not sufficient grounds to believe that different calculation procedures give fully coinciding protection coefficients. In some cases the radiation protection efficiency is characterized by the dose ratio $D_{abs}/(D_{abs})_0$ or yield ratio G_0/G.

5.5.1   Internal Protection

Internal radiation protection is due to dissipation of the radiation by fragments introduced into a macromolecule. Aromatic fragments are the most efficient in this respect; they have low-lying excited electronic states capable of radiative or nonradiative dissipation of excitation energy.[16,26]

Polystyrene may be considered an internally protected polyethylene with the protecting group introduced as monomer. A widely used approach is copolymerization of a protected monomer with styrene to obtain a copolymer containing units of both types.[21,26] Figure 5.4[73] shows the radiation-chemical yield of the cross-links as a function of the styrene content in its copolymer with butadiene. For 1:1 styrene-isobutylene copolymer, the radiation-chemical yield of degradation decreases by a factor of more than three compared to polyisobutylene. In the case of copolymerization of ethylene with styrene, at a 5% mol fraction of styrene, the protective effect of one styrene unit extends to three adjacent ethylene units, while at a 10% mol fraction of styrene, it extends to 8 to 10 adjacent ethylene units.[17] Finally, radiation resistance of a polymer can be enhanced by grafting a more stable compound; thus, the radiation resistance of polyethylene increases by a factor of more than two after grafting of acenaphthene.[20]

Data on the protective effect of benzene rings in poly(phenyl methacrylate) and poly(benzyl methacrylate) compared to nonaromatic methacrylates are contradictory.[29,74]

As stated in Makhlis,[29] reduction in the radiation-chemical yields of degradation, gas evolution, etc. on introduction of aromatic units into a copolymer is insufficient for concluding occurrence of the true internal radiation protection; however, this seems to be a matter of terminology. Indeed, in a polymer containing aromatic rings, the energy taken up by the ring increases, but intramolecular energy transfer from "weak" sites of the chain to aromatic rings (i.e., the true radiation protection) may be absent.

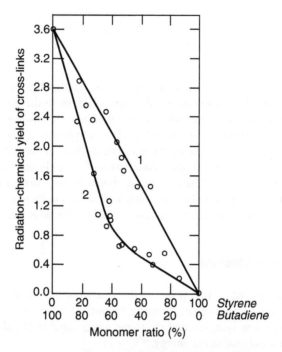

Figure 5.4 Radiation-chemical yield of cross-links on gamma irradiation in a vacuum as a function of the butadiene-styrene ratio (●) in the copolymer and (○) in a blend of the homopolymers. (From Witt, E., The effect of polymer composition on radiation-induced cross-linking, *J. Polym. Sci.*, 41, 507, 1959.)

Pankratova et al.[18] showed that dependency of the radiation-chemical yield $G(X)$ on the content of phenyl groups for blends of polydimethylsiloxane with polydiphenylsiloxane and for copolymers of dimethylsiloxane with diphenylsiloxane containing more than 1 mol% aromatic component can be described by the general exponential equation

$$G(X) = G_0(X)\exp\left[(-1.9 \pm 0.6)(N_{Ph}/N_{SiO})\right] \qquad (5.12)$$

where $G(X)$ and $G_0(X)$ are, respectively, radiation-chemical yields in the presence and in the absence of the aromatic component in the polymer or blend and N_{Ph}/N_{SiO} is the ratio of the numbers of phenyl groups and siloxane units (see Figure 5.5). A similar dependency is observed for the radiation-chemical yield of gaseous products (Figure 5.6[18]).

A simple exponential form of the dependencies led Pankratova et al.[18] to a suggestion that the observed first-order reaction is quenching with a phenyl group of the excitation transferred along the siloxane chain within a molecule or from one molecule to another. A similar dependency can be plotted for polyethylene–polystyrene copolymer (Figure 5.6), but the point corresponding to the polymer blend falls out from the general dependency.

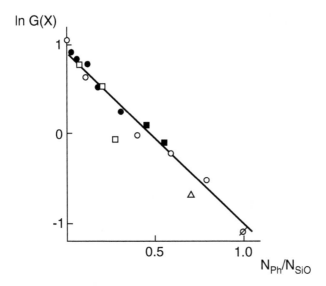

Figure 5.5   Logarithm of the cross-linking yield as a function of the number of phenyl groups per siloxane unit N_{Ph}/N_{SiO}: (●, Ø) copolymers of dimethylsiloxane with diphenylsiloxane (data from various sources); (○) blends of polydimethylsiloxane with polydiphenylsiloxane; (î, Δ) copolymers of methyl- and phenylsiloxanes with various degrees of substitution of one or two methyl groups by phenyl group). (From Pankratova, L.N., Bugaenko, L.T., and Revina, A.A, Influence of aromatic protectors on radiolysis of polyorganosiloxanes, *Khim. Vys. Energ.*, 34, 19, 2000 [in Russian].)

5.5.2   External Protection

With respect to the protection mechanism, radioprotectors are subdivided into two groups. Radioprotectors of the first group act as a "sponge," taking up the energy from the irradiated polymer and dissipating it by the nonradiative or radiative mechanism.[22,24] Typical representatives of this group of radioprotectors are aromatic hydrocarbons. The high radiation resistance characteristic of aromatic hydrocarbons is consistent with low quantum yields of their photochemical transformations and high relative intensity of their molecular ions (M^+) in the mass spectra.[38]

Ideally, the radioprotector should undergo no chemical transformations; however, in practice this is rare.[26] Aromatic hydrocarbons undergo radiation-chemical degradation to some extent, although their radiolysis products can partially preserve radiation-protecting properties.[22,36] In other cases, as shown by experiments with ^{14}C-labeled aromatic hydrocarbons, the radiolysis products of radioprotectors add to the polymer molecule, thus providing some internal protection.[22,26,38]

A more extensive group of protectors acts as "victims," accomplishing protection at the expense of decomposition.[24,26,37,74] For example, radioprotector (inhibitor) InH delivers hydrogen atom to macroradical R•, transforming into a relatively inactive radical In•:

$$R^\bullet + InH \rightarrow RH + In^\bullet$$

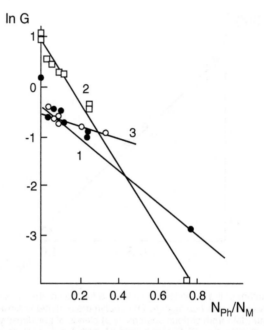

Figure 5.6 Logarithm of the radiation-chemical yield of gaseous products of polymer radiolysis as a function of the number of phenyl groups per phenyl-free monomeric unit N_{Ph}/N_M: (1) yield of hydrogen from polydimethylsiloxanes with methyl groups partially substituted by phenyl groups; (2) sum of logarithms of yield of methane and doubled yield of ethane from the same copolymers; (3) yield of hydrogen from ethylene-styrene copolymers. (From Pankratova, L.N., Bugaenko, L.T., and Revina, A.A, Influence of aromatic protectors on radiolysis of polyorganosiloxanes, *Khim. Vys. Energ.*, 34, 19, 2000 [in Russian].)

It is important that some radioprotectors of the "victim" type can neutralize ("cure") active centers formed in the polymeric chain, which would otherwise facilitate further degradation of the polymer. For example, the radical $In^{\bullet}$ can recombine with the macroradical, preventing cross-linking[17,74]:

$$R^{\bullet} + In^{\bullet} \rightarrow RIn$$

Radioprotector In can link two macroradicals, formally decreasing the fraction of the degraded polymer[17,74]:

$$R^{\bullet} + In + R^{\bullet} \rightarrow R\text{--}In\text{--}R$$

These facts suggest that there should be certain limiting doses at which a radioprotector of the "victim" type completely decomposes and becomes inefficient or poorly efficient.[74] These limiting values depend on the nature of the radioprotector, its concentration, and nature of the protected polymer.

Optimal concentrations of radioprotectors in the polymer range from several tenth fractions of wt% to several wt%,[37] usually to 1 to 2 wt%,[26] more seldom to 5 to 10 wt%,[21,26,37] and sometimes as high as 20 to 30 wt%.[22] An example of concentration dependences of the protective effect of radioprotectors is shown in Figure 5.7.[75]

Generally speaking, the lower limit of the efficient concentration of a radioprotector is determined by the mechanism of its action, and the upper limit is determined by the requirement to preserve the necessary service properties of the polymer (in some cases, even before external irradiation).[16] The experimentally observed saturation effect in the curves of protective effect vs. radioprotector concentration may be due also to limited solubility of the protecting additive in the polymer.[24]

Usually radioprotectors ensure more efficient protection from scission than from cross-linking; this trend is more pronounced in air than in an inert atmosphere.[21,17]

Radioprotectors can be introduced into polymers by several procedures: 1) addition into a polymer solution followed by solvent evaporation and drying, 2) joint grinding, 3) joint pressing, and 4) joint fusion. Different procedures ensure different uniformity of radioprotector distribution in the material, which can affect the protection efficiency in some cases.[22]

Even in the case of homogeneous distribution, radioprotector efficiency depends not only on its capability to deactivate excited states or scavenge free radicals but also on its compatibility with the polymer matrix and capability to be retained in this matrix without diffusion out. In this respect, covalently linked radioprotectors seem to be more reliable, though their introduction is more expensive.[21]

Absolute protection of polymers from radiolysis is impossible, but their resistance can be enhanced by a factor of two to five and, in some cases, ten.[26] There is some opinion that the radiation resistance of polymers can be enhanced by two orders of magnitude.[17] It should be noted that resistance enhancement by a factor

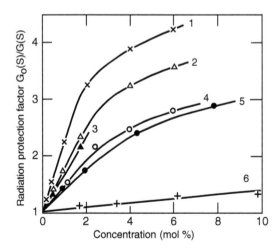

Figure 5.7   Efficiency of radiation protection of poly(methyl methacrylate) (as judged from yield of radiation-chemical scission) as a function of the radioprotector concentration: (1) benz[a]anthracene, (2) pyrene, (3) anthracene, (4) phenanthrene, (5) naphthalene, and (6) benzene. (From Wundrich, K., Effect of temperature and physical state on the inhibition by additives of radiation-induced degradation of poly(methyl methacrylate), *J. Polym. Sci., Polym. Phys. Ed.*, 12, 201, 1974.)

of 2 corresponds to the protection coefficient $p' = 0.5$, and enhancement by a factor of 100 corresponds to $p' = 0.99$. Of more than 300 values of p' for various polymer-radioprotector combinations,[22] 58% of p' values lie within $0.5 \geq p' > 0$, (i.e., the radiation resistance is enhanced by a factor of no more than 2); 39% of p' values lie within $0.9 \geq p' > 0.5$ (enhancement by a factor of 2 to 10); and only 3% of p' values lie within $0.97 \geq p' > 0.9$ (enhancement by a factor of 10 to 33).

Experiments showed that radioprotectors are more efficient for less radiation-resistant polymers. Proper choice of a polymer can lead to a better overall effect than attempts to enhance the radiation resistance of an unstable polymer with appropriate additives.[26] It is interesting that data are lacking on radioprotectors for polystyrene,[22,26,27,74,76] since polystyrene is sometimes used as a radioprotector for other polymeric materials.[22,24] There are only a few reports on the protective effect of luminophore additives on polystyrene plastic scintillators (see Section 5.5.3).

Principles of the action of radioprotectors in various stages of radiolysis and the main groups of radioprotectors used for a wide range of polymers are given in Table 5.9.

External protection is maximally efficient in all stages of radiolysis by suppressing undesirable chemical processes. Therefore, sometimes a mixture of several radioprotectors, each suppressing a particular radiolysis stage, is used for efficient protection. An alternative approach is combination in one molecule of fragments corresponding to radioprotectors of various classes. In some cases the protective

Table 5.9 Action of Radioprotectors in Various Stages of Radiolysis

Radiolysis Stage	Principle of Radioprotector Action	Main Groups of Radioprotectors
Intra- and intermolecular transfer of absorbed electronic excitation energy	Dissipation of electronic excitation energy in the form of heat or optical emission	Compounds containing aromatic fragments (especially hydrocarbons)
Ionization of polymeric molecule with formation of electron and organic (in particular, polymeric) ion; Formation of hyperexcited states and dissociation of polymer molecule	1) Electron transfer to organic ion (without subsequent excitation) 2) Electron acceptance (and thus decrease in probability of formation of excited macromolecules)	Aromatic amines and diamines, polysubstituted phenols and polyphenols Organic nitro compounds and organohalogen compounds, quinones, quinone imines
Bond cleavage, formation of hydrogen atoms and free radicals, reactions of free radicals	Acceptance of hydrogen atoms; transfer of hydrogen atoms to free radicals; reactions with free radicals to form less-active radicals	Aromatic amines and diamines, polysubstituted phenols and polyphenols, iodine, sulfur, sulfur-containing compounds, sterically hindered piperidine derivatives, spin traps (iminoxyl radicals)

Sources: Ivanov, V.S., *Radiation Chemistry of Polymers*, Khimiya, Leningrad, 1988 (in Russian); Milinchuk, V.K. and Tupikov, V.I., Eds., *Radiation Stability of Organic Materials Handbook*, Energoatomizdat, Moscow, 1986 (English translation, *Organic Radiation Chemistry Handbook*, Ellis Horwood, Chichester, 1989); Laricheva, V.P. and Suminov, V.I., *Problems of Radiation Protection of Polymers*, NIITEKHIM, Moscow, 1985 (in Russian); Makhlis, F.A., *Radiation Physics and Chemistry of Polymers*, Atomisdat, Moscow, 1972 (English translation, John Wiley & Sons, New York, 1975).

effect of a mixture exceeds the sum of those of the components; i.e., a synergistic effect is observed.

Monomolecular chemical reactions (dissociation, isomerization, etc.) of compounds excited to the singlet state occur with rate constants of 10^9 to 10^{13} sec^{-1}. Therefore, their suppression requires very high and practically inaccessible concentrations of radioprotectors. Since the lifetime of excited triplet states is much longer (up to 10^{-3} sec), their deactivation is possible at radioprotector concentrations on the order of 10^{-4} to 10^{-6} M.[16]

Electron transfer from excited macromolecules to radioprotector molecules occurs by the exchange or resonance mechanism. In both cases, the necessary condition is the overlap of the emission spectrum of the donor and absorption spectrum of the acceptor. By the exchange mechanism, both singlet–singlet and singlet–triplet energy transfer is possible, the necessary condition being overlap of the electronic shells of the excited donor molecule and ground-state acceptor molecule.[16] Energy transfer by the resonance mechanism requires that the lower vibration level of the donor molecule is higher than the lower vibration level of the acceptor molecule. The transfer is most efficient in the case of spin-allowed (singlet–singlet and triplet–triplet) transitions. The probability of the process occurring by the resonance mechanism is approximately two orders of magnitude higher than that of the process occurring by the exchange mechanism with a transfer rate constant of order 10^{11} l/(mol·sec).[16]

The acceptors of the electronic excitation energy should meet two requirements. They should deactivate the excited states of polymer macromolecules with high efficiency, and their own electronically excited states should exhibit low activity in chemical reactions. As already noted, the most efficient stabilizers of this type are aromatic compounds.[16]

To suppress the negative consequences of formation of radiolysis products by the ionic mechanism, it is necessary that the ion acceptors reduce chemical transformations, i.e., give less reaction products than the ions formed from the polymer. The lifetime of charges beyond Coulomb attraction is much longer than that of nonrecombined charges in pairs. Therefore, the acceptor concentration sufficient for accepting free radicals is about 10^{-8} M in liquids[16] and 10^{-3} to 10^{-1} M in solids. The charge transfer to an acceptor is feasible if the charge affinity of the acceptor is greater than that of the macromolecular ion. Good charge acceptors are aromatic amines, phenols, organic nitro compounds, organochlorine compounds, quinones, and quinone imines.

Decay of excited states, ion reactions, etc. result in generation of free radicals in a polymer. The majority of radicals form in spurs and tracks (see Section 5.3.5). Their local concentrations are fairly high, but such radicals rapidly recombine with each other. The rate of reaction of radicals with an acceptor becomes comparable to the rate of reaction of radicals with each other at acceptor concentration exceeding 10^{-6} M. The main aim of radical acceptors is to affect the direction of radical reactions or the yield of some products (e.g., gaseous). It is significant that when a polymer is irradiated in the absence of oxygen, the radical reactions are not catalytic. Therefore, scavenging of radicals with an acceptor does not alter the total amount of radicals formed but affects their composition and reactivity. Inhibitors of radical reactions exert their effect in the following ways:

1. Restoration of the initial macromolecule from which the radical has formed
2. Deactivation of the macroradical by formation of a new stable molecule
3. Formation of new, less-reactive radicals
4. Deactivation of radiolysis products and other substances that are potential sources of highly active free radicals[29]

The hydrogen atom bearing an unpaired electron in the valence shell is actually the simplest free radical. On the one hand, it is relatively readily generated; on the other, it readily binds with many substances exhibiting (in a broad sense) oxidizing properties. Therefore, introduction into a polymer of substances exhibiting efficient donor and acceptor properties with respect to hydrogen atoms can ensure efficient radiation protection. Transfer of a hydrogen atom from an additive to a formed macroradical restores the initial molecule. At the same time, the capability of the resulting low-molecular-mass radical of the additive to accept a hydrogen atom prevents the hydrogen atom's participation in further reactions of polymer degradation. Such highly efficient "stores" of hydrogen atoms are aromatic amines, phenols, aminoalkanethiols, thiourea derivatives, thiophenols, thiazolidines, and even elemental sulfur.[29]

$$H + S_n \rightarrow S_n{}^{\bullet}H$$

$$R^{\bullet} + S_n{}^{\bullet}H \rightarrow RH + S_n$$

Experiments with thiophenol tritium labeled at the SH group showed that, in the course of polyisobutylene irradiation, a tritium atom is transferred to the polymer matrix, with simultaneous decrease in degradation by 75%. Replacement of hydrogen atoms in –OH, –SH, and –NH_2 groups of the radioprotector by methyl groups, as a rule, reduces their efficiency.[29] Polynuclear aromatic hydrocarbons, along with protection by the "sponge" mechanism, can provide protection by binding hydrogen atoms through hydrogenation.[74]

On the whole, the mechanism of the action of most radioprotectors is complex and not limited to reactions with primary radiolysis products. Because of the large number of simultaneous reactions, a universal activity series of radioprotectors is complex.[16] The relative efficiency of stabilizers can vary, depending on their concentration, type of polymer matrix, and irradiation conditions.[21]

In the case of radiolysis of fluoro- and chloropolymers [polytetrafluoroethylene, poly(vinyl chloride), etc.], metal oxides (CaO, MgO, TiO_2) or salts of weak acids (sodium stearate, lead carbonate) are added[17,21] to chemically bind F_2, HF, and HCl that would otherwise cause chemical degradation of the polymers. Although such additives are sometimes classed with radioprotectors, they are chemical neutralizers. Also suitable as neutralizers are organic epoxy compounds that bind HCl at the epoxy ring.[24]

The main classes of chemical compounds used as radioprotectors have been considered; specific representatives of these classes will now be considered.[22,24,29,74] Since radioprotectors often contain several functional groups possessing protective functions, assignment of a particular substance to a particular class is, to a certain extent, arbitrary. Structure 5.6 shows aromatic hydrocarbons, Structure 5.7, aromatic heterocycles, Structure 5.8, quinones, and halogen rings.

> *Aromatic hydrocarbons*: naphthalene, alkyl- and arylnaphthalenes, di(2-naphthyl)methane, *p*-terphenyl, partially hydrogenated terphenyls, phenanthrene (XIV), fluoranthene (XV), pyrene, benz[*a*]anthracene (XVI), triphenylmethane, anthracene, tetralin (XVII), fluorene (XVIII), acenaphthene (XIX), etc.

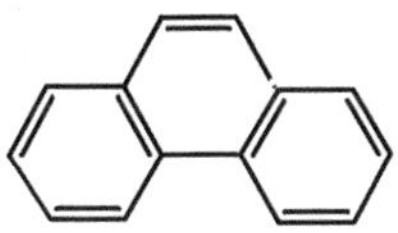

| XIV | 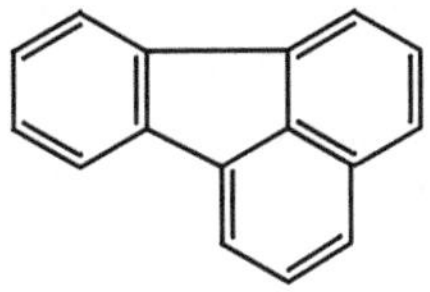 XV | 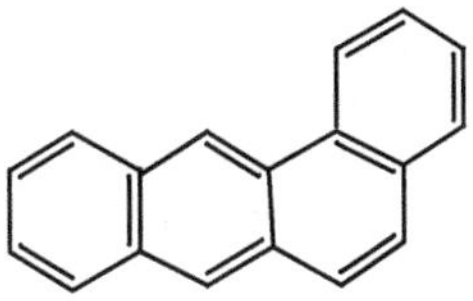 XVI |

| XVII | 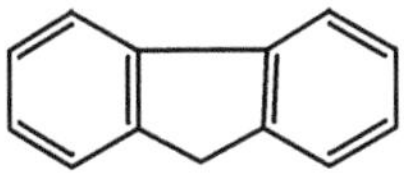 XVIII | 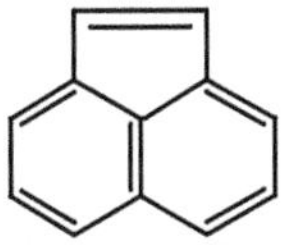 XIX |

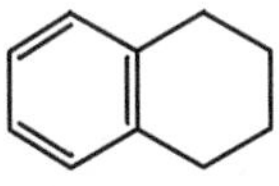

Structure 5.6

> *Aromatic heterocycles*: dibenzothiophene (XX), phenothiazine (XXI), 8-hydroxyquinoline (XXII), benzotriazole (XXIII), dibenzofuran, etc.

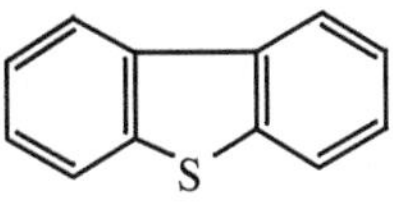

| XX | XXI | XXII | XXIII |

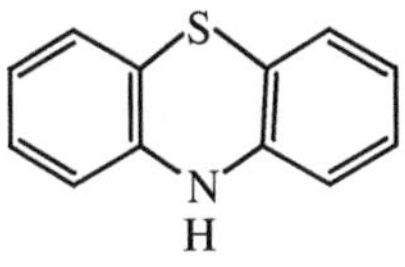

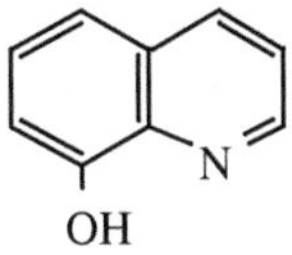

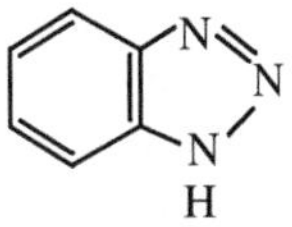

Structure 5.7

> *Quinones and quinone imines*: benzoquinone, anthraquinone (XXIV), 1,2- and 1,4-naphthoquinones, 3,3,5′,5′-tetra-*tert*-butyl-4,4′-diphenylquinone (XXV), phenylquinone imine, naphthylquinone imine, etc.
> *Organic nitro compounds*: 2,4- and 2,6-dinitrophenols, 4-nitrophenol, 2,4-dinitroaniline, etc.
> *Organohalogen compounds*: iodobenzene, haloacenaphthenes, chloranil (XXVI), etc.

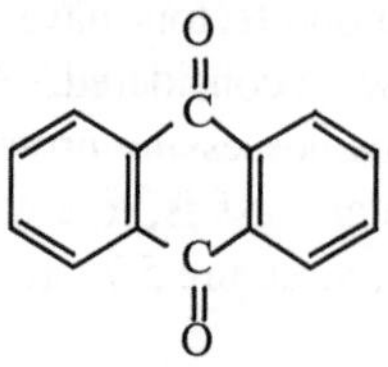

XXIV 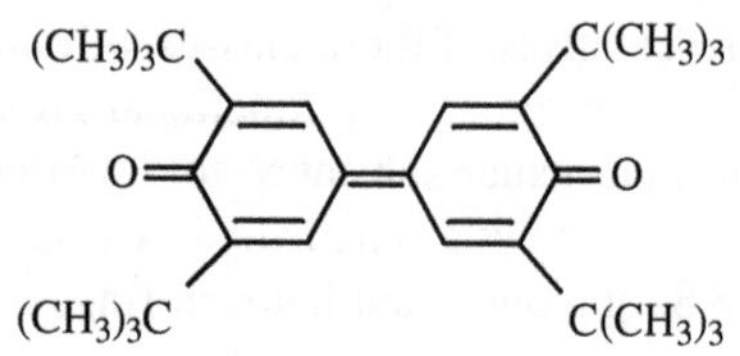   XXV 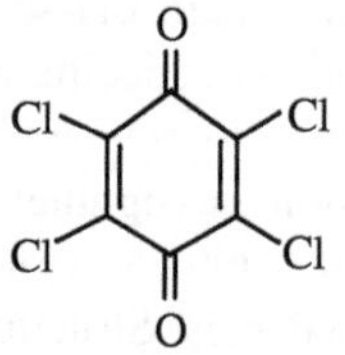   XXVI

Structure 5.8

Sulfur-containing compounds: alkyl- and dialkylsulfides, aryl- and diarylsulfides, diphenylthiourea (XXVII), allylthiourea, tetramethylthiuram disulfide (XXVIII), 2-mercaptobenzothiazole (XXIX), di(2-benzothiazolyl) disulfide (XXX), 4,4′-dihydroxydiphenyl disulfide, etc. Organoselenium compounds (alkyl selenides, etc.) can also be assigned to this group. (See Structure 5.9)

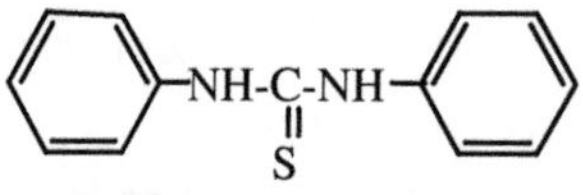

XXVII 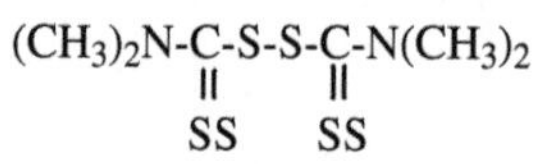   XXVIII

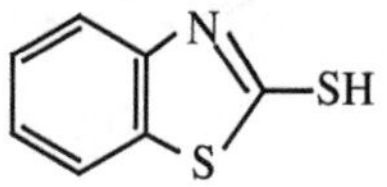

XXIX 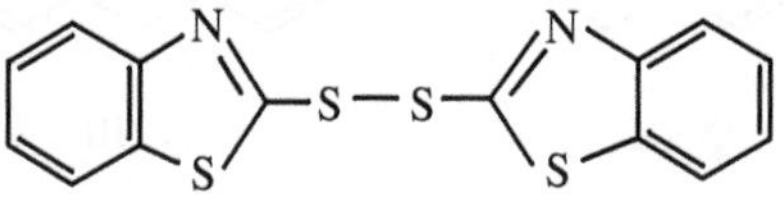   XXX

Structure 5.9

Aromatic amines: phenyl-1- and phenyl-2-naphthylamines (XXXI, XXXII), N,N′-diphenyl-*p*-phenylenediamine (XXXIII), diphenylamine, triphenylamine, etc. (See Structure 5.10)

Phenols and polyphenols: 2-naphthol (XXXIV), 2,6-di-*tert*-butyl-4-methylphenol (XXXV), 3,5-di-*tert*-butylpyrocatechol (XXXVI), 2,2′-methylenebis(4-ethyl-6-*tert*-butylphenol) (XXXVII), etc.

Sterically hindered piperidines: 2,2,6,6-tetramethylpiperidine (XXXVIII) and its derivatives.

Spin traps (stable iminoxyl radicals): 2,2,6,6-tetramethylpiperidine-1-oxyl (XXXIX) and its derivatives. (See Structure 5.11)

Theoretically the degree of radiation hardness of a polymer depends on many concomitant external factors: mechanical stresses, pressure, action of the light, etc.[77] However, specific polymeric articles considered in this book either are not subject to the action of these factors or their effects can be neglected.

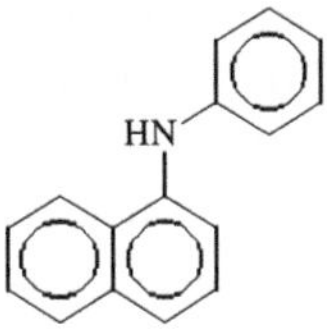

XXXI

phenyl-1-napthylamine

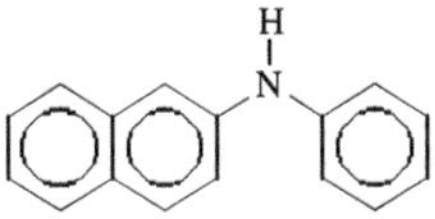

XXXII

phenyl-2-napthylamine

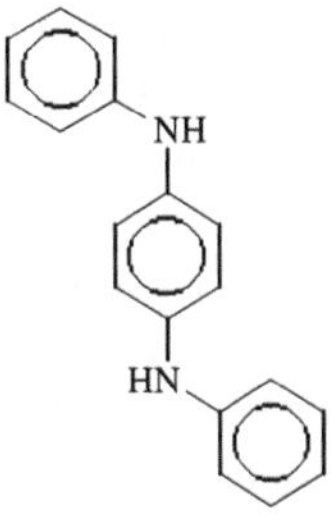

XXXIII

N,N'-dipenyl-p-phenylenediamine

Structure 5.10

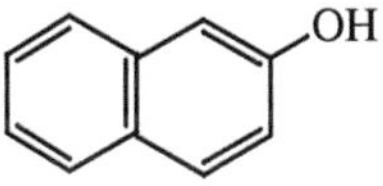

XXXIV

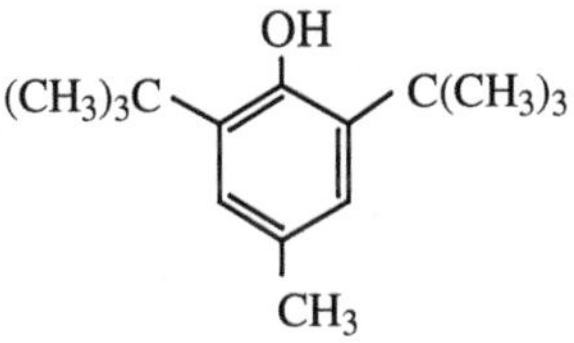

XXXV

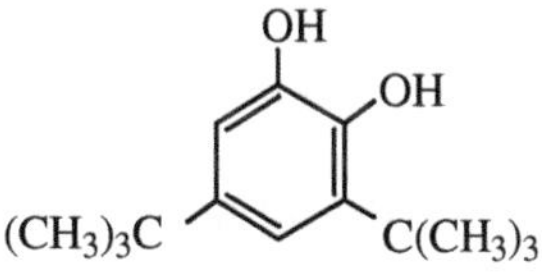

XXXVI

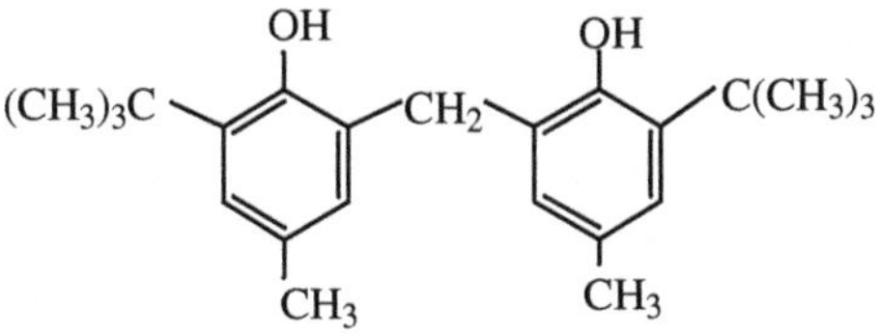

XXXVII

Structure 5.11

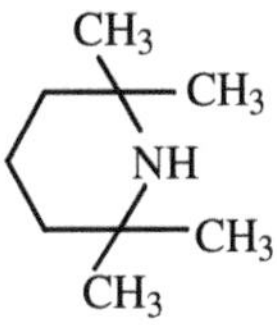

XXXVIII

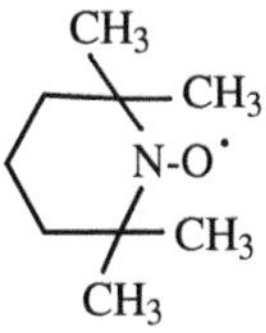

XXXIX

Structure 5.12

Under actual conditions, the only significant additional contribution to polymer degradation is oxidation processes. To protect the polymer from oxidation, it may be necessary to add antioxidants. Antioxidants were initially developed to protect polymers from thermal oxidative degradation, but certain common features of the mechanisms of thermal oxidative and radiation-induced oxidative degradation allow application of antioxidants in both cases.

If the term antioxidant refers merely to formal deceleration of radiation-induced oxidative degradation, then many radioprotectors inhibiting radiation-chemical formation of macroradicals $R^{\bullet}$ can be considered antioxidants. Specifically macroradicals $R^{\bullet}$ give rise to macroradicals $RO_2^{\bullet}$ responsible for further oxidation progress (see Section 5.3.1). Indeed, the ranges of antioxidants[40,41,78] and radioprotectors[21,22,24,37] appreciably overlap. In this connection, the antioxidative properties of fullerene h_{60} (without radiation), which compare well with standard antioxidants, are of interest. In accordance with kinetic studies in Zeinalov and Kossmehl,[79] the antioxidative action of fullerene is due to interception of $R^{\bullet}$ radicals.

If only substances affecting transformations of oxygen-containing macromolecules or macroradicals are classed with antioxidants, then the antioxidants can be subdivided into two groups: primary and secondary.[22,37] The primary antioxidants, represented by AH, terminate oxidation by hydrogen transfer to the peroxide radical with formation of hydroperoxide and weakly active radical $A^{\bullet}$:

$$RO_2^{\bullet} + AH \rightarrow ROOH + A^{\bullet}$$

Examples of primary antioxidants are aromatic amines and diamines, phenols and bisphenols (especially those with several bulky substituents), and also sterically shielded piperidine derivatives, nitroxyl radicals.[22,35] In this case, too, antioxidants and radioprotectors are similar, since the only difference is the nature of the species accepting the hydrogen atom: macroradical $RO_2^{\bullet}$ in the case of antioxidants and $R^{\bullet}$ in the case of radioprotectors.

Secondary antioxidants induce nonradical decomposition of hydroperoxides ROOH:

$$ROOH + AH \rightarrow RH + AH \text{ oxidation products}$$

The requirement for the nonradical character of hydroperoxide decomposition under the action of antioxidants is very significant, since radical decomposition causes propagation of the oxidation chain (see Section 5.3.1). A similar expression can be written for decomposition of peroxides:

$$RO_2^{\bullet} + AH \rightarrow RO^{\bullet} + AH \text{ oxidation products}$$

where $RO^{\bullet}$ is a relatively weak active radical. Typical antioxidants of this type are organic sulfides and disulfides, dialkyldithiocarbamates, organic phosphites, elemental sulfur, and complexes of various metals.[22,29,35] Secondary antioxidants are reductants; hydrogen donor radioprotectors are also reductants.

A number of binary and ternary synergistic mixtures of antioxidants that are efficient under irradiation conditions have been proposed. Mixtures of phenols with sulfides, phosphites, or amines appear useful. The structural fragments of antioxidants of both types can be combined in one molecule. Among such compounds are bisphenols with sulfide bridges, polyphenol phosphites, organotin derivatives of mercapto alcohols, etc. Instead of low-molecular-mass phenols, aromatic amines, or phosphites, less volatile oligomeric analogs or vinyl derivatives can be used since, in the course of synthesis or irradiation, they cross-link with the matrix.[22]

There are also less general mechanisms for some antioxidants such as direct reaction with molecular oxygen or binding in strong complexes of traces of variable-valence metals capable of catalyzing radiation-induced oxidation.[29] One of the mechanisms may be deactivation of excited singlet-state oxygen molecules. Quenching of singlet oxygen can be effected by tertiary amines, substituted phenols, metal chelates, and other substances.[23] A particular case of oxidation protection is protection from ozone formed under the action of radiation by using special additives called antiozonants; typical antiozonants are p-phenylenediamine and dihydroquinolines.[16] To prevent evaporation of stabilizers or their washout from the bulk of the polymer, stabilizers with molecular weight greater than 300 are usually used.[35]

As a rule, the discussion concerning the "limiting capacity" of a radioprotector is also relevant to antioxidants and substances combining both functions. There are some possibilities for regenerating antioxidants,[35,40] e.g., at the expense of a second, less active antioxidant present in the mixture.

In oxidative degradation, efficient stabilization of polymers is attained using the criterion

$$C_{In+O_2} \gg \Sigma\omega_I \tag{5.13}$$

where C_{In+O_2} is the rate of oxygen binding to form an inert product and $\Sigma\omega_I$ is the overall rate of polymer oxidation. As applied to a plate-shaped polymeric article, this criterion leads to the equation[35]

$$t = \frac{k'D_L^2[In]_0}{D_{O_2}[O_2]_0} + t' \tag{5.14}$$

where t is the time in which the article preserves the required set of properties in the presence of acceptor In; t' is the similar value for the article containing no acceptor ($t' < t$); k' is a constant; D_L is the plate thickness; D_{O_2} is the oxygen diffusion coefficient; and $[O_2]_0$ is the oxygen solubility. The above equality shows that stabilization of polymer films is short-term and inefficient. Efficient oxidation protection of polymers can be attained only within large pipes and parts.[35]

Another way of protection from oxygen and ozone is to coat the polymer surface with a protective layer such as wax.[16] Yet another approach is immersion in an inert gas atmosphere or in a vacuum.

It is known that ultraviolet light, that is, with a wavelength less than 400 nm, is the most active in photochemical degradation of polymers.[35] As already noted, the level of light fluxes attained in RLSs-T is too low for photostabilization to be a problem. Nevertheless, the mechanism of most photostabilizers consists in deactivating the photoexcited states of RH* or in scavenging macroradicals that R• formed from them.[25,78] Therefore, the ranges of radioprotectors and photostabilizers largely overlap. Two specific groups of substances used as photostabilizers for polymeric materials can be noted; however, they are not of interest as radioprotectors for tritiated polymers.[35,78] The first group includes inert diluents (e.g., carbon black). The second includes derivatives of o-oxybenzophenone acting by the light quantum absorption mechanism involving transition to the tautomeric quinoid state of a higher energy.

5.5.3 Enhancement of Radiation Stability of Plastic Scintillators

Plastic scintillators that are components of RLSs-T are subject to large doses of ionizing radiation. Improvement of the radiation stability of plastic scintillators[13-15,47,54,80,81] is an urgent problem in development of isotope sources of light and electric power.

In development of radiation-resistant plastic scintillators, proper choice of the plastic matrix and luminophore additives is very important. It is generally recognized that the plastic matrix is mainly responsible for distortion of optical properties through accumulation of color centers under the impact of ionizing radiation. Radiation-induced loss of light output of plastic scintillators is also due to reduction in the transmission of the polymer matrix in the wavelength range of the luminophore emission. Thus, one of the main lines in development of radiation-resistant plastic scintillators is the search for polymer matrices maximally resistant to radiation.[47,72,82-84]

The protection methods considered for polymeric materials in Sections 5.5.1 and 5.5.2 can be used for enhancement of the radiation stability of scintillators. However, some of these methods may distort optical characteristics of the material.

The role of the polymer matrix in radiolysis of plastic scintillators can be judged from the following experimental data. After irradiation of polystyrene and polyvinyltoluene scintillators of 0.03 Mgy, optical loss was 38%, whereas for phenyl-containing polyorganosiloxane scintillators, the loss did not exceed 1%, even after irradiation to a dose of 0.18 Mgy.[85] The first experience of using the polysiloxane polymer as a matrix for a scintillator[86] and subsequent studies[87,88] showed that radiation stability of plastic scintillators based on this material appreciably increases. Although polyorganosiloxane is best with respect to preservation of light output under irradiation, it has poor mechanical stability and luminophore additives have low solubility in it.

The effect of the polymer matrix on the radiation characteristics of plastic scintillators at various irradiation doses is well illustrated by the diagram in Figure 5.8 comparing the radiation stability of various materials used in scintillators.[89]

In controlling the radiation stability of plastic scintillators, it is important to properly choose the primary and secondary additives and their concentrations in the matrix; next some additional information is given on use of protective properties of luminophores in plastic scintillators.

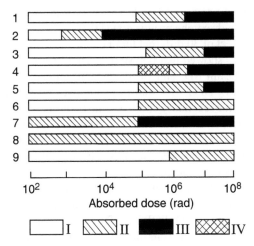

Figure 5.8   Radiation stability of some materials used in scintillators: (1, 2) poly(methyl methacrylate) at high and low dose rate, respectively; (3) polyvinyltoluene; (4, 5) plastic
scintillators based on polystyrene (PS + POPOP) and SCSN-38, respectively; (6)
GSI scintillation glass with addition of cerium; (7–9) scintillation crystals CsI(Tl),
$Bi_4Ge_3O_{12}$, and BaF_2, respectively. (I) Stable, (II) minor radiation damage, (III)
material is unsuitable, and (IV) no data. (From Stevenson, R.K. and Schoenbacher,
H., The feasibility of experiment of high luminosity of the large hadron collider,
Preprint CERN 88-02, Geneva, 1988.)

The majority of scintillation additives emit in the UV range; this radiation is
efficiently absorbed in the polymer matrix, which can additionally facilitate degradation of the scintillator. One of the ways to control radiation-induced loss of the
light output of plastic scintillators is to use additives emitting in the yellow-green
(> 500 nm) or even red (> 600 nm) regions of the spectrum, since at these wavelengths radiation damage for the polymer matrix is smaller.[82,90-92] Among the most
efficient spectrum shifters, as already noted, are 3-hydroxyflavone (3HF) and some
naphthyl derivatives (X31, X25). (Abbreviations accepted in the literature are used
for this discussion.)

Thus, a green scintillator with additions of DAT (di-*tert*-amyl-*p*-terphenyl, λ_{max}
= 375 nm) + 3HF (λ_{max} = 540 nm) at room temperature withstood a dose of 0.1
Mgy. The initial level of light output was restored in 25 days.[85] It should be noted
that annealing time can be reduced to 1 week by placing the sample in an oxygen
atmosphere.

Decreased radiation damage was also observed with OLIGO408 and K27 additives.[85] Light output decreased to 60% after uptake of 0.1 MGy but was restored to
90% in 13 days.

Vasil'chenko et al.[93] examined 70 plastic scintillators and made the following
conclusions on the stabilizing effect of luminophore additives. Such primary additives as 2,5-bis(4-styryl)oxadiazole (D11), 2-(4-dimethylaminophenyl)benzoxazole
(BO), and divinyl-*p*-terphenyl (DVpTP) ensure a high scintillation efficiency but do
not enhance radiation stability of the scintillator. The best stabilizing properties are

exhibited by the already mentioned additives 3HF, X25, X31, naphthalene (Nph), and 2-biphenyl-5-(2-hydroxyphenyl)thiazole (D12).

Analysis of the effect of 30 different luminophore additives on radiation stability of plastic scintillators[92] showed that polystyrene scintillators with the primary additives pTP, PPO, PBD, mPBD, and bPBD exhibit approximately equal radiation stability; introduction of BPO, BO, and PPD additives considerably decreases the radiation stability. As for secondary additives, the highest stabilizing effect is exhibited by X25, X31, 3HF, and M3HF. Thus, the light output of the scintillators (PS + 2.0% pTP + 0.025% X25) and (PS + 2.0% pTP + 0.025% X31) after irradiation with ^{137}Cs 3-quanta is 53 and 47%, respectively, whereas that of the scintillator containing no additives is as low as 23%.

Figure 5.9 shows the stabilizing effect of the 3HF additive on radiolysis of plastic scintillators based on polystyrene and polyvinyltoluene.[94]

The stabilizing effect of PPO, POPOP, pTP, and 1,1,3-trimethyl-3-phenylindane (TMPI) was also noted for plastic scintillators based on poly(methyl methacrylate).[95]

In enhancement of the radiation stability of plastic scintillators, it is important to optimize concentrations of primary and secondary luminophore additives. For example, it was found that for poly(methyl methacrylate) scintillators, the content of paramagnetic species decreases with increasing concentration of TMPI additive.[95] The highest radiation stability of this scintillator, with the high luminescence intensity preserved, is attained at a TMPI concentration of 0.98 M.

An increase in the concentration of a primary additive up to its solubility limit in a polymer enhances the radiation stability of plastic scintillators, which is probably due to substitution of the polymer by a more radiation-resistant component.[96] However, as the concentration of such additives is increased, the optical transmission of

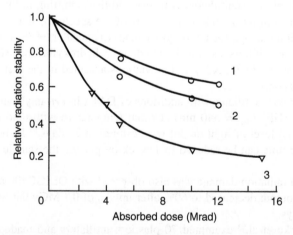

Figure 5.9 Effect of 3-hydroxyflavone (3HF) additive on the relative radiation stability of the plastic scintillators (1) PS + pTP + 3HF, (2) PVT + pTP + 3HF, and (3) PVT + pTP at various irradiation doses. (From Andryushchenko, L.A. and Grinev, B.V., Organosilicon materials for scintillator detectors of ionizing radiation, *Prib. Tekhn. Eksper.*, 4, 5, 1998 [in Russian].)

the initial samples decreases because the limit of solubility of the additives in the polymer matrix is achieved.

Zalyubovskii et al. studied how the ratio of the light outputs of irradiated L_L and nonirradiated $(L_L)_0$ samples of plastic scintillators depends on the content of various secondary additives and found that the highest radiation stability is attained at a concentration of secondary additives within 0.01 to 0.03%.[96] Thus, for polystyrene scintillators, the highest radiation stability is attained at a POPOP concentration of 0.02% $[L_L/(L_L)_0 = 57\%]$. POPOP concentrations above 0.1% show noticeably lower radiation stability $[L_L/(L_L)_0 = 44\%]$.

In some cases radiation stability of plastic scintillators can be enhanced by chemical modification of additives. This method was applied to 1,3,5-triaryl-Δ'^2-pyrazolines used in scintillation as activators and spectrum shifters. These additives exhibit high light output, low concentration quenching, and high solubility in the polymer matrix. However, their photochemical and radiation stability is poor. To overcome this drawback, Malinovskaya et al.[97] proposed a procedure for modification of pyrazolines by introducing fluorinated substituents, which allowed substantial enhancement of the radiation stability of plastic scintillators using these luminophore additives.

Since luminophore additives with energy transfer function exhibit protective properties, special antirad additives are seldom used in plastic scintillators; however, their use can be beneficial. Appreciable enhancement of radiation stability is seen in the standard scintillating formulation PS + 2% pTP + 0.2% POPOP by adding an organometallic compound designated CT-24. In Figure 5.10, $L_L/(L_L)_0$ is plotted against the content of the CT-24 stabilizer.[96] It is seen that the highest protecting effect is attained at a CT-24 content of 0.01%.

Gunder et al.[98] and Gunder and Koba[99] suggest that the nature of the polymer matrix can affect the stability of luminescent additives. In their subsequent study,[72] using IR and UV spectroscopy, they measured the degradation constants of luminescent additives (p-terphenyl and POPOP), holding concentration to 1 wt% and administering the same dose of 3- and electron irradiation using polystyrene, polyvinyltoluene, and polyvinylxylene. They found that the extent of radiation damage

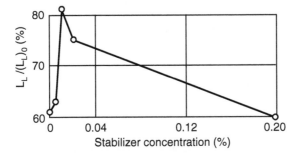

Figure. 5.10   Ratio of the light outputs $L_L/(L_L)_0$ of irradiated and nonirradiated scintillator as a function of concentration C of CT-24 stabilizer in polystyrene scintillator under irradiation at 0.028 Mgy. (From Zalyubovskii, I.I. et al., Ways to increase the radiation stability of polystyrene-based scintillators, *Prib. Tekhn. Eksper.*, 5, 76, 1995 [in Russian].)

of the additives was practically equal in the three polymer matrices. Thus, at a dose of 0.1 MGy in the three polymers, p-terphenyl degrades to 15% and POPOP to 35%. This result may reflect the very similar chemical nature of the tested polymers. At the same time, it was shown that the radiation stability of luminescent additives depends on the presence in the polymer matrix of secondary additives (spectrum shifters). Thus, the degradation constant of p-terphenyl in polystyrene with addition of POPOP considerably decreases and radiation-induced decomposition of the spectrum shifter grows several times. This effect may be due to transfer of electronic excitation energy from p-terphenyl to POPOP molecules, thus increasing stability of the primary additive and decreasing that of the secondary additive.[98]

A similar effect must be manifested in other mixed organic scintillators, since it is well known that, in such systems, the excitation energy usually migrates to the luminophore component with the longer-wave emission, e.g., from naphthalene to anthracene.[67]

An efficient way to enhance the radiation stability of plastic scintillators is introduction of diffusion intensifiers (enhancers), i.e., of substances enhancing the mobility of radiolysis products, thus accelerating their removal. Used as such substances are methylnaphthalene, 1-isopropylnaphthalene, m-diphenoxybenzene, polyphenyl oxide, dibutyl phthalate, etc.[93,96] Polyphenyl oxide appeared to be one of the best diffusion intensifiers. The ratios of the light output $[L_L/(L_L)_0]$ and transmission (i'/i'_0) vs. concentration of this additive are plotted in Figure 5.11.[96] The figure shows that, with increasing concentration of polyphenyl oxide to 20%, both $[L_L/(L_L)_0]$ and i'/i'_0 increase; i.e., radiation stability of the scintillator is appreciably enhanced. Further increase of the concentration of the diffusion enhancer results in decreased transmission of the initial sample, due to diffusion scattering, and in worse mechanical properties of the formulation.

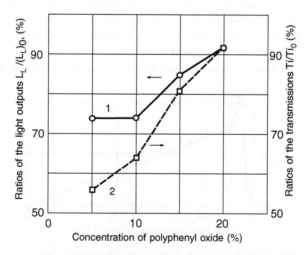

Figure 5.11   Ratios of (1) light outputs $[L_L/(L_L)_0]$ and (2) transmissions (i'/i'_0) as functions of the concentration of polyphenyl oxide in polystyrene scintillator under irradiation to 0.028 MGy. (From Zalyubovskii, I.I. et al., Ways to increase the radiation stability of polystyrene-based scintillators, *Prib. Tekhn. Eksper.*, 5, 76, 1995 [in Russian].)

Addition of 1-methylnaphthalene (1MN) enhances radiation stability so that the light output of the scintillator PS + 2% pTP + 0.05% POPOP + 20% 1MN decreases by no more than 10% after irradiation to 0.09 MGy. However, in high concentration, this additive appreciably softens the scintillator.

The most efficient enhancement of the radiation stability of plastic scintillators is attained by simultaneously introducing stabilizers and diffusion enhancers.[96] The general approach to enhancement of radiation stability of plastic scintillators is their use under oxygen-free conditions (in an inert atmosphere or in a vacuum). Below is only one of numerous examples of the effect of surrounding atmosphere on radiation stability of plastic scintillators: the radiation stability of the scintillator PS + 2% pTP + 0.02% POPOP + 0.01% CT-24 + 20% UD7 in a nitrogen atmosphere is, by a factor of approximately four, higher than in air.[96]

Bol'bit et al.[100] and Milinchuk et al.[101] suggested use of microporous forms of plastic scintillators. They showed that, with a large number of scintillator samples of diverse compositions, radiation stability of pore-structure forms of scintillators tends to increase several times compared to standard samples. These authors believe that porous scintillators owe their enhanced radiation stability to the action of oxygen that easily penetrates via pores to the macroradicals arising from irradiation and binds them. This prevents macroradicals from further participation in polymer degradation and luminescence quenching. This explanation seems somewhat unexpected, as it is inconsistent with traditional views on the role played by oxygen in origination and branching of chain reactions in radiation-induced degradation of polymers (see Section 5.3.1).

One condition necessary to ensure high radiation stability of plastic scintillators is high chemical purity of the polymer matrices and other ingredients.

In cases when plastic scintillators are used periodically, their radiation resource can be increased by alternating irradiation runs and pauses during which the system restores its properties.[102] It was found that during pauses, the lost scintillation properties are spontaneously restored, and this alternating mode allows the life of a plastic scintillator to be prolonged several times. Unfortunately, such an approach is not applicable to RLSs-T.

5.5.4   Deuteration Increases Radiation Stability of Polymeric Matrices

Incorporation of heavy hydrogen isotopes into molecules is known to change the physicochemical properties of compounds. Along with application of such isotopic effects in various branches of chemistry and physics, the heavy atom effect may find application in radioluminescent light sources by increasing radiation stability of polymeric matrices.

Ionization energy increases as it moves from protium to deuterium. The same is true for the binding energy of hydrogen isotope atoms in the hydrocarbon chain: for C–H and C–^{2}H, it is equal to 81.0 and 82.1 kcal/mol, respectively.[5] This accounts for a lower concentration of primary reactive species in radiolysis products of deuterated polymers. The radiation-chemical yield of paramagnetic particles on irradiating deuterated polyethylene and deuterated polypropylene is 1.5 to 2 times lower than for these polymers in the ordinary form.[28,103] The radiation-chemical yield

of gaseous hydrogen upon irradiation of [$^2H_8$]polystyrene is half that of the nondeu-
terated polymer.[104] Another effect of deuteration on polymers is on the structure of
the primary reactive species formed. For example, irradiation of ordinary polypropylene gives predominantly allyl radicals ~hd$_2$h·(hd$_3$)hd=h(hd$_3$)hd$_2$~, and irradiation
of deuterated polypropylene under identical conditions gives predominantly alkyl
radicals ~h^{2}d$_2$ h·(C^{2}d$_3$)C^{2}d$_2$~.[103,105] The difference in the structure of the radicals can
affect radiolysis consequences for ordinary and deuterated isotopic forms of polypropylene.

Of no less importance is the kinetic factor (kinetic isotopic effect), i.e., the rate
difference of the interaction of the primary reactive species of various isotopic
composition with the surrounding molecules as well as of ordinary species with
isotope-modified molecules.

An important role in radiation-induced degradation of polymers is played by the
reactions of atomic hydrogen yielded by breaking the C–H bond. These hydrogen
atoms readily react with surrounding macromolecules. On changing from the RH +
H to the R^{2}H + H reaction, a very large kinetic isotopic effect occurs. The bond is
broken up to 260 times as quickly in the protium case[28]; a probable reason is
hydrogen atom tunneling.[106]

Along with enhancing radiation stability, deuteration of polymers can be
employed for increasing their transparency.[107] As known, one factor responsible for
optical losses in polymers is stretching vibrations of the C–H bond. Replacement
of protium by deuterium changes the frequencies of these vibrations and, consequently, increases transparency.

Deuteration of a polymer (PMMA) for the purpose of increasing its transparency
was reported in 1977,[108] and later this effect was used for high-transparency polymers
intended for fiber optics application.[109-113]

Optical losses in polymer fibers are typically determined by the equation

$$T'' = \frac{10}{D_L} \lg \frac{I_L}{(I_L)_0} \qquad (5.15)$$

where i'' is the optical losses in dB/km; D_L is the light guide length in km; and $(I_L)_0$
and I_L are intensities of the incident and passed light, respectively.

In [$^2H_5$]PMMA, the optical losses decrease from 55 to 41 dB/km.[107] With increasing degree of deuteration ([$^2H_8$]PMMA), the polymer became more transparent.[110]
Light absorption also decreased significantly for deuterated polystyrene[109] and ther-
mostable polymer 2-fluoromethyl methacrylate.[111] Deuteration not only increases
the transparency of polymers but also shifts the transparency gap to greater wavelengths. For example, [$^2H_8$]PMMA exhibits high light transmission at 690 and 790
nm as opposed to 522 and 570 nm for ordinary PMMA.[108]

Synthesis of polymers combining deuterium and tritium labels is a fairly complex
task. Takagi et al. prepared [$^2H,^3H$]polystyrene by the Wilzbach method with a
specific activity of 150 Ci/g.[114] Despite synthesis difficulties, potential advantages
place such materials among the most promising for RLSs-T application. Deuteration
can be especially efficient in manufacturing tritium-containing light sources with a

plastic scintillator employed for converting the energy of beta particles emitted by the external source, such as labeled compounds dissolved in the polymer, gaseous tritium, krypton-85, etc. One can expect that deuterated luminophore additives will bring about improvements in radiation stability and optical properties.

REFERENCES

1. Evans, E., *Tritium and Its Compounds*, 2nd ed., Butterworths, London, 1974.
2. Bayly, R.J. and Weigel, H., Self-decomposition of compounds labeled with radioactive isotopes, *Nature*, 188, 384, 1960.
3. Tkachenko, S.E., Karpov, N.A., and Fedoseev, V.M., Autoradiolysis of labeled organic compounds, *Radiokhimiya*, 42, 123, 2000 [Russian language].
4. Belovodskii, L.F., Gaevoi, V.K., and Grishmanovskii, V.I., *Tritium,* Energoatomizdat, Moscow, 1985 [Russian language].
5. Gurvich, L.V. et al., *Chemical Bond Disruption Energies, Ionization Potentials, and Electron Affinity*, Nauka, Moscow, 1974 [Russian language].
6. Nefedov, V.D., Sinotova, E.N., and Toropova, M.A., *Chemical Consequences of Radioactive Decay*, Energoatomizdat, Moscow, 1981 [Russian language].
7. Cacace, F. and Speranza, M., *Techniques for the Study of Ion Molecule Reactions*, John Wiley & Sons, New York, 1988.
8. Speranza, M., Tritium for generation of carbocations, *Chem. Rev.*, 93, 2933, 1993.
9. Bethell, D. and Gold, V., *Carbonium Ions. An Introduction*, Academic Press, London-New York, 1967.
10. Khirnyi, Yu.M., Berezhko, P.G., and Kochemasova, L.N., On certain regularities in the ^{3}d| behavior in solid phases, *Dokl. Akad. Nauk SSSR*, 255, 1187, 1980 [Russian language].
11. Tyutnev, A.P. et al., Electron transport in polymers, *Vysokomol. Soedin.*, 40, 821, 1998 [Russian language].
12. Tolbert, B.M., Radiation decomposition of labeled compounds, *Adv. Tracer Methodol.*, 1, 64, 1963.
13. Vasil'chenko, V.G. and Solov'ev, A.S., Plastic scintillators with antirad additions and enhanced light output, *Prib. Tekhn. Eksp.*, 4, 46, 1998 [Russian language].
14. Protopopov, Yu.M. and Vasil'chenko, V.G., Radiation damage in plastic scintillators and optical fibers, *Nucl. Instr. Methods Phys. Res.*, 95B, 496, 1995.
15. Zorn, C., Studies in the radiation resistance of plastic scintillators: review and prospects, *IEEE Trans. Nucl. Sci.*, 37, 504, 1990.
16. Milinchuk, V.K., Klinshpont, E.R., and Tupikov, V.I., *Basics of Radiation Resistance of Organic Materials*, Energoatomizdat, Moscow, 1994 [Russian language].
17. Ivanov, V.S., *Radiation Chemistry of Polymers*, Khimiya, Leningrad, 1988 [Russian language].
18. Pankratova, L.N., Bugaenko, L.T., and Revina, A.A,. Influence of aromatic protectors on radiolysis of polyorganosiloxanes, *Khim. Vys. Energ.*, 34, 19, 2000 [Russian language].
19. Nikitina, T.S., Zhuravskaya, E.V., and Kuz'minskii, A.S., *Impact of Ionizing Radiations on Polymers*, GNTIKhL, Moscow, 1959 [Russian language].
20. Pikaev, A.K., *Modern Radiation Chemistry. Solids and Polymers: Applied Aspects*, Nauka, Moscow, 1987 [Russian language].

21. Clough, R.L., Radiation-resistant polymers, in *Encyclopedia of Polymer Science and Engineering*, 2nd ed., John Wiley & Sons, New York, 13, 667, 1988.
22. Milinchuk, V.K. and Tupikov, V.I., Eds., *Radiation Stability of Organic Materials. Handbook*, Energoatomizdat, Moscow, 1986 [English translation: *Organic Radiation Chemistry Handbook,* Milinchuk, V.K. and Tupikov, V.I., Eds., Ellis Horwood, Chichester, 1989].
23. Plotnikov, V.G. and Efimov, A.A., *Photostabilizers of Polymers. Photophysical Properties and Stabilizing Activity,* NIITEKhIM, Moscow, 1989 [Russian language].
24. Laricheva, V.P. and Suminov, V.I., *Problems of Radiation Protection of Polymers*, NIITEKhIM, Moscow, 1985 [Russian language].
25. Ränby, B. and Rabek, J.F., *Photodegradation, Photo-oxidation and Photostabilization of Polymers*, John Wiley & Sons, London, 1975.
26. Bolt, R.O. and Carrol, J.G., Eds., *Radiation Effects on Organic Materials,* Academic Press, New York, 1963.
27. Vekshin, N.A., *Excitation Transfer in Macromolecules*, VINITI, Moscow, 1989 [Russian language].
28. Milinchuk, V.K., Klinshpont, E.R., and Pshezhetsky, S.Ya., *Macroradicals*, Khimiya, Moscow, 1980 [Russian language].
29. Makhlis, F.A., *Radiation Physics and Chemistry of Polymers,* Atomisdat, Moscow, 1972 [English translation: Makhlis, F.A., *Radiation Physics and Chemistry of Polymers*, John Wiley & Sons, New York, 1975].
30. Zezin, A.A., Fel'dman, V.I., and Sukhov, F.F., Free radical formation mechanism and role of ionic processes in polystyrene radiolysis, *Vysokomol. Soedin., Ser. A-B*, 36, 925, 1994 [Russian language].
31. Fel'dman, V.I., Basics of selectivity of radiation-chemical processes in polymers, *Ross. Khim. Zh.*, 40, 90, 1996 [Russian language].
32. Chapiro, A., General consideration of the radiation chemistry of polymers, *Nucl. Instr. Methods Phys. Res.*, 105B, 5, 1995.
33. O'Donnel, J.H., Chemistry of radiation degradation of polymers, *ACS Symp. Ser.*, 475, 402, 1991.
34. Kalyazin, E.P. and Bugaenko, L.T., Inorganic and organic radiation chemistry: state and problems, *Zh. D. I. Mendeleev Vses. Khim. O-va*, 35, 551, 1990 [Russian language].
35. Gladyshev, G.P., Vasnetsova, O.A., and Mashukov, N.I., On polymer degradation and stabilization mechanisms, *Zh. D. I. Mendeleev Vses. Khim. O-va*, 35, 575, 1990 [Russian language].
36. Charlesby, A., *Atomic Radiation and Polymers*, Pergamon Press, New York, 1959.
37. Romantsev, M.F., *Chemical Protection of Organic Systems Against Ionizing Radiation*, Atomizdat, Moscow, 1978 [Russian language].
38. Swallow, A.J., *Radiation Chemistry of Organic Compounds*, Pergamon Press, Oxford, 1960.
39. Iskakov, L.I., *Radiation-Induced Gas Evolution from Polymeric Materials*, NIITEKhIM, Moscow, 1979 [Russian language].
40. Denisov, E.T., *Oxidation and Degradation of Carbon Chain Polymers*, Khimiya, Leningrad, 1990 [Russian language].
41. Shlyapnikov, Yu.A., Kiryushkin, S.G., and Mar'in, A.P., *Antioxidative Stabilization of Polymers*, Khimiya, Moscow, 1986 [English translation: Shlyapnikov, Yu.A., Kiryushkin, S.G., and Mar'in, A.P., *Antioxidative Stabilization of Polymers*, Taylor and Francis, London, 1996].

42. Laghari, J.R. and Hammond, A.H., A brief survey of radiation effects on polymer dielectrics, *IEEE Trans. Nucl. Sci.*, 37, 1076 ,1990.

43. Renschler, C.L. et al., Solid state radioluminescent lighting, *Radiat. Phys. Chem.*, 44, 629, 1994.

44. Mullins, D.F., Krasznai, J.P., and Mueller, D.A., Development of organic tritium light technology at Ontario Hydro, *Fusion Technology*, 21, 312, 1992.

45. Buontempo, S. et al., Influence of dissolved gas and temperature on light yield of new liquid scintillators, *Nucl. Instr. Methods Phys. Res.*, A425, 492, 1999.

46. Harrah, L.A. and Powell R.C., Dose rate saturation in plastic scintillators, in *Organic Scintillators and Liquid Scintillation Counting*, Horrocks, D.L., Ed., Academic Press, New York, 1976, 265.

47. Bross, A.D. and Pla-Dalmau A., Radiation damage of plastic scintillators, *IEEE Trans. Nucl. Sci.*, 39, 1199, 1992.

48. Decker, Ch., Mayo, F.R., and Richardson, H., Aging and degradation of polyolefins. III. Polyethylene and ethylene-propylene copolymers, *J. Polym. Sci., Polym. Chem. Ed.*, 11, 2879, 1973.

49. Clough, R.L. et al., "Accelerated-aging tests for predicting radiation degradation of organic materials," *Nucl. Safety*, 25, 238, 1984.

50. Shaw, M.T. and Yong–Ming, L., Modeling of thermal and radioactive aging of polymeric cable materials, *IEEE Trans. Nucl. Sci.*, 2, 638, 1996.

51. Akimov, Yu.K., *Scintillation Counters in High-Energy Physics*, Izd. Mosk. Gosud. Univ., Moscow, 1965 [Russian language].

52. Boyd, A.W., The radiolysis and pyrolysis of organic coolants, *J. Nucl. Mat.*, 9, 1, 1963.

53. Brooks, F.D., Development of organic scintillators, *Nucl. Instr. Meth.*, 162, 477, 1979.

54. Zorn, K. et al., Progress in the design of a radiation-hard plastic scintillator, *IEEE Trans. Nucl. Sci.*, 38, 194, 1991.

55. Yoshimura, Y. et al., Plastic scintillator produced by the injection-molding technique, *Nucl. Instr. Meth. Phys. Res.*, A406, 435, 1998.

56. Savinskii, A.K., Popov, V.I., and Kulyamin, V.A., LET Spectra and Quality Coefficients of Incorporated Radionuclides: Handbook, Energoatomizdat, Moscow, 1986 [Russian language].

57. McGlyn, S.P., Azumi, T., and Kinoshita, M., *Molecular Spectroscopy of the Triplet State*, Prentice–Hall, New York, 1969.

58. Ermolaev, V.L. et al., *Nonradiative Transfer of Electronic Excitation Energy*, Nauka, Leningrad, 1977 [Russian language].

59. Stopping powers for electrons and positrons, *ICRU Report 37*, Woodmont, 1984.

60. Henly, E. and Johnson, E., *The Chemistry and Physics of High Energy Reactions*, University Press, New York, 1969.

61. Mozumder, A. and Magee, J.L., Model of tracks of ionizing radiations for radical reaction mechanisms, *Radiat. Res.*, 28, 203, 1966.

62. Laustriat, G., The luminescence decay of organic scintillators, *Mol. Crystals*, 4, 127, 1968.

63. Bubnov, L.Ya. and Frankevich, E.L., Measurement of path lengths of kiloelectron-volt electrons in pyrene, *Khim. Vys. Energ.*, 19, 36, 1985 [Russian language].

64. Bubnov, L.Ya. and Frankevich, E.L., Changes in the energy exchange with the medium for kiloelectron-volt electron in thin film volume, *Khim. Vys. Energ.*, 30, 101, 1996 [Russian language].

65. Saraeva, V.V., Radiation-chemical liquid-phase oxidation of hydrocarbons, *Modern Problems of Physical Chemistry*, Topchieva, KV., Ed., Izd. Mosk. Gos. Univ., Moscow, 8, 367, 1975 [Russian language].

66. Renschler, C.L., Clough, R.L., and Sheppodd, T.J., Demonstration of completely organic, optically clear radioluminescent light, *J. Appl. Phys.*, 66, 4542, 1989.
67. Krasovitsky, B.M. and Bolotin, B.M., *Organic Luminophores*, 2nd ed., Khimiya, Leningrad, 1984 [Russian language].
68. Salmon, G.A., Some studies on the formation of excited states of aromatic solutes in hydrocarbons and other solvents, *Intern. J. Radiat. Phys. Chem.*, 8, 13, 1976.
69. Alfimov, M.V., Triplet states in radiolysis of solid solutions, *Int. J. Radiat. Phys. Chem.*, 8, 43, 1976.
70. Burton, M. and Magee, J.L., Eds, *Advances in Radiation Chemistry*, v. 2, John Wiley & Sons, New York, 1974.
71. Foldiak, G., *Radiation Chemistry of Hydrocarbones*, Akademiai Kiado, Budapest, 1981.
72. Koba, V.S. and Gunder, O.A., Radiation stability of the characteristics of a series of luminescent additions in various polymer bases, in *Monocrystals and Scintillators*, Collection of Works, VNII Monokristallov, Khar'kov, 1977, 102 [Russian language].
73. Witt, E., The effect of polymer composition on radiation-induced cross-linking, *J. Polym. Sci.*, 41, 507, 1959.
74. Tikhomirova, N.S. and Serenkova, V.I., *Polymer Protection Against Radiolysis*, NIITEKhIM, Moscow, 1977.
75. Wundrich, K., Effect of temperature and physical state on the inhibition by additives of radiation-induced degradation of poly(methyl methacrylate), *J. Polym. Sci., Polym. Phys. Ed.*, 12, 201, 1974.
76. Romantseva, O.N., Kirillova, E.I., and Nedotkina, K.S., *Stabilization of Polystyrene Plastics*, NIITEKhIM, Moscow, 1979.
77. Klinshpont, E.R. and Milinchuk, V.K., Polymer materials under combined action of ionizing radiation and other physical factors, *Zh. D. I. Mendeleev Vses. Khim. O-va*, 35, 618, 1990 [Russian language].
78. Voigt, J., *Die Stabilisierung der Kunstoffe gegen Licht und Warme*, Springer, Berlin, 1966.
79. Zeinalov, E.B. and Kossmehl, G., Fullerene C_{60} as an antioxidant for polymers, *Polym. Degrad. Stab.*, 71, 197, 2001.
80. Harmon, J.P., Gaynor, J.F., and Taylor, A.S., Approaches to optimize scintillator polymers for optical radiation hardness, *Radiat. Phys. Chem.*, 41, 153, 1993.
81. Markley, F. et al., Development of radiation hard scintillators, *Radiat. Phys. Chem.*, 41, 135, 1993.
82. Zorn, K. et al., Progress in the design of a radiation-hard plastic scintillator, *IEEE Trans. Nucl. Sci.*, 38, 194, 1991.
83. Ebdon, J.R. et al., Luminescence studies of polymer matrices. II. On the phosphorescence characteristics of 2-benzoylnaphtalene dispersed in various acrylic polymers, *High Performance Polymer*, 11, 49, 1999.
84. Britvich, G.I., Vasil'chikov, V.G., and Lapshin, V.G., Radiation stability of plastic scintillators, *Prib. Tekhn. Eksper.*, 1, 75, 1994 [Russian language].
85. Akimov, Yu.K., Plastic scintillator-based detectors of nuclear radiation, *Fiz. El. Chast. Atom. Yadra*, 25, 496, 1994 [Russian language].
86. Bowen, M. et al., A new radiation-hard plastic scintillator, *Nucl. Instr. Meth. Phys. Res.*, A276, 391, 1989.
87. Harmon, J. et al., Effects of ionizing radiation on the optical properties of polymers, in *Irradiation of Polymers: Fundamental and Technological Applications*, Clough, R.L. and Shalaby, S.W., Eds., Am. Chem. Soc., Washington, D.C., 1996, chap. 23.

88. Andryushchenko, L.A. et al., Organic scintillator-based detectors with enhanced performance characteristics, *Prib. Tekhn. Eksper.*, 6, 30, 1999 [Russian language].

89. Stevenson, R.K. and Schoenbacher, H., The feasibility of experiment of high luminosity of the large hadron collider, *Preprint CERN 88-02*, Geneva, 1988.

90. Rabin, N.V., Electron-photon calorimeters. Properties of the materials of calorimeter detectors, *Prib. Tekhn. Eksper.*, 6, 8, 1992 [Russian language].

91. Ambrosio, C.D. et al., Organic scintillators with large Stokes shifts dissolved in polystyrene, *Nucl. Instr. Meth.*, A307, 430, 1991.

92. Britvich, G.I. et al., Study of the radiation stability of polystyrene-based scintillators, *Prib. Tekhn. Eksper.*, 1, 109, 1993 [Russian language].

93. Vasil'chenko, V.G. et al., Study of radiation resistant plastic scintillators, *Prib. Tekhn. Eksper.*, 5, 85, 1995 [Russian language].

94. Andryushchenko, L.A. and Grinev, B.V., Organosilicon materials for scintillator detectors of ionizing radiation, *Prib. Tekhn. Eksper.*, 4, 5, 1998 [Russian language].

95. Leplyanin, G.V. et al., Influence of selected luminophores on the radiation stability of poly(methyl methacrylate), *Khim. Vys. Energ.*, 30, 424, 1966 [Russian language].

96. Zalyubovskii, I.I. et al., Ways to increase the radiation stability of polystyrene-based scintillators, *Prib. Tekhn. Eksper.*, 5, 76, 1995 [Russian language].

97. Malinovskaya, S.A. et al., Plastic scintillators containing pyrazolines with difluoromethylsulfonyl and trifluoromethyl groups, in *Monocrystals and Scintillators*, Collection of Works, VNII Monokristallov, Khar'kov, 1977, 94 [Russian language].

98. Gunder, O.A., Koba, V.S., and Ekkerman, V.M., Study of the radiation stability of polystyrene-based plastic scintillators, *Radiokhimiya*, 14, 913, 1972 [Russian language].

99. Gunder, O.A. and Koba, V.S., Photoresistance of a series of heterocylic and polyphenyl compounds in polymer bases, in *Scintillators and Organic Luminophores*, Collection of Works, VNII Monokristalllov, Khar'kov, 1972, 112 [Russian language].

100. Bol'bit, N.M. et al., Novel plastic scintillator with enhanced radiation stability, *Prib. Tekhn. Exper.*, 4, 31, 2000 [Russian language].

101. Milinchuk, V.K. et al., Patent WO9944378A, Russia, 1999.

102. Zaitsev, L.N. Problems of increasing the radiation service life of scintillator detectors for proton and ion colliders, *Fiz. El. Chast. Atom. Yadra*, 30, 1292, 1999 [Russian language].

103. Milinchuk, V.K., Klinshpont, E.P., and Vasilenko V.V., Effect of isotopic composition on free-radical reactions in polyolefins, *J. Polym. Sci., Polym. Chem. Ed.*, 14, 1419, 1976.

104. Wall, L.A. and Brown, D.W., γ-Irradiation of poly(methyl methacrylate) and polystyrene, *J. Phys. Chem.*, 61, 129, 1957.

105. Klinshpont, E.R., Milinchuk, V.K., and Pshezhetskii, S.Ya., Free radicals in irradiated deuterated polyethylene and polypropylene, *Vysokomol. Soedin.*, W15, p. 1963, 1973 [Russian language].

106. Dubinskaya, A.M. and Yusubov, N.N., Kinetics of low-temperature reactions of hydrogen atoms with polyisobutylene, *Izv. Akad. Nauk SSSR, Ser. Khim.*, 7, 1484, 1974 [Russian language].

107. Barashkov, N.N. et al., *Polymers in Fiber Optics: Review*, AO NII TEKhIM, Moscow, 1995 [Russian language].

108. Schleinitz, H.M., Ductile plastic optical fibers with improved visible and near infrared transmission, *Int. Wire and Cable Symp.*, 26, 352, 1977.

109. Kaino, T., Fujiki , M., and Nara, S., Low-loss polystyrene core-optical fibers,"*Polymer Prepr.* , 30, 544, 1981.

110. Kaino, T., Jinguji , K., and Nara, S., Low-loss poly(methyl methacrylate-d$_8$) core optical fibers, *Appl. Phys. Lett.*, 47, 567, 1983.
111. Baran, A.I. and Levin, V.M., *Physicochemical Basics of Synthesis and Processing of Polymers,* Izd. Gor'k. Univ., Gor'kii, 1984 [Russian language].
112. Barashkov, N.N. and Sakhno, T.V., *Optically Transparent Polymers and Materials Thereof,* Khimiya, Moscow, 1992 [Russian language].
113. Levin, M.B., Starostina, G.P., and Cherkasov, A.S., Efficiency of solar concentrators based on luminescent silicone glasses, *Zh. Prikl. Spektr.*, 46, 432, 1987 [Russian language].
114. Takagi, M. et al., Fabrication of deuterated-tritiated polystyrene shells for laser fusion experiments by means of an isotope exchange reaction, *J. Vac. Sci. Technol.*, A10, 239, 1992.

CHAPTER **6**

Silicon Voltaics for Direct and Indirect Radioactive Decay Energy Conversion into Electricity

V.B. Shuman

CONTENTS

6.1   FORMULATION OF THE PROBLEM

Semiconductor p–n junctions were first suggested in the 1950s for conversion of nuclear radiation into electricity.[1,2] Converters based on germanium and silicon p–n junctions were fabricated and irradiated with β- and γ-radiation from Sr[90]-Y[90] isotopes. The conversion efficiency was about 0.5% due to suboptimal isotope choice and poor semiconductor quality. Great progress has been achieved in improving the quality of silicon, and new classes of semiconductor materials have emerged with wider band gaps than silicon (amorphous silicon, W_3Y_5, W_2Y_6, etc.). Maximum conversion efficiency of nuclear radiation is expected with converters of high-band-gap materials. However, in view of the low cost of the starting material, wide use of silicon devices, and advanced development of their technology, the use of converters based on silicon technology should be estimated theoretically and experimentally. It is possible that mass production of nuclear batteries might favor less efficient but inexpensive and ecologically compatible silicon converters.

Operating under diverse environmental conditions, self-contained power sources with a very long service life would be useful for numerous sensor, communication, and control applications. Power source requirements can be defined in terms of average current consumption, voltage, and stored energy or charge, absolute and per unit volume. A preliminary estimate of such requirements provides the following specifications[3]:

Current:	1 to 10 µA
Voltage:	1.6 to 7 V
Power:	15 to 500 µW

Figure 6.1 shows required parameter values or their ranges for various applications, most of which can be provided with a 100 to 200 µW power unit.

Estimates of the specific power of existing chemical galvanic cells are 10^2 to 10^4 µW/cm³, with specific stored energy from 10^{-2} to a few Wh/cm³. Operational life is seldom longer than 2 years for chemical batteries.

As shown in the preceding chapters, tritium is a promising isotope for nuclear batteries due to ecological compatibility, low toxicity, relatively low cost, and suitable half-life. Publications on beta energy converters fabricated of semiconductors contain theoretical predictions of arrays for conversion of beta radiation from tritium that would have specific power of 10 to 100 µW/cm³ and specific energy[4] of 10

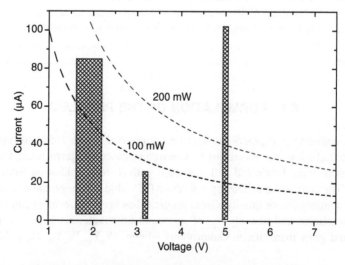

Figure 6.1 Application areas and corresponding parameters of desirable power cells; 100 and 200 µW. (From Trace Photonics, unpublished data, 2000.)

Wh/cm³. Semiconductor-based tritium beta converters are inferior to galvanic cells with regard to specific (instantaneous) power, but the greater half-life of tritium (12 years) results in greater lifetime and overall specific energy.

If the projected capacity of tritium beta converters is compared with demand for distributed power, the greater share of this demand (P ≤ 100 to 200 μW) could be satisfied by semiconductor-based tritium batteries with a volume < 20 cm³. Therefore, the development of semiconductor nuclear batteries is an important task.

Two methods of converting the power of beta radiation from tritium are known: direct and indirect. Direct conversion absorbs electrons emitted by tritium in semiconductor material, creating electron–hole pairs that are then separated by a p–n junction. Indirect conversion uses electrons to excite phosphors. These emit light in a narrow spectral range, which is absorbed by a semiconductor photovoltaic converter.

6.2 SILICON CELLS FOR DIRECT CONVERSION OF TRITIUM

The feasibility and efficiency of single crystal silicon cells for direct conversion of tritium beta radiation will now be determined. Direct conversion efficiency is determined by the energy of electron–hole pair formation and the band-gap width ($\backslash_g$). Electron–hole pair formation energy determines short-circuit current and the band gap determines open-circuit voltage. The basic energy conversion parameter is efficiency η. In the case of betavoltaic cells, the magnitude of η is defined by the expression

$$\eta = \frac{I_{sc} U_{oc} FF}{P_\beta} \tag{6.1}$$

where

I_{sc} = short-circuit current of betavoltaic
U_{oc} = open-circuit voltage of betavoltaic
FF = filling factor of current–voltage characteristic
P_β = β-radiation power incident onto the surface of betavoltaic

For a more comprehensive estimate of η, power generated by the betavoltaic is related to the full power of the radionuclide.

6.2.1 Short-Circuit Current I_{sc}: Estimates and Experiment

The magnitude of I_{sc} is given by

$$I_{sc} = q \frac{P_\beta}{\varepsilon}\left(1 - R_\beta\right) \times e^{-\mu_\beta \rho W_d}\left[\left(\frac{\mu_\beta \rho L_d}{1 + \mu_\beta \rho L_d} - 1\right) \times e^{-\mu_\beta \rho L_E} + 1\right] \tag{6.2}$$

where

ε	= energy required for the formation of electron–hole pair (in Si, $\varepsilon = 3.62$ eV)
R_β	= $\eta_\beta \cdot k$; the reflection coefficient of the β-radiation power
η_β	= backscattering factor (ratio of the scattered current to the primary current)
k	= $E_{\beta s}/E_0$, the ratio of the average scattered electron energy to the initial energy
μ_β	= mass factor of absorption of β-radiation from tritium in cm²/g
ρ	= absorbent density in g/cm³; $\rho_{Si} = 2.32$ g/cm³
W_d	= thickness of the dead layer of a betavoltaic, including the region of high surface and bulk recombination in the heavily doped part of the n'-layer, where the generated pairs recombine
L_E	= space-charge region of the n'-p junction and the concentration gradient field of the n'-region where carriers drift in the field instead of diffusing
L_d	= diffusion length of minority carriers (electrons in the neutral base)

Taking $\mu_\beta \cdot \rho_{Si} \geq 10^4$ cm⁻¹, we get $\mu_\beta \cdot \rho \cdot L_d \gg 1$; that is, the factor in brackets equals ≈ 1.

Now Equation 6.2 can be written in the form

$$\frac{I_{sc}}{P_\beta} = \frac{q}{\varepsilon}\left(1 - \eta_\beta \frac{E_{\beta s}}{E_0}\right)\exp\left(-\mu_\beta\rho_{Si}W_d\right) \qquad (6.3)$$

The first factor in this expression represents the current sensitivity per unit of efficiently absorbed β-radiation power. Since $\varepsilon = 3.62$ eV at 300 K in silicon, the first factor equals

$$q/\varepsilon = 0.276 \text{ A/W} \qquad (6.4)$$

The second factor concerns losses for reflection of backscattered and secondary electrons. As there does not appear to be a comprehensive theory of the secondary electron emission applicable to reflection of β-particles from real isotopes, only approximate estimates will be made. In particular, assuming isotropic angular distribution of β-radiation and using data on the reflection coefficient as a function of incidence angle,[5] the percentage of backscattered electrons in silicon can be estimated at ~30%. However, the average amount of energy lost in reflection is unknown. Energy loss due to genuine secondary electrons will probably make a much lower contribution since their energy is only a few electron volts. Thus, 30% can be considered an upper limit of loss for scattering of tritium β-particles from the silicon surface.

The third factor in Equation 6.3 defines the effect of losses in the "window" or the "dead" layer of thickness W_d. In the diffusion n'-p junction, W_d depends on the junction depth and the surface concentration level, i.e., the sheet resistance of diffusion layer. In experiments on the short-circuit current I_{sc} as a function of sheet resistance R_{sh} under irradiation at a known power level, from I_{sc} saturation as $R_{sh} \rightarrow \infty$, the maximum value of the ampere-watt sensitivity of a betavoltaic and the integrated reflection coefficient R_β can be determined using Equations 6.3 and 6.4.

To experimentally determine ampere-watt sensitivity of a betavoltaic and R_β, planar n'–p structures were fabricated by phosphorus diffusion into p-Si with a resistivity of 1 Ω cm. The sheet resistance was varied from 30 to 200 Ω by varying the diffusion time. The surface concentration of phosphorus was maintained at 2×10^{20} cm^{-3}. The area of the n'–p junction was about 0.56 cm^2. The sample was then irradiated with a TiT$_2$ source with a specific power of 0.23 μW/cm^2; the effective irradiated area was about 0.47 cm^2. The experimental results are shown in Figure 6.2.

The maximum value of I_{sc} = 21.5 nA corresponds to current density of about 46 nA/cm^2 and current sensitivity of about 0.2 A/W. Comparing this value with the calculated value of 0.276 A/W permits a power reflection coefficient estimate of R_β = 0.27. A more precise value of R$_\beta$ can be obtained from data measured in vacuum by plotting log I_{sc} vs. W_d and extrapolating the curve to W_d = 0. In this experiment W_d was not determined, so this value of R_β should be considered an upper limit.

It should be noted that p–n junctions with sheet resistance of 200 Ω/square on which a current sensitivity of 0.2 A/W has been measured are suboptimal with respect to reverse current and open-circuit voltage values. The highest values of these parameters have been obtained for R_{sh} = 60 to 80 Ω/square, with corresponding current sensitivity of about 0.16 A/W.

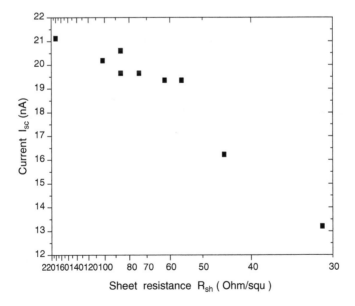

Figure 6.2   Current I_{sc} vs. sheet resistance R_{sh}.

Thus, from study of the short-circuit current:

- It is possible to obtain ampere-watt sensitivity of the betavoltaic with a planar $n'-p$ junction in single crystal silicon with irradiation from tritium of no less than 0.2 A/W.
- Total losses for reflection and inefficient absorption do not exceed 25%.
- Sensitivity can be increased by reduction of inefficiency losses, use of reflecting "mirrors" of materials with large atomic number, and use of an emission source with optimum self-absorption sandwiched between two silicon cells.

Current–voltage characteristics of improved Si $p–n$ junctions have been examined for betavoltaics. Silicon solar cells are designed to have large surface area and operate at high photocurrent densities (tens of mA/cm^2). Their $I–V$ characteristic is dominated by a diffusion current component with an ideality factor $n = 1$. In the worst case, a contribution to current from carrier generation in the $p–n$ junction space charge region is possible with an ideality factor $n = 2$. For solar cells, the small series resistance is important and leakage is insignificant. In silicon planar $p–n$ junctions of relatively small area designed for operation at low illumination levels, other factors should be taken into account related to the silicon–silicon oxide interface as well as leakage due to localized defects.

Components of the current through a $p–n$ junction determining the PV voltage at low levels of beta irradiation can be determined. In Figure 6.3 the photocell cross section is shown schematically.

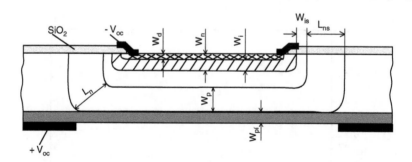

Figure 6.3 Schematic cross section of the photocell, where:

W_n	is n-region thickness
W_d	is thickness of the dead layer in n-region
W_i	is width of the space-charge region (W_{is} – at the surface)
L_n	is electron diffusion length in p-region (L_{ns} – at the surface)
W_p	is thickness of the (neutral) p-region
$W_{p'}$	is thickness of the p'-region reflecting minority carriers (BSF)
$+, - V_{oc}$	are wire leads from metal contacts

The current–voltage characteristic of the cell under irradiation is given by the formula

$$I = I_{sc} = I_p \left[\exp\left(+q(U - IR_s)/kT \right) - 1 \right] - I_{n'} \left[\exp\left(+q(U - IR_s)/kT \right) - 1 \right] -$$

$$I_i \left[\exp\left(+q(U - IR_s)/2kT \right) - 1 \right] - I_{ps} \left[\exp\left(+q(U - IR_s)/kT \right) - 1 \right] - \qquad (6.5)$$

$$I_{is} \left[\exp\left(+q(U - IR_s)/2kT \right) - 1 \right] - \frac{U - IR_s}{R_{sh}} - I_{def} \left[\exp\left((U - IR_s)/U_d \right) - 1 \right]$$

where

I, U	= current and voltage across the cell
I_{sc}	= photocurrent or short-circuit current
$I_{p,n'}$	= diffusion components of currents from p- and n'-regions, taking into account surface recombination velocity on the front and back surfaces (BSF)
I_i	= current due to carrier generation in the space-charge region (SCR)
I_{ps}	= current due to surface carrier generation at the boundary with oxide within a diffusion length L_{ns} from n'–p junction
I_{is}	= current due to carrier generation at space-charge region surface covered by the oxide
R_s, R_{sh}	= series and shunting ohmic cell resistances, respectively
I_{def}	= current flowing via structural defects such as decorated dislocations, stacking faults, and phase inclusions. Thermionic field or tunnel current can be represented with respective U_d values

In an abrupt, uniformly doped n^+–p junction, the total diffusion current through the bulk, $I_{dif} = I_p + I_{n'}$, is described by the formula

$$I_{dif} = q \frac{D_n}{L_n} n_{p0} \frac{Sh(W_p/L_n) + \eta_n Ch(W_p/L_n)}{Ch(W_p/L_n) + \eta_n Sh(W_p/L_n)} + q \frac{D_p}{L_p} p_{n0} \frac{Sh(W_n/L_p) + \eta_p Ch(W_n/L_p)}{Ch(W_n/L_p) + \eta_p Sh(W_n/L_p)}$$

$$(6.6)$$

where

$$L_n = \sqrt{D_n \tau_n}; \quad L_p = \sqrt{D_p \tau_p}; \quad n_{p0} = n_{ni_o}^\alpha / N_A; \quad p_{n0} = \exp\left(\Delta E_g / kT \right) n_{i_g}^2 / N_D;$$

ΔE_g = band gap narrowing at high doping levels ($\Delta E_g = 0.1$ eV at $N_D = 2 \times 10^{19}$ cm^{-3})

$L_{n,p}$ = diffusion length, with electrons n in p-region and holes p in n-region

$D_{n,p}$ = diffusion coefficient

$\tau_{n,p}$ = minority carrier lifetimes

n_{i0} = carrier concentration in the intrinsic semiconductor

N_A, N_D = acceptor and donor concentrations in p- and n-regions

η_n = $S_B L_n/D_n$; and $\eta_p = S_F L_p/D_p$, where S_B is the surface recombination velocity on the back surface and S_F is the surface recombination velocity on the front surface

An important parameter determining current magnitude is the minority carrier lifetime — τ_n (in p-region) and τ_p (in n-region). The resulting lifetime τ_r is determined by contributions from recombination via centers, radiative recombination, and Auger recombination.

Recombination via centers, that is, Shockley–Reed–Hall recombination with corresponding τ_{SRH}, is described by Equation 6.7:

$$\tau_{SRH} = \frac{\tau_p(n_0 + n_1 + \Delta n) + \tau_n(p_0 + p_1 + \Delta n)}{(p_0 + n_0 + \Delta n)} \tag{6.7}$$

where

n_0, p_0 = equilibrium concentrations;

Δn = excess concentration;

n_1 = $n_i \, exp(E_T - E_i)/kT$

p_1 = $n_i \, exp[-(E_T - E_i)/kT]$

E_i = the Fermi level in the intrinsic semiconductor

E_T = position in the bandgap of the recombination level at concentration N_T

$$\tau_n = \frac{1}{\sigma_n V_{th} N_t}; \quad \tau_p = \frac{1}{\sigma_p V_{th} N_t};$$

σ_n = capture cross section

V_{th} = thermal velocity

Radiative recombination is described by Equation 6.8, where $B = (0.75$ to $1.1)$ 10^{-14} cm^3/sec for silicon at 300 K.

$$\tau_{rad} = \left[B(p_0 + \Delta n)\right]^{-1} \tag{6.8}$$

Auger recombination, important at high concentrations, is described by Equation 6.9:

$$\tau_{Auger} = \left[C_p \left(p_n^2 + 2p_0 \Delta n + \Delta n^2 \right) + C_n \left(n_p^2 + 2n_0 \Delta n + \Delta n^2 \right) \right]^{-1} \qquad (6.9)$$

where

$$C_p = 10^{-31} \text{ cm}^6/\text{sec and } C_n = 2.8 \times 10^{-31} \text{ cm}^6/\text{sec}$$

The resulting recombination lifetime τ_r is defined by

$$1/\tau_r = 1/\tau_{SRH} + 1/\tau_{rad} + 1/\tau_{Auger} \qquad (6.10)$$

The generation–recombination current component from the space-charge region is described by Equation 6.11:

$$I_i = \frac{q n_i W_i}{2\tau_g} \qquad (6.11)$$

where

$$W_i = \sqrt{\frac{\varepsilon \left(V_{bi} - V \right)}{2\pi q N_A}}$$

is the space-charge region width, V_{bi} is the contact potential difference between n- and p-regions, ε is the dielectric constant, and

$$\tau_g = \tau_i \exp \left[\frac{|E_T - E_i|}{kT} \right]$$

After gettering, τ_g is approximately $(50 \text{ to } 100)\tau_r$.

The generation–recombination surface current component from the space-charge region, in contrast to the previous components, is not determined in the bulk, but at the semiconductor surface where the space charge contacts the oxide coat. This area is equal to the product of the cell periphery, P, and the width, W_{is}, of the exposed SCR surface. The corresponding current I_{is} is defined by

$$I_{is} = q \cdot S_i \cdot n_i \cdot P \cdot W_{is} \qquad (6.12)$$

where S_i is the surface recombination velocity at the boundary with oxide that can vary between 10^{-2} and 10^4 cm/sec.

The diffusion current component from the neutral surface at the boundary with the oxide over a distance from the SCR equal to the diffusion length is

$$I_{ps} = q \cdot S_i \cdot n_i^2 / p \cdot P \cdot L_{ns} \qquad (6.13)$$

Defects also can introduce an excess current component, depending on the presence of various types of defects influencing leakage currents, from ohmic-type to tunnel currents and currents due to carrier transfer over the top of the barrier. Predictions of currents through defects for particular samples are practically impossible.

We have conducted actual experiments to test betavoltaic designs. For operation at low short-circuit current densities, an n'–p structure has been fabricated using planar technology and p-type silicon with $\rho_p = 0.3$ ohm·cm. Chlorine and phosphorus gettering was employed and the structure had a guard ring; the effective area of the cell was about 0.5 cm².

In Figure 6.4a, a current-voltage characteristic of this cell is shown that was measured for the short-circuit current $I_{sc} = 18.1$ nA on a cell irradiated by a TiT$_2$ source of specific power 0.23 μW/cm². Measurement temperature was 23.5°C. The principal results are: $I_{sc} = 18.1$ nA; $U_{oc} = 162$ mV; $I_m = 13.7$ nA; $U_m = 110$ mV; $FF = 51.3\%$; and $\eta = 1.3\%$.

In Figure 6.4b a current-voltage curve of the same cell is shown under illumination by a luminescent lamp with short-circuit current of 900 nA. The following results have been obtained using light illumination of the cell: $I_{sc} = 900$ nA; $U_{oc} = 312$ mV; $I_m = 730$ nA; $U_m = 250$ mV; $FF = 65\%$; and $\eta = 3.1\%$. This output is 50 times greater than short-circuit current using the tritium source — approximately the ratio of the assumed maximum specific incident power from gaseous tritium (~11.5 μW/cm²) and the power value 0.23 μW/cm².

Direct conversion of tritium using silicon betavoltaics gives a short circuit current I_{sc} exceeding that of a wide band-gap semiconductor, with a value as high as 0.2 A/W. This is close to the theoretical limit and a direct consequence of the fact that the electron–hole pair creation energy in silicon semiconductors is lower than that in wide band-gap materials. Thus, energy conversion efficiency η of 1.3 and 3.0% for a solid tritiated titanium source of 0.23 μW/cm² and simulated gaseous tritium of 11.5 μW/cm², respectively, has been demonstrated.

6.3 SILICON CELLS FOR INDIRECT CONVERSION OF TRITIUM

Indirect conversion of beta radiation requires that it excite a phosphor and the phosphor then radiate light in a narrow spectral interval converted across a photovoltaic cell. This approach has the significant benefit that the semiconductor, or organic photovoltaic converter, is not exposed directly to ionizing radiation, which avoids degradation under a high radiation dose; it also introduces a new loss factor because of the phosphor efficiency, C_k. The maximum value of C_k for inorganic phosphors is 20 to 30%.[7] At the same time, the PV cell quantum efficiency for conversion of fairly monochromatic emission from the phosphor can be quite high, as the light reflection from the PV cell surface can be minimized by means of suitable antireflection coating. Internal and external quantum yield can reach unity (see Figure 6.5).

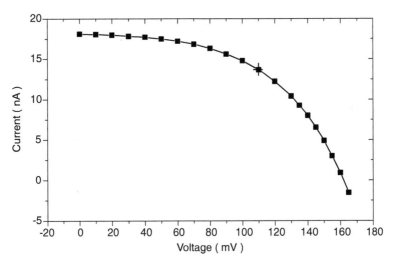

Figure 6.4a   Current-voltage characteristic of the monocrystalline Si-cell. Current I_{sc} corresponds to Ti-T$_2$ 0.23 µW/cm^2.

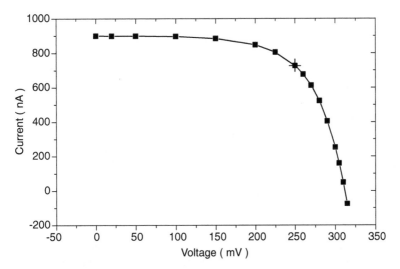

Figure 6.4b   Current-voltage characteristic of the monocrystalline Si-cell illuminated by a luminescent lamp. Current I_{sc} corresponds to T$_2$-gas.

For indirect conversion, both planar Si cells (Figure 6.3) and Si cells with vertical *p–n* junctions (Figure 6.6) can be used.[8] In the latter case, the technology is simpler and less expensive since no photolithography is required. Silicon cells with vertical *p–n* junctions have a number of substantial advantages over planar construction: no inconsistent requirements to the layer resistance of the emitter, spectral sensitivity, contact grid area, and so on. Since the front and back surfaces of such cells contain no metallization, the cells are bilateral. In contrast to planar junctions, cells with

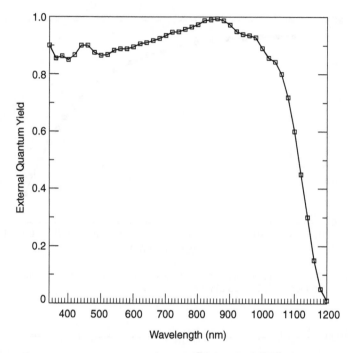

Figure 6.5 Spectral characteristic of an Si-based PV cell with vertical *p-n* junctions.

vertical *p–n* junctions generate higher voltage because they are connected in series and a lower current at the same power. In such structures no absorption occurs in the heavily doped emitter and little shading occurs with the contact grid, so they have good output current and spectral characteristics.

For these structures, the magnitude of j_{sc} under AM1.5D illumination (100 mW/cm^2) was 39.4 mA/cm^2. This is just 5% lower than the record value obtained for planar design. Structures with vertical *p–n* junctions have good spectral sensitivity in a wide range of wavelengths; internal quantum yield is close to unity in the wavelength range of 340 to 1080 nm. However, these structures were developed for operation with a concentrator; therefore, further work is needed to improve surface passivation and reduce the generation current.

In silicon structures with surface recombination velocity at the Si/SiO$_2$ interface of less than 1 cm/sec, the forward current-voltage characteristics at voltage higher than 100 mV have an ideality factor n = 1 (ideal diode). This is evidence that the generation component of the current is very low at current density above 10^{-10} A/cm^2. These results are quite attainable using current technology.[10] Obviously, ideal diode characteristics can be attained with appropriate structure geometry and surface passivation. In the range of low illumination levels and sharp visible phosphor emission, silicon structures have U_{oc} values somewhat below those of devices made of GaAs. In the case of indirect conversion, therefore, silicon is inferior to wide-gap semiconductors. For phosphors emitting over the relatively broad range of 800 < λ < 1100 nm, silicon is the best choice because of wide spectral response.

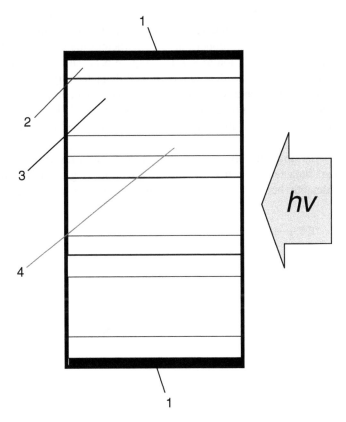

Figure 6.6 Schematic cross section of Si-array with vertical *p-n* junctions: (1) metal contacts; (2) *p*-Si; (3) *n*⁺-Si; (4) *p*⁺-Si.

6.4 MINIATURIZATION AND DESIGN OF TRITIUM BATTERY SILICON CELLS

From Figure 6.1 and the data of Section 6.2, the required battery current of 1 μA can be obtained with a minimum beta cell area of about 10 cm² using a solid tritium beta source. This means that miniaturization of betavoltaics is not a major problem. As the technology of nanoscale devices progresses, current consumption may drop two or three orders of magnitude. Therefore, this section explores the possibility of miniaturizing silicon betavoltaic cells.

Technological means of miniaturization are, in principle, the same as photovoltaic cells. However, one should bear in mind that, in contrast to solar energy, nuclear decay energy cannot be focused. In actual cells, the fraction of the efficient operating area of a cell diminishes with reduction of cell dimensions at least as the perimeter-to-area ratio, due primarily to edge insulation and contacts.

Physical limitations to miniaturization can arise, as well. In planar technology of silicon elements, the dark current density at the periphery can be higher than in the basic area (for example, because of different surface recombination velocities). In this instance, starting with some value of diameter D of the (round) cell, the dark current will be diminishing in proportion to ~D, whereas the photocurrent (or beta current) diminishes in proportion to ~D^2 (see Figure 6.7). This will result in open-circuit voltage U_{oc} dropping with reduction in cell diameter (Figure 6.8). In planar Si n^+–p test diodes of different diameters with experimental parameters as given in Figure 6.6 and surface recombination velocity under oxide of ~10^3 cm/sec, the drop of U_{oc} started at D ~ 1 to 2 mm. (The measuring device had an input resistance >$10^9 \, \Omega$.)

Consider the optimum dimensions and shape of an array of cells for direct conversion of beta radiation from tritium. Utilization of gaseous tritium in the cells is problematic because of large source dimensions or high pressures (tens of

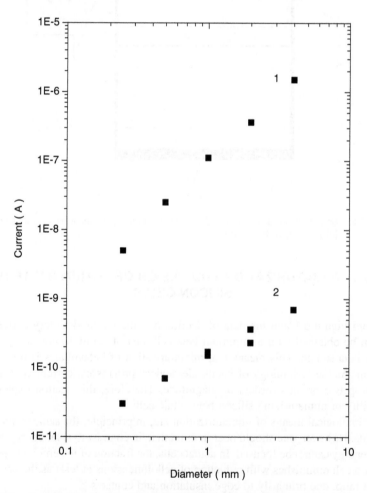

Figure 6.7   Current as a function of test diode diameter: 1) photocurrent (~D^2); 2) dark current (~$D^{1.15}$).

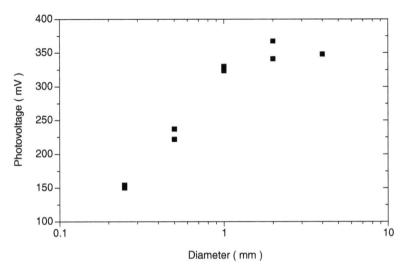

Figure 6.8 Photovoltage U_{oc} as a function of test diode diameter.

atmospheres) involved; therefore, use should be made of bound tritium, for example, in titanium. In this case, the source might have a specific power of ~0.25 to 1.5 $\mu W/cm^2$, conversion current $I_{sc} = 40$ to 240 nA/cm^2, and voltage $U_{oc} = 160$ to 240 mV (310 mV for gaseous tritium).

Comparing the range of requirements of potential consumers (2 $\mu A < I < 100$ μA; $1.5 < U < 6$ V), one can infer that the minimum operating area of the cell (or a row of cells in parallel connection) should amount to no less than 10 cm^2 with use of TiT_2 in order to obtain a minimum current of 2 μA.

For individual silicon cells, minimum size of the individual cell should be greater than 5 mm^2; in smaller cells, the conversion efficiency drops dramatically. For practical applications, the size of the cells should be greater than 1 cm^2 and a PV cell array is required. To get a semiconductor nuclear battery with an operating voltage of 1.6 V, it is necessary to connect 16 cells in series, assuming a TiT_2 solid-state source (0.25 to 1.5 $\mu W/cm^2$) is used. Seven to eight cells in an array may be needed if gaseous tritium is used as the source (11.5 $\mu W/cm^2$).

An optimum value of the specific power and packing density of a silicon-based nuclear battery could be achieved in a battery consisting of alternating Ti-T layers of thickness ~1000 Å and $n'-p-p'$ junctions of thickness ≤ 2 μm. Such a structure will have the following features:

• Optimum self-absorption losses
• Two-sided irradiation of the active region with beta particles that will double the short-circuit current
• Possible reduction of reflection losses (to less than ~25%)

Implementation of this structure will require additional technological research into ultrathin cells in the form of large-area membranes and methods of their assembly or a changeover to crystalline film structures deposited by various methods.

Subsequently, one more technique of fabricating single crystal beta-Si cells of high specific unit-volume power should be considered; this uses selective etching of (110)-oriented wafers and high-resolution, two-sided photolithography. Some designs hold promise of a 20- to 50-times-greater effective area.

In principle, it is possible to make small-diameter circular channels instead of slots; however, filling the channels with the isotope can be a problem. The idea of saturating porous silicon with tritium should possibly be reconsidered. Preliminary speculative estimates indicate the possibility of obtaining a several-times-greater flux of beta particles than from Ti-T, which is explained by lower self-absorption in porous silicon because of its lower specific weight. This approach will require some fundamental research.

The following parameters for direct-conversion silicon beta cells using tritium have been derived:

- Current sensitivity per unit incident power of beta radiation
  (theoretical prediction per unit absorbed power 0.27 A/W)   ~0.16 A/W
- Open-circuit voltage U_{oc}:
  for solid-state Ti-T source, 0.25 to 1.5 μW/cm^2 160 to 240 mV
  for gaseous tritium (simulation of 11.5 μW/cm^2 incident
  power of optical excitation)   up to 310 mV
- Conversion efficiency of the incident beta radiation
  from tritium:
  for solid-state Ti-T   up to 1.3%
  for gaseous tritium   up to 3%
- Minimum size of cell retaining the above efficiency   ~ 2 × 2 mm^2

6.5 INTEGRATED SILICON LOW-INTENSITY-RADIATION BETAVOLTAIC CELLS

The purpose of this section is to analyze and present data necessary for fabricating integrated OPVs based on single crystalline silicon. Main requirements for photovoltaic cells are 1) an integrated array of photovoltaic cells for operation at low light intensities, lower than 100 μW/cm^2; 2) optimized cell output for phosphor with wavelength of 400 to 700 nm; and 3) minimized cell size.

Assemblage of arrays of individual photovoltaics has always been an important stage in fabricating commercial arrays because parallel-series connections make the cost of the unit-produced power higher. In particular, this is due to the need to process each cell separately and by the loss of effective area. These problems are especially important for small-size photovoltaics of relatively low power used in consumer articles such as handheld computers, watches, and toys. At the present time these devices are powered by series arrays fabricated using hybrid integrated-circuit technology. In such arrays, individual PV cells are wired together with external connections.

It has long been recognized in integrated circuits that, when various functions can be integrated in the same silicon chip, a considerable increase in reliability can be obtained. This comes about in part because the number of interfaces between dissimilar materials can be reduced, thus reducing prime sources of product failures, such as cracks resulting from wafer cutting into squares, failure of wires, and so on. A further disadvantage of the conventional individual PV cell in these kinds of applications is that it requires the presence of back contact. The back contact makes the device unsuited for applications where it is desired to have the PV cell integrated with the circuitry it powers, a feature that will probably evolve in small instruments as well as in calculators, watches, and toys.

It is evident that a monolithic structure fabricated using conventional microelectronic batch techniques will share the yield characteristics of standard integrated circuits, following their cost "learning curves" without sacrificing the power conversion efficiency of standard photovoltaic batteries. Therefore, a monolithic array should have a reliability advantage over its hybrid counterpart.

In designing a monolithic series-array PV battery, cells must be designed for uniform response. Because the cells are connected in series, the lower current cell will fix the output current. Sometimes a poor cell can be strongly reverse-biased by the other cells in the array and act merely as a load rather than as a source. In the case under consideration, this condition should be fulfilled. Second, standard $p–n$ junction insulation is not sufficient because the blocking diode would be illuminated. Since the penetration depth of β-radiation and visible light is very small in comparison with sunlight, this problem is not significant. Manufacturing goals are to produce individual cells with the highest attainable output parameters (I_{sc}, U_{oc}, FF, η). Individual cells should be isolated with a view to eliminating stray couplings (shunting). Series connection of N individual cells is needed to obtain the required voltage: $V_{battery} = N \times U_{oc}$.

6.5.1 Optimized Si Parameters of Si-Based Photo- and Betavoltaic Cells

Choice of optimal parameters for photovoltaic cells is governed by the wavelength of incident radiation and its power per unit area. Taking into account that the best phosphors have spectral output in the wavelength region of 400 nm $< \lambda <$700 nm, one can assume, by way of example, an average wavelength of $\lambda_{\hat{a}\hat{a}} = 550$ nm corresponding to the green target phosphor fluorescence wavelength $\lambda_{\hat{a}\hat{a}} \sim 530$ nm for tritium.

The maximum watt-ampere $p–n$ junction sensitivity for radiation with $\lambda = 550$ nm is equal to 0.4435 A/W; the absorption coefficient in silicon[6] is K = 0.9×10^4 cm^{-1}. The thickness of a layer of Si absorbing 90% of light is $W_{0.90} = 2.55$ μm and $W_{0.99} = 5.1$ μm. For the red light with $\lambda = 630$ nm, the maximum watt-ampere sensitivity is 0.508 A/W, the absorption coefficient $K = 3.9 \times 10^3$ cm^{-1}, and the 99% absorption-layer thickness is 11.8 μm. Thus, for an Si photovoltaic cell with back surface field (BSF), the base thickness can be chosen not larger than 10 to 15 μm. (BSF is a reflecting barrier for minority carriers.)

Assume that the specific power of incident radiation is known. The maximum specific power can be determined from the condition for conversion of beta radiation from tritium gas[4] $P\beta_1 = 11.5$ μW/cm^2 with converter efficiency $h_k = 0.3$: $P_1 = 3.45$ μW/cm^2 of light emission and tritium in a solid Ti-T$_2$ source $P\beta_2 = 0.25$ to 1.5 μW/cm^2, corresponding to $P_2 = 0.075$ to 0.45 μW/cm^2 of light emission.

For estimating I_{sc} one can use the value of the maximum watt-ampere sensitivity, 0.4435 A/W, and for estimating the maximum value of U_{oc} and the optimum impurity concentration in the p-base, one can use data on the minority carrier lifetime as a function of impurity concentration.[10] An optimum value of the acceptor concentration in the p base was found to be $5 * 10^{17}$ cm^{-3} ($\rho_p = 0.1$ Ωcm) and the minority carrier lifetime $\tau_n = 1$ to 20 μsec.

Table 6.1 summarizes maximum values of I_{sc} and U_{oc}. In calculation of U_{oc}, the base thickness was assumed to be 10 μm, $\tau_n = 1$ μsec, $\tau_i = 20$ μsec, and the temperature 300 K. In the last three columns of Table 6.1, calculated values of the leakage current I_{sh}, shunting resistance R_{sh}, and series resistance R_s, which reduce the output power by ~10%, are given. In the calculation, rough criteria were used: $I_{sh} < 0.1$ I_{sc}; $R_{sh} > 10$ U_{oc}/I_{sc}; and $R_s < 0.1$ U_{oc}/I_{sc}.

It is interesting that the best experimental values of $I_{sc} = 1.8$ μA/cm^2 (gas source) and 0.036 μA/cm^2 (solid source, 0.23 μW/cm^2) for direct conversion are higher than maximum value for indirect conversion. Maximum experimental efficiency for direct conversion is 3.1% (see Section 6.3). However, if one takes into account the maximum values of $I_{sc} = 2.3$ μA/cm^2, $U_{oc} = 436$ mV, and $FF = 0.65$, the possible calculated efficiency would be about 6%.

6.5.2 Methods of Integrated Series Connection of Crystalline Si Cells

The solution to the problem of integrated arrays is simplest in the case of photovoltaic cells (PC) produced by thin-film technologies. In the first place, this is true of PCs based on amorphous silicon currently produced commercially on a large scale. The technology of depositing amorphous silicon films on insulating substrates allows making intercell connections with the same masks as those used in production of integrated circuits.[11]

Coutts and Meakin[12] applied epitaxy through shadow masks to fabricate an integrated-series connection of monocrystalline thin-film Si PV cells, a novel process that avoids trench etching and is therefore potentially cost effective. The p^+-, p- and n^+-type epitaxial Si layer are deposited on a textured crystalline Si substrate with porous Si on the surface through a shadow mask consisting of wires. The mask is translated along the surface, providing overlap and series connection of n^+- and p^+-layers. Then the stack is glued to a glass and split off from the substrate by applying mechanical stress. After removal of the residual porous Si, an integrated-series connection of monocrystalline and textured thin-film PV cells is formed. A mini-module consisting of six series-connected cells (8-μm-thick monocrystalline Si) on

Table 6.1 Maximum Values of I_{sc} and U_{oc} (in calculation of U_{oc}, base thickness was assumed to be 10 μm, $\tau_n = 1$ μsec, $\tau_i = 20$ μsec, and the temperature 300 K)

	P(μW/cm^2)	I_{sc} (μA/cm^2)	U_{oc} (mV)	I_{sh} (nA/cm^2)	R_{sh} (Ωcm^2)	R_s (Ωcm^2)
P_1	3.45	1.53	420	<150	>2.8 × 10^6	<2.8 × 10^4
P_2	0.075–0.45	0.033–0.20	250–330	<3	>8 × 10^7	<8 × 10^5

a glass has an open circuit voltage of only 1.6 V under one sun due to large defect density and an unoptimized doping profile. Note that, with the technological process, the value of U_{oc} can be obtained up to 623 mV.[13]

For PV cells based on monocrystalline silicon, attempts to achieve integrated design dating back to the late 1970s are known. Tayanava et al.[14] have described a monolithic array based on 30-μm-thick monocrystalline p-Si with an oxide insulation from a polysilicon substrate and isolating an n^+ diffusion layer between the cells. For five cells in a series connection at AM1, the following parameters have been obtained: $U_{oc} = 2.94$ V; $I_{sc} = 18.2$ mA/cm²; $FF = 0.75$; and $\eta = 8\%$. The low efficiency was related to series resistance and leaks. In 1981 fabrication was reported of a monolithic array with oxide isolation featuring $U_{oc} = 0.55$ V (per cell) and $\eta = 7.5\%$.[15] Attempts have been made at fabrication of monolithic arrays based on thin high-ohmic silicon with junctions at opposite sides of the wafer.[16,17]

A disadvantage common to all of these arrays is strong shunting of the base material; an efficient operation is possible only with concentrators of ≥ 5 sun AMI, when saturation sets in and $U_{oc} \approx 3$V (with six solar cells).

The highest conversion efficiency reported before 1999 for monolithically integrated cells, other than thin-film silicon solar cells, is $\eta = 11.4\%$ ($I_{sc} = 28.4$ mA/cm², $U_{oc} = 3.45$ V (six cells), $FF = 70.2\%$, 1 sun AM1.5, total area 21.1 cm²).[18] To achieve isolation of unit cells (UC), the authors used a rather simple approach: they removed the wafer material between the UCs by inserting trenches reaching from the cell front surface to the back surface, leaving narrow bridges to hold the device together. The main disadvantage was low isolation between the unit cells and fast drop of U_{oc} and η when illumination intensity decreased below 40 mW/cm².

A very interesting and promising novel multijunction PV cell has been proposed.[19] The new multijunction structure consists of alternating polarity n- and p-type layers, with like-polarity layers connected in parallel. The total thickness of the material is ~10 μm, and the calculated efficiency is $\eta = 10.7\%$ for single-junction design and $\eta = 17.3\%$ for multijunction design, both using poor-quality material of $\tau = 10$ nsec.

The fabrication sequence includes the following steps:

1. Deposition on dielectric substrate of alternating polarity multilayers
2. First laser grooving and groove-wall doping step
3. Second laser grooving and groove-wall doping step of opposite polarity, overlapping previous in some regions
4. Groove metallization

As a result, layers of the same type are connected in parallel, and the regions between the two widest metallization grooves are connected in series.

A novel approach has been described for potentially high-performance, low-cost, thin-film polycrystalline silicon PV cells. Unfortunately, experimental results are not given.

A process of contoured-oxide isolation applied in 1981 for producing a series-array PV battery[15] may prove useful. This is a widely known microelectronic "pocket" method for producing complementary transistors. Full area of the cell is ~10 × 1 mm²; the active area is 0.085 cm²; $U_{oc} = 0.55$ V/cell; $J_{sc} = 28.2$ mA/cm²;

$FF = 0.45$; and $\eta = 7.1\%$. Thickness of p-Si (100) is 58 μm ± 2 μm; thickness of polycrystalline Si is ~ 250 μm; and the thickness of insulating oxide is ~ 4800 Å. The I–V characteristics give evidence of large series resistance. U_{oc} is small because of shunting; poor isolation of Al contact from the polycrystalline Si is assumed to be the result of possible porosity of SiO_2.

In summary, it can be concluded that all reported designs of the integrated arrays can operate efficiently only at high illumination levels, 10^3 W/m² (1 sun) or higher. At lower illumination levels U_{oc} drops dramatically, due, apart from defects of material and technology, to considerable leakage between individual cells. Designers of integrated photovoltaic arrays for low illumination levels (1 W/m² and lower) should pay special attention to this problem. In the particular case of photovoltaic cells for a relatively narrow spectral range with high light absorption coefficients, isolation methods applied in microelectronics can be used.

In an integrated array, cells should be isolated from each other. Two isolation methods, using either a dielectric coating or a p–n junction, can be applied in a variety of ways; both techniques have their advantages and drawbacks. Dielectric insulation can be achieved using either the contoured monolithic oxide isolation (so-called "pocket") technique[15] or the well-known SOI (silicon on an insulator) technique. Both methods provide good dielectric insulation, but the former makes better use of the semiconductor wafer area. Nevertheless, the first technique can cause high mechanical stress, causing an undesirable increase of the p–n junction leakage. In the second, especially with direct bonding of the wafers, there is practically no stress or related leakage.

The second technique has a number of versions, all of which use a standard diffusion technology. However, the structure design should meet certain requirements. For example, the substrate should not be illuminated by light; the leakage current through the isolating p–n junction should be less than I_{sc}. This technique allows production of low-cost PV arrays; however, the problem to be anticipated is large leakage currents compared to those for the dielectric isolation technique. The same problem occurs for structures with vertical p–n junctions (discussed in Section 6.3).

Thus, silicon-based PV arrays design can be improved by means of intercell insulation. With indirect conversion of tritium beta radiation, the short-circuit current of silicon-based cells is highly competitive with wide band-gap semiconductor-based PV cells. Good agreement between theory and experimental values of I_{sc} is observed.

Direct-conversion short-circuit current of tritium beta radiation using silicon-based cells exceeds that of wide band-gap semiconductor-based beta cells and is as high as 0.2 A/W, approaching theoretical optimization. This is a direct consequence of the fact that e–h pair creation energy in Si is lower than that in wide band-gap materials. This allows energy conversion efficiency of 1.3% using a solid Ti-T_2 source of 0.23 μW/cm² and 3.0% using simulated gaseous tritium of 11.5 μW/cm².

Silicon-based cells have low cost and, in principle, may be integrated into monolithic arrays, which should give reliability advantages over hybrid counterparts. Si-based cells have lower conversion efficiency η as compared with wide band-gap compound-based cells because of lower open-circuit voltage U_{oc}. To improve the value of η up to the theoretical estimated value of 6%, it is necessary to increase the value of U_{oc} to the theoretical limit.

REFERENCES

1. Rappaport P., EMF across a *p–n* junction caused by bombardment with beta-particles, *Phys. Rev.*, 93, 246, 1954.
2. Pfann,W.G. and van Roosbroeck,W.J., Radioactive and photoelectric *p-n* junction power sources, *Appl.Phys.*, 25, 1422, 1954.
3. Trace Photonics, unpublished data, 2000.
4. Kherani, N.P. et al., Tritiated amorphous silicon for micropower application, *Fusion Technology*, 28, 1609, 1995; Walko, J. et al., *IEEE Proc. Int. Energy Convers. Eng. Conf.*, 6, 135, 1997.
5. Brodie, I. and Murray, J.J., *The Physics of Microfabrication*, Plenum Press, New York, 1982.
6. von Roos, O. and Landsberg, P.T., Effect of recombination on the open-circuit voltage of a silicon solar cell, *J. Appl. Phys.*, 57(10), 4746, 1985.
7. Mikhal'chenko, G.A., *Radioluminescentnye Izluchateli*, Energoatomizdat, Moscow, 1988, 152 [in Russian].
8. Voronkov, V. B. et al., Use of direct wafer bonding of silicon for fabricating solar cell structures with vertical *p-n* junctions, *Semiconductors*, 32(7), 789, 1998.
9. Aspnes, D.E. and Studna, A.A., Dielectric functions and optical parameters of Si, Ge, GaP, GaAs, GaSb, InP, InAs and InSb from 1.5 to 6 eV, *Phys. Rev.*, 27B, 2, 985, 1983.
10. Schroder, D.R., Minority carriers lifetime in silicon, *Solid State Phenomena*, 6–7, 383, 1989.
11. Altermatt, P. et al., The influence of a new band gap narrowing model on measurements of the intrinsic carrier density in crystalline silicon, *Tech. Dig. Int. PVSEC-11*, Sapporo, Hokkaido, Japan, 719, 1999.
12. Coutts, T.J. and Meakin, J.D., Eds., *Current Topics in Photovoltaics*, Academic Press, London, 1985, 139.
13. Brendel, R. and Oelting, S., A novel process for integrated series connection of crystalline thin-film silicon solar cell, *Tech. Dig. Int. PVSEC-11*, Sapporo, Hokkaido, Japan, 545, 1999.
14. Tayanava, H., Yamau, K., and Matsushita, T., Thin film crystalline silicon solar cell obtained by evaporation of porous silicon sacrificial layer, *Proc. 2nd World Conf. Photovoltaic Solar Energy Convers. Eur. Comm.*, Ispra, 1998, 272.
15. Warner, R.M., Murray, E.M., and Smith, W.K., *Conf. Record. 13th IEEE Photovoltaic Specialists Conf.*, Las Vegas, 1116, 1978.
16. Murray, E.M. and Warner, R.M., The contoured-oxide monolithic series-array solar battery, *J. Appl. Phys.*, 52, 6352, 1981.
17. Goetzberger, A., Integrated series-connected solar cell, U.S. Patent 4,330,680, 1982.
18. Kapoor, V.J. et al., High-voltage solar-cell chip, *J. Appl. Phys.*, 57, 1343, 1985.
19. Keller, S. et al., Progress in monolithic series connection of wafer based crystalline silicon solar cells by the novel "highvo" (high voltage) cell concept, *Tech. Dig. Int. PVSEC-11*, Sapporo, Hokkaido, Japan, 651, 1999; Keller, S. et al., Theoretical and experimental behavior of monolithically integrated crystalline silicon solar cells, *J. Appl. Phys.*, 87, 1556, 2000.
20. Green, M.A. and Wenham, S.R., Novel parallel multijunction solar cell, *Appl. Phys. Lett.*, 65, 2907, 1994.

CHAPTER 7

Nuclear Batteries Based on III-V Semiconductors

V.M. Andreev

CONTENTS

0-8493-0915-8/01/$0.00+$1.50
© 2002 by CRC Press LLC

7.1 MODELING THE EFFICIENCY OF III-V PHOTOVOLTAIC CELLS FOR RADIOLUMINESCENT LIGHTING

Photovoltaic (PV) conversion of a monochromatic light has been demonstrated with an efficiency of 50 to 60% at moderate illumination intensities of 10^{-1} to 10^{-2} W/cm^2 and reduces to 25 to 35% at low-intensity illumination of 10^{-5} to 10^{-6} W/cm^2. Modeling of conversion efficiency as a function of matched semiconductor PV cell band-gap energy and as a function of illumination intensity, as well as analysis of suitable semiconductors for PV conversion of radioluminescent light, will be described in this chapter.

III-V semiconductors are excellent materials for radiation-hard photovoltaic cell fabrication. GaAs, GaP and related compounds, and solid solutions (AlGaAs, GaInP, AlGaP) are suitable for indirect conversion of radioisotope decay energy into electricity. Their main advantages are good matching of band gap to radioluminescent light wavelength, relatively high efficiency at low-intensity light, and high quality of epitaxial materials, which can be fabricated by highly productive and cost-effective liquid phase epitaxy (LPE) and metal organic chemical vapor deposition (MOCVD).

7.1.1 Efficiency of Idealized Photovoltaic Cells

The processes taking place in a p–n junction under illumination[1-13] will now be discussed. Let the p–n junction be located near the illuminated semiconductor surface, free from surface states (Figure 7.1). Choose energy of the illuminating photons such that they would be absorbed in a semiconductor (for example, $hv = E_g$) and such that electron–hole pairs would arise only in the p-region at the distance from the p–n junction shorter than the electron diffusion length.

In exploiting a photoconverter as an electric energy source, an external load resistance must be connected to its terminals. Consider two ultimate cases: the load resistance $R = 0$ (the short-circuit regime) and $R = \infty$ (the open-circuit regime). Energy diagrams of the p–n junction in these regimes are depicted in Figure 7.1a and b. Photoelectrons generated in the immediate vicinity of the space-charge region are swept by the p–n junction electric field and collected in the n-region. All the remaining photoelectrons diffuse toward the p–n junction, trying to compensate for their diminution, and finally are collected in n-region, as well. From this region, electrons travel toward the back metal contact, and then they flow over into the

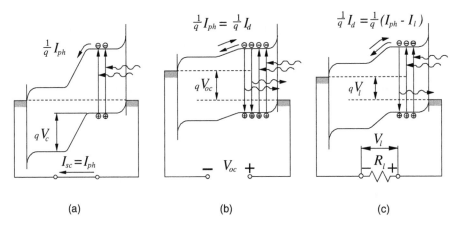

Figure 7.1   The band energy diagrams of a p–n junction under illumination: (a) in the short-circuit regime; (b) in the open-circuit regime; and (c) being connected to an external load resistance.

external circuit and into the contact to the p-region. Electrons reaching the interface between metal and p-region recombine there with the photogenerated holes.

Note the energy diagrams of the n- and p-regions in the vicinity of the contacts. They correspond to ideal nonrectifying (i.e., ohmic) contacts between a metal and a semiconductor when an energy barrier for the carrier flowover is absent. Such a situation is arranged by more heavy doping of the n- and p-regions in the vicinity of the contacts, so that $E_c - E_f$ and $E_f - E_v$ (E_c — conduction band bottom; E_f — Fermi level; E_v — conduction band top) would be equal to zero, and also by choosing metals (differing for contacts to the n- and p-regions) characterized by corresponding Fermi-level energy positions, similar to those in the semiconductor.

At an open external circuit of the p–n junction (Figure 7.1b), photoelectrons entering the n-region are collected and charge this region negatively. The excess holes remaining in the p-region charge this region positively. The potential difference arising is the voltage of the open-circuit regime V_{oc}. The polarity of V_{oc} corresponds to the forward bias of the p–n junction.

The magnitude of V_{oc} is always less than the contact potential difference V_c, which means that a little "step," dE, always exists on the p–n junction diagram. Potential energy of electrons situated near the conduction band bottom is lower in the n-region than in the p-region. This is the reason for effective extraction of photogenerated electrons from the p-region into the n-region. Thus, the photoelectron flux does not depend on the p–n junction bias voltage (positive and also negative up to the voltage, at which an avalanche multiplication of charge carriers starts). The light-generated carrier flux produces the photocurrent I_{ph}, the magnitude of which is equal to the number of photogenerated carriers passing through the p–n junction in a unit time:

$$I_{ph} = q \frac{P_r}{h\nu} \qquad (7.1)$$

where q is the electron charge and P_r is the absorbed light power. It is assumed that every absorbed photon with the energy $hv = E_g$ produces one electron–hole pair in a semiconductor. This condition is usually satisfied for photoconverters on the base of Si and GaAs.

At zero internal ohmic losses, the short-circuit regime in a PV cell (Figure 7.1a) is equivalent to a zero-bias voltage of the p–n junction. Therefore, the short-circuit current I_{sc} is equal to the photocurrent:

$$I_{sc} = I_{ph} \qquad (7.2)$$

Under open-circuit conditions (Figure 7.1b), the photocurrent is placed in equilibrium with the "dark" current; I_d is the forward current through the p–n junction arising at the forward bias voltage equal to the magnitude of V_{oc}. The open-circuit voltage at $I_{ph} \gg I_o$ is

$$V_{oc} \approx \frac{A_{id}KT}{q} \ln \frac{I_{ph}}{I_o} \qquad (7.3)$$

where A_{id} is the "ideality factor" ($A_{id} = 1$ if no recombination centers exist in the p–n junction); k is the Boltzmann constant; I_o is the saturation current; and T is temperature in Kelvin.

The dark current is accompanied by recombination of minority carriers (in this case, electrons in the p-region). At the recombination, potential energy of electron–hole pairs is released by means of photon emission with $hv0 = E_g$ or dissipated by means of heating the crystalline lattice. Radiative process is schematically shown by additional arrows in Figure 7.1b. Thus, the open-circuit regime of a photoconverter is equivalent to the forward bias regimes of rectifying diodes and light-emitting diodes.

To find a generalized expression for the I–V curve of an illuminated p–n junction, it is necessary to assume that a power source with a variable voltage is connected to it. At a positive bias voltage, the photocurrent I_{ph} is subtracted from the dark current of the p–n junction, and at a negative one, I_{ph} is summed with the dark curent. The expression for the I–V curve is written in the form

$$I = I_o \exp\left[\frac{qV}{A_{id}kT} - 1\right] - I_{ph} \qquad (7.4)$$

The I–V curve of the illuminated p–n junction can be derived graphically by means of moving down the whole dark I–V characteristic along the current axis by the value I_{ph}, as pictured in Figure 7.2. In the fourth quadrant of Figure 7.2, the direction of the current through the p–n junction is "opposite" to the applied voltage polarity; i.e., the illuminated p–n junction operates as an energy source. In this case, the point of zero-bias voltage of the p–n junction corresponds to the short-circuit regime, and the point of a zero current through the p–n junction corresponds to the open-circuit regime.

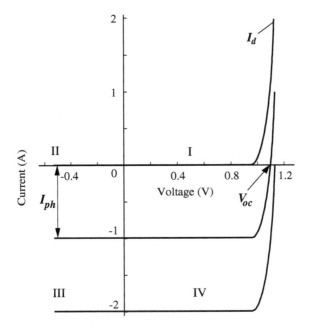

Figure 7.2   "Dark" and two "illuminated" *I–V* curves of the *p–n* junction in GaAs.

The segment of the *I–V* curve lying in the fourth quadrant can be constructed from measurements when a variable load resistance is connected to the *p–n* junction (Figure 7.1c). The direction of the current across the load always coincides with the direction of I_{ph}, and the load current I_l is equal to the resulting current through the *p–n* junction (see Equation 7.4). Taking the direction of I_{ph} as positive, one can write

$$I_l = I_{ph} - I_o \left[\exp \frac{qV_l}{A_{id}kT} - 1 \right]$$ (7.5)

Here V_l is the voltage drop across the load resistance equal to the voltage applied to the *p–n* junction. Equation 7.5 describes the *I–V* characteristic of the illuminated *p–n* junction ("illuminated" *I–V* curve). An illuminated *I–V* curve of the GaAs-based *p–n* junction for the photocurrent $I_{ph} = 1$A is shown in Figure 7.3a. This figure also shows the *I–V* curve of the load resistance for $R_{l1} = 0.1^*$, $R_{l2} = 1.026^*$, and $R_{l3} = æ\ 10^*$:

$$I_l = V_l / R_l$$ (7.6)

At known parameters of the illuminated *I–V* curve and a given value of R_l, magnitudes of I_l and V_l are found by successive approximation using Equations 7.5 and 7.6, solved simultaneously or solved graphically, as in Figure 7.3a. If R_l is small enough, intersection of the plots takes place on the horizontal segment of the illuminated *I–V* curve, where the dark current through the *p–n* junction can be neglected in comparison to the photocurrent. The current through the load resistance decreases with

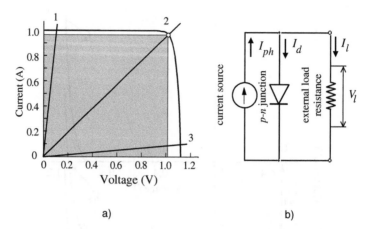

a)   b)

Figure 7.3   (a) Illuminated *I–V* curve of the *p–n* junction in GaAs and *I–V* characteristics of load resistance R_l for the magnitudes of 0.1 Ω (1), 1.026 Ω (2), and 10 Ω (3), and (b) the equivalent of the illuminated *p–n* junction with a load resistance.

increasing R_l because, with increase of the forward bias, the *p–n* junction begins to shunt the load resistance. Thus, in correspondence with Equation 7.5, the illuminated *p–n* junction can be presented in the form of the equivalent circuit of Figure 7.3b. The current source in the circuit simulates generation of a photocurrent independent of the *p–n* junction voltage, and the diode represents a darkened *p–n* junction. In varying R_l, the photocurrent is redistributed between load resistance and the *p–n* junction.

Output electric power utilized in the load resistance is determined by the following formula (neglecting the units in Formula 7.5):

$$P = I_l V_l = I_{ph} V_l - I_o V_l \exp\left(\frac{qV_l}{A_{id}kT}\right)$$ (7.7)

In the short-circuit and open-circuit regimes, $P = 0$, because either V_l or I_l is equal to zero. Apparently, some optimum magnitude of $R_l = R_{opt}$ exists for each illumination level, at which P reaches its maximum magnitude P_m. The maximum power condition can be determined from Equation 7.7, setting $dP/dV_l = 0$:

$$P_m \approx I_{ph}\left[V_{oc} - \frac{A_{id}kT}{q}\ln\left(\frac{qV_m}{A_{id}kT} + 1\right) - \frac{A_{id}kT}{q}\right]$$ (7.8)

In Figure 7.3a the area of the crosshatched rectangle is equal to P_m. Quality of the illuminated *I–V* characteristic is higher, the closer its shape is to rectangular. The fill factor of *I–V* characteristic *(FF)* is a parameter evaluating the quality of *I–V* characteristic:

$$FF = \frac{P_m}{I_{sc} V_{oc}} = \frac{I_m V_m}{I_{sc} V_{oc}}$$ (7.9)

Equation 7.8 can be rewritten in the form

$$P_m = I_{ph}\left(\frac{E_m}{q}\right) \tag{7.10}$$

where

$$E_m = q\left[V_{oc} - \frac{A_{id}kT}{q}\ln\left(\frac{qV_m}{A_{id}kT} + 1\right) - \frac{A_{id}kT}{q}\right] \tag{7.11}$$

is the energy calculated per one absorbed photon transferred into load resistance at an optimum electrical matching of the p–n junction and the external circuit.

The magnitude of the energy gap E_g defining the potential energy magnitude of one photogenerated electron–hole pair is the upper estimation limit for E_m. In principle, quanta of light energy as low as $hv = E_g$ can be absorbed in a semiconductor; however, energy of these quanta is not totally utilized in the optimum load resistance. The first cause of losses reflects the fact that the contact potential difference V_c is lower than the magnitude of E_g/q. These losses depend on the density of states in the valence and conduction bands of a semiconductor as well as on the majority carrier concentrations in the n- and p-regions of the p–n junction (recall that donor and acceptor atoms are totally ionized).

The second cause of losses reflects the fact that V_{oc} is lower than contact potential difference. These losses depend on the majority of carrier concentration and electrophysical parameters (mobility, diffusion lengths) of the minority carriers in n- and p-regions of the p–n junction. The third cause of losses reflects the fact that voltage V_m on the optimal load resistance is less than V_{oc}. Finally, the fourth cause of losses can be interpreted as losses due to the optimum current. Indeed, at the maximum power point on the I–V characteristic of the illuminated p–n junction, the current I_m is less than the photocurrent.

For further consideration of light energy conversion, it is important to consider photoelectrical properties of the p–n junction regarding the illumination level. Under conditions when a quantity of photogenerated charge carriers remains much less than a quantity of majority carriers, a variation of illumination level also means that photocurrent density changes in direct proportion to it. To carry out a corresponding analysis for the p–n junction in a GaAs-based photoconverter ($n = 10^{17}$cm^{-3}, $p = 10^{18}$cm^{-3}) at room temperature, $T = 297$ K. Let us begin in the range of $10^{-4} \le i_{ph} \le 100$ A·cm^{-2}, which covers the values of i_{ph} for such a type of photoconverter under illumination by optical power in the region of about $2 \cdot 10^{-4}$ to 200 W·cm^{-2}.

These intensities of illumination are much higher than those in the radioluminescent light sources. After consideration of high power efficiency, cell efficiency at the lower light intensities of 10^{-6} W·cm^{-2} will be considered, taking into account that this light power roughly corresponds to radioluminescent lighting. Illumination is supposed to be monochromatic, with $hv = E_g = 1.42$ eV. The dark current flow mechanism is a diffusion one ($A_{id} = 1$). It is necessary to perform calculations to evaluate the energy, calculating per one absorbed photon and utilizing the load resistance optimally matched with the $p–n$ junction at every new magnitude of i_{ph}. The results of calculations are accumulated in Figure 7.4 in the form of plots, which demonstrate energy losses in electron volts for each cause of loss. It is clear from Figure 7.4 that the output energy E_m increases with i_{ph}, i.e., with increase of the illumination level. This takes place at the cost of an increase of V_{oc} and, hence, of an increase of V_m.

The efficiency of the $p–n$ junction as a photovoltaic converter of monochromatic energy can be defined in the form of the ratio of E_m to the energy of one absorbed photon hv:

$$\eta = E_m/hv \tag{7.12}$$

The corresponding scale is on the right-hand side of Figure 7.4. The values of η defined by the E_m plot are maximum ones for a given $p–n$ junction, since the converted photon energy $hv = E_g$ has been chosen. Thus, the monochromatic efficiency of the $p–n$ junction increases noticeably with increasing photon current density, i.e., illumination level.

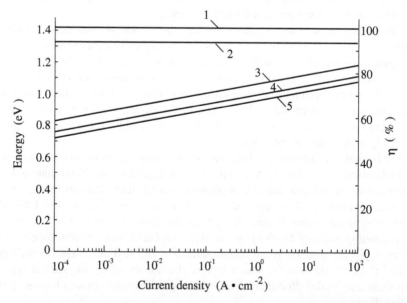

Figure 7.4   Plots of energy magnitudes (1) E_g, (2) qV_c, (3) qV_{oc}, (4) qV_m, and (5) E_m against the current density for an idealized $p–n$ junction in GaAs (left axis); the right axis is related to line 5 — dependence of conversion efficiency on photocurrent density at conversion of the monochromatic irradiation with quantum energy $hv = E_g$.

Radioluminescent light is not monochromatic. However, the width of the radiation band in this case is not as broad as sunlight. Therefore, it is possible to approximate the radioluminescent light as monochromatic without material changes in the results for PV cell efficiencies.

Photons with energy of $hv \geq E_g$ can be absorbed by PV cell material. Radiation with lower photon energies is not absorbed in a semiconductor and, hence, is useless from the point of view of photovoltaic conversion. Photons of energy $hv > E_g$ produce "hot" charge carriers which, in addition to excess potential energy E_g, gain excess kinetic energy equal to the difference $hv - E_g$. However, this kinetic energy is rapidly spent heating the crystalline lattice, and the photogenerated carriers reach thermal equilibrium with the lattice. Thus, only part of the energy of the absorbed photons converts into potential energy of the electron–hole pairs.

Maximum energy utilized in an electric load, calculated per one absorbed photon (E_m in Equation 7.11), rises with increasing E_g. To find the efficiency of PV cells, one should determine E_m. One way involves deducing Equation 7.11, where calculation of E_m requires the magnitudes of the electrophysical parameters (carrier mobility and diffusion lengths) of individual semiconductor materials characterized by different E_g magnitudes. The value of maximum efficiency obtained in this way is semiempirical, since it uses semiconductor technology achievements.

Another way of calculating E_m is based on general thermodynamic consideration of the problem.[8-10] A semiconductor photoconverter in thermodynamic equilibrium with its surroundings exchanges energy by means of radiant emission and absorption. Equilibrium black-body radiation that always exists inside a photoconverter material at a given temperature defines the lower limit of saturation current density, i_o, in the p–n junction and is expressed in the following way: [3]

$$i_o = \frac{q\left(n_o^2 + 1\right)E_g^2 kT}{4\pi^2 \hbar^3 c^2} \exp\left(-\frac{E_g}{kT}\right)$$ (7.13)

where n_o is the optical refraction index of the photoreceiver material and $\hbar = h/2\pi$ (h is the Planck constant). Physically this expression presents i_o as a photocurrent generated by the intrinsic black-body irradiation. A minimal calculated value of i_o for silicon at $T = 300$ K is $i_o^{min} \approx 1.4 \ 10^{-15}$ A·cm^{-2} and for gallium arsenide is $i_o^{min} \approx 2.2 \cdot 10^{-20}$ A·cm^{-2}.

In a similar manner one can find the values of i_o for other materials. Magnitudes of n_o for semiconductors usually range from 3 to 4. Knowing i_{ph}, it is possible to find V_{oc} from Equation 7.3, E_m from Equation 7.11, and monochromatic efficiency from Formula 7.12, at $hv = E_g$. These values of efficiency are depicted in Figure 7.5 by lines 1, 2, and 3 for three magnitudes of i_{ph}. In the figure, the abscissa is the wavelength, which corresponds to E_g for individual semiconductor material by the formula ($\lambda_g = 1.24/E_g$). For a chosen material, conversion efficiency magnitudes for light with wavelengths shorter than λ_g must be reduced in $\dfrac{\lambda_g}{\lambda}$ times, pictured in the figure by three tilted straight lines for four semiconductor materials and $i_{ph} = 10^{-6}$

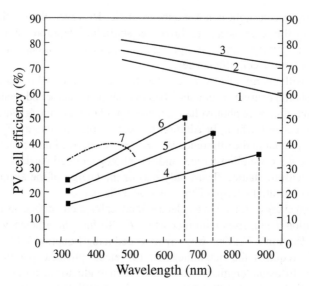

Figure 7.5   Lines 1, 2, and 3: plots of maximum magnitudes of the monochromatic efficiency of an idealized photoconverter for i_{ph} = 0.01, 0.1, and 1.0 A·cm^{-2}, correspondingly, in dependence on the boundary wavelength λ_g of the semiconductor material. Sloped lines 4, 5, and 6, and curve 7: spectral dependencies of conversion efficiency in the idealized photoconverter based on 4: GaAs, 5: Al$_{0.15}$Ga$_{0.5}$As, 6: Al$_{0.32}$Ga$_{0.68}$As, and 7: GaP for i_{ph} =10^{-6} A/cm^2. Optical and recombination losses were not taken into account.

A·cm^{-2}. Thus, the thermodynamically limited maximum value of the efficiency λ_{max} for an idealized light energy photoconverter on the base of each of the chosen materials rises with the increase of photocurrent density, i.e., with light power increase.

Now one formulates the question: what efficiency values are expected for single junction PV cells based on silicon in comparison with those for PV cells based on gallium arsenide at the up-to-date technology of these materials' productions? An answer depends on how close actual magnitudes of the saturation current density i_o are to the minimally limited magnitudes of $i_0^{min} \approx 1.4 \cdot 10^{-15}$ A·cm^{-2} for Si and $2.2 \cdot 10^{-20}$ A·cm^{-2} for GaAs. Analysis of the literature data[11-13] shows that the least value of i_o obtained by calculations using the best published electrophysical parameters[12] for Si is $i_o = 2 \cdot 10^{-14}$ A·cm^{-2}. At the same time, the ordinary value of i_o for GaAs is $i_o = 2 \cdot 10^{-19}$ A·cm^{-2}. Using these values, one may obtain calculated efficiency values at room and some higher temperatures.

Corresponding results are shown in Figures 7.5 and 7.6. Note that for GaAs-based PV cells, the calculated efficiency magnitudes η_{calc} at room temperature almost coincide with the efficiency magnitudes η_{max} of an idealized cell: η_{max} = 35% at i_{ph} = 10^{-6} A/cm^2 and $\eta_{max} \approx 60\%$ at i_{ph} = 10^{-2} A/cm^2. For Si-based cells, the efficiency magnitudes η_{calc} are lower. It is difficult to expect that η_{calc} and η_{max} will approach each other closely in the case of Si-based cells. This would require the radiative recombination probability in the indirect band-gap Si to be close to 100%, as is the case of direct band-gap GaAs.

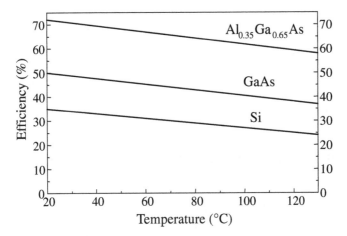

Figure 7.6   Temperature dependencies of the calculated magnitudes of efficiency for photovoltaic cells based on Si, GaAs, and $Al_{0.35}Ga_{0.65}As$ for conversion of light with
wavelength of 633 nm at moderate illumination intensity ($i_{ph} = 3 \cdot 10^{-2}$ A/cm²).

The smooth character of dotted curve 7 in Figure 7.5 reflects the fact of indirect
band-gap structure of GaP and lower value of quantum yield in the blue-green part
of spectrum. Therefore, the achievable efficiency of cells based on "indirect" GaP
under blue light (curve 7) is roughly the same as that in cells based on "direct"
$Al_{0.32}Ga_{0.68}As$ (curve 6) with lower band gap.

The advantage of wide band-gap semiconductors ($Al_{0.35}Ga_{0.65}As$ and GaP)
becomes more evident in the conversion of low-intensity light and at higher cell
temperatures. The temperature increase results in some decrease in E_g, which leads
to decrease in the cell operation voltage. Neglecting the weak temperature dependence of carrier mobility and diffusion lengths allows calculation of values of i_o (T)
in the case under consideration by:

for Si-based p–n junctions:

$$i_0(T) = 3.2 \times 10^{-37} \, Tn_i^2 \left[A \cdot cm^{-2} \right] \qquad (7.14)$$

and for GaAs-based p–n junctions:

$$i_0(T) = 2.1 \times 10^{-34} \, Tn_i^2 \left[A \cdot cm^{-2} \right] \qquad (7.15)$$

where T is temperature in degrees Kelvin and n_i is the intrinsic carrier concentration.

The values of efficiency of the PV cells based on gallium arsenide, $Al_{0.35}Ga_{0.65}As$,
and silicon at overheating by 100° above room temperature are presented in Figure
7.6. Thus, the AlGaAs-based cells are characterized not only by higher estimated
efficiencies but also by a weaker temperature dependency of efficiency in comparison
with GaAs- and Si-based cells, because $Al_{0.35}Ga_{0.65}As$ is a wider energy gap material.

All these efficiency magnitudes have been calculated on assumption of absence of the recombination component of the saturation current i_0, zero optical losses of light (i.e., zero reflection of light from the PV cell surface and zero shadowing of this surface by the contact grid), and zero recombination losses of photogenerated electron–hole pairs before their collection separation by the p–n junction. Effects connected with photogenerated carrier flow in the bulk of the p- and n -regions, as well as the majority carrier flow (ohmic losses), have not been taken into account. Further, the effects of numerous factors will be considered.

7.1.2 Efficiency of Photovoltaic Cells with Optical and Recombination Losses

Optical losses in PV cells are caused by reflection of some rays from the PV cell surface and by shadowing of this surface by the contact grid. A reduction of the reflection coefficient R_o is achieved by coating the cell surface with thin transparent layers of special materials affording interference and mutual suppressing of the light waves reflected from the layer surface and from the layer–semiconductor interface. By these means, the integral reflection coefficient can be decreased from $R_o \approx 35\%$ for an uncoated surface to $R_o \approx 8$ to 10% for application of single-layer antireflection coatings (ARCs) and to $R_o \approx 2$ to 3% for application of double-layer ARCs. Formation of a textured front surface, for example, by creation of regularly or irregularly spaced pyramids brings lowering of the reflection coefficient to about 12 to 15% for an uncoated surface and down to several percent for the application of a single-layer coating. These problems are considered in detail in a number of monographs.[11,12]

No specific features of reflection losses arise in cells operating with increase of the incident angle of rays. Thus, for a smooth cell surface (coated or uncoated), the reflection coefficient hardly increases when the incident angle of rays increases to 30 to 40°, but then it increases sharply, tending to unity at incidence angles close to 90°. In the case of isotropic radioluminescent lighting, the incidence angles may exceed 30 to 40°. Texturing the semiconductor surface and the cover glass surface allows one to decrease the angular dependence of the reflection coefficient significantly and therefore to reduce the integral reflection coefficient for isotropic light.

The second kind of optical loss in PV cells is due to shadowing the sensitive-to-light surface by an electrical contact to the p–n junction illuminated region. To reduce the ohmic losses, such a contact is usually made in the form of a metallic grid characterized by spacing from several millimeters (for operation at low power level) down to 0.1 to 0.3 mm (for operation at high power level). Losses due to shadowing in the case of low intensity of light are several percent. To lower the losses under consideration, special designs of PV cells have been proposed and developed.

A rather effective way of reducing losses caused by contact shadowing is the application of prismatic covers on the illuminated cell surface. The generator of the prismatic relief must be parallel to the contact strips, and the step must be equal to the contact grid step. The profile is chosen in such a manner that rays would be refracted by the outer cover surface and directed onto areas between contact strips.

For example, a cover profile may be presented as an array of cylindrical lenses near each other operating similar to raster-type optical systems. Made from transparent polymer materials and bonded to the cell surface by an appropriate adhesive, such covers can totally exclude losses due to contact shadowing.

Now consider recombination losses in PV cells. Electron–hole pairs in a real cell can be generated by light in both n- and p-regions, depending on the depth at which absorption of photons of given energy took place. Minority charge carriers (electrons in the p-region and holes in the n-region) move toward the p–n junction, the electric field of which executes a collection of electrons in the p-region and holes in the n-region. Part of the minority charge carriers may disappear due to recombination. The collection coefficient Q, equal to the ratio of the quantity of electron–hole pairs separated by the p–n junction to the total quantity of electron–hole pairs generated by light, is defined as a measure of efficiency of the photogenerated carrier collection process. The value of Q essentially depends on the irradiation wavelength due to spectral dependence of the semiconductor material absorption coefficient, so that carrier generation takes place at a different distance from the p–n junction.

In PV cells based on an indirect-gap semiconductor, a considerable portion of radiation generates electron–hole pairs deep in the base region. This results in the smooth spectral dependence of Q in the long-wavelength spectral range of the photoresponse, due to recombination of part of the generated carriers in the base bulk and on the back surface. In cells based on a direct-gap semiconductor, the spectral dependence of Q is characterized by a steeper rise in the long-wavelength range of the spectrum due to the sharper absorption edge. In both cases, the surface absorption takes place in a short-wavelength spectral range, resulting in a more or less strong drop in the magnitude of Q, with decreasing λ due to the beginning of the recombination of photogenerated carriers on the front surface.

The short-wavelength edge configuration of the photoresponse is defined mainly by collection of carriers from the front region, whereas the long-wavelength edge configuration is defined by collection from the base region. We shall carry out a brief analysis of the collection coefficient for a cell with a front illuminated region of n-type. For short wavelengths, when the conditions $\exp(-\alpha w) \ll 1$ and $KL_p \gg 1$ are met, the collection coefficient at the absence of an electric field in the n-region bulk is defined by the expression

$$Q = \frac{1 + \dfrac{s_o}{KD_p}}{\dfrac{s_o L_p}{D_p} sh\dfrac{w}{L_p} + ch\dfrac{w}{L_p}} \tag{7.16}$$

where s_o is the surface recombination velocity; L_p and D_p are the diffusion length and the diffusion coefficient for holes in the front n-region, respectively; w is the thickness of the front n-region; and K is the absorption coefficient of the semiconductor.

As follows from Equation 7.16, the magnitudes of Q in the short-wavelength spectral range are higher in a cell characterized by low surface recombination velocity, small thickness of the front region, and large diffusion length of minority charge carriers inside the front region material.

For the silicon PV cells, it is possible to lower the magnitude of s_o to 10^2 to 10^3 cm·s⁻¹ (sometimes to 10 cm·s⁻¹ or lower) using special treatment of the surface. This results in high photosensitivity in the short-wavelength range of the visible spectrum ($\lambda = 0.4 - 0.45$ μm) at the thickness of the front layer $w = 0.1$ to 0.2 μm.

In gallium arsenide the surface recombination velocity is much higher: $s_o \approx 10^6$ cm·s⁻¹. The growth of the thin wide-gap layer of the $Al_xGa_{1-x}As$ or $In_{0.5}Ga_{0.5}P$ solid alloys on the GaAs surface dramatically decreases the s_o- value on the heterointerface to about 10 to 10^3 cm·s⁻¹, owing to proximity of the lattice parameter for GaAs to those for AlAs and the solid alloys mentioned earlier. The corresponding interface is characterized by a reduced quantity of broken bonds in comparison to a free GaAs surface. The potential barrier on the heterointerface caused by a difference in band-gap magnitudes E_g of contacting materials inhibits charge carriers generated in the narrow-gap material from penetrating the surface vicinity of the wide-gap layer. In this case the value of the collection coefficient in the short-wavelength range of the spectrum is determined by the transmissivity properties of the wide-gap layer operating similar to a window. At thin layers (0.02 to 0.05 μm), keeping high photosensitivity is feasible even for the shortest-wavelength range of the spectrum, $\lambda \leq 0.4$ μm.

The presence of a built-in electric field in the vicinity of the illuminated PV cell surface also increases Q values in the short-wavelength range. In this case the photogenerated carriers are swept from the surface into the semiconductor. The surface recombination velocity is determined as an average velocity of the minority carriers migrating from the semiconductor bulk toward the surface. To suppress this migration it is necessary that the drift velocity of the carriers in the created electric field exceed s-magnitude, i.e., $E > s_o$. The electric field can be created in a semi-conductor bulk at the cost of tilting the conduction and valence band edges — for instance, in a crystal at spatially nonuniform doping or chemical composition. When the doping impurity concentration depends on coordinate, i.e., $N(x)$, the arising built-in electric field is

$$E = -\frac{kT}{q}\frac{d}{dx}[\ln N(x)] \qquad (7.17)$$

In silicon, when a poor surface recombination velocity $s_o \approx 10^4$ cm s⁻¹ takes place, the built-in field necessary to suppress the surface recombination must be on the order of 10^3 V·cm⁻¹. It can be created at the expense of the impurity gradient produced by the diffusion or ion implantation techniques. Such a field is achieved if, across the thickness $w = 1$ μm micrometer, the electron concentration changes from 10^{19} to 10^{17} cm⁻³. In the case of gallium arsenide ($s_o = 10^6$ cm s⁻¹), a substantially higher built-in field is required to suppress the surface recombination so that the drift velocity can approach a saturation value. The necessary built-in field can be realized in a front layer with the E_g gradient established by quiet variation of chemical composition of the semiconductor material in the surface vicinity. In this way it is possible to achieve, for instance, in the front AlGaAs-layer, the differential in E_g

more than 0.5 eV across the layer <1 μm thick, giving the built-in field $E > 5 \cdot 10^3$ V·cm^{-1}.

In general, there are also losses of photogenerated carriers due to bulk recombination, which affects the efficiency of carrier collection more strongly when carriers are generated in the cell base region. For long-wavelength solar irradiation, which passes the front region without absorption, the expression for the collection coefficient of charge carriers from the p-type base region has a form

$$Q = \frac{KL_n}{1 + KL_n} \qquad (7.18)$$

If light is absorbed in the base at a distance from the p–n junction much shorter than L_n, i.e., the condition $KL_n \gg 1$ is met, losses due to recombination are minimal and the collection coefficient Q is close to unity. With wavelength increase the carrier generation region shifts deeper into the base, resulting in increased losses through recombination. For silicon cells this corresponds to the spectral range 0.8 to 1.1 μm and for GaAs 0.85 to 0.9 μm, in which a smooth lowering of Q takes place with increasing λ.

One of the best ways to increase photosensitivity of a cell in the long-wavelength spectral range is to choose better-quality material with a higher electron diffusion length L_n. The recombination on the rear cell surface commonly decreases by increase of the surface impurity concentration, known as fabrication of the back surface field structure. To prevent additional optical losses, a reflection coating can be placed on the rear surface of the cell wafer and can double the distance passed in the base by the low-energy photons as well as increase the probability of their absorption. A similar effect of the light absorption increase is achieved by texturing front and rear surfaces of PV cells, which arranges the geometrical scattering and trapping of incident light inside the wafer. More comprehensive analysis of optical and recombination losses in various types of PV cells[12] have been completed.

In conclusion, recall that considered optical and recombination losses may be about 15% for monochromatic light. At low illumination intensities, additional losses caused by leakage currents and the recombination component of saturation current should be taken into account, as well. To obtain the achievable in practice efficiencies of radioluminescent radiation conversion, the values shown in Figure 7.5 must be multiplied by a factor of about 0.7 to 0.75.

Figure 7.7 shows calculated efficiencies of the PV cells, taking into account optical, ohmic, and recombination losses and current leakages. The efficiencies are shown at illumination intensity, which corresponds to photocurrent density of 10^{-6} A/cm^2. This value is close to the conditions of radioluminescent lighting. Therefore, the achievable-in-practice efficiency of about 35% may be received in a PV cell based on $Al_{0.32}Ga_{0.68}As$ and illuminated by red phosphor. Efficiency of about 28% is achievable in PV cells based on $Al_{0.32}Ga_{0.68}AsP$ illuminated by green phosphor and in GaP cells under blue phosphor illumination. Reducing photocurrent decreases the achievable efficiencies due to influence of leakage current.

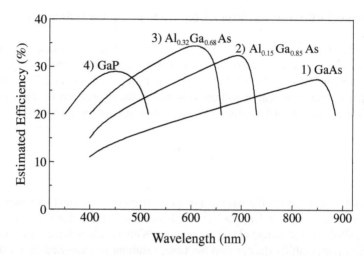

Figure 7.7 Achievable-in-practice estimated efficiencies vs. light wavelength for PV cells based on different semiconductors: 1) GaAs, 2) $Al_{0.15}Ga_{0.85}As$, 3) $Al_{0.32}Ga_{0.68}As$, and 4) GaP. Optical and recombination losses were taken into account in this calculation; $i_{ph} = 10^{-6}$ A/cm^2.

7.2 PHOTOVOLTAIC CELLS AND ARRAYS BASED ON AlGaAs–GaAs AND AlGaP–GaP HETEROSTRUCTURES

The use of III-V heterostructures for fabrication of photovoltaic cells provides high-conversion efficiency of radioluminescent light. Among the large number of heterojunctions estimated with regard to their applicability in these cells, AlGaAs–GaAs heterojunctions may find the widest application due to the well-matched lattice constants of aluminum-gallium arsenide and gallium arsenide and because aluminum-gallium arsenide has a near-optimal band gap (E_g = 1.7 to 1.95 eV) for effective conversion of red, yellow, green, and blue phosphor fluorescence wavelength of 700 to 400 nm.

The first photovoltaic cells based on AlGaAs–GaAs heterojunctions were produced at the Ioffe Physico-Technical Institute.[14,15] The basic low-band material was GaAs and a wide-gap window was made of $Al_xGa_{1-x}As$ with $x > 0.6$. Such a heterostructure is illuminated through the window; the light with photon energy $hv > 1.4$ eV exceeding the band-gap value of GaAs is absorbed in it, while the minority carriers are separated by the p–n junction field. Because of the close lattice parameters of the contacting materials, the interface in $Al_xGa_{1-x}As$–GaAs heterojunctions has a low density of surface states, providing a highly effective accumulation of carriers. Thin layers of $Al_xGa_{1-x}As$ solid solutions close in composition to aluminum arsenide (x = 0.8 to 0.9) are almost completely transparent to visible light, making PV cells very sensitive in the spectral region λ = 0.4 to 0.9 μm. Extensive investigations have resulted in fabrication of AlGaAs heterocells exhibiting high performance characteristics in high and in low intensities of illumination.[16-20] Before analyzing the photovoltaic cells, properties of heterojunctions in the AlGaAs system will be considered.

7.2.1   Properties of AlGaAs–GaAs Heterojunctions

The lattice parameters of gallium arsenide ($a' = 0.5653$ nm) and aluminum arsenide ($a' = 0.566$ nm) differ by 0.12% at room temperature. Owing to the different thermal expansion coefficients[5,21] of GaAs ($K_T = 6.5 \cdot 10^{-6}$ deg^{-1}) and AlAs ($K_T = 5.2 \cdot 10^{-6}$ deg^{-1}) at epitaxial growth temperatures of solid solutions ($T = 600$ to 900°C), the lattice parameters of the substrate and layers become still closer, thereby facilitating growth of high-quality $Al_xGa_{1-x}As$ solid solutions free from misfit dislocations on the GaAs substrate. However, as the temperature is reduced to room temperature, elastic strain arises because of different K_T values. Reducing the thickness of the $Al_xGa_{1-x}As$ layer can reduce the strain in the active region of an AlGaAs–GaAs cell.

Aluminum arsenide is a material sensitive to corrosion. However, when doped with a stabilizing component (gallium) to form a solid solution with AlAs, its stability becomes much higher in a humid atmosphere. Addition of 10 atomic percents of gallium makes $Al_xGa_{1-x}As$ solid solutions with $x = 0.9$ applicable for the production of PV cells and other devices possessing stable characteristics.

Aluminum arsenide is an indirect semiconductor, with $E_g = 2.17$ eV. The energy gap between the direct minimum of the conduction band and the top of the valence band in it is about $E_g = 3.0$ eV. The relationship between the band gap and the composition of $Al_xGa_{1-x}As$ solid solutions[20] can be obtained by interpolating direct and indirect energy band gaps of AlAs and GaAs. In the range of $Al_xGa_{1-x}As$ compositions from $x = 0$ (GaAs) to $x < 0.4$ and of band-gap energy $E_g = 1.40$ to 1.95 eV (300 K), this material has a direct energy band structure, but it is indirect in the range $x > 0.4$.

The $Al_xGa_{1-x}As$ layer functioning as a window in AlGaAs–GaAs PV cells should be transparent to the light. In direct $Al_xGa_{1-x}As$ solid solutions, the absorption edge $K = f(9)$ is nearly as abrupt as in GaAs.[22] In indirect materials with absorption coefficient $K < 10^4$ cm^{-1}, K rises with increasing $h\nu$ at a rate that is lower, the higher the AlAs content. The absorption coefficient at $x = 0.8$ initially ($h\nu < 2.5$ eV) increases slowly and only at $h\nu > 2.6$ eV begins to increase rapidly to (1 to 2)$\cdot 10^4$ cm^{-1}. This means that the absorption edge in indirect solid solutions is largely determined by the direct energy gap. For fabrication of PV cells from these materials, the important conclusion is that almost complete transparency of $Al_xGa_{1-x}As$ layers to short-wavelength light ($\lambda = 400$ to 500 nm) may be attained with fairly thin layers (0.1 μm) and x values in the range 0.8 to 0.9 (solid solutions with $x > 0.9$ cannot be used because of their poor resistance to corrosion).

Another parameter important for designing antireflection coatings is the refractive index n_o of the front layer. In $Al_xGa_{1-x}As$ solid solutions, the refractive index varies nearly linearly from $n_o = 3.6$ for GaAs to $n_o = 3.0$ for AlAs at $\lambda = 0.9$ μm.[23] Hence, the refractive index at $x = 0.8$ to 0.9 and $0 = 0.9$ μm is $n_o = 3.1$ to 3.05 and increases with decreasing λ. Thus, with λ varying from 0.9 to 0.7 μm, the refractive index of $Al_{0.4}Ga_{0.6}As$ rises from 3.34 to 3.5.

The band diagram of an abrupt heterojunction differs from that of a homo p–n junction in the appearance of offsets in the valence and conduction bands (ΔE_v and ΔE_c, respectively). The band diagrams shown in Figure 7.8 refer to an ideal heterojunction containing no surface states at the interface. Considering the small difference in lattice parameters of contacting materials, $Al_xGa_{1-x}As$–GaAs heterojunctions may be placed into the category of structures, which exhibit a low recombination rate at the interface.

The experimental values of ΔE_c obtained for direct AlGaAs solid solutions fall in the range $\Delta E_c = 0.55$ to $0.85\ \Delta E_g$.[24,25] The dependence of ΔE_v on the parameter x for $Al_xGa_{1-x}As$–GaAs heterojunctions for the entire composition range is given in Batey and Wright[25]: $\Delta E_v = 0.55x$. Using these data, the conduction-band offset for direct solid solutions is $\Delta E_c = 0.9x$ and the contributions to the net offset from ΔE_c and ΔE_v amount to 60 and 40%, respectively. In the range of indirect compositions, the value of ΔE_c drops with increasing x. For example, according to Batey and Wright,[25] $\Delta E_c = 0.2$ eV and $\Delta E_v = 0.5$ eV for $Al_{0.9}Ga_{0.1}As$–GaAs; in other words, most of the offset will be due to the valence band in this case.

As seen in Figure 7.8, the appearance of the band offsets in abrupt N–p and P–n heterojunctions (the first symbol indicates the type of conductivity in the wide-gap material) gives rise to additional potential barriers (the kinks in the figures) that present an obstacle to minority carrier separation. A barrier in the conduction band

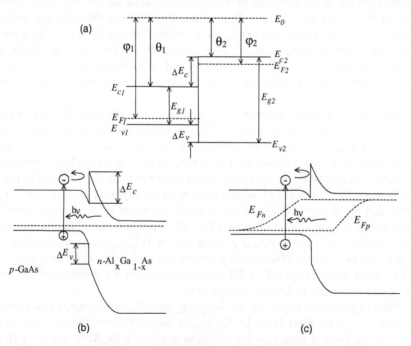

Figure 7.8 Band diagrams of a p-GaAs–n-$Al_xGa_{1-x}As$ heterojunction: (a) prior to formation of heterojunction; (b) zero bias voltage; (c) forward bias.

of a *N–p* heterojunction impedes the separation of electrons (Figure 7.8), and a barrier in the valence band of an *P–n* heterojunction does the same to holes generated in the low-band-gap material. High-energy carriers can overcome the barrier or tunnel through it. The efficiency of these processes increases with the doping level of the wide-band-gap material due to the barrier width decrease. It also decreases in heterojunctions operating in reverse bias, when the probability for a carrier to overcome the barrier increases. Photovoltaic cells operate with a *p–n* junction in forward bias equal to the voltage at the point of optimum load. In this case, the potential barriers for the separation of minority carriers will be higher (Figure 7.8c).

One way of eliminating these barriers is to incorporate, in the heterojunction, intermediate layers with a composition gradually varying over a thickness d_n (Figure 7.9a, c). In these graded heterojunctions, provided the buffer layers are of sufficient thickness, there are no barriers to minority carriers not only in reverse or zero bias but also in forward bias, corresponding to the operating regime of the PV cell. An important problem here is optimization of the thickness (d_n) of the graded layer necessary for complete elimination of the potential barrier to the minority carriers. The required thickness *d* of the layers increases with increase of the space-charge region width in the *p–n* junction, i.e., with the doping level decrease in the contacting materials. For the majority of carrier concentrations typical for Al$_x$Ga$_{1-x}$As–GaAs

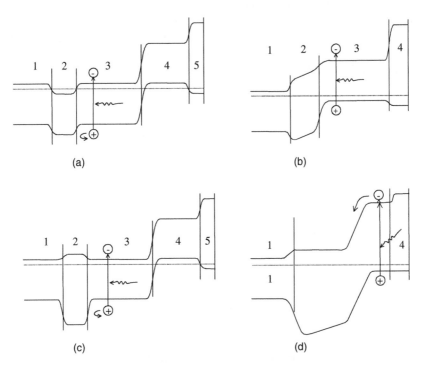

(a)   (b)

(c)   (d)

Figure 7.9   Band diagrams of AlGaAs heterostructures for photovoltaic cells with back-surface field (a, c); of AlGaAs heterostructures for photovoltaic cells with a thin wide-band gap layer in the space-charge region (b); and of a structure with Al$_x$Ga$_{1-x}$As (E_g = 1.5 to 1.9 eV) in the photoactive layer (d).

PV cells, the graded layer thickness must be about 5 nm. Buffer layers of such a thickness can be obtained in $Al_xGa_{1-x}As$–GaAs heterostructures grown by liquid-phase epitaxy, using isothermal mixing of melts containing different amounts of aluminum. This can also be done by submelting the substrate and subsequent crystallization of the solid solution of graded composition, as well as by high-temperature diffusion of the heterojunction components Al and Ga.

A simple and common way of eliminating potential barriers to minority carriers in PV cells based on AlGaAs–GaAs heterojunctions consists of displacing the p–n junction from the interface into gallium arsenide (Figure 7.9a, c) by a distance smaller than the electron diffusion length in p-GaAs. In such a structure, electron–hole pairs generated near the p–n junction in n-GaAs and in p-GaAs meet no obstacles on their way to the space-charge region of the junction, while the potential barrier in the isotype heterojunction p-AlGaAs–p-GaAs does not hinder the movement of majority carriers (holes).

7.2.2 AlGaAs–GaAs Heterostructures for Photovoltaic Cells

Figure 7.9 shows some possible types of AlGaAs heterostructures for conversion of radioluminescent light. The band diagrams relate to heterostructures with graded buffer layers incorporated into AlGaAs–GaAs heterojunctions. The simplest type of heterostructure is that with the composition n-GaAs–p-GaAs–p-AlGaAs. The basic material for their fabrication is n-GaAs with majority carrier concentration (1 to 5)$\cdot 10^{16}$ cm^{-3}, grown mostly by liquid-phase epitaxy. The p-GaAs layer ($p = 10^{18}$ to 10^{19} cm^{-3}) of 0.5 to 2.0 μm thick is grown either epitaxially or by zinc or beryllium diffusion during the growth of $Al_xGa_{1-x}As$ solid solution doped with one of these impurities. The diffusion produces a quasi-electric field arising from the acceptor concentration gradient, which enhances the effective diffusion length of electrons generated by light in the p-GaAs layer if the acceptor concentration in p-GaAs is high enough (up to 10^{19} cm^{-3}) near the interface to reduce sheet resistance in the front layers of the cells.

Higher conversion efficiencies of the short-wavelength part of the spectrum have been obtained in structures with the wide-band-gap layer of smaller thickness. To reduce the contact resistance, an additional p^+-GaAs layer is sometimes grown onto the solid solution layer and subsequently etched off in places free from metallization. Another way of enhancing the short-wavelength photosensitivity is to use, for the front layer, a solid solution of graded composition with the band gap increasing toward the illuminated surface.[26,27] The field due to E_g gradient significantly enhances the effective electron diffusion length and suppresses the surface recombination of electron–hole pairs generated near the surface by short-wavelength light.

As noted earlier, introduction of a potential barrier at the back of the photoactive region improves collection efficiency of minority carriers generated by low-energy photons in the base. This barrier is made by growing buffer layer 2 (Figure 7.9a) of n-GaAs,[28,29] doped to a level exceeding that in active layer 3 or of n^+-$Al_xGa_{1-x}As$ layer 2 (Figure 7.9c).

Thin wide-band-gap layer 2 (n-$Al_xGa_{1-x}As$) was introduced in the structure shown in Figure 7.9b to separate the p- and n-layers.[30] This buffer layer ensures that the *p–n* junction effectively separates electron–hole pairs generated in layer 3. The band gap in the space-charge region of the *p–n* junction significantly increases and the recombination component of the reverse saturation current is lowered by several orders of magnitude. The structure with wide-band-gap (E_g) layer in the photoactive region (Figure 7.9d) is optimal for conversion of the visible light, taking into account the increase of the output voltage with the E_g increase.

Among the structures discussed, heterostructures with a thin p-Al$_x$Ga$_{1-x}$As window ($x = 0.8$ to 0.9) have found a very wide application. Solar cells based on these structures have high efficiencies at 1 sun and concentrated sunlight conversion. Therefore, fabrication of these structures and characteristics of low-intensity PV cells made from them will be covered in detail next.

7.2.3   Fabrication of AlGaAs–GaAs Photovoltaic Cells

The principal methods used for the fabrication of photovoltaic cell heterostructures in the Al-Ga-As system are liquid-phase epitaxy (LPE) and metallorganic chemical vapor deposition (MOCVD).

Consider three main processes involved in the fabrication of n-GaAs–p-GaAs–p-AlGaAs heterostructures by liquid-phase epitaxy: the growth of the n-GaAs base region, the formation of the p-GaAs layer by diffusion or epitaxy, and the growth of p-AlGaAs. The growth of structures for PV cells is carried out in a purified hydrogen atmosphere by cooling the Ga + As melt saturated with arsenic to produce n-GaAs and the Ga + Al + As + Zn (Be) melt to produce p-Al$_x$Ga$_{1-x}$As. To obtain the majority carrier concentration in the range of $n \cong 10^{16}$ cm^{-3}, the melt annealing is carried out during 5 to 10 h at a temperature of $T_0 = 850$ to $800°$C.

Crystallization of Al$_x$Ga$_{1-x}$As solid solution from the liquid phase has been studied quite extensively.[2,5,31-37] A fragment of the phase diagram for the compositions $x > 0.65$ used in PV cells is shown in Figure 7.10. Usually, a AlGaAs layer with $x = 0.7$ to 0.9 is grown from a thin (0.5 to 1.0 mm) Ga + Al + As melt layer at a temperature (T_0) below 700°C at the start of the epitaxial growth. Cooling from $T_0 = 700°$C to room temperature produces layers of about 1.0 to 0.3 μm thick. Thinner layers, necessary for extending the photoresponse spectrum to the shorter wavelengths, are obtained by lowering T_0 and melt thickness, thus reducing the arsenic content in the melt.[38] For example, 0.1 μm layers can be produced at $T_0 = 600°$C by reducing the melt thickness to 0.1 mm or by interrupting the cooling and then removing the unused melts.

As mentioned earlier, high short-wavelength photosensitivity can be obtained in AlGaAs heterostructures with a quasi-electrical field in the top layer produced by the band-gap gradient. Structures with a thin graded band-gap layer can be grown using the following modification of liquid-phase epitaxy.

Thin (< 0.3 μm) graded band-gap layers have been crystallized[27] under equilibrium conditions between a GaAs substrate and a Ga + Al + As melt nonsaturated with arsenic. When the melt is brought into contact with the substrate, the latter melts to some extent and its thin layer near the interface becomes rich in gallium. Then follows the epitaxial growth of the layer, in which the aluminum arsenide

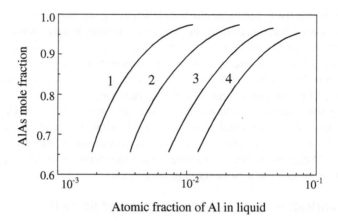

Atomic fraction of Al in liquid

Figure 7.10 Dependence of AlAs content in $Al_xGa_{1-x}As$ solid solutions on Al concentration in the melt at initial growth temperatures (T_0): 600° (1), 700° (2), 800° (3), and 900°C (4).

content gradually increases due to aluminum diffusion into the gallium-rich layer of the melt near the substrate.

$Al_xGa_{1-x}As$ solid solutions with a high AlAs content ($x = 0.8$ to 0.9) are usually doped with zinc, magnesium, or beryllium. In the case of zinc, the ionization energy ΔE_{Zn} of the acceptor level varies from 0.02 eV in GaAs to 0.07 to 0.09 eV in $Al_{0.8}Ga_{0.2}As$,[39,40] which accounts for a low zinc concentration in the solid solution despite a high zinc content in the melt. In shallower magnesium or beryllium levels ($\Delta E_{Be} = 0.05$ eV in $Al_{0.8}Ga_{0.2}As$), the free carrier concentration will be higher and the front layer resistance will be lower.

Since the late 1970s, AlGaAs–GaAs heterostructures have been produced by metalorganic chemical vapor deposition (MOCVD), using metalorganic compounds of Group III elements and hydrides of Group V elements.[41,42] The MOCVD process is carried out in reactors, with the substrate holder heated to 650 to 700°C at atmospheric or lower pressure and with hydrogen as the carrier gas. An advantage of MOCVD is the possibility of fabricating multilayer structures in high-yield reactors with layers of specified composition and thickness varying from 1 to 10 nm to several microns.

Because MOCVD is capable of producing GaAs epitaxial layers on silicon and germanium substrates, it has potential for fabrication of low-cost, high-efficiency AlGaAs photovoltaic cells on these substrates. Of course, a successful application of the MOCVD technique necessitates use of up-to-date automated equipment and high-purity carrier gases and materials. In addition, special precautions should be taken when handling toxic metalorganic compounds and hydrides.

An important phase in photovoltaic cell fabrication is the formation of stripe contacts to the thin-front layer. Multilayer or composite coats are deposited for better adhesion and lower contact resistance: Au/Zn, Ag/Mn, Ag/Zn, Cr/Ni, Pd/Ni, and Pd/Zn/Au. The Au/Ge eutectic alloy is typically used to make contacts to the back surface of n-GaAs substrate.

Stripe contacts in structures with a thick AlGaAs layer are applied directly to the solid solution (Figure 7.11a) subjected to Zn diffusion from the gaseous phase to increase the acceptor density at the surface; recessions for the stripe contacts are etched out to improve the adhesion. In structures with a thin wide-band-gap layer, the contacts are made by one of the methods illustrated in Figure 7.11: directly to the AlGaAs layer (b), to the p-GaAs layer inside the stripe windows etched in the p-Al$_x$Ga$_{1-x}$As layer (c), or to a p^+-GaAs layer grown onto the solid solution and subsequently etched off in areas free from contacts (d).

Antireflection coating for PV cells based on AlGaAs structures is made by anodic oxidation of the Al$_x$Ga$_{1-x}$As surface or by deposition of thin films of Si$_3$N$_4$, ZnS, Ta$_2$O$_5$, and others. With the Si$_3$N$_4$ antireflection coat (74 nm), the reflection loss is about 8%. A two-layer coat composed of, for example, Ta$_2$O$_5$ (53 nm) and SiO$_2$ (76 nm) can reduce reflection loss to 4%.[43] The same result can be obtained for a two-layer coat of ZnS and MgF$_2$.

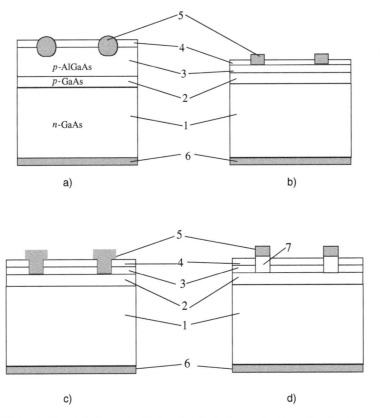

Figure 7.11   Different designs of n-GaAs–p-GaAs–p-Al$_x$Ga$_{1-x}$As PV cells; 1) n-GaAs; 2) p-GaAs layer; 3) p-Al$_x$Ga$_{1-x}$As layer; 4) antireflection coat; 5) stripe contacts; 6) back contact; 7) cap p-GaAs layer.

7.2.4 Spectral Characteristics and Photocurrent of AlGaAs–GaAs Photovoltaic Cells

The photoresponse spectra of cells based on heterostructures of the n-GaAs–p-GaAs–p-AlGaAs type vary with the collection efficiency of carriers generated in n-GaAs and p-GaAs and with the transmission spectrum of the wide-band-gap solid solution layer. The first factor determines the shape of the long-wavelength photosensitivity edge and the spectral variation of photoresponse, while the second determines shape and position of the short-wavelength edge of photosensitivity.

The short-wavelength edge of photoresponse depends on the solid solution composition and thickness. The AlAs content in the $Al_xGa_{1-x}As$ layer is usually chosen in the range of $x = 0.8$ to 0.9. As seen from Figure 7.12, as the $Al_xGa_{1-x}As$ layer thickness decreases from 10 to 0.05 µm, the short-wavelength photoresponse edge shifts from $9 = 0.55$ µm to $9 = 0.4$ µm, which means that the wide-band-gap layer of 0.05 to 0.1 µm is almost completely transparent to visible light. Maximum short-wavelength photoresponse values have been obtained in cells with wide-band-gap layer thickness of $D = 0.03$ to 0.05 µm. These ultrathin layers were grown either by liquid-phase epitaxy[27,43-45] or by MOCVD epitaxy.[20,29,46]

In addition to composition and thickness of the $Al_xGa_{1-x}As$ layer, the short-wavelength spectrum is strongly influenced by the antireflection coat, especially around $\lambda = 0.4$ µm. The maximum value of external quantum yield at $\lambda = 0.4$ µm to 0.5 µm has been obtained for solar cells with $D = 0.03$ µm coated with an Si_3N_4 layer (Figure 7.12, curve 1) or with $ZnS + MgF_2$ ARC layers (Figure 7.13, curve 1).

In cells based on n-GaAs–p-GaAs–p-AlGaAs heterostructures in which the p-GaAs layer thickness d is greater than 0.5 µm, the photocurrent is determined mainly by the relation between layer thickness and electron diffusion length (L_n).

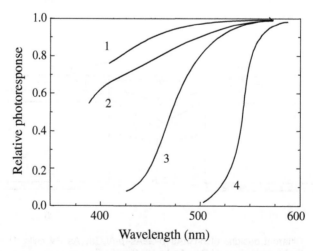

Figure 7.12 Experimental quantum yield in the short-wavelength range of photovoltaic cells based on an n-GaAs–p-GaAs–p-Al$_x$Ga$_{1-x}$As structure at different thicknesses (D) of the wide-band-gap Al$_x$Ga$_{1-x}$As layer; µm: 1) 0.03; 2) 0.07; 3) 0.5; 4) 10.

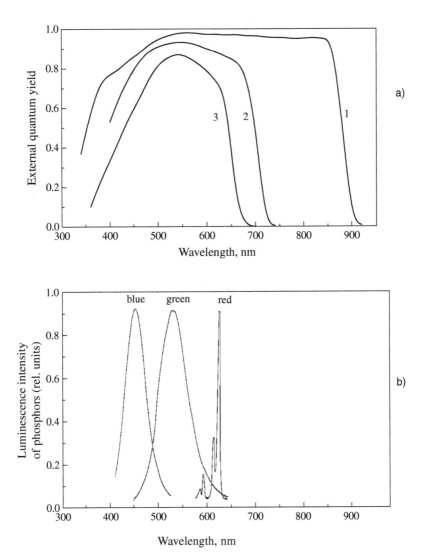

Figure 7.13   a) Spectral response of advanced PV cells based on $Al_xGa_{1-x}As$ with different band-gap energy in photoactive region: 1) 1.4 eV; 2) 1.72 eV, 3) 1.93 eV; b) emission spectra of blue, green, and red phosphors in tritium-fueled lamps.

The photocurrent at L_n = 3 to 6 μm was found to be close to its maximum value at d (p-GaAs) equal to 0.5 to 1.5 μm. For L_n = 5 to 6 μm, d (p-GaAs) may be increased to 3 μm with photocurrent values still remaining fairly high. High-efficiency radiative recombination effects may give rise to secondary photons that may be reabsorbed, leading to larger diffusion lengths. This photon recycling effect may increase the photoresponse in cells with a thicker active region.

Figure 7.13a shows the spectral responses of AlGaAs–GaAs cells in which the external quantum yield increases owing to decrease of the optical and recombination losses. The cells with spectral response shown by curves 1, 2, and 3 are very effective in converting the phosphor light from tritium-fueled lamps (Figure 7.13b).

7.2.5 Current-Voltage Characteristics of AlGaAs–GaAs Photovoltaic Cells

Due to high probability of radiative recombination at photocurrents higher than 10^{-1} A/cm^2 in GaAs p–n junctions, the saturation current amounts to $i_o = 10^{-19}$ A/cm^2, while the minimum value is $i_o = 10^{-20}$ A/cm^2. The close values of the actual and theoretical currents are a necessary condition for achieving open-circuit voltage and fill-factor values close to their theoretical limits. Current through the load connected in parallel with an illuminated PV cell can be represented as a difference between the photocurrent and the dark current (i_d) in the junction. For $i_d \gg i_o$, the latter represents the sum of the injection and recombination components: $i_d = i_{01}$ exp $(qU/kT) + i_{02}$ exp $(qU/2kT)$, where i_{01} and i_{02} are the injection and recombination components of the saturation current, respectively.

At a current density of $I \leq 3 \cdot 10^{-2}$ A/cm^2, photovoltaic cells based on GaAs p–n junctions exhibit a combined mechanism of current production (Figure 7.14, curve 1), with the total saturation current equal to 10^{-13} to 10^{-15} A/cm^2. Minimum values of i_o in this case are found in structures fabricated from high-quality n-GaAs grown by liquid-phase epitaxy or MOCVD. At a current density of $I < 10^{-5}$ A/cm^2 corresponding to the radioluminescent lighting, the cells exhibit a recombination mechanism of current production.

The recombination component of the saturation current is proportional to the intrinsic carrier concentration, which decreases with the band-gap energy increase. In n-GaAs–n-Al$_{0.4}$Ga$_{0.6}$As–p-GaAs–p-Al$_{0.8}$Ga$_{0.2}$As structures with a thin Al$_{0.4}$Ga$_{0.6}$As layer in the space-charge region, the value of recombination current saturation is

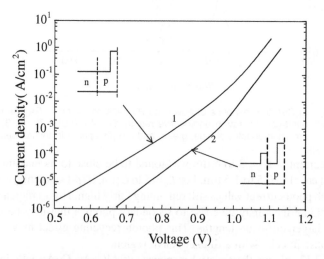

Figure 7.14 Experimental dark current–voltage characteristics of PV cells based on (1) a heterostructure with GaAs p–n junction and (2) a structure with Al$_{0.4}$Ga$_{0.6}$As layer in the space-charge region.

appreciably lower — actually below $i_{02} = 10^{-15}$ A/cm^2. The transition from the recombination to the injection mechanism of current flow occurs here (Figure 7.14, curve 2) at a current density two orders of magnitude lower than in a GaAs p–n junction structure. As a result, the respective PV cells with increased band gap in the space-charge region allow obtaining of a higher open-circuit voltage and fill factor. The difference between curves 1 and 2 in Figure 7.14 increases with the current decrease. Therefore, the advantage of structures with a wide-band-gap semiconductor in the space-charge region is more pronounced under low intensity of illumination.

The saturation-current component due to injection component decreases and V_{oc} increases for higher doping levels in the n- and p-regions. However, leakage and tunnel currents are more likely at higher doping levels. For these reasons, the optimum free-carrier concentrations near the p–n junction are $n = (1 \text{ to } 3) \, 10^{17}$ cm^{-3} and $p = (1 \text{ to } 2) \, 10^{18}$ cm^{-3} and the maximum V_{oc} values of 1.04 to 1.05 obtained in GaAs-based cells for $i_{ph} = 3 \cdot 10^{-2}$ A/cm^2.

Figure 7.15 illustrates the open-circuit voltage V_{oc} vs. the photocurrent for different values of injection (i_{01}) and recombination (i_{02}) saturation currents for GaAs at $T = 300$ K. Curve 1 corresponds to $i_{01} = 10^{-20}$ A/cm^2 and $i_{02} = 10^{-12}$ A/cm^2, values achievable in GaAs PV cells and with a space-charge region located in a semiconductor with a higher band gap than in GaAs. Curves 3 and 5 demonstrate that the higher saturation currents (lower values of L_D and $\square$) due to poor material quality and the higher concentrations of deep levels due to impurities and defects can appreciably reduce V_{oc}, especially at $i_{ph} < 0.1$ A/cm^2. At $i_{ph} > 1$ A/cm^2, the open-

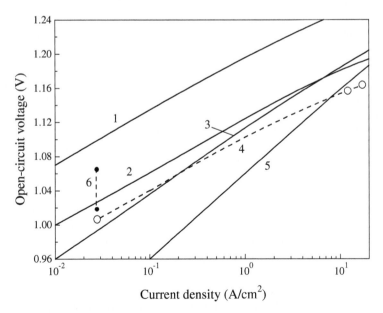

Figure 7.15   Theoretical (1,2,3,5) and experimental (4) dependencies at 300 K of the open-circuit voltage vs. photocurrent density for different values of the injection (i_{01}) and recombination (i_{02}) components of saturation currents i, A/cm^2: 1) $i_{01} = 10^{-20}$, $i_{02} = 10^{-12}$; 2) $i_{01} = 10^{-19}$, $i_{02} = 10^{-11}$; 3) $i_{01} = 10^{-19}$, $i_{02} = 10^{-10}$; 4) experiment; 5) $i_{01} = 10^{-19}$, $i_{02} = 10^{-19}$, $i_{02} = 10^{-9}$; 6) measured V_{oc} values at $i_{ph} = 3 \cdot 10^{-2}$ A/cm^2.

circuit voltage varies mainly with i_{01} because, despite a fairly large recombination component of the saturation current ($i_{02} = 10^{-10}$ A/cm^2), the injection-current mechanism becomes dominant. The experimental dependence $V_{oc} = $ f(i_{ph}) in Figure 7.15 (curve 4) can be fitted with the calculation, assuming $i_{01} = 10^{-19}$ A/cm^2 and $i_{02} = 10^{-10}$ A/cm^2.

At current density of $i \leq 10^{-6}$ A/cm^2, the optimal doping level in the n-region is $n \leq 10^{16}$ cm^{-3}, owing to a lower value of leakage current in the structures with lower doping level.

The dependence of V_{oc} as a function of band-gap energy of AlGaAs in the active region of the fabricated PV cells under two illumination intensities is shown in Figure 7.16. The V_{oc} values in the cells using a green LED at $i_{ph} \cong 1$ μA/cm^2 (Figure 7.16, curve 1) increase from 0.5 V in the cell based on GaAs to $V_{oc} = 0.9$ V in the cell based on Al$_{0.35}$Ga$_{0.65}$As ($E_g = 1.9$ eV). The V_{oc} values are lower under a green tritium lamp at $i_{ph} \cong 100$ nA/cm^2 (Figure 7.16, curve 2): $V_{oc} = 0.36$ V in a GaAs-based cell and 0.8 V in a Al$_{0.35}$Ga$_{0.65}$As-based cell.

The fill factor (FF) decreases with higher saturation current and lower photocurrent value. Figure 7.17 (curves 1 to 3) illustrates the effect of the injection (i_{01}) and recombination (i_{02}) components of the reverse saturation current on FF for concentrated sunlight (neglecting the ohmic losses). For example, an increase in i_{02} from 10^{-11} to 10^{-10} A/cm^2 causes FF to decrease from 0.86 to 0.8 at $i_{ph} = 3 \cdot 10^{-2}$ A/cm^2. At lower photocurrents ($i_{ph} < 10^{-2}$ A/cm^2), the difference between these FF values grows, demonstrating a stronger influence of reverse saturation current on FF. Experimental curves 4 to 5 in Figure 7.17 are visibly affected by ohmic losses, but at $i < 10^{-2}$ A/cm^2, this influence is insignificant.

The influence of Al$_x$Ga$_{1-x}$As band-gap value in the p–n junction region on the density of the reverse dark currents (i_d) has been studied in the cells. In Figure 7.18 the reverse branches of the dark current–voltage characteristics of the cells were plotted with $x = 0$, 0.13, 0.3, and 0.62 (curves 1 to 4, respectively) obtained at room temperature. It is evident that major changes in the values of reverse current occurred as a result of increasing the AlAs content in the Al$_x$Ga$_{1-x}$As near the p–n junction. We then observed reduction in the reverse-current value by two orders of magnitude with increasing x in the range of 0 to 0.3 to values $i_d < 10^{-12}$ A/cm^2 (curve 3) and subsequent increase in i_d with an increase in $x = 0.62$ (curve 4). The increase in the dark current on transition to the indirect band-gap compositions indicated that the crystal quality of the solid solutions with high AlAs concentration was worse.

Figure 7.19 shows the current-voltage characteristics of three PV cells based on Al$_x$Ga$_{1-x}$As ($x = 0.34$ to 0.37) at photocurrent densities of about 10^{-6} A/cm^2 and $2 \cdot 10^{-7}$ A/cm^2. Open-circuit voltages 0.92 and 0.93 and fill factor of 0.84 to 0.86 have been achieved in the cells based on Al$_{0.36}$Ga$_{0.64}$As and Al$_{0.37}$Ga$_{0.63}$As at $i_{ph} \cong 10^{-6}$ A/cm^2 (curves 1 and 2). $V_{oc} = 0.83$ and $FF = 0.81$ were measured in one of the cells at $i_{ph} = 2 \cdot 10^{-7}$ A/cm^2 (curve 4).

Figures 7.20 and 7.21 show the illuminated I–V curves under the low-intensity green and blue LED lighting. It will be demonstrated at the end of Section 7.2 that tritium-powered lamps generate photocurrent density of 100 to 150 nA/cm^2 and output electrical power density of 50 to 100 nW/cm^2 in developed cells based on AlGaAs.

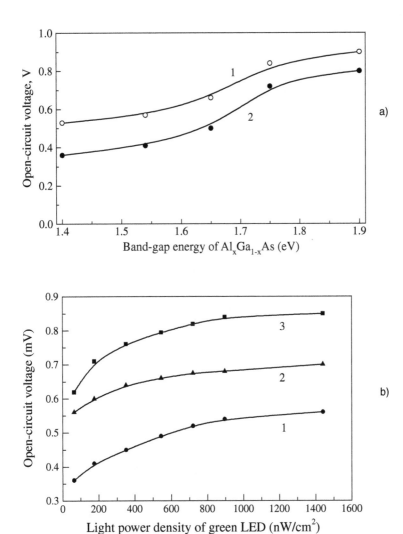

Figure 7.16 Dependencies of open-circuit voltage a) as a function of AlGaAs band gap in the photoactive region of PV cells at photocurrent densities: 1) $i_{ph} \cong 10^{-6}$ A/cm^2 (under green LED), 2) $i_{ph} \cong 10^{-7}$ A/cm^2 (green tritium lamp); b) as a function of light power density from green LED in the cells with a different band-gap energy in photoactive area: 1) 1.43 eV, 2) 1.82 eV, 3) 1.9 eV.

7.2.6 Temperature Dependencies of AlGaAs–GaAs PV Cell Parameters

As the operating temperature of PV cells rises, efficiency drops, mainly due to reduced V_{oc} (Figures 7.22 and 7.23). The larger band gap of Al$_x$Ga$_{1-x}$As and GaAs, relative to Si, accounts for better temperature stability of AlGaAs–GaAs cell parameters. According to theoretical concepts, the temperature coefficients of V_{oc} and FF increase with decreasing photocurrent. Calculated values of the temperature

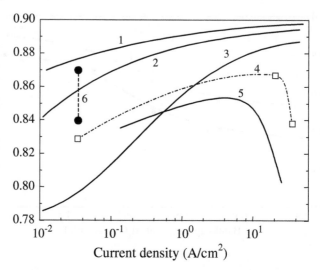

Figure 7.17   Calculated (1 to 3) and experimental (4 to 6) dependencies of the fill factor on the photocurrent density for GaAs PV cells at different values on injection (i_{01}) and recombination (i_{02}) saturation currents, A/cm²: 1) $i_{01} = 10^{-20}$, $i_{02} = 10^{-12}$; 2) $i_{01} = 10^{-19}$, $i_{02} = 10^{-11}$; 3) $i_{01} = 10^{-19}$, $i_{02} = 10^{-10}$; 4, 5) experiment; 6) range of maximum measured FF values at $i_{01} = 3 \cdot 10^{-2}$ A/cm².

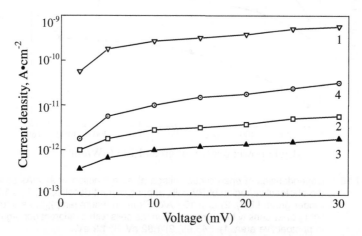

Figure 7.18   Reverse branches of the current-voltage characteristics of $(n\text{-}p)$-Al$_x$Ga$_{1-x}$As–p-Al$_{0.85}$Ga$_{0.15}$As heterostructures with different values of x: 1) 0, 2) 0.13, 3) 0.3, 4) 0.62.

coefficient of $dV_{oc}/V_{oc}dT$ in GaAs are $1.5 \cdot 10^{-3}$/°C and -1.5×10^{-3}/°C at $i_{ph} = 10^{-2}$ A/cm² and 1 A/cm², respectively. The experimental $dV_{oc}/V_{oc}\, dT$ values are typically about $-2 \cdot 10^{-3}$/°C, and the temperature coefficient of the fill factor $dFF/FFdT$ is about $-1 \cdot 10^{-3}$/°C at these excitation levels.

The temperature coefficients of V_{oc} and the efficiency are lower in PV cells on structures with an electric field produced by the holes' concentration gradient in

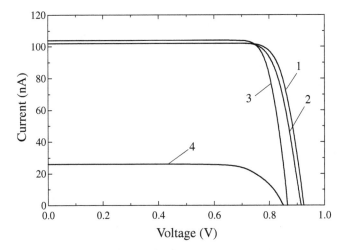

Figure 7.19   Illuminated current-voltage characteristics of three PV cells (S = 0.12 cm²) based on (1, 4) $Al_{0.37}Ga_{0.63}As$, (2) $Al_{0.36}Ga_{0.64}As$, and (3) $Al_{0.34}Ga_{0.66}As$ at two illumination intensities.

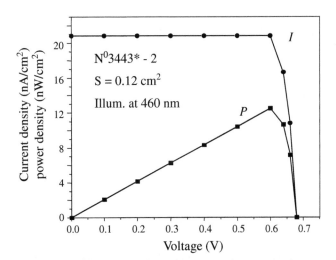

Figure 7.20   Illuminated *I–V* curve and electrical power density curve under blue (λ_{max} = 460 nm) LED light in the cell based on $Al_{0.33}Ga_{0.67}As$.

p-GaAs. Hole concentration can be significantly increased ($p = 10^{19}$ cm^{-3}) by zinc diffusion from the vapor phase to raise quasi-electric field strengths to $E = 2500$ V/cm. This improves the temperature stability of the fill factor and especially of the open-circuit voltage (Figure 7.22a). The minimum values of the respective temperature coefficients at $i_{ph} = 10$ A/cm² are $dFF/FFdT = -6 \cdot 10^{-4}/°C$ and $dV_{oc}/V_{oc} dT = -1.3 \cdot 10^{-3}/°C$. Figure 7.22 shows the *I–V* characteristics of the cell based on $Al_{0.37}Ga_{0.63}As$ at two temperatures and photocurrent density of 10^{-6} A/cm². It is seen that V_{oc} reduces from 0.93 V at 20°C to 0.87 V at 60°C. Fill factor in these conditions

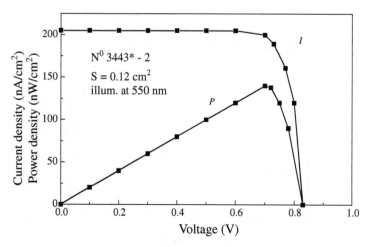

Figure 7.21 Illuminated current-voltage and output power–voltage curves under green (λ_{max} = 550 nm) LED in the cell, based on $Al_{0.33}Ga_{0.67}As$.

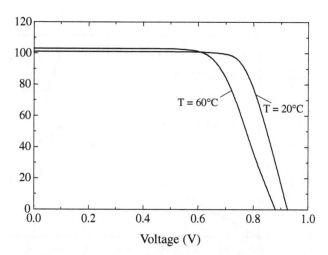

Figure 7.22 Illuminated current-voltage characteristics of PV cell based on $Al_{0.37}Ga_{0.63}As$ at two temperatures, 20°C and 60°C, and at photocurrent density of 10^{-6} A/cm².

decreases from 0.78 to 0.7. Temperature dependencies of V_{oc} are linear up to 100°C (Figure 7.23). In the best cells, based on $Al_{0.37}Ga_{0.63}As$, the V_{oc} reduces from 0.93 V at 20°C to 0.8 V at 100°C.

The efficiency of the PV cell in Figure 7.24a (high excitation level) decreases from 39% at 25°C to 34% at 100°C. Figures 7.24b and c show the temperature dependencies of photovoltaic parameters of the PV cell based on GaAs and $Al_{0.32}Ga_{0.68}As$ photocurrents of 10^{-2} and 10^{-6} A/cm² under light with λ = 633 nm. The efficiency of 26% at 20°C decreases to 18% at 100°C at photocurrent of 10^{-6}

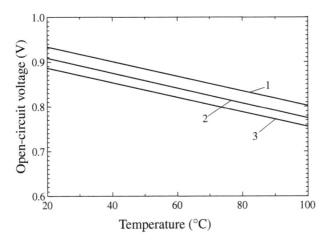

Figure 7.23 Temperature dependencies of open-circuit voltage for three PV cells based on (1) $Al_{0.37}Ga_{0.63}As$, (2) $Al_{0.36}Ga_{0.64}As$, and (3) $Al_{0.34}Ga_{0.66}As$ at photocurrent density of 10^{-6} A/cm².

A/cm² (low excitation level) in PV cells based on $Al_{0.32}Ga_{0.68}As$. The main reason for this dependence is the decrease of V_{oc} from 0.87 V at 20°C to 0.72 V at 100°C.

7.2.7   Photovoltaic AlGaAs Arrays for Radioluminescent Lighting

Heterostructures (Figures 7.25 and 7.26) of low-intensity photovoltaic arrays (LIAs) fabricated by liquid-phase epitaxy consist of n-GaAs substrate ($n = 10^{18}$ cm⁻³, Sn), n-GaAs buffer layer ($n = 10^{16}$ cm⁻³, undoped), $n-Al_xGa_{1-x}As$ base layer with thickness of 2 to 6 μm ($n = 10^{16}$ cm⁻³, undoped), $p-Al_xGa_{1-x}As$ (0.5 to 1 μm) - emitter ($p = 10^{18}$ cm⁻³, Mg) and $p-Al_{0.85}Ga_{0.15}As$ (0.03 to 0.05 μm, $p = 10^{17}$ cm⁻³, Mg) – window layer. Antireflection coating (ARC) an anodic oxide. The front ohmic contact is fabricated to cap the p-GaAs layer ($p = 10^{19}$ cm⁻³, Mg), which is etched off in areas free from contacts.

Photovoltaic arrays with square of 0.82 to 3.45 cm² were assembled on bases made of metalized ceramic or fiberglass. Each array consisted of 2 to 10 series-connected cells with cell-designated illumination areas of 0.2 × 0.5, 0.45 × 0.45, 0.4 × 1.2, and 0.9 × 0.9 cm and with array-designated illumination areas of 0.82, 1.0, 2.54, 3.36, and 3.45 cm² (Table 7.1). Some arrays were protected by Ce-doped glass.

Expected output power from radioluminescent batteries is approximately the same for blue, green, and yellow tritium lamps. Therefore, the AlGaAs beta cells have been optimized for operation in these parts of the spectrum. High external quantum yield of 0.8 to 0.9 at λ = 500 to 650 nm (green, yellow, and red colors) and 0.7 to 0.8 at λ = 450 to 500 nm (blue phosphor) were measured in the developed photovoltaic arrays (Figure 7.27). Illuminated current-voltage and output power-voltage characteristics show enough high fill factor of 0.7 to 0.85 under tritium lamps at light densities of 300 to 1100 nW/cm² and under white and green light at density of 3 to 10 μW/cm² simulating the concentrated radioluminescent lighting.

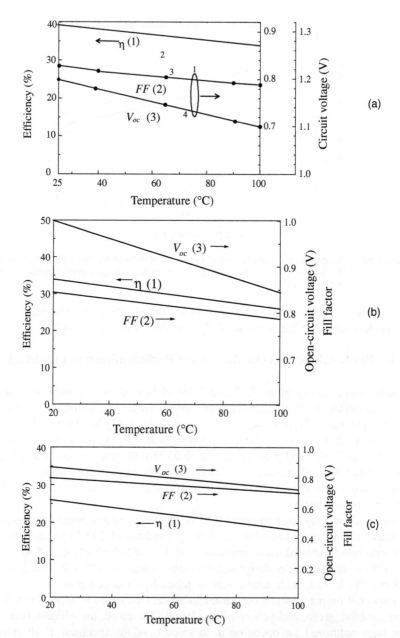

Figure 7.24 Temperature dependencies of efficiency (1) η, (2) FF, and (3) V_{oc} in PV cells based on different A^3B^5 semiconductors: (a, b) GaAs, (c) $Al_{0.32}Ga_{0.68}As$, and at different excitation levels under illumination with $\lambda = 633$ nm: (a) GaAs, $i_{ph} = 10$ A/cm²; (b) GaAs, $i_{ph} = 3 \cdot 10^{-2}$ A/cm²; (c) $Al_{0.32}Ga_{0.68}As$, $i_{ph} = 10^{-6}$ A/cm².

Photovoltaic arrays with dimensions of 1×1 cm² (type "E," Table 7.1) were assembled on bases made of metallicized ceramic (Figures 7.28, 7.29, and 7.30). Each array consists of 10 series-connected cells with dimensions of 2×6 mm with

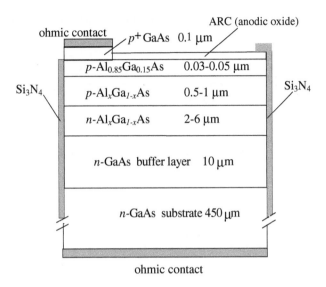

Figure 7.25   Cross section of the photovoltaic cell based on AlGaAs–GaAs heterostructure for low-intensity arrays.

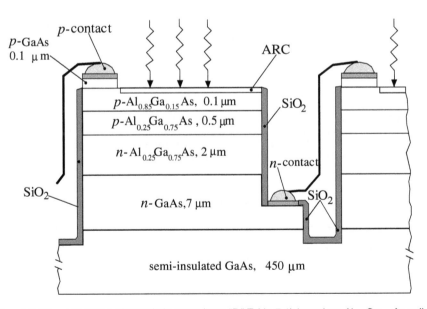

Figure 7.26   Integrated photovoltaic array (type "D," Table 7.1) based on $Al_{0.25}Ga_{0.75}As$ cells fabricated on the semi-insulated GaAs substrate.

cell-designated illumination areas of 2×5 mm and with array-designated illumination areas of 10×10 mm (Figures 7.29 and 7.30). The array is protected by Ce-doped glass (Figure 7.28).

Table 7.1 Dimensions of Fabricated PV Arrays for Illumination by Tritium-Powered Lamps

Array Design	Array Type				
	"A"	"B"	"C"	"D"	"E"
Array dimensions, cm	1.86 × 1.86	0.9 × 2.72	0.45 × 1.83	1.2 × 2.8	1 × 1
Array square, cm^2	3.45	2.54	0.82	3.36	1.0
Array photoactive area square, cm^2	3.24	2.43	0.8	3.22	0.95
Cells quantity in array, pieces	4	3	4	7	10
Cells dimensions in array, cm	0.9 × 0.9		0.45 × 0.45	0.4 × 1.2	0.2 × 0.5

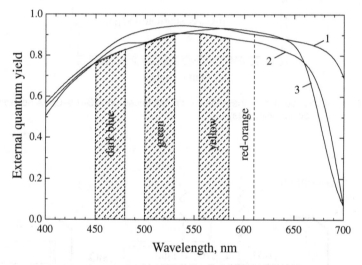

Figure 7.27 Photoresponse spectra of photovoltaic cells (0.9 × 0.9 cm^2) based on $Al_{0.17}Ga_{0.23}As$ (N° 863-03, curve 1), and $Al_{0.25}Ga_{0.75}As$ (N°872-01, curve 2) and (N°P3728-01, curve 3). Shaded bands correspond to the spectral position of radiation from available tritium-powered lamps.

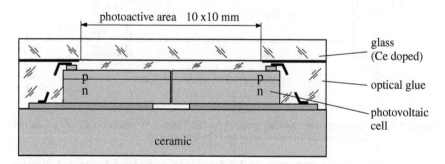

Figure 7.28 Cross section of photovoltaic array based on AlGaAs cells for radioluminescent batteries.

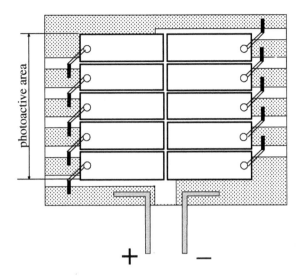

Figure 7.29 Photovoltaic array of type "E" with photoactive area of 10 × 10 mm based on 10 series-connected 2 × 6 mm $Al_{0.1}Ga_{0.9}As$ cells.

Figure 7.30 Top view of the photovoltaic low-intensity array of type "E" based on ten cells connected in series.

Illuminated current-voltage characteristics (Figures 7.31 and 7.32) show enough high fill factor values of 0.8 to 0.84 at illumination intensities of 10^{-7} to 10^{-5} W/cm^2. Table 7.2 demonstrates the performance of fabricated arrays LIA 3 and LIA 5 under red light at different illumination intensities.

Figures 7.32 and 7.33 show the increase of open-circuit voltage in the developed photovoltaic arrays from 3.5 to 5 V at photocurrents of 10^{-8} A to 5 to 6 V at $I_{ph} = 10^{-7}$ A. Output voltage in the point of optimal load is 80 to 85% from the value of V_{oc}. Cell efficiency of 19 to 20% and array LIA-3 efficiency up to 18.1% have been

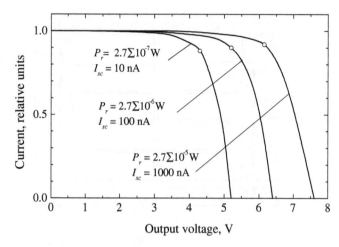

Figure 7.31 Illuminated current-voltage characteristics of PV array ($S = 1$ cm²) based on ten series-connected AlGaAs cells at three intensities of light with wavelength of 633 nm.

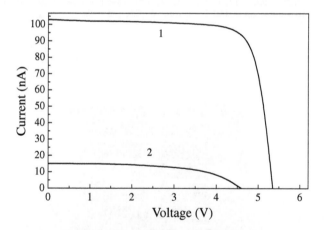

Figure 7.32 Illuminated current-voltage characteristics of photovoltaic low-intensity array LIA-3 ($S = 1$ cm²) made of ten series-connected AlGaAs–GaAs cells at two illumination intensities of the light from green LED (curve 1) and from tritium green lamp (curve 2).

measured under red light ($\lambda = 633$ nm) at output electric power of $5 \cdot 10^{-7}$ W (see Table 7.2). Array power output of $7 \cdot 10^{-8}$ W/cm² was received at radioluminescent lighting from tritium-phosphor green lamp. Photocurrent of 15 nA and output voltage of 4.6 V (Figure 7.32, curve 2) have been measured in these conditions of lighting.

The constructions of four different arrays (see Table 7.1) are shown in Figure 7.34. Performance of these arrays under tritium lamps is described in Tables 7.3 and 7.4 as well as in the next discussion concerning nuclear battery prototypes based on established photovoltaic arrays and tritium lamps.

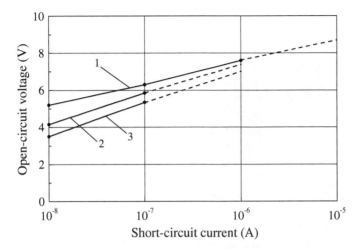

Figure 7.33   Dependence of the open-circuit voltage as a function of short-circuit current in photovoltaic arrays (type "E") LIA-1 (curve 1), LIA-3 (curve 2), and LIA-5 (curve 3).

Table 7.2   Parameters of Arrays LIA-5 and LIA-3 of Type "E" (see Table 7.1) Measured at Two Illumination Intensities

Array	I_{sc}, nA	V_{oc}, V	I_m, nA	V_m, V	P_m, nW	FF, %	Eff, %
LIA-5	103	5.35	97	4.5	436.5	79.2	15.6
LIA-5	9.2	3.5	7.9	3.0	23.7	73.6	9.5
LIA-3	103	5.85	101	5.0	505	83.8	18.1
LIA-3	9.7	4.15	8.6	3.5	30.1	74.8	11.5

Note: Efficiencies are calculated for the light wavelength of 633 nm.

7.2.8   Batteries Based on Tritium Lamps and Photovoltaic AlGaAs Arrays

Constructions of radioluminescent batteries based on tritium lamps and developed arrays are shown in Figure 7.35. Various approaches are used to increase the output power of batteries, such as using the mirror concentrator (Figure 7.35a, b) and batteries based on two or four arrays surrounding the lamps (Figure 7.35c, d). It was found that maximum specific electric power is ensured in the batteries shown in Figure 7.35a, b and maximum specific power output (expressed in nW/Ci) in the batteries shown in Figure 7.35c, d.

Table 7.4 shows the parameters of various tritium radioluminescent lamps that have been used for assembling nuclear batteries. The light densities of the comparable construction are approximately the same for green and yellow tritium lamps. The light density on the surface of PV arrays can be increased approximately two times in battery constructions with mirror concentrator or reflector (Figure 7.35a, b).

Figure 7.36 illustrates an example of optical efficiency measurements in photovoltaic arrays operating with tritium lamps. The following formulas allow calculation of the optical efficiency of arrays.

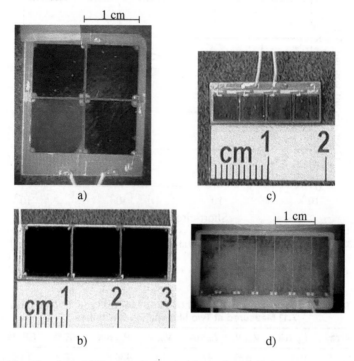

Figure 7.34 Top view of PV arrays designated for nuclear batteries.

Table 7.3 Performance of PV Arrays Fabricated for Illumination by Tritium Lamps

N	Name of Parameter	Magnitude of Parameter
1	Spectral region of array photosensitivity	from 350 nm to 650 to 800 nm
2	Light densities from tritium lamps	320 to 670 nW/cm² (without mirror) 480 to 1140 nW/cm² (with mirror concentrator)
3	Maximum array responsivity under green tritium lamp	0.33 A/W
4	Arrays output voltage under tritium lamps	1.5 to 3.5 volts
5	Fill factor of *I–V* curves of arrays under tritium lamps	0.8 to 0.84
6	Maximum array optical efficiency at illumination by tritium lamps	10 to 12%
7	Maximum relation of the array's electric power to tritium activity in the lamps	150 to 220 nW/Ci in batteries, based on array situated on one side of tritium lamp

Table 7.4 Parameters of Tritium Radioluminescent Lamps (RLS) Available for Assembling of Tritium-Powered Batteries with Photovoltaic Arrays

Lamp Number	Lamp Type	λ_{max}, nm	Lamp Diameter, mm	Lamp Length, mm	Tritium Activity in Lamp, Ci	Specific Activity, Ci/cm²	Light Density, nW/cm²
RLS1	Green tube	510	3	30	0.165	0.23	320
RLS2	Green tube	510	5	49	1.0	0.48	650
RLS3	Yellow tube	570	5	47	1.0	0.5	600
RLS4	Assemblage — four green tubes	510	3	30	0.66	0.23	320
RLS5	Assemblage — four green tubes	510	5	31	1.52	0.32	390

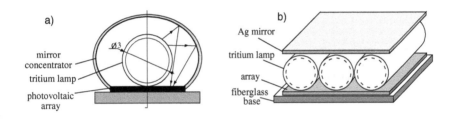

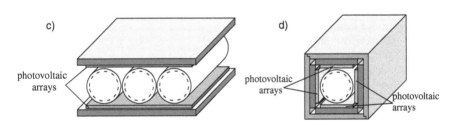

Figure 7.35 Design of tritium batteries based on green ³H₂ lamps and photovoltaic arrays consisting of two to ten series-connected cells: a) with mirror concentrator, b) with mirror reflector, c) with two arrays situated on two sides of tritium lamps, and d) with four arrays surrounding the tritium lamp from four sides.

$$\text{Efficiency } (\eta) = \frac{g_{el}}{g_{lum}} \qquad (7.19)$$

$$g_{el} = I_{max} V_{max} = I_{sc} \cdot V_{oc} \cdot FF \qquad g_{lum} = \frac{n I_{sc}}{Q} h v_{max} = \frac{n I_{sc}}{Q} \frac{1.24}{\lambda_{max}[\mu m]}$$

$$FF = \frac{I_{max} V_{max}}{I_{sc} V_{oc}}$$

$$\text{Efficiency } (5) = \frac{0.806}{n} \lambda_{max} \cdot Q \, V_{oc} \cdot FF \qquad (7.20)$$

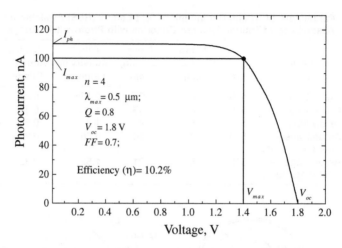

Figure 7.36   Example of load voltage-current characteristics of nuclear battery based on PV array (total area of 4 cm², four cells connected in series) with green tritium lamps with light density of 340 nW/cm².

where

I_{sc} – short-circuit current in the array
V_{oc} – open-circuit voltage
R_{opt} – optimal load of array
I_{max} – photocurrent through the optimal load
U_{max} – voltage on the optimal load
P_{el} – output power from array
P_{lum} – input power of light from tritium lamp
Q – external quantum yield of the array at $9 = 9_{lum}$
hv_{max} – energy (in eV) of maximum fluorescence from tritium lamp
λ_{max} – wavelength (in µm) of light from tritium lamp
FF – fill factor of illuminated current-voltage characteristics of the array
n – quantity of cells connected in series in the array

Formula 7.19 was used in cases of array illumination by a tritium lamp with a preliminary measured value of input light power (P_{lum}). Formula 7.20 was used for illumination of arrays with preliminary measured external quantum yield from photoresponse spectra, e.g., at $n = 3$, $\lambda_{lum} = 0.51$ µm, $Q = 0.85$, $V_{oc} = 1.6$ V, $FF = 0.8$, and optical efficiency $\eta = 14.9\%$.

Figures 7.37 and 7.38 and Table 7.5 show the output current-voltage and output power-voltage curves for four tritium-fueled batteries of different design. The maximum optical efficiency of 10.7% was received in battery N1 with the output electric power of 147 nW. The maximum output power of 234 nW was generated by the battery N5, based on a PV array with square of 6.92 cm² under four tritium lamps with total tritium activity of 1.3 Ci. The relation of output power to tritium activity $K = 180$ nW/Ci was observed in this battery. Figure 7.39 demonstrates the operation of display, with battery N2 generating the output power of 90 nW.

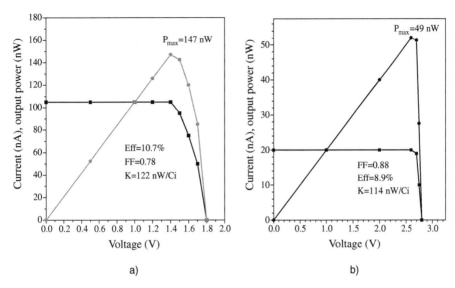

Figure 7.37   Current-voltage and output power-voltage curves of tritium batteries: a) battery N1 (see Table 7.5) and b) battery N3 based on PV arrays with green tritium lamps.

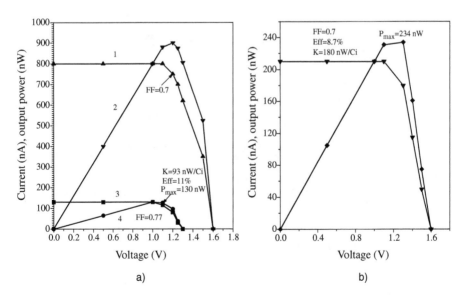

Figure 7.38   Current-voltage and output power-voltage curves of tritium batteries: a) battery N4 (see Table 7.5) with two green tritium lamps of total activity A = 1.4 Ci (curves 3, 4) and under white light (without tritium lamps) simulating the concentrated lighting (curves 1, 2); b) I–V curves of nuclear battery N5 (see Table 7.5) based on PV array with area of 6.92 cm² and tritium lamps with activity A = 1.3 Ci.

Table 7.5 Performance of Experimental Tritium Batteries Based on AlGaAs PV Arrays and Tritium-Powered Lamps

Battery N	PV Array Type	PV Array Square, cm^2	Tritium Activity, Ci	Light Density, nW/cm^2	Battery Performance					
					V_{oc}, V	J_{sc}, nA	FF	P_{max}, nW	Opt. Eff, %	K, nW/Ci
1	"A"	3.45 (1.86 × 1.86)	1.2	390	1.8	105	0.78	147	10.7	122
2	"D"	3.36 (2.8 × 1.2)	0.5	480	3.5	45	0.54	90	5.6	179
3	"C"	0.82 (0.45 × 1.83)	0.46	670	2.8	20	0.89	49	8.9	114
4	"B"	2.54 (0.9 × 2.28)	1.4	640	1.3	130	0.77	130	8	93
5	"B"	6.92 (1.86 × 1.86)	1.3	390	1.6	210	0.7	234	8.7	180
6	"E"	1 (1 × 1)	0.32	390	3.4	13	0.74	32.5	8.3	102

Figure 7.39 Display operation with tritium battery N2 (see Table 7.5) based on three tritium lamps (total tritium activity of 0.5 Ci) and photovoltaic array (square of 3.36 cm^2) with output electric power of 90 nW.

7.2.9 Photovoltaic Cells Based on AlGaP–GaP for Blue Phosphor Lighting

Devices based on semiconductors with band gaps of more than 2 eV are very promising for conversion of energy from short-wavelength radioluminescent lamps. GaP devices compared with AlGaAs, GaAs, and Si devices have the following advantages: better thermal stability and radiation resistance, lower dark-current densities, longer service life, lower noise levels, and higher voltage response that can ensure higher efficiency.

Until now, interest in semiconductor devices based on GaP were focused primarily on red-green light-emitting diodes and on low-noise ultraviolet and blue photo detectors. The low leakage current of GaP p–n junctions and wide band gap ($E_g =$ 2.25 eV, 300 K) influence a number of successful high-temperature devices, which can operate at temperatures above 300°C. Liquid-phase and vapor-phase epitaxies were developed during the last decades to improve GaP material quality, increase minority carrier diffusion lengths, decrease concentration of impurities, and decrease leakage currents, taking into account that photocurrents in the case of tritium-fueled batteries are about 10^{-7} to 10^{-6} A/cm^2. The photoresponse maximum of GaP photodiodes is in the range of 400 to 460 nm, corresponding to high sensitivity to radiation from blue phosphors. Using AlGaP window layer allows for increasing the quantum yield in the near UV spectrum, which is matched to common scintillators and offers an alternative to photomultiplier tubes for reading pulses from ionizing radiation.

The p-GaP–n-GaP (Figure 7.40a) and p-AlGaP–p-GaP–n-GaP (Figure 7.40b) heterostructures were grown on n-GaP (111) and (100) oriented substrates with a donor concentration of 10^{17} to 10^{-18} cm^{-3}. GaP cells were fabricated by Zn-diffusion into n-GaP layer or by epitaxial growth of n- and p-GaP layers. The p-Al$_x$Ga$_{1-x}$P window layers with an acceptor concentration up to 10^{18} cm^{-3} were grown in the epitaxial AlGaP–GaP heterostructures; window-layer thickness was varied from several micrometers to 0.07 μm in these structures. During growth of the Al$_x$Ga$_{1-x}$P window layer, the p-GaP region (Figure 7.40b) with thickness of 0.3 to 0.5 μm was formed by Zn or Mg diffusion into the n-GaP.

In some heterostructures the p-GaP region was epitaxially grown, as well. The AlP concentration in the window-layer Al$_x$Ga$_{1-x}$P was varied from $x = 0.3$ to $x = 0.8$. Samples with square of 0.1 to 2 cm^2 were cleaved out and ohmic contacts were fabricated.

Figure 7.41 shows spectral responses of GaP- and AlGaP–GaP-based devices. The photoresponse maximum is located in the spectral range of 450 to 460 nm in cells with p-GaP emitter thickness of 1 to 1.5 μm. The reduction of the cap p-GaP

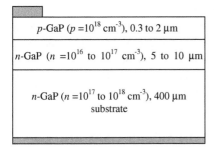

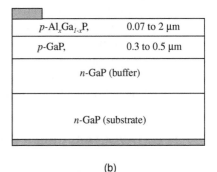

(a)   (b)

Figure 7.40 Cross section of developed PV cells based on (a) GaP and (b) AlGaP–GaP heterostructures.

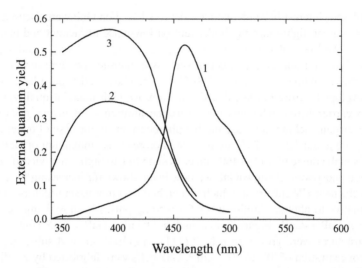

Figure 7.41 Spectral photoresponse of p–n GaP cells: 1 – p-GaP front-layer thickness of 1.4 μm; 2 – p-GaP layer thickness of 0.3 μm; 3 – p-Al$_{0.7}$Ga$_{0.3}$P–p-GaP–n-GaP structure with AlGaP layer thickness of 0.07 μm.

and p-Al$_x$Ga$_{1-x}$P layer thickness allows for increasing the UV photosensitivity with photoresponse maximum shifted to 400 nm. This means that these cells can also ensure higher conversion efficiency of betas from tritium absorbed mainly on the thickness of about 0.5 μm from the irradiated surface.

Figure 7.42 shows the dark I–V curves for two p–n GaP cells fabricated on GaP substrates with different quality and substrate orientation. The typical forward and reverse current densities at 0.1 V are in the range of (2 to 4) 10^{-12} A/cm^2. Shunt resistance of $5 \cdot 10^{10}$ Ω was measured in the developed cells at $V = 10$ mV. The forward current of 10^{-6} A/cm^2 takes place at voltage of about 1 V, which is more than in the case of AlGaAs-based cells. This demonstrates the possibility to receive the higher value of V_{oc} at generated currents of 10^{-7} to 10^{-6} A/cm^2 expected in GaP PV cells designated for radioluminescent lighting.

Developed GaP PV cells were tested under blue LED with illumination intensity in the range of 1 to 10 μW/cm^2. Open-circuit voltage of 0.97 V, $FF = 0.77$, and output power $P_{el} = 750$ nW/cm^2 were measured at photocurrent density of 1 μA/cm^2. Open-circuit voltage $V_{oc} = 0.78$, $FF = 0.67$, and $P_{el} = 52$ nW/cm^2 were received at $i_{ph} = 100$ nA/cm^2, simulating the blue tritium lamp illumination, as shown in Figure 7.43.

7.3 BETAVOLTAIC CELLS BASED ON AlGaAs–GaAs AND GaP

Selecting a beta emitter involves consideration of isotope half-lives as well as the effect of radiation damage of semiconductor converters. To create a long-living battery, it is desirable to utilize isotopes with long half-lives. However, since beta flux from an isotope is inversely proportional to half-life, the power in the beginning of life (BOL) is inversely proportional to the half-life. The maximum energy of beta

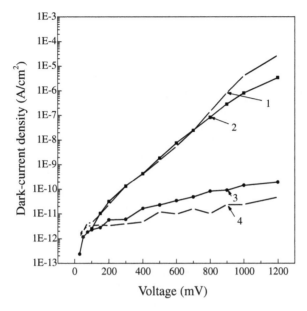

Figure 7.42   Forward (curves 1, 2) and reverse (3, 4) dark *I–V* curves of *p–n* GaP PV cells: (1, 4) substrate orientation (100) and (2, 3) substrate orientation (111).

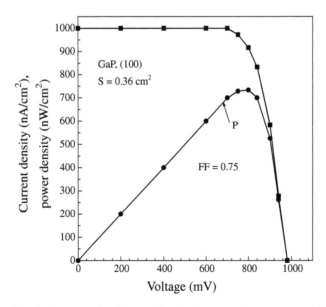

Figure 7.43   Illuminated current-voltage and output power-voltage curves of GaP PV cell under blue (λ_{max} = 460 nm) LED.

radiation (E_{max}) in relation to the semiconductor damage threshold (E_{th}) must be taken into account, as well.

On the basis of an analysis of radiation characteristics of radioactive isotopes and their safety, availability, and costs, tritium is one of the best beta sources for these types of batteries. The half-life of tritium is 12.3 years; it radiates low-energy $(E_{max} = 18.6\ \text{KeV})$ beta particles that make batteries using tritium safe from the point of view of safety. The last feature of tritium is very important, also, to ensure a long lifetime of semiconductor converters because low-energy (lower than damage threshold) radiation from tritium does not generate defects in semiconductors. Tritium is low cost and available in gaseous form, in the form of tritiated metal (titanium) foils, and in the tritiated organic materials form. The surface power density from gaseous tritium is about 15 μW/cm^2 and about 1.5 μW/cm^2 from the Ti-T$_2$ surface, taking into account the self-absorption of beta particles in the titanium.

Theoretical analysis shows that efficiency of betavoltaic conversion increases with increase of the band gap (E_g) of semiconductor converters (Figure 7.44). The theoretical maximum of betavoltaic conversion efficiency is about 25% for GaP and SiC and 22% for semiconductors with $E_g = 1.9$ eV (Al$_{0.35}$Ga$_{0.65}$As) and decreases to 15% for Si. The theoretical maximum of tritium-based betavoltaic converter efficiency reduces to about 15% for semiconductors with $E_g \cong 1.9$ eV and to 10% for Si-based cells, owing to low photocurrent density ($\leq 10^{-6}$ A/cm^2) under tritium betas.

These values of efficiency are considerably less than efficiencies of photovoltaic conversion of monochromatic light, for several reasons. The effective ionization energy $M =$ (average amount of beta particle energy expended to create one electron–hole pair), determined by the semi-empirical relation $M = 2.8E_g + (0.5\ \text{to}\ 1.0\ \text{eV})$, is about three times more than the photon energy expended to create one electron–hole pair with energy approximately equal to E_g. The photocurrent value in

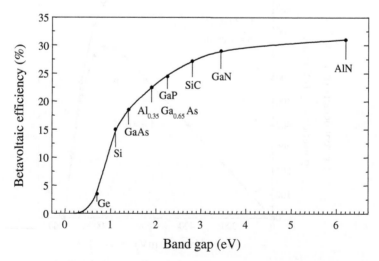

Figure 7.44 Theoretical maximum of betavoltaic efficiency vs. semiconductor band gaps (300 K): Ge — 0.7 eV; Si — 1.1 eV; GaAs — 1.4 eV; Al$_{0.35}$Ga$_{0.65}$As — 1.9 eV; GaP — 2.25 eV; SiC — 2.8 eV; GaN — 3.44 eV; AlN — 6.2 eV. (From Olsen, L.C., *Proc. 12th Space Photovoltaic Res. Technol. Conf.*, Cleveland, 1992, 256.)

tritium betavoltaic cells will typically be in the range of 10^{-7} to 10^{-6} A/cm^2, whereas in solar cells the photocurrent is typically higher than 10^{-2} A/cm^2. The lower photocurrent value in betavoltaic cells causes lower values of V_{oc} and of efficiency. The reduction of the efficiency in Si-based cells at reducing photocurrent reducing and temperature increasing is stronger than in the case of wide-band-gap *III-V* semiconductors. Reasons for this effect were considered earlier in the discussion concerning theoretical modeling of low-intensity PV cells based on Si and III-V semiconductors.

Among available semiconductor materials, Al$_x$Ga$_{1-x}$As solid solution with $x =$ 0.1 to 0.35 is the best candidate for fabrication of betavoltaic cells based on tritium. AlGaAs–GaAs heterostructure PV cells were developed in the Ioffe Institute and studied under the tritium beta particles.[77-79] These heterostructures consist of *n*-GaAs substrate, *n*-Al$_x$Ga$_{1-x}$As base layer, *p*-Al$_x$Ga$_{1-x}$As, and *p*-Al$_{0.85}$Ga$_{0.15}$As front layer. The main advantage of this structure for conversion of tritium beta radiation to electricity follows.

Since the wide-band-gap AlGaAs front layer is thin (20 to 30 nm thick), the tritium beta particles are weakly absorbed in this layer and separation of minority carriers generated by low-energy beta radiation is highly efficient. It was found from theoretical modeling that less than 10% of total electron–hole pairs created by beta particles from tritium are generated uselessly in the Al$_{0.85}$Ga$_{0.15}$As layer with thickness of 10 nm; about 80% are generated in the part of the active *p*-Al$_x$Ga$_{1-x}$As layer about 250 nm thick (Figure 7.45).

Among the other wide-band-gap III-V semiconductors, GaP is also available for betavoltaic device fabrication. GaP has the highest theoretical efficiency of beta particle conversion (see Figure 7.44).

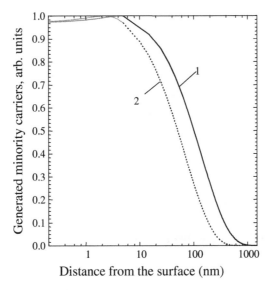

Figure 7.45 Calculated quantity of minority carriers generated by tritium betas in Al$_{0.1}$Ga$_{0.9}$As for (1) normal and (2) isotropic distribution of irradiation from tritium to the cell.

7.3.1 Betavoltaic Cells Based on AlGaAs–GaAs Heterostructures

AlGaAs–GaAs heterostructures for beta cells were developed and grown by liquid-phase and MOCVD epitaxy. These heterostructures (Figure 7.46a) consist of n-GaAs substrate ($n = 2 \cdot 10^{18}$ cm^{-3}); n-Al$_x$Ga$_{1-x}$As ($x = 0.1$ to 0.35) base layer, $n = 10^{16}$ cm^{-3}; p-Al$_x$Ga$_{1-x}$As layer, $p = 10^{18}$ cm^{-3}; and p-Al$_{0.85}$Ga$_{0.15}$As front layer, 0.02 to 0.05 μm thick.

Characteristic features of the structure shown in Figure 7.46a are:

1. Since the wide-band-gap front layer is thin, separation of current carriers generated by radiation from the radioactive source near the surface of the heterostructure is highly efficient. This conclusion is demonstrated by the high value of internal quantum photosensitivity, which is realized in the violet region in such structures (Figure 7.47).
2. There is a composition gradient in the n-Al$_x$Ga$_{1-x}$As layer; band-gap energy increases from the p–n junction. This gradient sets up a quasielectric field capable of improving the collection of carriers generated beyond the p–n junction. It can also reduce the effect of radiation-induced defects that lower the diffusion length of minority carriers (holes). As a result, radiation and thermal stability are improved, with the implication that converters of the ionizing radiation have a long service life.

Another type of cell structure with internal Bragg reflector (Figure 7.46b) has been developed for beta cells. The Bragg reflector (BR) is widely used in lasers and other optical devices.[20,65-67] The principles of multilayer dielectric reflectors are well known. By using a multiple layer composed of materials with different refractive indexes, nearly 100% reflectance can be achieved over a restricted wavelength range.

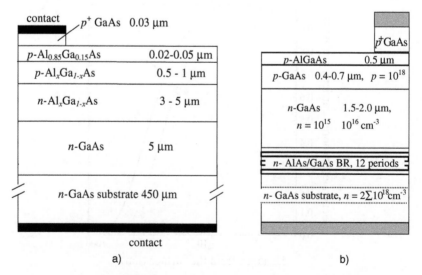

Figure 7.46 Cross section of betavoltaic cells based on AlGaAs–GaAs heterostructures: (a) structures fabricated by liquid-phase epitaxy; (b) structure with Bragg reflector fabricated by MOCVD.

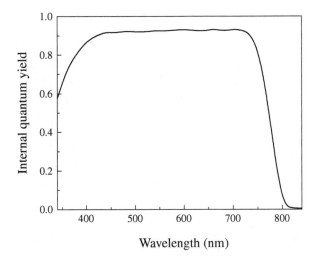

Figure 7.47 Spectral response of developed betavoltaic cells based on $Al_{0.12}Ga_{0.88}As$.

The thickness of each of the two materials is chosen for quarter-wavelength reflection for the given wavelength. This multilayer dielectric stack will selectively reflect reradiated photons (created in the process of recombination of electron–hole pairs generated by betas) with energy near the GaAs band gap, providing a second pass through the photoactive region and thereby increasing the induced current.

Epitaxial (MOCVD) Bragg reflectors were designed for pairs of AlAs and GaAs layers. As the number (N) of pairs is made larger, the BR reflectance monotonically increases, asymptotically tending to unity, and reaches 96% at $N = 12$. This type of reflector increases the effective absorption of photons and allows for making the n-GaAs base layer thinner. Cell efficiency in this case should be more tolerant to lower diffusion lengths of holes. A cell structure with a Bragg reflector is shown schematically in Figure 7.46b. The heterostructures were grown in Ioffe Institute by MOCVD using equipment with a horizontal reactor at low pressure. The solar cell consists of a top p-GaAs contact layer, 0.03 to 0.05 µm p-AlGaAs window layer, 0.5 µm p-GaAs photoactive layer, 1.5 to 2.0 µm n-GaAs base, and a Bragg reflector grown on n-type GaAs substrate. The BR was optimized for reflectance in the 800 to 900 nm spectral region and consists of 12 pairs of AlAs and GaAs layers with thickness of 72 nm for AlAs and 59 nm for GaAs.

The photoresponse of a cell with internal Bragg reflector with a 1.5 to 2 µm n-GaAs layer is nearly the same as for a cell with a 3 to 4 µm n-GaAs layer without a BR that improves the radiation resistance, as will be demonstrated next.

Beta cells of 1×1 cm^2 and 2×2 cm^2 based on p–n-$Al_xGa_{1-x}As$–p-$Al_{0.85}Ga_{0.15}As$ with different contents of AlAs in the active layers ($x = 0.1$ to 0.35) were fabricated (Figure 7.48). The maximum value of the current generated by betas from tritium can be received in AlGaAs heterostructures with passivating wide-band-gap layer $Al_{0.85}Ga_{0.15}As$ thickness reduced to 20 to 30 nm. Figure 7.49 shows spectral responses of three beta cells based on n–p-$Al_{0.25}Ga_{0.75}As$–p-$Al_{0.85}Ga_{0.15}As$ with different thicknesses of the top passivating layers. It is seen that UV sensitivity is higher in cells

Figure 7.48 Beta cells based on Al$_x$Ga$_{1-x}$As with dimensions of 2 × 2 cm² (three pieces) and 1 × 1 cm² (five pieces).

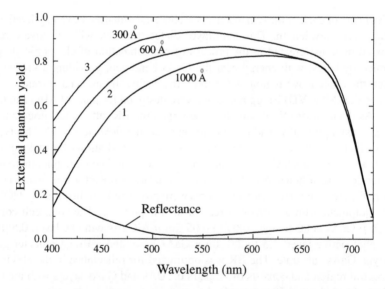

Figure 7.49 Spectral responses of three n–p-Al$_{0.25}$Ga$_{0.75}$As–p-Al$_{0.85}$Ga$_{0.15}$As beta cells with different thicknesses of top p-Al$_{0.85}$Ga$_{0.15}$As layer: 1) 0.1 μm; 2) 0.06 μm; 3) 0.03 μm. Typical reflectance spectrum is plotted, as well.

with lower thickness of the Al$_{0.85}$Ga$_{0.15}$As layer (30 nm) because useless absorption of UV radiation is reduced in this cell. The same conclusion can be made concerning reduction of nonactive absorption of betas from tritium in the samples with lower thickness of the passivating layer.

It is seen from Figure 7.50 that cells based on active layers with different band gaps (E_g = 1.7 to 1.8 eV) demonstrate enough good photosensitivity in the short wavelength part of spectrum. Cells with higher band gap ensure higher predicted

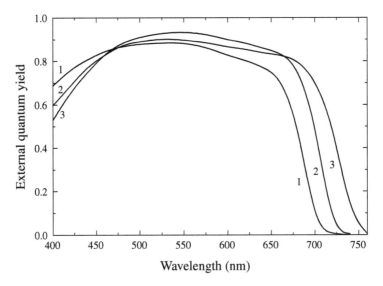

Figure 7.50   Spectral responses of three n–p-Al$_x$Ga$_{1-x}$As–p-Al$_{0.85}$Ga$_{0.15}$As beta cells with various contents of AlAs in the p–n junction region: curves 1) $x = 0.28$; 2) $x = 0.23$; 3) $x = 0.2$ and with top Al$_{0.85}$Ga$_{0.15}$As layer thickness of 0.03 to 0.05 μm.

efficiency of beta conversion. Therefore, fabricated beta cells based on Al$_x$Ga$_{1-x}$As ($x > 0.2$) with good UV photosensitivity would ensure high conversion efficiency.

Figure 7.51 shows the dark current-voltage characteristics of beta cells with square of 0.1 cm^2. The leakage current $i = 10^{-12}$ A/cm^2 at $V = 10$ mV was measured in this cell. Shunt resistance as high as 10^{10} Ω has been received at $V = 10$ mW in the developed beta cells based on Al$_x$Ga$_{1-x}$As.

Figure 7.52 shows that open-circuit voltage in Al$_{0.15}$Ga$_{0.85}$As beta cells is about 0.5 V at photocurrent density of 10^{-7} A/cm^2 and 0.5 to 0.65 V at 10^{-6} A/cm^2. These values of I_{sc} and V_{oc} are typical in betavoltaic cells based on Al$_{0.15}$Ga$_{0.85}$As–GaAs heterostructures with tritium beta radiation sources. Increasing band-gap energy in the active region leads to increase of open-circuit voltage up to 0.9 V at $i_{sc} = 10^{-6}$ A/cm^2.

Two modifications of tritium beta sources were used for experiments with beta cells based on Al$_x$Ga$_{1-x}$As ($x = 0.15$ and 0.35): tritiated Ti foils and capsules filled with gaseous tritium (pressure of 1 atm.) containing betavoltaic cells. Open-circuit voltage of 0.65 to 0.91 V at short-circuit current of $(0.75$ to $1) \cdot 10^{-6}$ A/cm^2 and fill factor of 0.8 to 0.85 were achieved in cells situated in the capsule with tritium gas. The maximum output specific power of 0.55 μW/cm^2 and maximum conversion efficiency of 5.8% were received in Al$_{0.35}$Ga$_{0.65}$As cells in tritium gas (see Figure 7.53 and Table 7.6).

Solid radioactive sources are of greater interest from a practical standpoint. The available tritium-titanium sources generate a considerably lower electrical output power than tritium gas. The current density increases from $i_{sc} = 40$ nA/cm^2 in cells based on Al$_{0.25}$Ga$_{0.75}$As with passivating layer thickness of 100 nm to $i_{sc} = 55$ to 60 nA/cm^2 in cells with layer thickness of 30 nm under the T$_2$-Ti source with $P_2 = 0.48$

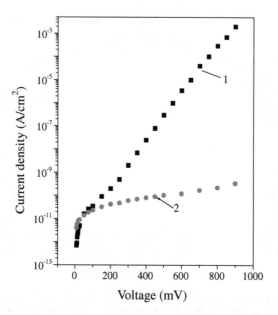

Figure 7.51 Dark current-voltage characteristics of $Al_{0.1}Ga_{0.9}As$-based betavoltaic cell with square of 0.1 cm^2: 1) forward bias, 2) reverse bias.

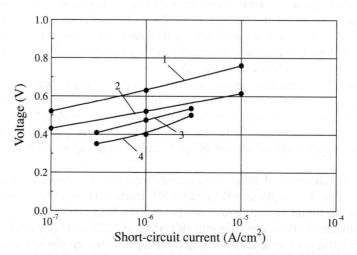

Figure 7.52 Dependencies of open-circuit voltage (curves 1, 3) and output voltage in the point of the optimal load (curves 2, 4) as a function of short-circuit current for two betavoltaic cells: based on $Al_{0.15}Ga_{0.85}As$ (curves 1, 2) and on $Al_{0.1}Ga_{0.9}As$ (curves 3, 4).

µW/cm^2 (Figures 7.54 and 7.55b). These current densities were received in the cells based on the structures without antireflection coatings. Antireflection coating made of ZnS with total thickness of about 50 nm is deposited on the structure with top passivating $Al_{0.85}Ga_{0.15}As$ layer thickness of 30 nm. This leads to reduction of the

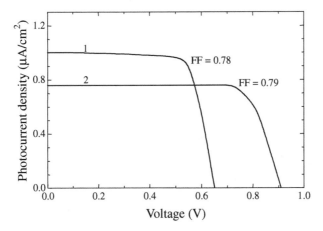

Figure 7.53   *I–V* curves of AlGaAs beta cells under betas from tritium gas at 1 atm.: 1) $Al_{0.15}Ga_{0.85}As$ beta cell, 2) $Al_{0.35}Ga_{0.65}As$ beta cell.

Table 7.6   Performance of Betavoltaic Cells Based on $Al_{0.35}Ga_{0.75}As$ under Different Sources of Tritium Betas

Tritium Fuel	Generated Current Density, nA/cm²	Open-Circuit Voltage, V	Output Power, nW/cm²
Tritium-titanium	58	0.62	27
Tritium gas, 1 atm	760	0.91	550
Tritium green lamp	120	0.78	74

current density of 15 to 20 nA/cm², owing to additional useless absorption of betas in the ARC layers (Figure 7.55a). Specific output electrical power of 15 to 27 nW/cm² and efficiency of 5.6% was received in $Al_{0.25}Ga_{0.75}As$ cells under the available T_2-Ti source (Figure 7.55b).

In spite of low output power in these experiments, tritium-powered solid sources of betas are of great interest because of the possibility of increasing tritium contents in titanium or in the other solid materials. The volumetric output power of the isotope batteries can be increased substantially, also, by using thin-film III-V heterostructure beta cells; these can be assembled in multilayer battery with systematic alternation of semiconductor converters and layers of the solid radioisotope beta source.

7.3.2   Radiation Resistance of AlGaAs–GaAs Beta Cells

Betavoltaic cells operate under the nuclear decay energy of beta-emitting radio-isotopes. It is known that the threshold of electron energy for defect generation in Si and GaAs is about 200 keV. Taking into account the low energy of betas from tritium, one can suppose the low rate of cell degradation caused by radiation from tritium. However, considering the possible use of other radioisotopes (promethium, strontium) with much higher beta energy, the results of radiation resistance studies

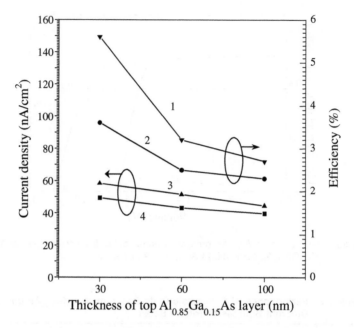

Figure 7.54 Dependencies of generated current density (curves 3, 4) and beta cell efficiency (curves 1, 2) as a function of the thickness of the top $Al_{0.85}Ga_{0.25}As$ layers in p–n-$Al_{0.25}Ga_{0.75}As$–p-$Al_{0.85}Ga_{0.15}As$ beta cells under the $^3H_2$-Ti source with output beta power of 0.48 µW/cm²: (1, 3) structures without antireflection coating, (2,4) structures with antireflection coating with thickness of 50 nm.

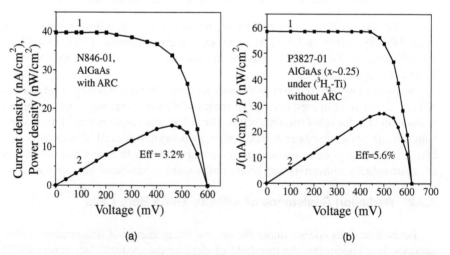

(a)   (b)

Figure 7.55 (1) I–V and (2) power-voltage curves of AlGaAs beta cells, based on $Al_{0.25}Ga_{0.75}As$ under (T_2-Ti) beta source with output beta power of 0.48 µW/cm². The thickness of top passivating layer is 30 nm. Antireflection coating: (a) ARC with thickness of 50 nm and (b) without ARC.

of AlGaAs–GaAs cells are presented here. These energies of electrons are usually used for estimating radiation resistance of semiconductor solar cells in order to predict their lifetimes in space environment conditions.

The extent of radiation damage of cells based on n-GaAs–p-GaAs–p-AlGaAs heterostructures strongly depends on the depth (d_p) of the p–n junction. Analysis of the theoretical model shows that cells with a thinner p-GaAs layer should exhibit higher radiation resistance for the same recombination parameters of the semiconductor. This conclusion is supported by results[66-69] of an experimental investigation on irradiation with 1 and 3.7 MeV electrons in cells with different thicknesses (d_p) of the p-GaAs layer. Cells based on GaAs with $d_p < 0.5$ μm possess better radiation resistance than silicon cells.

The results of experimental studies of radiation damage in GaAs and respective cells indicate that temperature increase up to 150 to 250°C enhances radiation resistance. The same effect is observed with "injection" anneal. It was demonstrated that direct current density of 3 A·cm^{-2} during proton irradiation resulted in a substantial increase in the output power.

Figure 7.56 shows the spectral response curves for GaAs PV cells with Bragg reflector and thin base (1.1 to 1.3 μm) after $3\cdot10^{15}$ cm^{-2} electron fluence as a function of base doping. Until the diffusion length is somewhat higher than the absorption depth of the reflection light, the effect of Bragg reflector of long-wavelength photons will stay sufficient. Lower diffusion length in the high base doping ($N_d=1\cdot10^{17}$ cm^{-2}) after electron exposure predetermines the impossibility of collecting the carriers efficiently from the base region (Figure 7.56, curve 4). The cells with low doped thin base ($N_d=1\cdot10^{15}$ cm^{-3}) and internal Bragg reflector demonstrate higher radiation resistance in photocurrent and efficiency (Figures 7.56, curves 2 and 3, and 7.57) than in the case of high base ($N_d=1\cdot10^{17}$ cm^{-3}) doping.

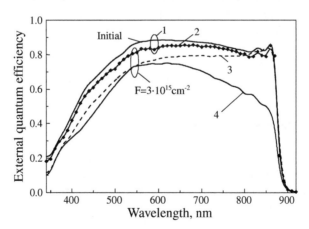

Figure 7.56 Spectral response curves before (curve 1) and after (curves 2 to 4) 3.7-MeV electron fluence ($F = 3\cdot10^{15}$ cm^{-2}) for cells with Bragg reflector and with different base doping levels. The best radiation resistance was measured for GaAs cell with $N_d =1\cdot10^{15}$ cm^{-3}. The cell parameters are: 1, 2, 3) low doped ($N_d = 10^{15}$ cm^{-3}) base n-GaAs, 4) moderate doped ($N_d = 10^{17}$ cm^{-3}) base.

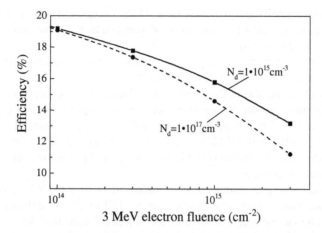

Figure 7.57   Comparison of efficiency degradation in AlGaAs–GaAs cells with Bragg reflector under solar spectrum as a function of 3-MeV electron fluence at base doping $N_d = 10^{15}$ cm^{-3} and $N_d = 10^{17}$ cm^{-3}.

Thus, reducing the base thickness to 1.1 to 1.5 μm as well as base doping to N_d $=1\cdot10^{15}$ cm^{-3} in GaAs cells with internal Bragg reflector demonstrates the increase of radiation resistance. Comparison of these results with degradation of Si cells leads to a conclusion that AlGaAs–GaAs cells ensure an increase lifetime about two times in the same conditions of irradiation.

Presented results allow for comparing cells of different types by their radiation resistance under 1 to 3.7 MeV electron irradiation. The right experiments of long-lifetime cells irradiated by beta particles from different radioisotopes must be carried out in order to obtain information about beta cell lifetime under tritium betas or under other beta sources (^{63}Ni, Pm, Sr, ^{55}Fe) that can be used in betavoltaic batteries.

Figure 7.58 demonstrates the changes of forward and reverse *I–V* curves of AlGaAs beta cell before and after Ti layer deposition on the front surface of the cell and structure exposure in tritium. SiO$_2$ layers protected the *p–n* junction on the lateral surface of the cells. A small increase in dark current was observed after exposure of this structure with deposited Ti in the tritium gas.

7.3.3   Beta Cells Based on GaP

The structures of these cells were similar to those discussed in the Section 7.2.7 concerning PV cells based on GaP. The results of experiments with these beta cells under T$_2$-Ti betas are shown in Figure 7.59. The worst efficiency was received in the cell with thick (1.4 μm) *p*-GaP layer corresponding to worst UV sensitivity of these cells (see Figure 7.41, curve 1) and low current density of 7 nA/cm^2 generated by betas from solid T$_2$-Ti source with 0.48 μW/cm^2 output power. The higher generated current density of 22 nA/cm^2 and higher efficiency of 2% (Figure 7.59a) were received in the cell with thinner *p*-GaP layer. The highest photocurrent density of 41.5 nA/cm^2, output electric power of 21 nW/cm^2, and efficiency of 4.4%

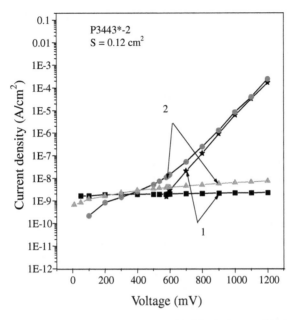

Figure 7.58   Dark forward and reverse *I–V* curves in AlGaAs beta cell before Ti deposition (curve 1) and after Ti deposition and exposure in T_2-capsule (curve 2).

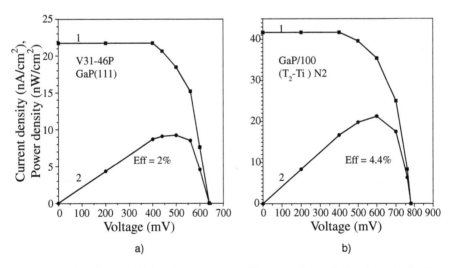

a)   b)

Figure 7.59   Irradiated *I–V* (1) and output power (2) curves of *p–n*-GaP cells under the T_2-Ti source with output beta power density of 0.48 μW/cm². The thickness of *p*-GaP layer in the cells is a) 1.4 μm, b) 0.3 μm.

(Figure 7.59b) were achieved in the cell with *p*-GaP layer thickness of 0.3 μm. A similar efficiency increase was received in beta cells based on heterostructures AlGaP–GaP with thin wide-band-gap passivating layer.

The following possible improvement of parameters would allow increasing efficiency up to about 10% in AlGaP–GaP beta cells under the tritium betas: 1) increasing the external quantum efficiency to 80%; 2) increasing the fill factor of I–V curves to 0.85; and 3) increasing V_{oc} to 1.1 V.

7.3.4 Conceptual Design of Betavoltaic Batteries Based on III-V Beta Cells

Increase of volumetric power output can be obtained in the betavoltaic battery by using two photosensitive surfaces. For example, in the p–n–p cells designed for this purpose, photosensitive p–n junctions are made on both sides of the substrate. The total power output of this three-terminal cell is twice as large as that of a similar one-sided cell.

Figure 7.60 shows the tritium battery based on bifacial AlGaAs and GaP beta cells on Ge (a) and Si (b) substrates. The first structure (a) consists of: n-Ge substrate of 115 μm thick, n^+ Ge p^+ Ge tunnel junction structure, p-Al$_{0.3}$Ga$_{0.7}$As n-Al$_{0.3}$Ga$_{0.7}$As photoactive region of 2 μm thick and n-Al$_{0.8}$Ga$_{0.2}$As passivating layer of 0.02 μm thick. On the back side of substrate, the similar structure without tunnel junction is grown. The titanium-tritium with thickness of 1 μm is sputtered on both sides of the structure. This bifacial tandem beta cell with tunnel junction ensures series connection of such structures of the required quantity in the battery.

The thickness of each cell in these batteries is about 125 μm. Output voltage will be about 1.5 V based on AlGaAs tandem cells (a) and about 2 V based on GaP tandem (b), taking into account an increase of voltage of two times in tandem cells in comparison with single-junction cells. Output voltage will be 10 to 15 V in a tritium-powered battery 1 mm thick and 100 to 150 V in a battery 1 cm thick. At

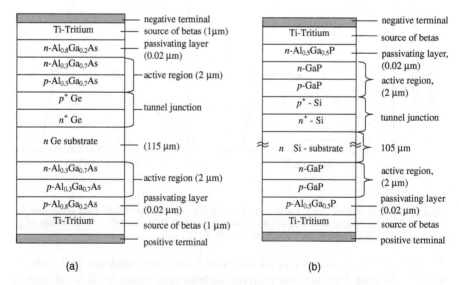

(a)   (b)

Figure 7.60   Conceptual design of tritium batteries based on (a) bifacial AlGaAs–Ge–AlGaAs and (b) GaP–Si–GaP tandem beta cells.

expected output current density of about 10^{-7} A/cm^2, the predicted output-specific volumetric power from this battery is 10^{-5} W/cm^3.

The structures shown in Figure 7.60 may be grown by MOCVD technology. The similar single junction and tandem structures on Ge substrates (a) are produced in high scale for space arrays.[72,73] Technology of AlGaP–GaP heterostructures (b) is available for betavoltaic device fabrication, as well. These heterostructures for beta cells may be grown on cheap Si-substrates, taking into account lattice matching AlP and GaP with Si.

Much higher output electric power density per unit volume may be obtained in a betavoltaic battery based on thin-film (about 2 μm thick) bifacial cells without substrates (Figure 7.61). The specific output power of bifacial cells can be increased by removing substrates in cells with a thin low-band-gap layer situated between two higher-band-gap layers. If the low-band-gap layer thickness is less than the minority carrier diffusion length, a bifacial sensitivity to betas can be achieved in the cells based on heterostructure with a single p–n junction. Recombination losses in beta-voltaic cells based on similar p–n heterostructures can be reduced by confining the carrier generation region by potential barriers and owing to reduction of low-band-gap layer thickness, taking into account that complete absorption of tritium betas occurs in a thin layer of about 1 μm.

This structure may be fabricated in the following way. At first, the four-layer heterostructure is grown on Ge-substrate (Figure 7.61). Then this structure is bonded to the Ti-tritium source fabricated in the output terminal, the substrate is removed from the heterostructure by selective etching, and Ti-tritium source of betas is sputtered on the surface of heterostructure. This thin Ti-T layer serves as a positive terminal. The process may be reiterated to obtain the necessary output electric power in the battery. The volumetric specific power of the battery based on thin-film bifacial

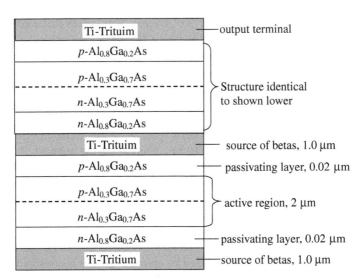

Figure 7.61   Cross section of tritium-powered battery based on two thin-film bifacial betavoltaic cells.

cells may be increased by some ten times in comparison with the batteries shown in Figure 7.60 based on cells with the Ge, Si substrates.

The decrease in beta cell and titanium layer thickness and an increase in tritium concentration in the titanium would allow increasing the battery power density to 0.1 mW/cm^3 in the beginning of life. This specific output electric power might be ensured in thin-film batteries at volumetric tritium activity of about 50 Ci/cm^3. The 12-year half-life betavoltaic battery with power density of 0.1 mW/cm^3 has an electrical power capacity of about 10 W-h/cm^3, which is more than in chemical batteries.

Volumetric tritium density and beta surface activity may be increased by the involvement of new materials. Single-walled carbon nanotube (SWCNT) layers are promising for tritium storage in addition to with T_2-Ti sources. Pores of molecular dimensions can absorb large quantities of hydrogen, owing to enhanced density of the absorbed material inside the pores, a consequence of the attractive potential of pore walls. Estimates[74] show that volumetric hydrogen density of 50 kg H_2m^{-3} can be received in single-walled carbon nanotubes with diameters of 1.5 to 3.0 nm; this hydrogen density is approximately two times more than in metal hydrides.

The additional advantage of SWCNT for tritium storage is simplification of fabrication battery technology because of lower temperature of tritium absorption in the case of SWCNT in comparison with fabrication of T_2-Ti. After annealing the SWCNT at 450 to 500°C in vacuum, the absorption of tritium would be effective at room temperature. In addition, surface activity of betas in SWCNT can be higher due to probable lower self-absorption of betas in SWCNT in comparison with metal hydrides, including the tritiated titanium.

Tritium is available from by-products of modern nuclear reactors, which makes it a cheap radioactive source. Therefore, the price of the unit area of semiconductor for beta cells is more important, taking into account that the output power density from tritium-fueled beta cells is about 10^{-7} W/cm^2. This means that a large area of beta cells is required to obtain acceptable power. III-V semiconductors have a higher price than silicon, but epitaxial thin-film technology on cheap Ge and Si substrates allows for reducing the price of III-V beta cells. For example, lattice-matched AlGaAs–GaAs heterostructures on Ge are widely used for high-scale production (more than 10^3 m^2 per year) of space PV arrays[72,73]; high-quality GaP heterostructures may be grown on Si, as well.

The typical thickness of Ge in modern space arrays is about 125 μm. Current Ge wafers may sell for $1.00 to $1.50/cm^2; if price can be reduced, the scale of wafer production can be increased. The cost of Si wafers for terrestrial solar cells is less than $0.10/cm^2. The epitaxial process is necessary to grow III-V heterostructures on Ge and Si substrates. There are productive MOCVD installations for such structure production. For example, the productivity of AIX-3000 MOCVD installation allows for fabricating approximately 2000-cm^2 cell structures in one 4 hour run, which means that about 3·10^2 m^2 cell structures can be fabricated during a year. Therefore, the increase of cell cost caused by epitaxial growth is not significant. The thickness of III-V heterostructure in a beta cell may be about 2 μm and the quantity and cost of A^3 and B^5 materials in the structure are very small. Thus, the expected cost of beta cells made of AlGaAs on Ge or GaP on Si would be about $1/cm^2 or less at a high scale of production. Taking into account the advantages of III-V beta cells, one can conclude that it is possible to use III-V compounds in profitable production of beta cells.

7.4 THERMOPHOTOVOLTAIC III-V CONVERTERS POWERED BY RADIOISOTOPE HEAT SOURCES

Thermophotovoltaic generators (TPV) convert broadband infrared radiation into electricity by means of infrared-sensitive photovoltaic converters. The TPV principle has been known for more than 25 years. Recently, interest has been renewed in TPV generators because of development of high-efficiency, low-band-gap PV cells based on GaSb and InGaAs materials developed as bottom subcells for mechanically stacked tandem solar cells.[4,75-77]

Thermophotovoltaics, which provide heat and electricity cogeneration, are valuable for families living off electric power grids because they use a propane or natural gas-fired source rather than the sun. For homes on the electric grid, this product has potential, as well. If electric power is lost, a TPV cogenerator of electricity and heat provides power. Excess electric power can also be utilized to operate a television, a computer, or lights. Other applications of TPV generators are for cabins, cars, and boats.

In radioisotope thermophotovoltaic (RTPV) generators, the fuel that provides the heat to a TPV emitter can be a Pu-based radioisotope. An RTPV system can be used for space missions where solar flux is too low for conventional solar PV arrays to be practical. The long half-life of radioisotope fuel would ensure long duration of these missions.

Figure 7.62 shows an example of an RTPV generator consisting of a Pu-based radioisotope heat source, matched emitter, selective filters, four PV arrays, and cooling fins. The PV cells directly convert emitted radiation into electric power output. Input power to the RTPV system is provided by a heated emitter that maintains a high source temperature. A temperature range of 1000 to 1200°C is dictated by practical considerations: lower temperatures result in low RTPV generator electric power output, while the opposite is true for higher temperatures. The source may emit a selective (narrow-band) or a blackbody spectrum. Common emitter materials, for example, silicon carbide, radiate near-blackbody (gray body) spectra. PV semiconductor converters can convert photons with energies greater than their band gap. For blackbody spectra, a large fraction of the radiation is contained in the long-wavelength part of the spectrum with photon energies lower than the PV cell band gap. Therefore, an optical element can be usefully included in the RTPV system to return sub-bandgap-energy photons to the emitter for reheating. This optical element may be made as dielectric stacks or as a metallic reflector at the back surface of the PV cells.

One of the main reasons for the new level of interest in TPV is the availability of high-performance cells based on semiconductors from III-V compounds.[78-89] GaSb and InGaAs, with band gaps of 0.7 to 0.75 eV, and InGaAsSb (E_g = 0.5 to 0.6 eV) are near optimal for emitters with temperatures of 1000 to 1200°C. Figure 7.63

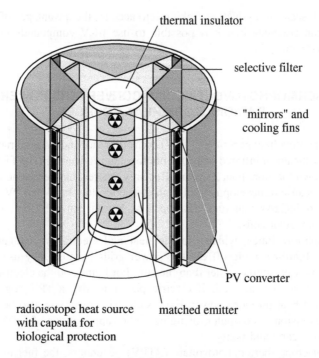

Figure 7.62 Radioisotope thermophotovoltaic generator concept drawing.

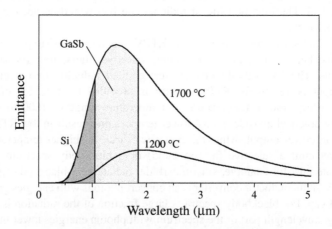

Figure 7.63 Blackbody emittance as a function of wavelength for two emitter temperatures: 1200 and 1700°C. The parts of radiation that may be absorbed and converted by Si and GaSb cells are shown.

shows emittance as a function of wavelength of a blackbody at 1200 and 1700°C. GaSb, with a band-gap energy of 0.7 eV, converts much more energy available from such blackbody emitters than Si. Figure 7.64 shows the calculated electric power density output for single-junction TPV converters operated at 50° C as a function

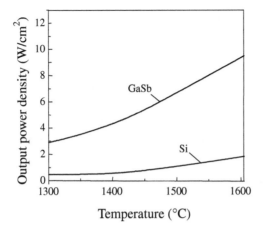

Figure 7.64   Output power density for GaSb- and Si-based cells as a function of blackbody emitter temperature.

of emitter temperature. Output power density is about five times greater for GaSb cells in comparison to silicon-based cells. This advantage of GaSb cells increases with emitter temperature decrease to 1000 to 1200°C.

This section presents some results of recent development of PV cells prepared by Zn-diffusion into "bulk" GaSb wafers[80-89] as well as into epitaxial GaSb and InGaAsSb[85] layers grown by liquid-phase epitaxy. This research will extend the state of the art through improved material matching and efficiency in the RTPV system.

The main technological approach to producing GaSb TPV cells is the use of bulk wafers as a base material. These wafers are prepared from ingots grown by a Czochralski technique and then Te-doped. At the first stage, GaSb wafers are exposed for the first zinc diffusion procedure in a "pseudo–closed box" technique to form a shallow p–n junction in the photoactive area of a cell. Anodic oxidation and selective etching are employed for precise thinning of the diffused p-GaSb layer. At the second stage, a deep p–n junction (1 to 1.5 μm) is formed by a spatially selective diffusion process; this avoids current leakage under contact grid fingers. A blanket Au (Ge) ohmic contact is then deposited on the back surface by vacuum evaporation and alloyed at 300°C. A Cr/Au/Ni/Au contact is deposited on the front surface and alloyed at 250°C. The front contact grid is thickened by electroplated Au up to 3 μm. A two-layer ZnS–MgF$_2$ antireflection coating is applied to minimize reflection losses in a spectral range of 700 to 2200 nm. The cell area is 1×1 cm^2 or 2×1 cm^2 and contact grid spacing is 0.1 or 0.2 mm. The scheme of a bulk GaSb cell is given in Figure 7.65.

Figure 7.66 shows the spectral response of cells based on bulk GaSb with different p-layer thicknesses. As can be seen from this figure, maximum photocurrent can be obtained from a cell with p-region thickness in the range of 0.1 to 0.3 μm, which is why the second selective diffusion process is necessary. Epitaxial growth of Te-doped GaSb layers is carried out[83,84] by liquid-phase epitaxy (LPE) in attempts to prepare material of a higher crystal quality. Ga-, Sb-, and Pb-rich melts are used in the LPE processes. Figure 7.67 shows the spectral responses of the epitaxial TPV

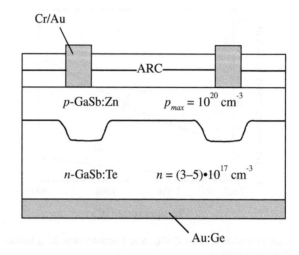

Figure 7.65   Cross section of a "bulk" GaSb TPV cell.

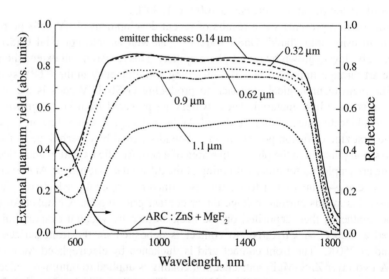

Figure 7.66   Spectra of photoresponse (on the area including contact grid) for "bulk" GaSb cells with different p-emitter thicknesses. Spectrum of reflectivity in the case of ZnS+MgF$_2$ antireflection coating is shown, as well.

cells. A maximum of external quantum yield was obtained in the cells, which were prepared using epilayers grown from Ga-rich melt.

Photocurrent density as high as 54 mA/cm^2 under AM0 spectrum was obtained in GaSb cells, and 58 mA/cm^2 was obtained in InGaAsSb cells. Efficiency of 19.1% (under the part of AM0 spectrum with $\lambda < 900$ nm) has been achieved (see Figure 7.68).

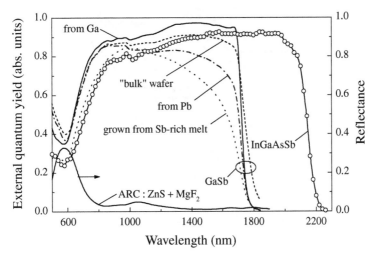

Figure 7.67   Spectral responses (on the photoactive area) of TPV cells using the epitaxial layer of GaSb and InGaAsSb grown from Ga-, Pb-, and Sb-rich melts. Reflectance from a cell with ZnS+MgF$_2$ antireflection coating is shown, as well.

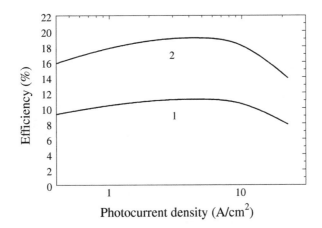

Figure 7.68   Efficiencies of the GaSb TPV cell as a function of photocurrent under concentrated sunlight: 1) AM0 spectrum, 2) part of AM0 spectrum with $\lambda > 900$ nm.

The cells of 1 or 2 cm^2 in the area (see Figure 7.69) were developed for TPV generators and are able to generate photocurrent up to 2 to 10 A (CW-operation) at I–V curve fill factor values of about 0.7 (see Figure 7.70). The deviations of V_{oc} and FF in the 1-cm^2 cells fabricated using advanced technology are in the following ranges: V_{oc} = 0.42 to 0.45 and FF = 0.70 to 0.72 at photocurrent of 1 A. Cells 2 cm^2 in area generate photocurrent up to 9 A at V_{oc} =0.52 V.

Small fuel-fired TPV demonstration systems have been built in different laboratories.[80,88-96] Possible applications of such systems are as stand-alone generators of heat and electricity able to charge an electrochemical battery. TPV system design

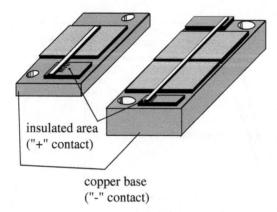

insulated area
("+" contact)

copper base
("-" contact)

Figure 7.69   TPV cell design: for lower generated power (left); for higher generated power (right).

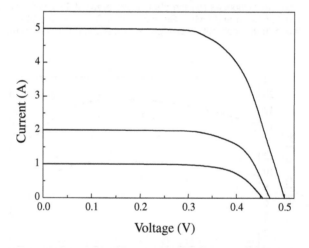

Figure 7.70   Illuminated current-voltage characteristics of a 1-cm² GaSb TPV cell.

can possibly be improved through use of more efficient TPV cells and better matching of the TPV cell and emitter through use of selective filters and back surface mirrors. The simplicity of the device is its greatest asset. The overall system efficiency depends on how the three key components — PV cells, radiant emitter, and heat source — operate as a system.

GaSb cells are well characterized with respect to external quantum efficiency of photoresponse, open-circuit voltage, and fill factor of the illuminated *I–V* curve. In addition, the fabrication process for them is relatively simple, which is why such cells have been employed in the Ioffe Institute experimental TPV generator.[92] Passive air cooling of the cells was used in that configuration. For this, each cell of 1×2 cm² in area (or two cells of 1×1 cm² in a parallel connection) was soldered to a copper prism equipped with two aluminum fins. Some of the cells were mounted on a special holder to allow alternative cooling methods by water or by air (with blower, or passively). High temperature–resistant metals as the emitter material can be used in the TPV generators. Metallic alloys of Fe/Cr/Al/Co/C are proper materials for emitter operating up to a temperature of 1250°C.[92] Attractive features of such a material are the spectral selectivity of emission; resistance against thermoshock; and transparency for exhaust gas, but solid for heat. The latter feature may be realized in practice if the emitter of a cylindrical geometry has a view of a "basket." Figure 7.71 shows measured spectral distribution of emissive power of the basket-type metallic emitter heated by flame up to 1200 to 1250°C.

In Figure 7.72, a schematic of one of the fuel-fired TPV generators built in the Ioffe Institute[92] is shown. A burner with gas inlet 0.15 mm in diameter is connected (directly) with a butane tank or through a pressure-reducing unit with a propane tank. Gas stream force and natural convection process are employed for gas–air premixing and transport to the combustion zone; the premixing zone is separated from the combustion zone by a wire screen. The basket-type metallic emitter has an external diameter of 3.6 cm and height of 1.8 cm, and the combustion process is carried out directly at its surface. There is a quartz window 5 cm in diameter arranged around the stream of exhaust gas. Fifteen GaSb cells of 1×2 cm^2 each are situated around the emitter and window. The heat sink of each cell consists of copper prism and two aluminum fins (Figure 7.73). Output electric power of 6.2 to 7 W was received in this generator (Figure 7.74), which may be considered a prototype for RTPV generators for space missions far from the sun.

A small GaSb PV cell–based RTPV system fueled by a ^{238}PuO$_2$ radioisotope has been developed recently.[96] This was designed for a space mission to Pluto, in the vicinity of which solar illumination is too small to permit solar arrays to be used. It was shown that replacing the radioisotope thermoelectric generator typically used

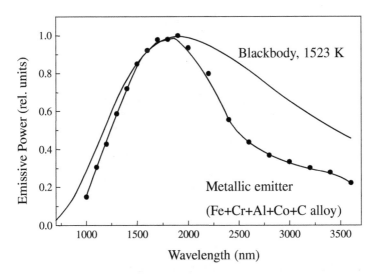

Figure 7.71   Emissive power spectra for the basket-type metallic emitter (experimental) and for blackbody at T = 1523 K (calculated).

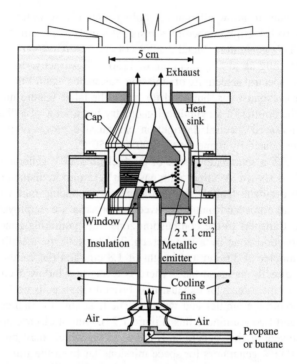

Figure 7.72 Schematic of portable fuel-fired TPV generator based on 15 GaSb 1 × 2 cm² cells.

Figure 7.73 Seven mounted 1 × 2 cm² GaSb cells on heat sink (copper prisms and aluminum fins) for TPV generator.

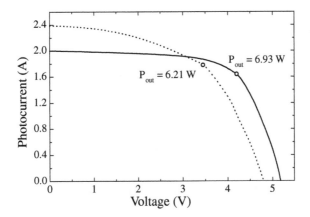

Figure 7.74 *I–V* curves for fuel-fired TPV generator based on 15 GaSb TPV cells at intensive (solid line) and convective (dotted line) air cooling.

on such missions with the RTPV generator triples the efficiency and specific power of the power-generating system.

ACKNOWLEDGMENTS

The author wishes to thank Zh. Alferov for helpful discussions and support and V.D. Rumyantsev, A.G. Kavetsky, V.S. Kalinovsky, V.P. Khvostikov, V.R. Larionov, V.M. Lantratov, and M.Z. Shvarts for their contribution to this work.

REFERENCES

1. Sze, S.M., *Physics of Semiconductor Devices*, John Wiley & Sons, Inc., New York, 1981, Part 1.
2. Andreev, V.M., Dolginov, L.M., and Tret'iakov, D.N., Liquid phase epitaxy in semiconductor devices technology, Sov. Radio, Moscow, 1975 [in Russian].
3. Ioffe, A.F., *Physics of Semiconductors*, Moscow, Leningrad, 1957 [in Russian].
4. Andreev, V.M., Grilikhes, V.A., and Rumyantsev, V.D., *Photovoltaic Conversion of Concentrated Sunlight*, John Wiley & Sons, Chichester, 1997, chap. 1 and 3.
5. Casey, H.C. and Panish, M.B., *Heterostructure Lasers*, Academic Press, New York, 1978, Part B.
6. Ryvkin, S.M., *Photovoltaic Phenomena in Semiconductors*, Fizmatgiz, Moscow, 1963 [in Russian].
7. Partain, L.D., Solar cell fundamentals, in *Solar Cells and Their Application*, Partain, L.D., Ed., John Wiley & Sons, Chichester, 1995, chap. 1.
8. Henry, C.H., Limiting efficiency of ideal single and multiple energy gap terrestrial solar cells, *J. Appl. Phys.*, 51, 4494, 1980.
9. Shockley, W. and Queisser, H.J., Detailed balance limit of efficiency of *p–n* junction solar cells, *J. Appl. Phys.*, 32, 510, 1961.

10. Vasil'ev, A.M. and Landsman, A.P., Semiconductor photoconverters, Sov. Radio, Moscow, 1971 [in Russian].

11. Hovel, H.J., *Semiconductors and Semimetals*, vol. 11, *Solar Cells*, Willardson, R.K. and Beer, A.C., Eds., Academic Press, New York, 1975, chap. 1.

12. Luque, A., *Solar Cells and Optics for Photovoltaic Concentration*, Adam Hilger, Bristol-Philadelphia, 1989.

13. Fahrenbruch, A.L. and Bube, R.H., *Fundamentals of Solar Cells*, Academic Press, New York, 1983.

14. Alferov Zh. I. et al., Photovoltaic properties of heterojunction AlGaAs-GaAs, *Fiz. i Tekhn. Polupr.*, 3, 1633, 1969; translated in English in *Sov. Phys. Semicond.*, 3, No. 11, 1969.

15. Alferov, Zh. I. et al., Solar cells based on heterojunction p-AlGaAs–n-GaAs, *Fiz. i Tekhn. Polupr.*, 4, 2378, 1970; translated in English in *Sov. Phys. Semicond.*, 4, No. 12, 1970.

16. Andreev, V.M. et al., Investigation of high efficiency AlGaAs-GaAs solar cells, *Fiz. i Tekhn. Polupr.*, 7, 2289, 1973; translated in English in *Sov. Phys. Semicond.*, 7, No. 12, 1973.

17. Andreev, V.M. et al., Comparison of different heterophotoconverters for achievement of highest efficiency, *Fiz. i Tekhn. Polupr.*, 8, 1328, 1974; translated in English in *Sov. Phys. Semicond.*, 8, No. 7, 1974.

18. Hovel, H.J. and Woodall, J.M., High-efficiency AlGaAs-GaAs solar cells, *Appl. Phys. Lett.*, 21, 379, 1972.

19. Woodall, J.M. and Hovel, H.J., Outlook for GaAs terrestrial photovoltaics, *J. Vac. Sci. Technol.*, 12, 1000, 1975.

20. Andreev, V.M. et al., High-efficiency AlGaAs-GaAs solar cells with internal Bragg reflector, *Conf. Rec. 1st World Conf. Photovoltaic Energy Convers.*, Hawaii, 1994, p. 1894.

21. Ettenberg, M. and Paff, R.J., Thermal expansion of AlAs, *J. Appl. Phys.*, 41, 3926, 1970.

22. Monemar, B., Shih, K.K., and Pettit, G.D., Some optical properties of the $Al_xGa_{1-x}As$ alloy system, *J. Appl. Phys.*, 47, 2604, 1976.

23. Casey, H.C., Sell, D.D., and Panish, M.B., Refractive index of $Al_xGa_{1-x}As$ between 1.2 and 1.8 eV, *Appl. Phys. Lett.*, 24, 63, 1974.

24. Dingle, R., Wiegmann, W., and Henry, C.H., Quantum states of confined carriers in very thin $Al_xGa_{1-x}As$-GaAs-$Al_xGa_{1-x}As$ heterostructures, *Phys. Rev. Lett.*, 33, 827, 1974.

25. Batey, J. and Wright, S.L., Energy band alignment in GaAs: (Al,Ga)As heterostructures: the dependence on alloy composition, *J. Appl. Phys.*, 59, 200, 1986.

26. Alferov, Zh. I. et al., Solar cells on the basis of AlGaAs-heterostructures with graded layers, *Pis'ma v Zh. Tekn. Fiz.*, 4, 305, 1978; translated in English in *J. Techn. Phys. Lett.*, 4, No. 6, 1978.

27. Woodall, J.M. and Hovel, H.J., An izothermal etchback-regrowth method for high efficiency AlGaAs-GaAs solar cells, *Appl. Phys. Lett.*, 30, 492, 1977.

28. Hamaker, H.C. et al., 26% efficient magnesium doped AlGaAs/GaAs solar concentrator cells, *Appl. Phys. Lett.*, 47, 762, 1985.

29. Werthen, J.G. et al., 21% (one sun, air mass zero) 4 cm² GaAs space solar cells, *Appl. Phys. Lett.*, 48, 74, 1986.

30. Andreev, V.M. et al., Heterojunction solar cells with low value of saturation current, *Fiz. i Tekhn. Polupr.*, 19, 276, 1985; translated in English in *Sov. Phys. Semicond.*, 19, No. 2, 1985.

31. Alferov, Zh. I. et al., Liquid phase epitaxy of $Al_xGa_{1-x}As$-GaAs heterostructures, *Kristall und Technik.*, 10, 103, 1975.

32. Alferov, Zh. I. et al., Investigation of new LPE method of obtaining Al-Ga-As heterostructures, *Kristall und Technik.*, 11, 1013, 1976.

33. Andreev, V.M., Heterostructure solar energy converters, in *Optoelectronic Materials and Devices*, Herman, M.A. Ed., Polish Sci. Publ., Warsaw, 1983, 479.

34. Small, M.B. et al., Al diffusivity as a function of growth rate during the formation of (GaAl)As heterojunctions by liquid phase epitaxy, *J. Appl. Phys.*, 52, 814, 1981.

35. Alferov, Zh. I. et al., Preparation and investigation of epitaxial layers of $Al_xGa_{1-x}As$ solid solutions and heterojunctions in the AlAs-GaAs system, *Kristall und Technik.*, 4, 495, 1969.

36. Panish, M.B. and Sumski, S., Ga-Al-As: phase, thermodynamic and optical properteies, *J. Phys. Chem. Solid.*, 30, 129, 1969.

37. Andreev, V.M., Liquid-phase epitaxy of electroluminescent AlGaAs heterostructures, *Czechoslov. J. Phys.*, B30, 262, 1980.

38. Andreev, V.M. et al., Photoelectric properties of AlGaAs-GaAs heterostructures with a tunnel-thin wide-gap window, *Sov. Phys. Semicond.*, 23, 74, 1989.

39. Masu, K., Konagai, M., and Takahashi, K., Acceptor energy level for Zn in $Ga_{1-x}Al_xAs$, *J. Appl. Phys.*, 51, 1060, 1980.

40. Vassilieff, G. and Saint–Cric, B., Zn incorporation in $Ga_{1-x}Al_xAs$ grown by liquid phase epitaxy and its electrical properties, *J. Appl. Phys.*, 54, 4581, 1983.

41. Dupuis, R.D. et al., High-efficiency GaAlAs/GaAs heterostructure solar cells grown by metalorganic chemical vapor deposition, *Appl. Phys. Lett.*, 31, 201, 1977.

42. Nelson, N. J. et al. Organometallic-sourced VPE AlGaAs/GaAs concentrator solar cells having conversion efficiencies of 19%, *Appl. Phys. Lett.*, 33, 26, 1978.

43. Sahai, R., Edwall, D.D., and Harris, J.S., High-efficiency AlGaAs-GaAs concentrator solar cells, *Appl. Phys. Lett.*, 34, 147, 1979.

44. Andreev, V.M. et al., High-efficiency (24.6%, AM0) LPE grown AlGaAs/GaAs concentrator solar cells and modules, *Conf. Rec. 1st World Conf. Photovoltaic Energy Convers.*, Hawaii, 1994, p. 2096.

45. Khvostikov, V.P. et al., Space concentrator solar cells vased on multilayer LPE grown AlGaAs/GaAs heterostructure, *Proc. 4th Eur. Space Power Conf.*, Poitiers, 1995, 359.

46. Tobin, S.P. et al., Advanced in high-efficiency GaAs solar cells, *Conf. Rec. 21st IEEE Photovoltaic Specialists Conf.*, Las Vegas, 1990, 158.

47. Sims, P. E. et al., High efficiency GaP power conversion for betavoltaic applications, *Proc. 13th SPRAT Conf.*, Cleveland, 1994.

48. Sims, P. E. et al., Gallium phosphide energy converters, *Proc. 14th SPRAT Conf.*, Cleveland, 1995, 231.

49. Hughes, R.C. et. al., Gallium phosphide junction with low leakage for energy conversion and near ultraviolet detectors, *J. Appl. Phys.*, 69, 6500, 1991.

50. Prutskij, T.A., Andreev, V.M., and Larionov, V.R., GaP-AlGaP heterojunction photosensors with photosensitivity maximum at 320 to 420 nm, *Sol. Energy Mater. Sol. Cells*, 37, 380, 1995.

51. Ehrenberg. W. et al., The electron voltaic effect, *Proc. R. Soc.*, 64, 424, 1951.

52. Rappaport, P., The electron-voltaic effect in *p–n* junctions induced by beta particle bombardment, *Phys. Rev.*, 93, 1953, 246.

53. Pfan, W.G. and Van Roosbroeck, W., Radioactive and photoelectric *p–n* junction power sources, *J. Appl. Phys.*, 25, 1954, 1422.

54. Rappaport, P., Loferski, J.J., and Linder, E.G., The electron-voltaic effect in germanium and silicon *p–n* junctions, *RCA Rev.*, 17, 1956, 100.

55. Flicker, H., Loferski, J.J., and Elleman, T.S., Construction of a promethium-147 atomic battery, *IEEE Trans.*, ED-11, 2, 1964.
56. Olsen, L.C., Betavoltaic energy conversion, *Energy Convers.*, 13, 117, 1973.
57. Olsen, L.C., Advanced betavoltaic power sources, *Proc. 9th Intersociety Energy Convers. Eng. Conf.*, 1974, 754.
58. Balonov M.I., *Dozimetrija and Normirovanie of Tritium*, Energoatomizdat, Moscow, 1983 [in Russian].
59. Kherami, N.P. and Shmayada, W.T., Radioluminescence using metal tritides, *Zeitshrift fur Physikalische Chemie*, Bd. 183, S. 453, 1994.
60. Birks, L.S., *Electron Probe Microanalysis*, Interscience Publishers, New York, 1963.
61. Olsen, L.C., Review of betavoltaic energy conversion, *Proc. 17th Space Photovoltaic Res. Technol. Conf.*, Cleveland, 1992, 256.
62. Andreev, V.M. et al., Behavior of AlGaAs/GaAs photodiodes subjected to soft ionizing radiation, *Semiconductors*, 28, 207, 1994.
63. Andreev, V.M. et al., Betavoltaic cells based on tritium and AlGaAs/GaAs heterostructures, *Techn. Dig. Int.* PVSEC-11, Sapporo, 1999, 817.
64. Andreev, V.M. and Rumyantsev, V.D., A3B5 based solar cells and concentrating optical elements for space PV modules, *Sol. Energy Mater. Sol. Cells*, 44, 319, 1996.
65. Tobin, S.P. et al., Enhanced light absorption in GaAs solar cells with internal Bragg reflector, *Conf. Rec. 22nd IEEE Photovoltaic Specialists Conf.*, Las Vegas, 1991, 147.
66. Shvarts, M.Z. et al., Effect of structure parameters on radiation resistance of AlGaAs/GaAs solar cells with internal Bragg reflector, *Proc. 5th Eur. Space Power Conf.*, Tarragona, 1998, 513.
67. Andreev., V.M. et al., High efficiency radiation stable AlGaAs/GaAs solar cells with internal Bragg reflector, *Proc. 4th Eur. Space Power Conf.*, Poitiers, 1995, 367.
68. Loo, R., Knechtly, R.C., and Kamath, G.S., Enhanced annealing of GaAs solar cell radiation damage, *Conf. Rec. 15th IEEE Photovoltaic Specialists Conf.*, New York, 1981, 33.
69. Walker, G.H. and Conway, E.J., Short circuit current changes in electron irradiated GaAlAs/GaAs solar cells, *Prog. Int. Solar Energy Soc. Congr.*, Pergamon Press, New York, 1978, 575.
70. Friedman, D.J. et al., GaInP/GaAs monolithic tandem concentrator cells, *Conf. Rec. 1st World Conf. Photovoltaic Energy Convers.*, Hawaii, 1994, 1829.
71. Friedman, D.J. et al., On-sun concentrator performance of GaInP/GaAs tandem cells, *Conf. Rec. 25th IEEE Photovoltaic Specialists Conf.*, Washington, D.C., 1996, 73.
72. Chiang, P.K. et al., Large area GaInP/GaAs/Ge multijunction solar cells for space application, *Conf. Rec. 1st World Conf. Photovoltaic Energy Convers.*, Hawaii, 1994, 2120.
73. Chiang, P.K. et al., Experimental results of GaInP$_2$/GaAs/Ge triple junction cell development for space power systems, *Conf. Rec. 25th IEEE Photovoltaic Specialists Conf.*, Washington, D.C., 1996, 183.
74. Dillon, A.C. et al., Storage of hydrogen in single-walled carbon nanotubes, *Nature*, 386, 377, 1997.
75. Fraas, L.M. et al., Over 35% efficient GaAs/GaSb tandem solar cells, *IEEE Trans. Electron Devices*, 37, 443, 1990.
76. Fraas, L.M., Concentrator modules using multijunction cells, in *Solar Cells and Their Applications*, Partain, L.D., Ed., John Wiley & Sons, 1995, 301.
77. Fraas, L.M. et al., Tandem gallium concentrator solar cells, *Proc. 11th Photovoltaic Sol. Energy Conf.*, Montereux, 1992, 135.

78. Wanlass, M.W. et al., Ga$_x$In$_{1-x}$As thermophotovoltaic converters, *Proc. IEEE 1st World Conf. Photovoltaic Energy Convers.*, Hawaii, 1994, 1685.

79. Coutts, T.J. et al., A review of recent advances in thermophotovoltaics, *Proc. 25th IEEE PVSC*, Washington, D.C., 1996, 25.

80. Fraas, L. et al., A thermophotovoltaic electric generator using GaSb cells with a hydrocarbon burner, *Proc. IEEE 1st World Conf. Photovoltaic Energy Convers.*, Hawaii, 1994, 1713.

81. Khvostikov, V.P., Sorokina, S.V., and Shvarts, M.Z., GaSb based solar cells for concentrator tandem application, *Proc. 13th Eur. Photovoltaic Sol. Energy Conf.*, Nice, 1995, 61.

82. Andreev, V.M. et al., Tandem solar cells based on AlGaAs/GaAs and GaAs structures, *Proc. 23rd ISCS*, St. Petersburg, 1996, 425.

83. Andreev, V.M. et al., GaAs and GaSb based solar cells for concentrator and thermophotovoltaic applications, *Proc. 25th IEEE PVSC*, Washington, D.C., 1996, 143.

84. Andreev, V.M. et al., GaSb based PV cells for solar and thermophotovoltaic applications, *Proc. 14th Eur. Photovoltaic Sol. Energy Conf. Exhibition*, Barcelona, 1997, 1763.

85. Andreev, V.M. et al., Tandem GaSb/InGaAsSb thermophotovoltaic cells, *Proc. 26th IEEE Photovoltaic Specialists Conf.*, Anaheim, 1997, 935.

86. Bett, A.W. et al., GaSb-based (thermo)photovoltaic cells with Zn diffused emitters, *Proc. 25th IEEE PVSC*, Washington, D.C., 1996, 133.

87. Bett, A.W. et al., Large-area GaSb photovoltaic cells, *Proc. 3rd NREL Conf. Thermophotovoltaic Generation Electr.*, Colorado Springs, 1997, 41.

88. Fraas, L.M. et al., Low cost high power GaSb photovoltaic cells, *Proc, 3rd NREL Conf. Thermophotovoltaic Generation of Electr.*, Colorado Springs, 1997, 33.

89. Fraas, L., Commercial GaSb and circuit development for the midnight sun TPV stove, *Proc. 4th NREL Conf. Thermophotovoltaic Generation Electr.*, Denver, 1998, 480.

90. Schubnell, M., Gabler, H., and Broman L., Overview of European activities in thermophotovoltaics, *3rd NREL Conf.*, Colorado Springs, 1997, 3.

91. Gabler, H., Hein, M., and Xenker, M., A propane-fueled thermophotovoltaic energy converter using low-bandgap photovoltaic cells, *Proc. 2nd World Conf. Photovoltaic Sol. Energy Convers.*, Vienna, 1998.

92. Rumyantsev, V.D. et al., Portable TPV generator based metallic emitter and 1.5-Amp GaSb cells, *Proc. 4th NREL Conf. Thermophotovoltaic Generation Electr.*, Denver, 1998, 384.

93. Fraas, L. et al., Development status on a TPV cylinder for combined heat and electric power for the home, *Proc. 4th NREL Conf. Thermophotovoltaic Generation Electr.*, Denver, 1998, 371.

94. Ferguson, L. and Fraas, L., Matched infrared emitter for use with GaSb TPV cells, *Proc. 3rd NREL Conf. Thermophotovoltaic Generation Electr.*, Colorado Springs, 1997, 169.

95. Stone, K.W. et al., Operation and component testing of a solar thermophotovoltaic power system, *Proc. 25th PVSC*, Washington, D.C., 1996, 1421.

96. Counts, T. and Fitzgerald, M.C., Thermophotovoltaics, *Sci. Am.*, September, 1998, 90.

CHAPTER 8

Wide-Band Semiconductors for Direct-Conversion Nuclear Batteries

Y.S. Shreter, Y.T. Rebane, and N.I. Bochkareva

CONTENTS

8.1   MATERIAL SPECIFICATIONS FOR SEMICONDUCTOR NUCLEAR BATTERIES

The idea to use semiconductor p–n junctions to convert the energy of nuclear decay into electrical current is as old as the junctions themselves. However, effective nuclear batteries based on Ge, Si, and GaAs semiconductor p–n junctions have not yet been successful. This is due mainly to high degradation of conventional semiconductors under ionizing radiation and low carrier separation efficiency by p–n junctions at the low radiation intensity typical for ecologically acceptable nuclear radiation sources.

Use of radiation hard semiconductors with wide band gaps such as GaN, AlN, SiC, and diamond provides a possible way to overcome these problems. Until recently the quality of these semiconductors was poor, but new growth methods developed in the last decade for wide-band-gap semiconductors improves the prospect of their application in nuclear batteries.

The use of metal-semiconductor Schottky barriers for carrier separation seems to be the most effective way to produce nuclear batteries based on wide-band-gap microcrystalline thin-film semiconductors. This chapter provides an analysis of nuclear battery designs based on thin-film wide-band-gap microcrystalline semiconductors covered with thin metallic films to form Schottky barriers for carrier separation.

8.1.1   Choice of Radioactive Isotope

The key issue for nuclear battery design is choice of the radioactive isotopes and semiconductors. There are two types of nuclear battery applications: for industrial, space, and military purposes such as aerospace electronics and environmental sensors, and for general consumer utilization such as power sources for PC memory and watches. The first type of application allows consideration of isotopes that produce gamma radiation, because ecological requirements are not as strict since people will not be directly exposed to ionizing radiation. Radioisotope batteries for mass utilization are commercially more attractive since their market potential is so much more significant; however, they must meet very strict ecological requirements and be absolutely safe. These strict requirements favor tritium as the best radioactive power source for semiconductor nuclear batteries.

Tritium produces only low-energy β-radiation (mean value ~ 6 keV), with a very small penetration depth of less than 1 μm in solids. After decay it transmutes into a safe inert gas, helium. From the technical point of view, tritium has another advantage: soft β-radiation produces the least possible damage to semiconductor devices used for current generation and its diffusion into solid films allows construction of compact solid-state radioactive sources. The price of tritium is also reasonable for mass production of semiconductor nuclear batteries.

8.1.2 Choice of Semiconductor Material

The shallow penetration depth of tritium β-electrons dictates the choice of the semiconductor material. Thin-film semiconductors provide the most effective utilization of tritium β-radiation.

Another factor that strongly affects semiconductor nuclear batteries is degradation of the semiconductor material under tritium β-electrons. In spite of the low energy of tritium β-electrons, which is below the threshold for direct generation of point lattice defects in single collisions of tritium β-electrons with host atom nuclei, tritium β-electrons can promote lattice defects indirectly through creation of excess electron–hole plasma along their path in the host semiconductor. This indirect creation of lattice defects is less effective in radiation-hard wide-band-gap materials like GaN, AlN, SiC, and diamond, where formation energy of the point lattice defects and energy needed for generation of the electron–hole pair are higher.

In these wide-band-gap semiconductors, the dark current is much lower than in conventional Si and GaAs semiconductors, which is also favorable for effective carrier separation under low β-irradiation intensity. The use of high-quality monocrystalline wide-band-gap semiconductor epitaxial films on crystalline substrates in nuclear batteries is expensive, so cheap and robust microcrystalline films on thin flexible substrates are attractive. Another possibility is to use cheap hydrogenated or tritiated amorphous semiconductor films in the nuclear batteries, but they have low carrier mobility and instability under β-radiation, probably related to low hydrogen isotope diffusion activation energy in amorphous semiconductors.

8.2 ENERGY LOSS AND GENERATION OF ELECTRON–HOLE PAIRS BY BETA ELECTRONS

8.2.1 Maximal Penetration Length of Beta Electrons

Energy losses of nonrelativistic beta electrons in semiconductors are mainly related to excitation of individual atoms or the generation of plasmons. Both these excitations then decay with creation of electron–hole pairs. The corresponding energy losses per unit path length for elemental semiconductor are given by Equation 8.1[1,2]:

$$-\frac{dE}{dx} = \frac{4\pi Ne^4}{mv^2} \ln\left(\frac{2E}{E_{ion}}\right) \tag{8.1}$$

where N is the total electron density in the semiconductor, v and m are the electron velocity and mass, and E_{ion} is the mean ionization energy of the semiconductor atoms.

It can be seen from this equation that the square of the beta electron energy with the logarithmic accuracy decays linearly with the electron path:

$$-\frac{dE^2}{dx} = 4\pi Ne^4 \ln\left(\frac{2E}{E_{ion}}\right) \tag{8.2}$$

The total electron density in an elemental semiconductor is related to its mass density ρ by the equation $N = \rho Z/M$, where Z and M are the charge and mass of the semiconductor atom, respectively. For a compound semiconductor, Equation 8.3 and the Thomas–Fermi approximation, $E_{ion,i} \sim Z_i \cdot Ry$, is used to estimate the mean ionization energy of the i-atom:

$$-\frac{dE^2}{dx} = 4\pi e^4 \rho \sum_i Z_i \ln\left(\frac{2E}{Z_i Ry}\right)\bigg/ m_p \sum_i M_i \tag{8.3}$$

where Z_i and M_i are the charge and mass of the i-atom nucleolus, and m_p is the proton mass.[2]

From Equation 8.3, the beta electron maximum penetration length into the semiconductor can be estimated with logarithmic accuracy:

$$L_{max} = E^2 m_p \sum_i M_i \bigg/ 4\pi e^4 \rho \sum_i Z_i \ln\left(\frac{2E}{Z_i Ry}\right) \tag{8.4}$$

Because the square of the beta electron energy decays linearly, the electron–hole pair generation rate G per unit path length is nonuniformly distributed along the beta electron path, increasing with the path length L_p as

$$G = G_0\bigg/\sqrt{1 - L_p/L_{max}} \tag{8.5}$$

where L_{max} is the maximal penetration length for beta electrons.

8.2.2 Optimal Thickness of Metallic Schottky Contact for Betavoltaic Cells

As can be seen from Equation 8.5, the generation rate goes to infinity at L_{max}, although its integral giving the number of electron–hole pairs created is finite. Therefore, most electron–hole pairs are generated in the vicinity of L_{max}. In the near-surface region, $L_p < 0.1\, L_{max}$, so only 5% of the total electron–hole pairs are generated there.

The maximal penetration lengths of the tritium beta electrons with energy ~ 6 keV calculated on the basis of Equation 8.4 for several semiconductors and metals are shown in the Table 8.1. As can be seen, the penetration length of tritium β-electrons into semiconductors is less then 0.5 μm; therefore, only thin-film semiconductor devices should be used for carrier separation to provide maximal volume efficiency of the nuclear battery. Microcrystalline thin films deposited on flexible substrate seem to be most promising to provide compact packaging and cheap mass production of nuclear batteries.

The use of thin metallic layer, with thickness L^{met} smaller than one-tenth of L_{max} for the corresponding metal ($L^{met} \leq 0.1\ L_{max}$) and deposited on microcrystalline semiconductor thin film to form a Schottky barrier for carrier separation, is the cheapest and most effective way to produce nuclear batteries. As can be seen from the table, the corresponding thickness L^{met} should be in the region of 100 to 400 Å.

8.3 CARRIER SEPARATION BY THIN-FILM MS AND MIS DIODES

8.3.1 Band Structure and Current-Voltage Characteristics of MS and MIS Diodes

The general band structure scheme of metal–semiconductor (MS) diodes for n-type semiconductor is shown in Figure 8.1. Figure 8.2 shows the band structure of metal–intrinsic semiconductor–semiconductor (MIS) diode for the case of p-type semiconductor.

The corresponding I–V characteristics include the three main current components. These are the thermionic current J_{therm} resulting from carrier thermal activation over the Schottky barrier, the tunneling current J_{tunnel} resulting from the carrier tunneling through the Schottky barrier, and the nuclear radiation–generated current

Table 8.1 Maximal Penetration Lengths of the Tritium Beta Electrons in Different Materials

Semiconductor	M_{cation}	Z_{cation}	M_{anion}	Z_{anion}	ρ, g/cm³	L_{max}, μm
Si	28.08	14			2.33	0.48
Ge	72.59	32			5.32	0.30
C	12.01	6			3.51	0.26
GaAs	69.72	31	121.75	51	5.4	0.33
GaP	69.72	31	30.97	15	4.14	0.34
GaN	69.72	31	14	7	6.1	0.23
AlN	26.98	13	14	7	3.13	0.34
SiC	28.08	14	12.01	6	3.21	0.33
Metal						
Ti	47.9	22			4.5	0.30
Pd	106.4	46			12.02	0.15
Au	197.0	79			19.32	0.12
Pt	195.1	78			21.45	0.11
Al	26.98	13			2.69	0.42
Ni	58.7	28			8.91	0.16

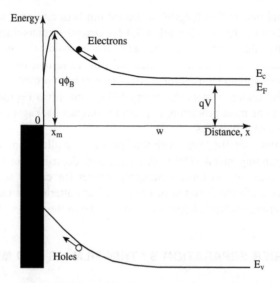

Figure 8.1 Energy-band diagram of MS Schottky diode for *n*-type semiconductor under beta radiation. Beta-generated electrons and holes are separated by built-in electric field in the narrow depletion region $0 < x < W$ and produce current J_{gen} directed against applied forward bias V.

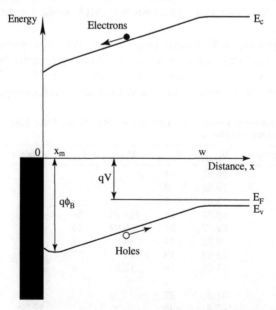

Figure 8.2 Energy-band diagram of MIS Schottky diode for *p*-type semiconductor under beta radiation. Beta-generated electrons and holes are separated by the built-in electric field in the wide depletion region $0 < x < W$ and produce current J_{gen} directed against applied forward bias V.

J_{gen}, which results from separation of the electron–hole pairs by the electric field built into the Schottky barrier.

$$J = J_{gen} - J_{therm} - J_{tunnel}$$ (8.6)

The thermionic current J_{therm} has the form[3,4]

$$J_{therm} = A^*T^2 \exp\left(-\frac{q\varphi_B}{kT}\right)\left[\exp\left(\frac{qV}{kT}\right) - 1\right]$$ (8.7)

where A^* is the effective Richardson constant $A^* = 4\pi q m^* k^2/\hbar^3$, m^* is the charge carrier effective mass, $m^* = m_e$ for n-type semiconductor and $m^* = m_h$ for p-type semiconductor and m_e and m_h are electron and hole effective masses, respectively. In a more rigorous treatment,[3,5] A^* should be substituted by A^{**}, which in most cases is slightly different from A^*.

The tunneling current in the quasi-classical approximation can be written as

$$J_{tunnel} = A^*T^2 \exp\left(-2\int_0^W dx\sqrt{\frac{2m_e\left(E_c(x) - qV\right)}{\hbar^2}}\right)\left[1 - \exp\left(-\frac{qV}{kT}\right)\right]$$ (8.8)

for n-type semiconductor and

$$J_{tunnel} = A^*T^2 \exp\left(-2\int_0^W dx\sqrt{\frac{2m_h\left(-E_v(x) - qV\right)}{\hbar^2}}\right)\left[1 - \exp\left(-\frac{qV}{kT}\right)\right]$$ (8.9)

for p-type semiconductor, where $E_c(x)$ and $E_v(x)$ are space-depending positions of the conduction and valence band edges (Figures 8.1 and 8.2).

For MS diodes, Equations 8.8 and 8.9 give

$$J_{tunnel} = A^*T^2 \exp\left(-\frac{q(\varphi_B - V)}{E_{00}}\right)\left[1 - \exp\left(-\frac{qV}{kT}\right)\right]$$ (8.10)

where

$$E_{00} = \frac{e\hbar}{2}\sqrt{\frac{N_D}{\varepsilon\varepsilon_0 m_e}}$$ (8.11)

for n-type semiconductor, and

$$E_{00} = \frac{e\hbar}{2}\sqrt{\frac{N_A}{\varepsilon\varepsilon_0 m_h}} \qquad (8.12)$$

for p-type semiconductor.

For MIS diodes, Equations 8.8 and 8.9 yield

$$J_{tunnel} = A^* T^2 \exp\left(-\sqrt{\frac{q(\varphi_B - V)}{E_{01}}}\right)\left[1 - \exp\left(-\frac{qV}{kT}\right)\right] \qquad (8.13)$$

where

$$E_{01} = \lambda\hbar^2/4m^* w^2 \qquad (8.14)$$

and $w \approx L_{int}$. L_{int} is the thickness of intrinsic semiconductor layer.

8.3.2 Efficiency of MS and MIS Beta Cells

Generally, the efficiency of the Schottky diode is given by the conventional equation used for PV cells

$$\eta = J_{sc}V_{oc}\,FF/P \qquad (8.15)$$

where J_{sc} is short-circuit current, V_{oc} is open-circuit voltage, FF is filling factor characterizing the shape of I–V characteristic, and P is incident power of beta radiation.

For heavily doped semiconductor layer, the internal resistance of the Schottky barrier cell is negligible; $V = 0$, $J_{therm} = 0$, and $J_{tunnel} = 0$ for the short-circuit case. Thus, from Equation 8.6, $J_{sc} = J_{gen}$ and is given by the equation

$$J_{sc} = qP(1-R)\left(1 - \sqrt{1 - L_{diff}/\left(L_{max}^{sem} + L_{diff}\right)}\right)\sqrt{1 - L^{met}/L_{max}^{met}}/E_{pair} \qquad (8.16)$$

where P is incident power of beta electrons, R is reflection coefficient of beta electrons, L_{diff} is carrier diffusion length in semiconductor, L_{max}^{met} and L_{max}^{sem} are maximal penetration length of beta electrons in metal and semiconductor, respectively, L^{met} and w are metal layer thickness, and E_{pair} is the energy spent by beta electrons for generation of one electron–hole pair.

Equation 8.16 is valid for the case $w < L_{diff}$, which is typical for MS and MIS diodes with narrow intrinsic layers. In the opposite case, $w < L_{diff}$, which can take place for MIS diodes with wide intrinsic layers, an additional factor related to an incomplete separation of the carriers in the depletion region in Equation 8.16 becomes

$$J_{sc} = qP(1-R)\left(1 - \exp\left\{-q\varphi_B L_{diff}/wkT\right\}\right)$$

$$\left(1 - \sqrt{1 - L_{diff}/\left(L_{max}^{sem} + L_{diff}\right)}\right)\sqrt{1 - L^{met}/L_{max}^{met}}\Big/E_{pair} \tag{8.17}$$

Equations 8.16 and 8.17 are written for an incident beta electron beam perpendicular to the surface. For randomly oriented beta electrons, averaging over incident angles should be performed. In a first approximation, such averaging can be performed by substitutions:

$$L_{max}^{sem} \rightarrow \frac{2}{\pi} L_{max}^{sem} \quad \text{and} \quad L_{max}^{met} \rightarrow \frac{2}{\pi} L_{max}^{met} \tag{8.18}$$

Equations 8.16 and 8.17 are written for the straight beta electron path; however, momentum relaxation of the beta electron via elastic and inelastic scattering has the same or higher rate as its energy relaxation. This can lead to a further reduction of L_{max} by a factor of two or more.

It can be seen from Equations 8.16 and 8.17 that J_{sc} is maximal at $R<<1$, $L^{met} << L_{max}^{met}$, $L_{diff} >> L_{max}^{sem}$, and $L_{int} \approx w \leq q\varphi_B L_{diff}/kT$. Physically, this represents a maximally transparent metal layer, good-quality semiconductor material with large diffusion length, and a wide depletion region in which generation of all electron–hole pairs occurs.

Also, it can be seen from Equations 8.10 and 8.14 that, in MIS diodes, the contribution of the tunnel current J_{tunnel} to the total current J is smaller compared to MS diodes. This is especially important for semiconductor nuclear batteries with direct conversion of radioactive energy which, in contrast to indirect-conversion batteries, cannot use concentrators of incident power and need to work at low incident power fluxes. In the very low incident power flux limit, the nuclear battery efficiency given by Equation 8.15 becomes

$$\eta = J_{sc}^2 R_{diff}(0)/4P \tag{8.19}$$

where $R_{diff}(0)$ is differential resistance of MS or MIS diodes at zero bias.

It can be seen from Equation 8.19 that the efficiency of the MS and MIS nuclear battery is proportional to $R_{diff}(0)$. Explicit analytical expressions for $R_{diff}(0)$ can obtained from Equations 8.7 and 8.14:

$$R_{diff}(0) = \frac{k}{qA^*T\left\{\exp\left(-\frac{q\varphi_B}{kT}\right) + \exp\left(\frac{q\varphi_B}{E_{00}}\right)\right\}} \tag{8.20}$$

for MS diodes and

$$R_{diff}(0) = \frac{k}{qA^*T\left\{\exp\left(-\dfrac{q\varphi_B}{kT}\right) + \exp\left(-\sqrt{\dfrac{q\varphi_B}{E_{01}}}\right)\right\}} \tag{8.21}$$

for MIS diodes.

It is found from Equations 8.11, 8.12, and 8.14 that, usually, for MS and MIS diodes, $kT \ll E_{00} \ll E_{01}$, and E_{01} exponentially increases with increasing L_{int}.

This can also be seen from experimental data for Al–undoped polycrystalline diamond–heavily boron-doped diamond MIS diodes[6] listed in Table 8.2. Therefore, according to Equation 8.19, an increase of L_{int} should lead to exponential increase of the efficiency of MIS diode–based nuclear batteries. However, this is true only for $L_{int} \leq L_d$ where L_d is carrier diffusion length, since for $L_{int} > L_d$, Equation 8.6 for thermionic current should be replaced by one for thermionic-diffusion current[3]; this will lead to significant reduction of J_{sc}. Thus, one can conclude that MIS diodes fabricated of heavily doped wide-band-gap semiconductor polycrystalline films covered by a metal layer with thickness $L^{met} \leq 0.1\ L_{max}^{met}$ over a reasonable-quality intrinsic layer with carrier diffusion length $L_{diff} \gg L_{max}^{sem}$ and thickness $L_{int} \approx w \leq q\varphi_B L_{diff}/kT$ seem to be most promising for betavoltaic applications.

8.4 FORMATION MECHANISMS OF SCHOTTKY CONTACTS TO WIDE-BAND SEMICONDUCTORS

8.4.1 Technological Aspects Of Schottky Contact Formation To Wide-Band Semiconductors

The technology of the Schottky contact formation to wide-band semiconductors includes three main stages: surface preparation, deposition of the metal layer, and annealing of the contact. While preparing Schottky contacts for nuclear batteries with direct conversion, most attention should be devoted to reducing the density of interface defects, since they can enhance the tunnel current by orders of magnitude. The high height of the Schottky barrier is much less important since the batteries operate at low voltages, usually much less than φ_B. Also, considering the metallic contact formation to wide-band-gap semiconductor films such as diamond, GaN, AlN, and SiC films, it is important to take into account the adverse possibility of

Table 8.2 Differential Resistances at Zero Bias of Al/Polycrystalline i-Diamond–B-Doped Diamond MIS Schottky Diodes with Different Thicknesses of Intrinsic Layers L_{int}

L_{int}, μm	$R_{diff}(0)$, kΩ
0	5.5
0.1	1,000
0.2	4,600
0.3	15,600

metal reacting chemically with the semiconductor to form metallic carbides, nitrides, or silicides. The formation of metallic carbides, nitrides, or silicides at the annealing stage of the contact formation can lead to a high density of defects at the interface between metal and semiconductor films; this increases tunneling currents and suppresses the Schottky contact rectifying properties.

8.4.2   Interface Band Structure Model

The basic model of the interface band structure formation for wide-band-gap semiconductors' surface defects has been illustrated with the case of diamond.[7] The model suggests division of contact metals into three groups according to their chemical reactivity with respect to the semiconductor surface and division of semiconductor surfaces into two groups according to their quality.

Thus, six different cases are possible:

1. Inert metal is deposited on a high-quality surface; the resulting contact is rectifying with low tunnel current. After medium-temperature annealing, the rectification ratio decreases and tunnel current inceases.
2. Inert metal is deposited on a low-quality surface; the resulting contact is ohmic as deposited. After medium-temperature annealing, the contact becomes rectifying with high rectification ratio and low tunnel current.
3. Slightly chemically active metal is deposited on a high-quality surface; the resulting contact is rectifying. After annealing, the contact remains rectifying.
4. Slightly chemically active metal is deposited on a low-quality surface; the resulting contact is ohmic. After annealing, the contact remains ohmic.
5. Highly chemically active metal is deposited on a high-quality surface; the resulting contact is rectifying. After annealing, the contact converts into ohmic.
6. Highly chemically active metal is deposited on a low-quality surface; the resulting contact is ohmic. After annealing, the contact remains ohmic.

This model is supported by experimental data on the metallic contact formation of boron-doped polycrystalline diamond films with Au that forms no carbides, Al found to be a weak carbide former, and Ti, which is a strong carbide former.[7] The corresponding data are summarized in Table 8.3.

In addition to these six cases, it is possible to make contacts by direct deposition of metallic carbides, nitrides, or silicides. Such contacts are expected to be most stable at high temperatures since the metal–semiconductor chemical reaction is

Table 8.3   Rectifying Properties of Different Metal Contacts to Diamond

Metal	Perfect Surface as Deposited	Perfect Surface after Annealing	Damaged Surface as Deposited	Damaged Surface after Annealing
Au (inert)	rectifying	rectifying	ohmic	rectifying
Al (low reactive)	rectifying	rectifying	ohmic	ohmic
Ti (high reactive)	rectifying	ohmic	ohmic	ohmic

already completed. Experimental data on TiC contact to boron-doped polycrystalline diamond films show that this contact is rectifying with reverse leakage current of 0.4 μA as deposited, and it remains rectifying after annealing at 850°C with slight leakage current increase to 0.7 μA.

Thus, one can conclude that metals that do not form carbides, nitrides, or silicides are preferable for Schottky contact formation, whereas those forming carbides, nitrides, or silicides are preferable for ohmic contacts. Also, for the betavoltaic cell operating at high temperatures, the Schottky contact of deposited carbides, nitrides, or silicides is preferable.

8.5 III-NITRIDES AS NEW MATERIALS FOR BETAVOLTAIC CELLS

8.5.1 Recent Progress in Technology of III-Nitride Films

Great progress has been achieved in the growth of high-quality III-nitride epitaxial layers for light-emitting diodes, semiconductor lasers, and high-temperature electronics. These materials demonstrate high-quality electronic, optical properties combined with very low degradation in spite of high concentration of threading dislocation, stacking faults, and other growth defects.

To extend the applications of III-nitrides to low-cost and large-area electronic devices, it is desirable to grow the crystals on cheap substrates such as glass, transparent conductive oxides, silicon, or metal films. From this point of view, microcrystalline III-nitride films are the most promising material for large-area nuclear batteries with direct conversion of radioactive energy.

A significant achievment in this field is the growing of high-quality hydrogenated microcrystalline GaN (GaN:H) films by plasma-enhanced metalorganic chemical vapor deposition (p-CVD) in the mixtures of NH_3 and trimethylgallium.[8] The films are deposited at temperatures between 100 and 300°C on aluminum, fused silica, and Si (100) substrates and have a stoichiometric composition with 17 to 30% of hydrogen.

Using an analogy with hydrogenated amorphous silicon (α-Si:H), InP, and GaAs where hydrogen atoms passivate dangling bonds, one can expect that hydrogen atoms passivate dangling bonds, dislocations, and defects in amorphous and microcrystalline nitrides, as well. Since the bond energies of Ga-H (66 kcal/mol) and N-H (85 kcal/mol) are comparable to the Si-H bond energy of 76 kcal/mol, hydrogenation of the nitrides is possible by a low-temperature growth method as for α-Si:H.

The n-GaN:H films on Al substrates exhibit UV photosensitivity. The dark current and photocurrent of microcrystalline GaN:H films are two orders of magnitude larger than those of amorphous films. The photoresponse of a microcrystalline GaN:H film is 0.012 A/W at 330-nm monochromatic light of incident intensity of 5 μW/cm^2 with a 3 V bias. Excellent photoelectrical properties are reported for Mg-doped hydrogenated p-GaN (GaN:H) films grown at 380°C.[9] These films are fabricated using dual remote-plasma metallorganic chemical vapor deposition under hydrogen-rich conditions. The peak responsivity is 0.11 A/W at 360 nm with the dark current of 10^{-11} A at –1 V bias. The corresponding $R_{diff}(0)$ in Equation 8.19 is very high — $\sim 10^{11}\,\Omega$ — making these films very promising for effective nuclear batteries under direct conversion.

8.5.2   Theoretical Models of Schottky Barrier Formation Mechanism to III-Nitride Films

To explore this potential, a technology is needed for formation of stable Schottky contacts with high Schottky barrier height and low leakage current; this is crucial for realization of the GaN-based nuclear batteries with direct conversion. However, not much is known at present about controlling Schottky barrier height on GaN or about the underlying Schottky barrier formation mechanism.

The barrier height φ_B is the energy distance between the Fermi level E_F and valence-band maximum E_v for p- and conduction-band minimum E_c for n-type semiconductors shown in Figures 8.1 and 8.2. The famous Schottky–Mott rule predicts barrier height to be equal to the difference between the electron affinity (χ) of the semiconductor and the work function of the metal. Deviation from the Schottky-Mott rule prediction is usually explained by interface states.[3] A continuum of intrinsic interface states can arise from wave function tails of the metal electrons.[10] The energy position of the Fermi level at a metal–semiconductor interface and Schottky barrier height depend on the occupation of the metal-induced gap states.

Also, deviation from the Schottky–Mott rule can result from the charge transfer across the metal–semiconductor interface. A generalization of the Pauling model describing the ionicity of diatomic molecules by the difference in the atomic electronegativities[12] relates the charge transfer across metal–semiconductor interfaces with the differences $X_m - X_s$ of the electronegativities of the metal and semiconductor in contact.

The metal-induced gap states model and the electronegativity model predict barrier heights to vary as $\varphi_B = \Phi_{cnl} + S\,(X_m - X_s)$ on semiconductors doped n-type, where Φ_{cnl} is the charge neutrality level, $S = \partial\varphi_B / \partial X_m$. The metal-induced gap states model represents the primary mechanism that determines barrier heights in ideal semiconductor contacts. Deviations from what is predicted by this model are caused by secondary mechanisms such as interface dipoles and defects.

It is often said that, due to the substantial ionic component of the bonds in III-V nitrides, the Fermi level at the nitride surface and at the metal–nitride interface should be unpinned; the barrier height should consequently depend on the work function or electronegativity of the contacting metal. Schottky barrier heights of a variety of elemental metals including Au, Ti, Pt, Pd, Ni, and Cr have been reported and are summarized in Table 8.4.[13-15] Note that Schottky barrier heights on n-GaN indeed vary with the metal work function within experimental scattering. Thus, for formation of ohmic contacts on n-GaN, a metal with a small work function such as Ti can be used; Schottky barriers using a metal with a large work function such as Pt are most suitable. However, some data indicate that the electric properties of GaN Schottky contacts depend on the interface formation process, so it is not clear at present whether the metal-induced gap states model is applicable to the metal–GaN systems.

Table 8.4   Summary of Experimental Characteristics of
n-GaN Schottky Diodes

Diodes	Metal Work Function Φ_m, eV	Schottky Barrier Height Φ_B, eV
Ti/n-GaN	4.4	0.57
Au/n-GaN	5.1	0.88
Pd/n-GaN	5.12	1.11
Ni/n-GaN	5.15	0.95
Pt/n-GaN	5.65	1.13

In contrast to the ideal Mott–Schottky limit, real metal–semiconductor interfaces are known to exhibit a strong tendency for the metal Fermi level aligning with a characteristic energy position, E_0, of the semiconductor. This makes the value of Schottky barrier height independent of or weakly dependent on the metal work function, causing the so-called "Fermi level pinning."

Major models to account for the origin of the Fermi level pinning include the unified defect model, metal-induced gap states model, disorder-induced gap states model, and effective work function model.[15] In spite of a long history of research, the main mechanism for Fermi level pinning at the metal–semiconductor surface is not well established.

The physical pinning position E_0 is the energy position of the acceptor-type discrete deep level for n-type materials in the unified defect model and the common anion metallic pinning position in the effective work function model. On the other hand, in the metal-induced gap states and disorder-induced gap states models, E_0 has the physical meaning of the charge neutrality energy level of the entire band of semiconductor. The position of E_0 for GaN was recently proposed to lie at 1.1 eV below the conduction band edge. The dependence of the barrier height, φ_B, on the metal work function, Φ_m, is usually expressed phenomenologically in terms of the slope factor, S, defined by $S = d\varphi_B/d\Phi_m$, which allows the Schottky barrier height for n-type material to be expressed as

$$\varphi_B = S(\Phi_m - \chi_s) + (1 - S)(E_c - E_0) \tag{8.22}$$

where χ_s is the electron affinity. $S = 1$ corresponds to the completely pinned limit (Bardeen limit).

For the metal-induced gap states model, the screening capability of metal-induced gap states against contact potential difference depends on density and penetration depth of these states.

The semiempirical theoretical expression is given by

$$S = \left[1 + \frac{q^2}{\varepsilon_i \varepsilon_o} D_{MIGS}(E_o)\delta_i\right]^{-1} = \left[1 + 0.29\left(\frac{(\varepsilon_\infty - 1)^2}{\varepsilon_i}\right)\right]^{-1} \tag{8.23}$$

where ε_o is the dielectric constant of vacuum, ε_i is the dielectric constant of the interface, and ε_∞ is the dielectric constant of the semiconductor. $D_{MIGS}(E_o)$ is the metal-induced gap states density at E_o and δ_i is the penetration depth of the these states.

On the other hand, the disorder-induced gap states model gives the following expression for S:

$$S = \text{sech}(\delta/\lambda) \qquad (8.24)$$

where δ is the thickness of the disordered layer and λ is the disorder-induced gap states screening length given by

$$\lambda = \left[\varepsilon_o \, \varepsilon_\infty / q^2 \, N_{DIGS}(E_o)\right]^{1/2} \qquad (8.25)$$

Here, $N_{DIGS}(E_o)$ is the volume density of disorder-induced gap states at E_0.

8.5.3 Physical Characteristics of GaN Schottky Diodes

Schottky diodes formed by vacuum depositions after NH_4OH surface pretreatment show nearly ideal thermionic emission characteristics with Schottky barrier height values weakly dependent on the metal work function with $S \approx 0.1$.[15] Schottky diodes formed *in situ* electrochemically also show nearly ideal thermionic emission *I–V* characteristics but demonstrate strong metal work function–dependent Schottky barrier height values. The slope factor can be as large as $S = 0.49$. This indicates that the electronic properties of GaN Schottky contacts depend strongly on the interface formation process. The observed strong dependence of Schottky barrier height on processing can be related to disorder-induced gap states or metal nitride formation.

8.6 RADIATION RESISTANCE OF III-NITRIDES

High stability of electrical and optical properties of III-nitrides with respect to induced structural defects makes these materials good candidates for application in nuclear batteries. Little information is available on radiation stability of GaN and the effect of electron irradiation on its physical properties. Recently, several groups have reported investigation of irradiation-induced defects in GaN and related compounds. Most of the results are still not fully understood, but some useful conclusions can be made.

In all investigated structures, which include *n*-type, compensated, and *p*-type GaN epilayers on 6H SiC or Al_2O_3 substrates grown by MOCVD or HVPE technique, electron irradiation performed at room temperature by 2.5-MeV electrons with a dose up to 10^{17} electrons/cm² has a minor effect on photoluminescence. This indicates the very high radiation stability of GaN.[16,17]

As-grown Mg-doped GaN films usually show high electrical resistance and low luminescence intensity. It has been proposed by Nakamura and Fasol[18] that Mg dopants are passivated by atomic hydrogen. To activate p-type conductivity and enhance luminescence intensity, post-growth treatment by low-energy electron beam irradiation or thermal annealing in N_2 ambient is required. The electron beam generates free electrons and holes that stimulate the breaking of acceptor-H bonds, creating active acceptors and increasing the p-doping level.[19]

Investigations of low-energy electron beam irradiation of Mg-doped GaN support this model.[20,21] With electron beam energy of 15 keV for a penetration depth of 0.7 μm, electron beam current 6.8 nA, and 0.52 μm in diameter, the enhancement of luminescent intensity initially occurs rapidly and then proceeds gradually until saturation after about 30 msec of irradiation at room temperature.[20] Considering the e-beam spot diameter and large electron beam current density in this experiment, irradiation over a period of 50 msec is equivalent to 1 month of betavoltaic cell exposure to tritiated titanium.

Experimental investigation of hydrogenated microcrystalline GaN films has shown that there are no changes in the conductivities of GaN:H films after 5 h of irradiation with 500 mW/cm^2 UV light. However, α-SiH shows severe degradation related to the Steabler–Wronski effect. The higher stability of GaN:H films can result from higher binding energy of the Ga–N chemical bond compared to the Si–Si bond since the binding energy of the Ga–H bond is lower than that of Si–H. The high radiation stability of GaN:H makes it a good material for direct-conversion nuclear batteries.

8.7 BETAVOLTAIC CELLS BASED ON GALLIUM NITRIDE

8.7.1 Optimal Betavoltaic GaN Cells

Optimized parameter goals for betavoltaic GaN cells can be estimated with use of the mean energy needed per pair generation by beta electron $E_{pair} \sim 10$ eV.[22]

For tritium betavoltaic cells, average beta electron kinetic energy is 5.7 keV and maximal possible short-circuit current density per curie is $I_{sc,max} = 1.6 \cdot 10^{-19}$ C $\times$ $3.7 \cdot 10^{10}$ s$^{-1} \times 5.7$ keV/10 eV = $3.37 \cdot 10^{-6}$ A/Ci. For p–n diode cells, the corresponding maximal possible output power per curie is $P_{max} = I_{sc,max} \times E_g = 10$ μW/Ci. For n-GaN MS and MIS diode cells, the maximal output power is $P_{max} = I_{sc,max} \times \varphi_B = 3$ μW/Ci for typical barrier heights of 1.1 eV for Pt, 0.91 to 1.15 eV for Au, 0.94 eV for Pd, and 0.66 to 0.99 eV for Ni[26], for p-GaN MS and MIS diode cells, the maximal output power is $P_{max} = I_{sc,max} \times \varphi_B = 8$ μW/Ci for barrier height values of 2.38 eV for Au on p-GaN[27] and 2.4 eV for Ni.[28] The corresponding maximal efficiencies of nuclear energy conversion into electricity are 30, 10, and 24% for p–n diode, Schottky diode to n-GaN, and Schottky diode to p-GaN, respectively.

For real beta cells, the efficiency is much lower. From Equation 8.15 (cell with semiconductor material of high quality or with high content of radioactive isotope) and from Equation 8.19 (cell semiconductor material of low quality or with low content of radioactive isotope), cell efficiency can be calculated.

8.7.2 Experimental Parameters of *n*-GaN Schottky Diodes

The best results on *n*-GaN Schottky diodes obtained to date are reported by Hasegawa et al.[15] Nearly ideal thermionic-emission *I*–*V*-characteristics with a Schottky barrier height of 1.03 eV and ideality factor close to unity were realized on Au Schottky diodes formed on NH_4OH-treated surfaces. At forward current density, $1.12 \cdot 10^{-6}$ A/cm^2 forward voltage for vacuum deposition–based Au–*n*-GaN Schottky diodes[12,15] is equal to 0.33 to 0.788 V. Therefore, the maximum possible open-circuit voltage that can be achieved in *n*-GaN Schottky beta detector exposed to beta radiation from tritium gas is 0.788 V.

Accordingly, maximum possible power P_{max} per unit area (1 cm^2) on an *n*-GaN beta cell can be 0.37 to 0.88 μW/cm.^{2} Energy-conversion efficiency, defined as P_{max} per cell area divided by the total power density of beta radiation on the surface of cell-exposed beta radiation from tritium gas (11.24 μW/cm^2), can be 3.24 to 7.65% and 2.06 to 4.1% exposed beta radiation from tritiated titanium (Ti^{3}H$_2$) (0.23 μW/cm^2), respectively.

8.7.3 Experimental Parameters of *p*-GaN Schottky Diodes

The best reported results on *p*-GaN Schottky diodes obtained to date on hydrogenated microcrystalline films of *p*-GaN[9] exhibited forward currents of $1.12 \cdot 10^{-6}$ A/cm^2 and 23 nA/cm^2 at 2.36 and 0.6 V, respectively. Accordingly, the maximum possible power P_{max} on a *p*-GaN beta cell is 2.67 μW/cm^2 using tritium gas and 0.68 μW/cm^2 with tritiated titanium. This is an energy-conversion efficiency from tritium gas of 23.2% and 5.9% from tritiated titanium (Ti^{3}H$_2$) (0.23 μW/cm^2).

GaN–Schottky diodes have been fabricated in this laboratory using *n*-GaN films grown by metalorganic chemical vapor deposition (MOCVD) on (0001) sapphire substrates. A thin GaN buffer layer was grown prior to growing the 1.12-μm-thick Si-doped GaN main layer. The films were of *n*-type, sheet resistance of about 35 Ω/square, (accordingly, ρ = $3.92 \cdot 10^{-3}$ Ωcm), and carrier concentration of about 10^{18} cm^{-3} at room temperature. The Schottky metal contacts were formed by evaporating semitransparent Au layers in vacuum through the mask. The device structure consisted of an array of Au hexagons (with area 0.4 mm^2).

The diodes display rectifying current-voltage (*I*–*V*) characteristics in dark and under quartz mercury lamp illumination (see Figure 8.3). The leakage current increases almost linearly with reverse bias to reach 0.1 nA at a reverse bias of –0.4 V at room temperature. In the range from 0.5 to 5 V, superlinear increase of reverse current with voltage up to 160 nA was observed.

8.7.4 Testing of GaN Schottky Diodes under Quartz Mercury Lamp Illumination

For testing Schottky diodes, a quartz mercury lamp generating radiation with ~ 260 < λ < ~ 400 nm has been used. Incident power, measured using Si photodiode with photoresponse down to 200 nm, was $1.75 \cdot 10^{-5}$ W/cm^2. Short-circuit current observed under illumination with mercury lamp was 2.41 nA and superlinear,

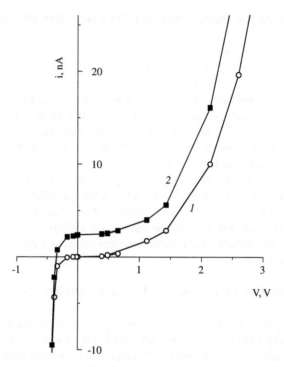

Figure 8.3 *I–V* characteristics of Au–n-GaN Schottky photodiode (1) in the dark and (2) illuminated by 1.75·10⁻⁵ W/cm² with a quartz mercury lamp and UV filter.

increasing with voltage in the range from 2 to 5 V up to 90 nA at 5 V. Open-circuit voltage was 337 mV.

The photoresponse of the Schottky diode in the UV region is 0.019 A/W. In general, the excess (leakage) current across a GaN p–n junction is expected to be lower than that across a Schottky barrier, due to a larger barrier height.

The device structure consists of $Al_2O_3(0001)$ substrate, GaN–nucleation layer, a 0.24-μm-thick layer of undoped GaN, a 2.1-μm-thick layer of GaN:Si, and a 0.34-μm-thick layer of GaN:Mg. From separate measurements of specific resistance of GaN p- and n-layers with resolution of 1 mm, it was found that the mean concentration of electrically active impurities (Si and Mg) is relatively uniform over the layer area, with spatial variation of no more than 10%.

After growth, the mesa p–n junction diodes were processed by patterning in hexagons with areas of 0.4 mm². The surface of the p-type GaN was partially etched until the n-type GaN layer was exposed. Next, an Ni–Au contact was evaporated onto the p-type GaN layer and a Ti/Al contact onto the n-type GaN. The devices, fabricated in a set of 40 elements, were examined for I–V characteristics. These characteristics of each p–n junction sample were measured in the dark and under illumination with a quartz mercury lamp through the UV filter ($\lambda < 400$ nm) with intensity of 1.75·10⁻⁵ W/cm². Representative forward-biased I–V curves are shown in Figures 8.4 and 8.5.

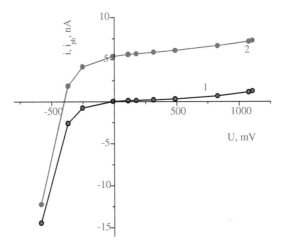

Figure 8.4 *I–V* characteristics of *p–n* GaN photodiode (1) in the dark and (2) illuminated by 1.75·10⁻⁵ W/cm² with a quartz mercury lamp with UV filter.

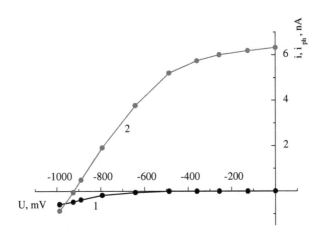

Figure 8.5 *I–V* characteristics of *p–n* GaN photodiode (1) in the dark and (2) illuminated by 1.75·10⁻⁵ W/cm² with a quartz mercury lamp with UV filter, $S = 4 \cdot 10^{-3}$ cm².

The p–n GaN junction photodetectors have shown typical zero-bias rsponsivities of ~ 0.1 A/W and open-circuit photovoltage, U_{oc}, ~ 400 to 700 mV. The best p–n GaN junction diodes exhibited zero-bias responsivity of ~ 0.1 A/W and open-circuit photovoltage $U_{oc} = 930$ to 1000 mV (Figure 8.5).

An increase in the open-circuit photovoltage U_{oc} with short-circuit current is approximately logarithmic (Figure 8.6). This dependence allows one to estimate possible photovoltage of beta cell–exposed beta radiation or light generated with organic phosphors in an indirect conversion photon battery.

The estimated value of U_{oc} is about 570 and 320 mV (at short-circuit photocurrent density of 1.75 and 0.03 μA/cm², respectively) for intensity of UV radiation equal

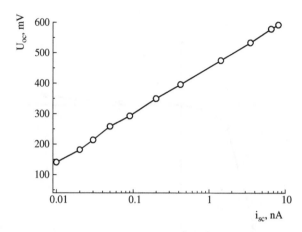

Figure 8.6 Open-circuit photovoltage vs. short-circuit photocurrent of p–n-GaN junction photodiode illuminated with a quartz mercury lamp with UV filter.

to 11.5 $\mu W/cm^2$ and 0.23 $\mu W/cm^2$, corresponding to an intensity of beta radiation from tritium gas and tritiated titanium (Ti^3H_2), respectively, and 490 and 230 mV for illumination level of indirect-conversion photon battery with conversion efficiency $C_k = 0.3$.

Accordingly, the maximum possible power P_{max} of 1 $\mu W/cm^2$ and 0.01 $\mu W/cm^2$, respectively, can be expected in n–p–GaN beta cell-exposed beta radiation from tritium gas and from tritiated titanium (Ti^3H_2) (0.23 $\mu W/cm^2$).

8.7.5 Testing Radiation Resistance of GaN Schottky Diodes under Beta Radiation

Radiation resistance of photovoltaic cells is crucial for realization of the GaN-based betavoltaic battery. In order to confirm applicability of Schottky and p–n junction diode formation processes to betavoltaic cells based on GaN, an attempt was made to characterize the degradation process by means of measuring the main parameters and I–V characteristics of Schottky and p–n junction diodes before and after exposure to beta radiation from tritium gas inside an experimental setup under tritium pressure of 0.8 atm.

Both open-circuit voltage and short-circuit current of Schottky diodes decrease over a period of 400 h. The I–V curves obtained before and after 407 h of tritium radiation are displayed in Figure 8.7. As can be seen, a short-circuit current decreased from 2.42 to 2.23 nA while an open-circuit voltage decreased from 337 to 295 mV. Accordingly, the relative decrease of open-circuit voltage was 12.4%, while that of short-circuit current was 6%. This degradation may take place due to decrease of Schottky barrier height under tritium action.[28]

The p–n junction diodes demonstrated high radiation resistance; I–V curves obtained before and after 200 h of tritium radiation are displayed in Figure 8.8. As can be seen, short-circuit current and open-circuit voltage remained practically unchanged after exposure to beta radiation from tritium gas. In contrast to the

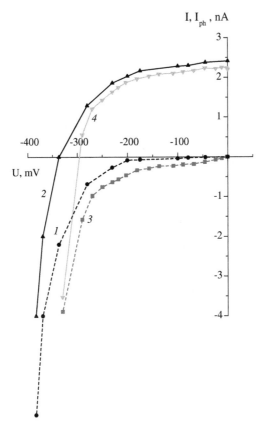

Figure 8.7 *I–V* characteristics of an Au–*n*-GaN Schottky ultraviolet photodiode in the dark (1) and (3), and illuminated by 1.75·10⁻⁵ W/cm² with a quartz mercury lamp with UV filter (2) and (4), before (1) and (2), and after (3) and (4), the exposure to beta radiation from tritium gas over a period of 400 hours.

Schottky diodes, the *I–V* curves showed insignificant degradation at low voltages (U < 0.3 to 0.5 V). At higher voltages (U > 0.5 V), current was found to degrade.

Tritium is the most appropriate source for commercial semiconductor nuclear batteries. The small penetration depth of tritium β-electrons dictates the choice of semiconductor material. Practically, the choice is restricted to thin-film semiconductors, which provide the most effective utilization of the tritium β-radiation.

The use of metal–semiconductor Schottky barriers for carrier separation seems to be the most effective way to produce nuclear batteries based on wide-band-gap microcrystalline thin-film semiconductors.

MS and MIS diodes fabricated of heavily doped wide-band-gap semiconductor polycrystalline films covered by the thin metal layer over intrinsic layer seem to be suitable for betavoltaic applications. This point is supported by experiments with GaN-based MS beta cells. Schottky photodiodes on *n*-type GaN as well as *p–n* junction GaN photodiodes were fabricated, demonstrating applicability of the present GaN photodiode formation process for low-intensity photo- and betavoltaic

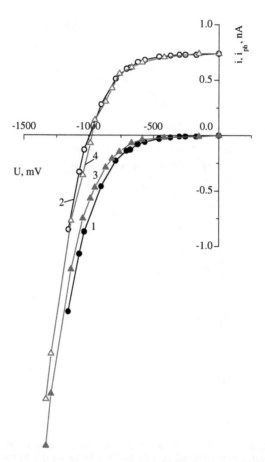

Figure 8.8 *I–V* characteristics of a *p–n*-GaN ultraviolet photodiode in the dark (1) and (3), and illuminated by 1.75·10⁻⁵ W/cm² with a quartz mercury lamp with UV filter (2) and (4), before (1) and (2), and after (3) and (4), the exposure to beta radiation from tritium gas over a period of 200 hours.

applications. However, the fabrication process has not been optimized. *p–n* junction GaN photodiodes showed highest efficiency, with an output voltage close to 1 V under an illumination level of $2 \cdot 10^{-5}$ W/cm⁻².

The main problem of GaN-based Schottky barriers and *p–n* junction photodiodes is the high density of defects in GaN epitaxial films. These defects are found to critically limit large-area GaN photovoltaic and betavoltaic cells with low dark-leakage currents. With improved growth technologies and techniques of grain boundary passivation, this material may become suitable as low-cost material for photovoltaic and betavoltaic low-level power-conversion applications. Many companies are working to improve GaN material quality and fabrication of GaN-based light-emitting diodes and ultraviolet photodetectors for medicine. For these purposes, defect passivation technique improvements are especially important. In the future, substantial efforts must be directed to investigation and development of microcrystalline GaN, AlN, and AlGaN on aluminum foil, which are promising candidates for cheap betavoltaic cells for conversion of low-intensity nuclear power into electricity.

REFERENCES

1. Landay, L.D. and Lifshitz, E.M., *Theoretical Physics*, vol. VIII, Nauka, Moscow, 1982.
2. Landay, L.D. and Lifshitz, E.M., *Theoretical Physics*, vol. VIII, Nauka, Moscow, 1974.
3. Sze, S.M., *Physics of Semiconductor Devices*, 2nd ed., John Wiley & Sons, New York, 1981.
4. Crowell, C.R. and Sze, S.M., *Solid State Electron.*, 9, 1035–1048, 1966.
5. Andrews, J.M. and Lepselter, M.P., *Solid State Electron.*, 13, 1011, 1970.
6. Miyata, K., Kobashi, K.K., and Dreifus, D.L., *Diamond Related Mater.*, 2, 1107–1111, 1993.
7. Tachibana, T. and Glass, J.T., *Diamond Related Mater.*, 2, 963–969, 1993.
8. Yagi, S., Ultraviolet photoconductive hydrogenated amorphous and microcrystalline GaN, *Jpn. J. Appl. Phys.*, 38, L792, 1999.
9. Yagi, S., Highly sensitive ultraviolet photodetectors based on Mg-doped hydrogenated GaN films grown at 380°C, *Appl. Phys. Lett.*, 76, 345, 2000.
10. Tersoff, J., Schottky barrier heights and the continuum of gap states, *Phys. Rev. Lett.*, 52, 465, 1965.
11. Pauling, L.N., *The Nature of the Chemical Bond,* Cornell University Press, Ithaca, NY, 1939.
12. Yu.Koyama, Y., Hashizume, T., and Hasegawa, H., Formation process and properties of Schottky and ohmic contacts on n-type GaN for field transistor applications, *Sol. St. Electron.*, 43, 1483, 1999.
13. Wang, L. and Nathan, M.I., High barrier height GaN Schottky diodes: Pt/GaN and Pd/GaN, *Appl. Phys. Lett.*, 68, 1267, 1996.
14. Schmitz, A.C. et al., Schottky barrier properties of various metals on n-type GaN, *Semicond. Sci. Technol.*, 11, 1464, 1996.
15. Hasegawa, H., Koyama, Y., and Hashizume, T., Properties of metal-semiconductor interfaces formed on n-type GaN, *Jpn. J. Appl. Phys.*, 38B, 2634, 1999.
16. Polenta, L., Fang, Z.Q., and Look, D.C., On the main irradiation defect in GaN, *Appl. Phys. Lett.,* 76, 2086, 2000.
17. Buyanova, I.A. et al., Photoluminescence of GaN: Effect of electron irradiation, *Appl. Phys. Lett.*, 73, 2968, 1998.
18. Nakamura, S. and Fasol, G., *The Blue Laser Guide. GaN Based Emitters and Lasers*, 1st ed., Springer, Berlin, 1997, chap. 7.
19. Nakamura, S. and Fasol, G., *The Blue Laser Guide. GaN Based Emitters and Lasers*, 1st ed., Springer, Berlin, 1997, chap. 5.
20. Li, X, and Coleman, J.J., Time-dependent study of low energy electron beam irradiation of Mg-doped GaN grown by metalorganic chemical vapor deposition, *Appl. Phys. Lett.*, 68, 1605, 1996.
21. Inamori, M. et al., Direct patterning of the current confinement structure for p-type column-III nitrides by low-energy electron beam irradiation treatment, *Jpn. J. Appl. Phys.*, 34, 1190, 1995.

22. Kozlovsky, V.I. et al., Electron beam pumped MQW InGaN/GaN laser, *MRS Internet J. Nitride Semicond. Res.*, 2, 38, 1997.
23. Osinsky, A. et al., Low noise *p-π-n* GaN ultraviolet photodetectors, *Appl. Phys. Lett.*, 71, 2334, 1997.
24. Asif Khan, M. et al., Schottky barrier photodetector based on Mg-doped *p*-type GaN films, *Appl. Phys. Lett.*, 63, 2455, 1993.
25. Chen, Q. et al., Schottky barrier detectors on GaN for visible-blind ultraviolet detection, *Appl. Phys. Lett.*,70, 2277, 1997.
26. Pearton, S.J. et al., GaN: Processing, defects, and devices, *J. Appl. Phys.*, 86, 1, 1999.
27. Kuznetsov, N.I. et al., Schottky barriers on GaN, *Mater. Res. Soc. Symp. Proc.*, 395, 837, 1996.
28. Shiojima, K., Sugahara, T., and Sakai, S., Large Schottky barriers for Ni/*p*-GaN contacts, *Appl. Phys. Lett.*, 74, 1936, 1999.

CHAPTER 9

Organic Photoconducting Materials

T.A. Iourre, L.I. Rudaya, and N.V. Klimova

CONTENTS

0-8493-0915-8/01/$0.00+$1.50

389

9.1 INTRODUCTION

This review summarizes the state of the art in organic photovoltaic cells (OPVs). While organic materials are readily ionized and have not functioned adequately in radioisotope decay energy direct-conversion prototypes, the ongoing effort to miniaturize electronics, sensors, and actuators may well use organic materials as light, cheap, and flexible materials of construction. Applications in solar voltaics, chemical batteries, hydrogen fuel storage, and indirect radioisotope decay energy converters are promising. OPV designs have been analyzed and data have been collected on electronic organic materials for both OPVs and organic light-emitting diodes (OLEDs). In the course of this work, ecological, economic, and technological aspects have been considered concerning application of organic photoconducting materials and means of enhancing efficiency. The major types of OPV materials considered in this survey are:

- Systems simulating photosynthesis in which conversion of light into electrical energy occurs in porous oxides sensitized with Grätzel complexes
- Macrocycles of phthalocyanines, porphyrins, and chlorophylls
- Fullerenes
- Conducting polymers and polyheteroarylenes

9.2 ORGANIC-BASED PHOTOVOLTAIC CELLS SIMULATING PHOTOSYNTHESIS

By analogy with photosynthesis, organic-based photovoltaic cells (OPVs) use light energy to drive charge-carrier transport processes. An attractive strategy is the development of systems that simulate natural photosynthesis in the conversion of light energy for the fixation of CO_2. Based on the processes used by plants, molecular photovoltaic devices have been constructed with overall efficiency for solar energy conversion to electricity of 10 to 11%.

The purpose of this section is to review recent research and select the best systems for light conversion into electricity based on nanocrystalline oxide films and transition metal charge-transfer sensitizers. In porous films, short-circuit currents are four to five orders greater than those in nonporous films. Photogeneration of holes takes place in a pigment coating, the thickness of which is limited by the light penetration depth. Their excellent performance will promote acceptance of renewable energy technologies, not least by setting new standards of economy.

The most striking successes are based on molecular engineering of ruthenium complexes such as cisX₂bis(2,2′-bipyridyl-4,4′-dicarboxylate)ruthenium(II), where X = Cl⁻, Br⁻, I⁻, CN⁻, or SCN⁻. Among these compounds, cis(dithiocyanato)bis(2,2′-bipyridyl-4,4′-dicarboxylate)ruthenium(II), sometimes referred to as Grätzel complex, performs best as a solar light absorber and charge-transfer sensitizer.[1] A broad range of visible light absorption and long-lived excited states renders it an attractive sensitizer for homogenous and heterogeneous reactions. As a component in a photovoltaic cell, the Ru complex is anchored to the surface of a wide-band oxide semiconductor such as nanoporous TiO_2. Due to its large band gap (3.2 eV), the semiconductor does not absorb visible radiation. The Grätzel complex sustained over 5×10^7 redox cycles without loss of performance, corresponding to about 10 years of continuous operations in natural sunlight. Other organic sensitizers photobleach after fewer than 10^4 redox cycles. The required amount of ruthenium for the OPV is 10^{-3} mol/m², corresponding to U.S. \$0.07/m² for the noble metal.[1]

9.2.1   Parameters Determining OPV Efficiency

The absorption coefficient of dye used in sensitized OPVs is a major factor in the total energy efficiency of the cell. The absorption coefficient of the Grätzel ruthenium(II) complex can be greatly increased by having the dye absorbed on silver islands.[2] The quantum efficiency is strongly dependent on the presence of oxygen.[3]

Ruthenium(II) complexes were prepared and characterized with respect to their absorption, luminescence, and redox behavior.[4] They act as efficient charge-transfer sensitizers for nanocrystalline TiO_2 films (thickness 8 to 12 μm) of very high internal surface area (roughness factor ca.1000) prepared by sintering 15 to 30 nm colloidal titanium particles on a conducting glass support. The performance of the Grätzel complex is unmatched by any other sensitizer. Nanocrystalline TiO_2 films coated with a monolayer of this compound collect visible light very efficiently, with an absorption threshold around 800 nm.

Conversion of incident photons into electric current was demonstrated over a large spectral range. These films were incorporated in a thin-layer regenerative PV cell equipped with a light-reflecting counter electrode. Short-circuit photocurrents exceeding 17 mA/cm² were obtained in simulated AMI.5 sunlight using lithium iodide/triiodide in acetonitrile or acetonitrile/3-methyl-2-oxazolidone mixture as a redox electrolyte. The open-circuit photovoltage was 0.38 V and increased to 0.72 V by treating the dye-covered film with 4-tert-butylpyridine. A light-to-electric energy-conversion efficiency of 10% was attained with this system. The effect of the temperature on the power output and long-term stability of the dye was also investigated.

The frequency-dependent photocurrent response of dye-sensitized TiO_2 cells to modulated illumination has been analyzed.[5] Analytical expressions were derived that describe generation, collection, and recombination of electrons in a thin-layer nanocrystalline PV cell under conditions of steady illumination and with superimposed small-amplitude modulation. The analysis considers illumination from the contact side and from the counter electrode side. Differences in the intensity-modulated photocurrent response are predicted for the two cases.

A solid-state dye-sensitized OPV was fabricated by photodepositing polypyrrole onto the surface of a porous TiO_2 electrode anchored to the photosensitizing Grätzel complex. Polypyrrole in the porous medium successfully works as a hole transport layer connecting the sensitizer dye molecules to a counter electrode. Doping density of the polypyrrole-conducting layer was optimized to improve the efficiency of the system.[6]

The mechanism of instability in Ru-based dye-sensitized OPVs has been considered.[7] *In situ* infrared attenuated total reflection studies of the Ru-complex OPV reveal that the Ru-complex (both tri- and mononuclear) attached to TiO_2 is irreversibly consumed under conditions of insufficient iodide or oxide regeneration. The ruthenium sensitizer $((Ru(bipy)_2(CN)_2)_2Ru(bpca)_2)^{2-}$, where bipy is 2,2'-bipyridine and bpca is 2,2'-bipyridine-4,4'-dicarboxylate, decomposes into fragments. One of the fragments is $Ru(bpy)_2(CN)_2$. For the sensitizer $Ru(bpca)_2(SCN)_2$, it has been shown that a molecular fragment absorbing at a wavelength of 4.96 μm diffuses from the nanostructured TiO_2. Due to correlation with the photocurrent density, it is identified as a product of the oxidized sensitizer.

A photovoltaic device based on thin films (10 to 20 μm) of TiO_2 nanoparticles in contact with nonaqueous liquid electrolyte has been proposed.[8] The cell is simple to fabricate. In principle its color may be tuned through the visible spectrum, from transparent to black, by changing the absorbing characteristics of the dye. The photovoltage of the cells was not determined by the threshold energy for light absorption (band gap) as in conventional semiconductors. The highest efficiency was 11%. The cell has potential to become a low-cost photovoltaic option. Unique applications include photovoltaic and photoelectrochromic windows.

Another solid-state OPV has been fabricated by photoelectrochemical polymerization of pyridine (Py) on porous nanocrystalline TiO_2 sensitized by the Grätzel dye, or a newly synthesized cis(dcb)$_2$(pyrrole-1-yl-methyl)pyridine ruthenium.[9] Poly(pyrrole) is a hole transport carrier that improves cell characteristics.

Some OPVs are based on dye-sensitized nanoporous oxides.[10,11] An SnO_2/perylene 3,4-dicarboxylic acid-9,10-(5-phenanthroline)carboximide structure had the following photovoltaic parameters: I_{sc} = 3.26 mA/cm^2, V_v = 0.45 V, and overall cell efficiency (/) of 0.89% at one sun intensity[10]; for a quasi-solid-state TiO_2 photoelectrochemical cell dye-sensitized polymer gel: $I_{sc} \geq 3$ mA/cm^2 under 30 mW/cm^2 white light illumination, V_v = 0.6 V, / < 3 to 5%.[11]

TiO_2 acts as an efficient electron acceptor for MEH–PPV in the excited state. An Al/ultrathin nanocrystalline TiO_2-poly(p-phenylenevinylene) (PPV)/ITO structure has been studied.[12] The results indicate that most photogenerated excitons are charge separated at the interface between the polymer and the nanocrystals (larger than 20% by weight of TiO_2). Electron collection efficiency in photovoltaic devices is limited by fast recombination. The transport model indicates that, even for a relatively long recombination time and well-ordered nanocrystals, most of the collected charge originates from the first several nanocrystal layers of the electrode, rather than from the sides throughout the film, due to recombination.

9.2.2 Solid-State Dye-Sensitized Meso- and Nanoporous TiO₂ OPVs

OPVs based on dye-sensitized mesoporous films are a low-cost alternative to conventional solid-state devices. Impressive light-to-electrical energy-conversion efficiencies have been achieved with such films when used in combination with liquid electrolytes. Practical advantages may be gained by the replacement of the liquid electrolyte with a solid charge-transport material. The dye-sensitized heterojunction of TiO_2 with the amorphous organic hole-transport material 2,2',7,7'-tetrakis(N,N-di-p-methoxyphenyl-amine)-9,9'-spirobifluorene is described.

Photoinduced charge-carrier generation at the heterojunction is very efficient. An OPV based on these materials converts photons to electrical current with a high yield within a narow wavelength range (33%).[13] The effect of blended or layered TiO_2 nanoparticles on charge-transfer processes in conjugated polymers is observed in different systems.[14-16] The light is absorbed by a monolayer of Ru complex tethered to the TiO_2 surface. In such systems, efficiency of up to 10% is reached.[15] The number of organic ligands that can be used in Ru complexes is sufficiently large to give the material engineer a variety of wavelength absorption options. Device data (photovoltaic characteristics) are limited, but new information is being accumulated in this promising applied research area.[17-30]

9.2.3 Nanoporous Solid-State OPVs Sensitized with Macrocyclic Dyes

A fully solid-state dye-sensitized nanoporous n-TiO₂/copper chlorophylline/p-Cu cell was fabricated by depositing p-Cu onto a nanoporous film of TiO_2 coated with Cu chlorophylline. The photoexcited pigment is found to inject electrons into n-TiO₂ and holes into p-Cu, generating a photocurrent and photovoltage significant for a solid-state dye-sensitized cell.[31]

An OPV cell made from a porous film of a mixture of Sn(IV) and Zn oxide sensitized with a ruthenium bipyridyl complex suppresses recombination of the photogenerated electrons and dye cations, generating a short-circuit photocurrent of ca. 22.8 mA/cm² and an open-circuit voltage of ca. 670 mV in direct sunlight of 900 W/m² with an efficiency of 8%.[32]

The nanocrystalline TiO_2 film is modified simultaneously with porphyrin (PPh) and phthalocyanine (Pc) by chemical absorption. Since PPh and Pc dyes have different absorption bands in the visible range, TiO_2 sensitized with a combination of these dyes can be better matched to the solar energy spectrum, and the photoresponse of the electrode is extended to cover most of the visible range. This kind of electrode exhibits significant energy-conversion efficiency improvement.[33,34]

Co-adsorption complexes of gallium or zinc tetrasulfophthalocyanine (GaTsPc or ZnTsPc) on porous TiO_2 improve photovoltaic properties of the electrodes.[35-37] This is attributed to the formation of Pc–PPh heteroaggregates on the electrode, resulting in a reduced concentration of GaTsPc dimer and providing a low-lying charge-transfer state.

A photoelectrochemical cell utilizing flavonoid anthocyanine dyes extracted from blackberries, along with colloidal TiO_2 powder, has been shown to convert sunlight to electrical power with an efficiency of 0.56 % under full sun. Fluorescence quenching has been observed for the excited state of the TiO_2–adsorbed anthocyanine dye, cyanine. The photocurrent spectrum correlates well with the optical absorption of the cyanine-sensitized TiO_2 nanocrystalline film. The incident photon–to-current efficiency of 19% at the peak of the visible absorption band of the dye, open-circuit voltage of 0.5 to 0.4 V, and short-circuit photocurrent of 1.5 to 2.2 mA/cm^2 are remarkable for such a simple system and suggest efficient charge-carrier injection.[38]

Self-assembled functional molecules have been introduced into mesoporous materials. Such molecules are synthesized directly or by co-assembly with dye-bound surfactants of ferrocenyl trimethyl ammonium bromide with silicates or Pc molecules doped with C_{16} trimethyl ammonium bromide micelles on V_2O_5, WO_3, or SiO_2. The process provides a well-organized molecular doped mesoporous structure by a direct and simple procedure.[39]

An interesting variant is the cell $ITO/TiO_2/PbPc/Au$.[40] Transparent TiO_2 films were deposited onto a base electrode comprising an ITO glass substrate using the (alkoxide) sol-gel technique. Lead phthalocyanine was vacuum sublimed onto the TiO_2 surface. The resulting heterojunction had photovoltaic properties; however, the power-conversion efficiency was low.

When nanostructured TiO_2 was sensitized by monochromophore (E)-N-hexadecyl-4-(2-(4-diethylamino)phenyl) ethenyl)pyridium stearate, the photocurrent onset extended to 800 nm. Under the illumination of 100 mW/cm^2 simulated solar light, over 1 mA/cm^2 photocurrent was obtained with the dye-loaded nanocrystalline TiO_2 electrode. This is a very high value among pure organic dye sensitizers.[41]

Complexes of Ru (II) with nitrogen-containing ligands coating meso- or nanoporous TiO_2 are promising candidates as photosensitizers in dye-sensitized OPVs. Ru complexes of this type are not produced on a commercial scale. However, since OPV performance to cost is superior to conventional photovoltaics, their commercial development can be expected.

9.3 APPLICATION OF PHTHALOCYANINES IN PV ELECTRONIC DEVICES

Due to chemical and thermal stability and semiconductor properties, metalocyanines represent one of the most promising classes of molecular electronic materials. Photoconductivity in the visible spectral range permits their development as solar voltaics and for indirect conversion of radioactive decay energy. Composite structures with other organic materials provide additional flexibility in available electrophysical properties.

9.3.1 Structure and Properties of Phthalocyanines

Phthalocyanines (Pc) and porphyrines are characterized by a 16-membered macrocycle built of four pyrrole groups bound together by nitrogen atoms. These are typically planar aromatic molecules, the tetrazaporphyn molecule containing 38 and phthalocyanines 54 "aromatic" electrons. Because of phthalocyanine's extended conjugated electronic system, its color is deep red, with $\lambda_{max} = 703$ nm.

Porphynes of various degrees of substitution and hydrogenation are called porphyrins. Chlorophyll is a porphyrin (Figure 9.1). Porphyrins may have been formed on the primeval earth by ablation of pyrrole compounds. Pyrroles can be formed from ammonia, acetylene, and formaldehyde — components in early earth atmosphere and hydrosphere. Porphyrins and their complexes with metals served as photosensitizers during biological evolution, which attests to the ecological safety of these organic macrocyclic dyes.

Central (imine) hydrogen atoms in these compounds are easily replaced by metal cations forming chelate compounds. Mostly stable, these chelates sublime in vacuum without decomposition and are resistant to acids and weakly soluble in organic solvents.[42] The stability of metalophthalocyanines depends on the central metal atom. Many of these compounds, such as PcH$_2$, PcNi, PcCu, and PcZn, are very stable in the presence of oxygen, with a few exceptions. In the synthesis of PcFe, oxygen can not be excluded totally, leading to the formation of (PcFe)$_2$O.[43] CuPc is thermally stable up to 200°C, sublimates in vacuum without decomposition at 550 to 580°C, and is moderately soluble when boiled in chloronaphthalene or quinoline.

More then 70 elements have been used as the central metal atom in the phthalocyanine moiety to control the oxidation potential of the phthalocyanato compounds, leading to very different electrical properties. Mono- and di-anions of porphyrins have a much greater proton–acceptor ability compared to anions of phthalocyanines.

9.3.2 Methods of Producing Phthalocyanines and Porphyrins

Phthalocyanines (Pc) have been produced commercially since the 1930s. The basic method of production is by heating phthalonitrile with salts of metals (e.g., Cu$_2$Cl$_2$) in the presence of NaCl. This process is exothermic, with the temperature rising to 300°C. Pc can be obtained by condensation of phthalic anhydride, urea, and salts of metals in the presence of H$_3$BO$_3$, or As$_2$O$_3$ in trichlorobenzene solution, among other methods. 1-amino-3-iminoiso-indolenine in the presence of metal ions and soft reducing agents at elevated temperatures forms a corresponding phthalocyanine.[44] Metal Pc compound is prepared by treating 1,3-diiminoisoindolenine and a metal salt in a polar aprotic solvent. A photodetector using Pc prepared by this method showed high sensitivity.[45]

Water-soluble Pc are prepared by direct sulfonation of phthalocyanine macrocycles. Phthalocyanines can also be prepared by ring-insertion reaction from subphthalocyanines leading to asymmetrically substituted phthalocyanines.[46] Oxotitanium Pc, used as a photoconductive and laser printing material, is prepared by reaction of o-C$_6$H$_4$(CN)$_2$ with TiCl$_4$ in Ph$_2$O followed by hydrolysis.[47] Metal Pc exhibit high electron transfer abilities but are insoluble in most organic solvents, which has inhibited their commercial use.[48]

4-(polyfluoroalkoxy)phthalonitriles were prepared with high yield by nucleophilic substitution of the nitro group of 4-nitrophthalonitrile by polyfluoroalkoxy

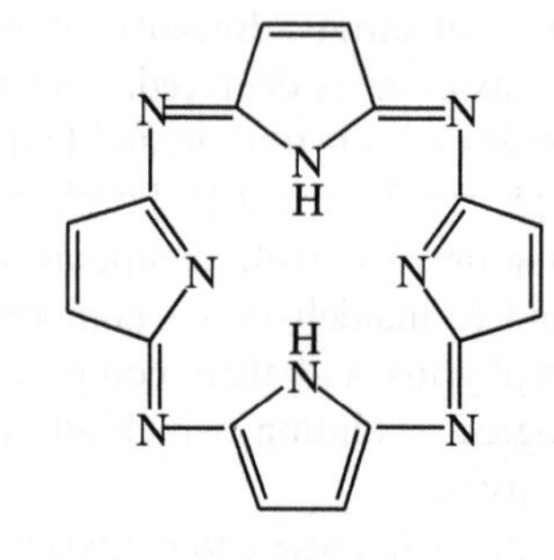

I

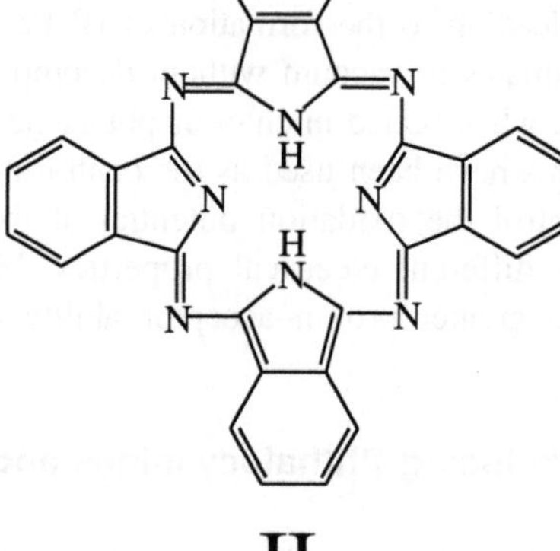

II

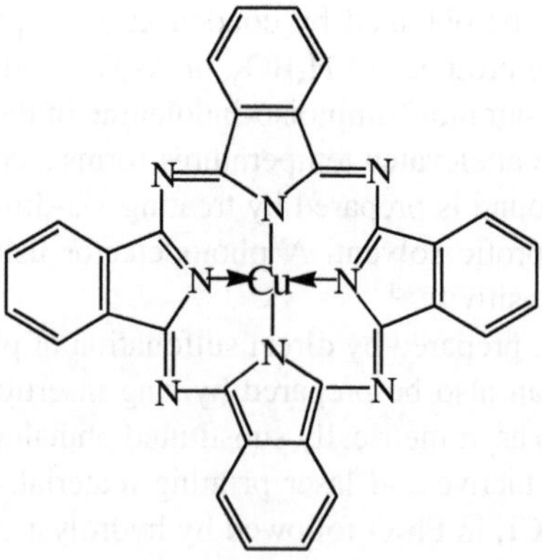

III

Figure 9.1 The structures of tetrazaphorphyn (I), tetrabenztetrazaphorphyn (II), and CuPc (III).

groups at room temperature. These nitriles were cyclotetramerized in the presence of Co or Cu salts into Pc complexes.[49] Several methods of purification of photoconductive phthalocyanines have been proposed. Metal-free Pc is dissolved in 85 wt% H_2SO_4, diluted with water to deposit α-type Pc, and milled to produce photoconductive Pc.[50] Water-soluble Pc are purified by dissolving them in water, boiling the filtrates, and filtering again.[51] Synthesis of conducting bridged macrocyclic metal complexes is described next. Since Pc compounds can be synthesized by simple methods, their cost can be greatly reduced through mass production for solar voltaics and other electronic devices.

9.3.3 Electrical Properties of Phthalocyanine

Metal Pc are molecular crystals with narrow band gaps of about 2 eV.[52] Theoretical conversion efficiency for the OPV can be derived either through a thermodynamic argument involving energy and entropy balance or through the detailed balance of carrier generation in a semiconductor.[53] H_2Pc single crystals grown by vacuum sublimation were investigated for their conductivity in dark and light. The applied electric field ranged from 0.8 kV/cm to 6 kV/cm; the temperature range was 300 to 570 K. The crystal was found to be photoconductive.

Based on activation energies calculated from temperature-dependent photoconductivity, an energy level scheme for crystalline H_2Pc was proposed. The model consists of two trapping levels within the forbidden gap. One is 0.4 eV below the conduction band edge from which electrons are thermally excited into the conduction band, and the other acts as a recombination center 0.3 eV above the valence band edge. The band gap is calculated to be 1.4 eV. The proposed model is in agreement with an earlier investigation, indicating that H_2Pc as a semiconductor is thermally stable, even at relatively high temperatures.[54]

Charge-transfer reactions are photoinduced by exciting strongly coupled donor–acceptor pairs with UV and visible light. The systems investigated are composed of porphyrin (PPh) or Pc mixed dimers. In this case proton transfer competes with electron transfer.[55]

The influence of electric field and humidity on photoconduction and fluorescence has been studied on particles of highly photoconductive TiOPc dispersed in a poly(vinylbutyral) matrix. Integrated fluorescence quenching showed a linear dependence at low applied fields and a linear correlation with the carrier generation efficiency. Time-resolved fluorescence decay was analyzed by fitting the data to a sum of two exponentials representing fast and slow fluorescence compounds. Results indicate that carrier generation in TiOPc originates from relaxed and nonrelaxed intrinsic excited singlet states, while the trapped excitons do not lead to significant carrier production. The field dependence of integrated fluorescence quenching supports the existence of a carrier precursor state with charge-transfer character.[56] A photoconductor of TiOPc in a new crystal form was prepared using a crystal-transform technique by microporous membrane percolating; the ratio of photoconduction to the dark conduction[57] reached 10^4.

The photovoltaic action spectra of Al/VOPc/ITO at different modulation frequencies of illuminating light were measured. The VOPc film acts as a p-type semiconductor and photovoltage is mainly generated at the Al–VOPc interface. The charge-transfer excitation and Frenkel excitation were revealed by electric modulation reflecting spectroscopic technique.[58] Potential barriers at interfaces of organic semiconductors belonging to linear polyacenes and Pc with SnO_2 and CuI transparent electrodes, as well as spectral dependence of photogalvanic properties of thin-film SnO_2/OS/CuI sandwich structures illuminated on both sides, were studied. Their correlation with spectra of the optical absorption coefficients was analyzed. Crystal and energy structure of organic semiconductor thin films depend essentially on preparation conditions which influences critical interface properties.[59]

Excitation of surface plasmon polaritons in CuPc OPVs was achieved by means of spectral attenuated total reflectance.[60]

Photoelectric properties of Pc powders, films, and Pc compounds on a variety of substrates have been the subject of numerous investigations.[61,62] Photoconductivity is affected by the nature of a Pc substrate as well as by its crystal modification, doping agents, and the way the substrate is prepared. Unsubstituted CuPc shows higher sensitivity in the near-IR region than substituted CuPc complexes. This is explained by substituent disruption of the stacked arrangement characteristic of unsubstituted CuPc.

OPV conversion efficiency is expressed by four terms: photon collection factor, voltage output factor, average quantum efficiency of photocarrier generation, and fill factor (FF).[63] CuPc absorbs light around the blue-green region in the visible spectrum. Poly(aniline), which has conjugated bonds, is photosensitive. In poly(aniline) thin films prepared with CuPc powder, not only does the absorption intensity increase but the absorption band also broadens over a region between 1.7 and 2.3 V and above 3.0 eV of the visible spectrum. Thus, CuPc has been shown to increase the efficiency of OPVs made of conducting polymers such as poly(aniline). The use of iodine along with CuPc in preparation of thin films was shown to smooth the broadening effect in absorption of OPVs. Attempts are underway to increase absorption on both ends of the visible spectrum, using suitable dopants.[64]

CuPc thin films were deposited on crystalline Si (111) substrates with surface dangling bonds terminated with hydrogen. The deposited films were contaminated with oxygen and showed p-type conduction. CuPc/Si p–n heterojunction diodes exhibited rectification in their I–V characteristic with relatively large reverse leakage current. This is associated with trap-assisted tunneling via traps at the CuPc–Si interface resulting from incomplete hydrogen termination of Si dangling bonds. A strong blue-green light is emitted from CuPc film exposed to UV light.[65]

The dependence upon applied voltage of the transient photoconductivity response in the nanosecond range at different excitation light intensities in an Al/SnPc/ITO sandwich cell has been reported. Charge-transfer excitons and Frenkel excitons have been revealed by the technique of electronic modulation reflection spectroscopy. The photoelectric properties of this cell might result from dissociation of the charge transfer excitons and the Frenkel excitons via multistep thermalization and ionization.[66]

Metal-substituted Pc are a class of weakly semiconducting organic dyes. Many of these materials were studied for molecular-switching, gas-sensing, and photovoltaic properties. In particular, incorporation of a strong electron acceptor such as iodine or tetracyanoquinono-dimethane (TCNQ) substantially increases conduction.

Sandwich structures of aluminum/α-nickel Pc/silver were fabricated by chemical vacuum deposition. TCNQ spin coating on an aluminum electrode prior to NiPc deposition is reported.[67] Spin coating is a simpler technique for thin-film deposition than CVD, and performances in which TCNQ was deposited by co-sublimation with NiPc were similar. The effect of exposure to air increases the conductivity by an order of magnitude. Conductivities of the devices incorporating TCNQ are about 20 times those of devices without TCNQ. Doping level of devices without TCNQ increased throughout a 94-h measurement period, while the devices with a TCNQ layer were fully doped within 2 h.

Current density–voltage characteristics obtained for thin films with NiPc particles dispersed in a polymer binder sandwiched between ohmic Au contacts showed ohmic conduction in the lower voltage range and space-charged limited conduction controlled by an exponential distribution of traps above the valence band edge for higher voltage ranges. The following values were obtained: hole mobility $\mu = 1.0 \cdot 10^{-5}$ m^2/V·S, room temperature hole concentration $\rho = 1.2 \cdot 10^{14}/m^3$, concentration of traps per unit energy range at the valence edge $\rho_0 = 2.2 \cdot 10^{44} J^{-1} m^{-3}$, temperature parameter of trapping distribution $T_1 = 820$ K, and total trapping concentration $N_t(e)$ $= 1.8 \cdot 10^{24} m^{-3}$.[68]

Measurement of direct current as a function of inverse temperature at constant applied voltage for H_2Pc with gold electrodes yielded a hole mobility value $\mu = 6.25 \cdot 10^{-7} m^2 \cdot V^{-1} s^{-1}$. At room temperature,[69] hole concentration was $\rho = 8 \cdot 10^{18}$ m^{-3}. Data for MPc ($CoPc$[70] and $PbPc$[71]) give values similar to those for H_2Pc. The charge transport in LiPc and iodinated compounds ($LiPcI$)[72] and $ClAlPc$[73] was investigated. Tunneling of charge carriers is the dominant conduction process in both compounds up to room temperature, yielding intrinsic DC conductivities of $\sigma_{dc} = 5.3 \cdot 10^{-4} \Omega^{-1} cm^{-1}$ ($LiPcI$).[72]

9.3.4 Phthalocyanine-Based Molecular Electronic Devices

A number of investigations suggest that dye-sensitized OPVs represent efficient solar converters for power generation.[74,75] The OPV consists of an n-type layer containing fine crystalline particulate organic pigments (e.g., Pc compounds) dispersed in a binder resin, a p-type layer containing evaporated organic pigments (e.g., perylene compounds), and counter electrodes on transparent electrode substrates. These OPVs having organic p–n heterojunctions and a polymer resin substrate such as polyethylene terephthalate, polyethylene naphthalate, polyestersulfone, polyimide, or aramide have high efficiency and can be manufactured at low cost.[76,77]

A semiconductor fire detector has been proposed.[78] On an insulating substrate, Pc films were vacuum deposited in such proportions as to enable detection of fires with or without flames. Optimization of the thickness of ZnPc in a p–n cell gave an excitation diffusion length of 30 to 40 nm.[79] Good correlation between thickness and photocurrent yield has been shown. Thin films of ClAlPc of 100 to 300 nm thickness sublimated onto a glass substrate in high vacuum gave good agreement with findings for CuPc, PbPc, NiPC, and CoPc.[80] Self-assembled functional molecules in mesostructured materials were synthesized directly, either by co-assembly of dye-bound surfactant or ferrocenyl TMA with silicate or by Pc doped with

C_{16}TMA micells assembled with VO_x, MoO_3, WO_3, and SiO_2 oxides to produce Pc-doped mesostructured material.[81] Reasonable properties were reported for a system formed of n-type material, di-pyridyl-perylene tetracarboxylicdiimide, and CuPc or pentacene.[82]

The influence of chloranil doping on photovoltaic performance of a semitransparent indium electrode and a CuPc binder layer was investigated.[83] The undoped OPV showed moderate photovoltaic performance. Chloranil doping increased the current in the cell but decreased the open-circuit voltage. In two-layer organic devices fabricated from Al/CV/NiPc/ITO where NiPc is p-type and Crystal Violet is the n-type semiconductor, the ITO transparent electrode has been found to form an ohmic contact with p-type NiPc, while the aluminum electrode forms an ohmic contact with Crystal Violet.[84]

Quantitative data obtained by major research groups on short-circuit photocurrent, quantum yield (φ), open-circuit voltage (V_{op}), fill factor (FF), energy-conversion efficiency (η), and short-circuit current (I_{sc}) of dye-sensitized OPVs are tabulated in Table 9.1.[85-103]

NiPc(AsF_6)$_{0.5}$ is a double-chain system — a metal and macrocyclic chain within the same molecular column.[104] The photovoltaic effect has been observed for ITO/TiO_2/ZnPc/Au.[105] Photocurrent flows from ZnPc to TiO_2 in the external circuit when the cell is illuminated, with photocurrent varying directly and photovoltage varying logarithmically with light intensity. The Al/dye/Au sandwich using 5,10,15,20-tetraphenylporphirinatozinc (ZntpPPh) and 3-ethyl–5(3-ethyl-2-(3H)-benzothiazolyliden)-2-thioxo-4-thiazolidinone (MC) exhibits a larger energy-conversion yield due to a better spectral match compared to pure ZntpPPh or an MC solid.[106] OPVs based on the electrochemical oxidation of PcLi$_2$ using LiPc on an ITO substrate were demonstrated but with no appreciable improvement in photovoltaic characteristics.[107]

Photoexcitation of a Zn-tetraphenyl Pc complex leads to electron transfer with very long lifetimes of the charge-separating pairs.[108] Blends of porphyrins have been proposed for use in OPVs. The photovoltaic properties of meso-tetraphenyl porphyrin (PPh), Cr(III), and Mn(II) with solid thin films between two optically transparent electrodes have been studied. Due to differing electronic configurations, mixed metal porphyrins with different central ions exhibit a photovoltaic effect under an external electrical field.[109] Similar PPh blends for OPVs have been patented in Japan,[110,111] as well as a film representing a homogenous mixture of PPh derivatives and Ph$_3$CH derivatives, and others.[112] The contribution of electric field–induced photocurrent in metal-free PPh is described.[113] The heterojunction ITO/chlorophyll-a/TPyPh/Al OPVs were prepared by electrodeposition of chlorophyll-a onto a conducting ITO electrode. Two-layer cells show better results[114]; the typical parameters of these OPVs under light density of 20 μW/cm^2 at 470 nm are given.[115]

Elaboration of OPVs based on rare-earth Pc complexes is described.[116] The preparation of novel transition-metal Pc-type compounds (Me$_8$Pc)RuL$_2$ with axial ligands capable of binding semiconductors such as TiO_2 has also been described.[117] Preferred compounds offer absorption of solar radiation in the near IR and might be used in solar voltaic windows. Pc have been successfully used in electroluminescent devices,[118,119] as electrochromic materials,[120] in photodetectors,[121,122] and in photoelectrochemical cells.[123]

Table 9.1 Characteristics of Thin-Film PV Cells Comprising Porphyrins and Phthalocyanines

PV Cell Structures	I_{sc} :A/cm²	V_{op} V	η,%	FF	Notes	Ref.
Al/C$_{60}$/OEP/OOPPV/	1.4	0.15	0.01	0.3	ITO side	85
ITO	1.1	0.35	6	—	Al side	85
TiO$_2$(H$_2$Pc)/NiPc	0.56	0.24	0.02	0.64	—	86
Al/CuPc/ITO	0.25	0.4	9	—	—	87
Al/mixtDye/Au	—	0.90	0.73–0.82	0.14	φ = 14.7%, 440 nm, 14.7 :W/cm², Al side	88
Al/mixtDye/Au	—	0.90	0.72	0.19	φ = 9.1%, 570 nm, 32.4 mW/cm²	88
ITO/MPcl/TiOPc/Au	—	—	0.7	—	400 to 720 nm	89
	—	—	0.35	—	white light 8 to 130 mW/cm²	
TiO$_2$/BPl/ZnPc	1.5·10³	—	—	—	φ = 18%, under 1 sun condition	90
ITO/TiO$_2$sensZnTSPc	—	—	2.1	—	700 nm	91
	—	—	1.7	—	35 mW/cm²	91
ITO/di-Me/ZnPc/Au	3.5	0.68	0.43	0.52	white light 100 mW/cm²	92
Au/α-TiOPcPerylene/ITO	—	—	—	—	cryst. α-TiOPc film	93
SnO$_2$/CuPc/BP/Ag	0.5	0.47	0.66	—	white light 750 W/m²	94
Al/HD/Me/Au	—	—	—	—	φ = 39%, 445 nm, white light 21.5 :W/cm²	95
TiO$_2$/ZnTs PPh+GaPc (0.61:0.39)	—	—	—	—	φ = 32.2%, 700 nm, monochrom, at 138 :W/cm²	96
ITO/C$_{60}$/H$_2$Pc/Au	89	0.18	0.03	0.25	white light, 12.5 mW/cm²	97
ITO/CuPc/(CuPc +TpyPh)/TPyP/Al	—	—	0.35	—	monochrom light 520 nm, at 30 mW/cm²	98
ITO/CuPc/(CuPc +TpyPh)/TPyPh/Al	—	—	0.9	0.32	monochrom light 430 nm, at 10 µW/cm²	99
PcuPc/n-Si doped Perylene dye	0.75	210	3.9	0.22	light intensiv. 0.88 mW/cm²	100
Al/(H$_2$ PPh + Zn PPh)	—	0.79	1.21	0.23	—	101
SnO$_2$/CuPc/BP/Ag	0.21	0.47	0.33	0.19	at 3 mW/cm²	102
ITO/CuPc + Perylene/Al	—	—	9.1	—		103

9.3.5 Conducting Bridged Macrocyclic Metal Complexes

One of the best investigated types of co-facially linked stacked macrocyclic metal complexes is polyphtalocyaninato metalloxanes [PcMO]$_n$ (M = Si, Ge, or Sn). Iodine-doped compounds have stoichiometries [(PcMO)I$_y$]$_n$ where y is 1.1, at most. The electrical conductivity of [(PcGeO)I$_y$]$_n$ at room temperature is 1.1·10⁻¹ S/cm but 2.2·10⁻¹⁰ S/cm for the undoped sample. Conductivity of [(PcSiO)I$_y$]$_n$ is 6.7·10⁻¹ S/cm, while undoped conductivity is 5.5·10⁻⁶ S/cm. [(PcSnO)I$_y$]$_n$ has a conductivity of 1.1·10⁻⁶ S/cm, while the undoped sample conductivity is 1.2·10⁻⁹ S/cm.[46]

A stacked arrangement of phthalocyaninato and naphthalocyaninato transition metal compounds yields coordination polymers. The macrocycle, central metal atom, and bridging ligand can be varied systematically. The stacking is achieved by biaxially connecting the central transition metal atoms of the macrocycles with bidentate bridging ligands. The bridging ligands are linear π-electron-containing organic molecules such as [pyrazine(pyz),p-diisocyanobenzene(dib),tetrazine(tz)]. If the oxidation state of the central metal atom is +3 (e.g., Co^{+3}, Fe^{+3})-charged bridging ligands such as cyanide (CN^-) or thiocyanate (SCN^-), a cyano-bridged intrinsic semiconducting Pc complex can be synthesized by displacement of the axial anion (e.g., Cl^-) by CN^- in a coordinatively unsaturated PcMX. Room temperature electrical conductivities for polynuclear cyanometal complexes are, for $[PcFe(CN)]_n$, $6 \cdot 10^{-3}$; for $[PcCr(CN)]_n$, $3 \cdot 10^{-6}$; and for $[PcMn(CN)]_n$, $1 \cdot 10^{-5}$ S/cm.[124]

Polymeric macrocyclic compounds $[PPh(M)\text{-}R'\text{-}C\equiv C\text{-}C\equiv C\text{-}R]_n$ are known that have Si, Ge, or Al as a central atom covalently bonded to bridging groups of the polymeric compound. When doped with $FeCl_3$, these compounds have high conductivity of 0.2 to 70 S/cm.[125]

Another type of polymeric Pc is mesogenic Pc.[126,127] Columnar mesophases derived from suitable substituted triphenylene or Pc derivatives are of great interest due to their proven anisotropic electronic conductivity and photoconductivity. In particular, the study of Pc-based columnar mesogens is relevant due to the widespread use of Pc photoconductors and their high electronic conductivity when doped.[128] They combine the advantage of high stability with interesting electrical and optical properties.

Ladder polymers, which include Pc units, are difficult to synthesize. The most suitable method is repetitive Diels–Alder. A convenient macrocyclic metal complex for repetitive Diels–Alder is a hemiporphyrazine system. It is easier to synthesize hemiporphyrazine compounds, which always exhibit D_{2h} symmetry.

Thin films of poly(titanyloxo-phthalocyanine) (pTiOPc), in which phthalocyanini rings are bridged with carbonyl, sulfur dioxide, or alkylene groups, show photocurrents of 2 to 10 :A/cm² upon excitation with a light above 560 nm with an applied potential of 0.6 V (vs. Ag/AgCl). The intensity of the photocurrent was highly dependent on the polymer structure. Thin films of pTiOPc with carbonyl bridging groups showed the highest photocurrent generation, whereas hexafluoroisopropyliden groups showed lower photocurrent.[129] Resulting oligomers must then be converted into an extended π-electron conjugated system to achieve electrical conductivity.[46] Ladder polymers may prove to be very promising.

Dye-sensitized OPVs are very promising solutions to cheap solar cells. Among the dyes, electron-rich Pc systems are the best understood. Their ecological safety, low cost, ease of synthesis, and amazing electrical properties make possible effective use in electronic devices. The commercial potential of phthalocyanine systems has not been fully explored, and it can be expected that electronic properties of devices using these systems will be greatly improved.

9.4   OTHER DYES USED IN PHOTOVOLTAIC ELECTRICAL DEVICES

The photogalvanic effect was studied in cells containing Safranine as photo-sensitizer and EDTA, glucose, and NTA as reductants in different systems. The effects of different reductants on the electric output of the cell were observed and a mechanism for the generation of photocurrent in this kind of photogalvanic cell has been proposed.[130] Photopotential and photocurrent generated were (EDTA) 760 mV and 50 µA, (NTA) 415 mV and 35 µA, and (glucose) 373 mV and µA.[131] The conversion efficiency was 0.26% and the maximum power output of the cell was 38.0 λW.[132]

Mono-N-alkylquinacridone pigments exhibit light bleaching stability, high temperature stability, and high migration fastness, as well as hypsochromic shift fluorescence and high photoconductivity. They are high-molecular-weight organic materials for use as light-emitting materials of organic electrolumenescence elements as well as photoreceptors in electrophotography. Enhancement in power output of photovoltaic cells is reached by mixed dyes of phenazine, thiazine, xanthene, and acridine.[133]

A novel compound, 3-N,N′-dipropyl-1,6,7,12-tetraphenoxy,4,9,10-perylenedi-carboximide (PTC), has been synthesized. This compound is a p-type semiconductor material.[134] Two types of OPVs, (Al/PTC/Ag) and (ITO/PTC/Ag) and a light-emitting diode having a two-layer structure (ITO/PTC/Znq$_2$/Al) were fabricated by vacuum deposition. The first cell exhibited the photovoltaic effect, but the second did not. The LED emitted at $\lambda = 550$ nm, a yellow-green light.

Azo compounds were found to have high photovoltaic efficiency. Nitro- and thio-groups showed electron withdrawing properties, while alkyl, amino, and hydroxy groups showed electron donor properties. The compounds are very stable in air.[135]

Highly stable organic luminescent devices based on vapor-deposited tris(8-hydroxyquinolinato)aluminum (Alq) thin films have been demonstrated. Improvements in stability are achieved with multilayer thin-film structures using a CuPc stabilized hole–injection contact, a hole transport diamine layer using naphthyl-substituted benzidine derivatives, and an alternating current source. These emissive devices showed an operational half-lifetime of 4000 h with an initial luminance of 510 cd/m^2.[136]

Fabrication and performance of photovoltaic cells using organic p-type (pentacene) and n-type (perylene) have been described.[137] Different complex spiro compounds have been found suitable as charge-transfer materials for PV devices.[139-141] Seven analogues of bishemicyanine dye chromophore cations, ionically combined with the bivalent zinc anion bis-(2-thione-1,3-dithiol-4,5-dimercapto)zinc demonstrate that enhancement of the electron-donating ability or electron-accepting ability of hemicyanine dye promotes intramolecular charge-transfer reaction and can improve photocarrier generation performance.[142] The best results were obtained for cells made of crystalline pentacene. The illuminated current-voltage characteristics of such OPVs based on bromine-doped pentacene with total area of 4.6 mm^2 and active area of 4.15 mm^2 had an energy-conversion (AMI.5) efficiency of 4.5%, for a carrier concentration of 10^{15} cm^{-3}. The open-circuit voltage V_{oc} was 900 mV, short-circuit current density I_{sc} was 7.7 mA/cm^2, and the fill factor (FF) was 66%.[143]

The structural, electronic, and spectroscopic properties of polymers of the form $\{[M(dmb)_2]Y\}_n$ (where M = Cu, Ag; dmb is 1,8-diisocyano-p-methane, and Y = BF_4^-, NO_3^-, PF_6^-, or ClO_4^-) were described.[138] The replacement of Y by BF_4(tetra-cyanoquinolinato-methane anion) (TCNQ-) produced electrical insulating materials which, upon doping with neutral TCNQ, become conducting. Photovoltaic cells using these materials have been designed.

Photoenhanced current dependence on temperature and light intensity with pentacene and tetracene thin films was studied.[144] The exciton charge-carrier rate constants depend on temperature.

Oxygen effect on photovoltaic properties of sandwich structure based on pentacene (Pn) films and Pn/CdSSe heterostructures has been investigated. The sensitivity to oxygen of the heterostructure is determined mainly at the $p-n$ interface.[145] Different dyes have been successfully used to selectively modify various photovoltaic parameters.

9.5 FULLERENES

One of the promising ways of fabricating cheaper photovoltaic cells involves the use of fullerenes with conducting organic compounds. Fullerenes are a new allotropic modification of carbon discovered in the mid-1980s. Early in the 1990s, a simple method of producing C_{60} fullerene in gram amounts was developed, thus giving impetus to research on these materials. In 1996, Curl, Smalley, and Kroto were awarded the Nobel Prize for their research of fullerenes.

The C_{60} molecule has a polyhedral form, possesses icosahedral symmetry (group Ih), and consists of 20 six- and five-membered carbon rings. All 60 carbon atoms in the fullerene molecule are equivalent, which is proven by the presence in the ^{13}C NMR spectrum of a single signal. Bonding of carbon atoms in C_{60} is close to sp^2 hybridization. Because of the spherical shape of the molecule, the carbon atoms are not located in the same plane. Their geometric arrangement is pyramidal, which causes significant strain. This changes the nature of the π-orbitals by admixing a σ-component to them. Thus, by altering their electronic properties, the pyramidal structure of fullerenes makes their electron affinity high.

Fullerenes are attractive with regard to their acceptor properties; they can be considered π-acceptors. Significant distinctions from other acceptor systems exist: large size, unique electronic structure, and spherical shape, which yields materials with surprising physical properties.[146] Fullerenes form various noncovalent compounds of the donor–acceptor type that are bonded by relatively weak van der Waals forces or by donor–acceptor charge transfer. Charge transfer gives these compounds new physical and chemical properties.

9.5.1 Conduction in Fullerenes

Overlapping of π-orbitals creates new valence and conduction bands. The energy gap between these bands in C_{60} is estimated at 1.5 to 1.8 eV; therefore, the crystalline fullerene is a semiconductor.[147] Calculation of the charge transfer state for C_{60} indicates a band gap of 3.1 eV.[148] It has been demonstrated that higher temperature and oxygen depletion increase the photocurrent of fullerene OPVs. The photocurrent is increased by an order of magnitude (to 6.4 μA/cm^2 under 100 mW/cm^2 illumination) by an increase of the aqueous cell temperature from 24 to 82°C. Adsoprtion range is outstanding at 720 nm (1.7 eV) to 560 nm (2.2 eV), and a substantial photocurrent response peak occurs at 395 nm (3.1 eV), falling off at shorter wavelengths.[149]

Little is known about the forbidden gap states and the recombination center in C_{60} films. Oxygen centers play a major role in forming the electronic structure of solid C_{60}. These acceptor-like centers compensate the shallow donors and act as recombination centers with levels 0.3 eV below the conduction band edge. Oxygen suppresses phase transition that occurs around 260°C by preventing the expansion of the lattice with increasing temperature.[150]

The electrical properties of C_{60} change as pressure increases from 1.5 to 10 GPa at temperatures between 100 to 1000°C. Pressure-induced polymerization leads to a change in the band structure and electrical conductivity. An increase of temperature at constant pressure increases the degree of polymerization, reduces the band gap, and increases the conductivity. A polymer phase obtained above the transition temperature of C_{60} has metallic conductivity.[151]

At highest temperatures, the mechanism of conductivity of C_{60} and C_{70} films changes from hopping conductivity to the mechanism of thermal activation with activation energies $E_a = 0.3$ and 0.6 eV, respectively. In the whole temperature range, oxygen-exposed C_{60} and C_{70} films exhibit the conductivity of a semiconductor with a constant activation energy $E_a = 0.6$ eV.[152]

9.5.2 Fullerenes for OPVs

Application of C_{60} 100-nm thin film to OPV fabrication has met with success. The active area of these cells was about 1 cm^2. A xenon lamp with about 400 W/m^2 was used as an illumination source; the light was incident on the film side. Open-circuit voltage was $V_{oc} = 200$ mV and short-circuit current was $I_{sc} = 300$ μA; the calculated power efficiency was 2 to 3%. These values correspond to relatively low light intensity. With an illumination source of 1000 W/m^2, the expected values could be as high as 400 to 500 mV for open voltage and about 400 to 600 μA for short-circuit current.

Organic light-emitting diodes (OLEDs) showed I–V curves similar to those of conventional p–n junctions with an electrical breakdown voltage over 50 V. The photogenerated current of the photodiodes was about 20 mA and quantum efficiency was higher than unity. Results obtained from OPVs and OLEDs are very promising for commercialization.[153]

The crystal structure and photovoltaic properties of single crystal C_{60}, polycrystalline film of C_{60}, systems C_{60}/Ag, C_{60}/Si, and C_{60}/n-GaAs were studied.[154-156] Solid C_{60}–n-GaAs heterojunctions have been fabricated by deposition of solid C_{60} film on (100)-oriented epitaxial n-type GaAs substrates and their electric characteristics have been measured. The rectification ratio was greater than 10^6 at a bias of ± 1 V. The current for a fixed forward bias was an exponential function of the reciprocal temperature, from which the effective potential barrier height of the heterojunction, 0.58 eV, was determined. A trap with an energy level 0.35 eV below the conduction band of GaAs at C_{60}–GaAs interface has been observed by a deep level transient spectroscopy technique.[157]

OPVs may have a sandwich structure: the first transparent conductive layer, a second fullerene layer, a third organic p-type semiconductor layer, and a fourth metal electrode whose work function is greater than 4.5 eV. The layers are formed successively on a substate, where at least the second through the fourth layers are formed by vapor deposition at less than 10^{-5} torr vacuum. The organic layer is preferably a porphyn derivative, $(\eta^2\text{-}C_{70})$ $Pd(PPh_3)_2$ complex, and the fullerenes are C_{60} or C_{70}.[158,159]

9.5.3 Dye-Sensitized Fullerene Layers for OPVs

Dye-sensitized fullerene thin films represent a new class of OPV materials. The correlation of microporous fullerene structure with electrical properties provides evidence of the hopping conduction charge-transfer mechanism via extended surface states of interconnected fullerene superclusters.[160]

The influence of the electrode and insulating layers on the photovoltaic effect of a C_{60}–oxotitanium phthalocyanine (Pc) junction was studied. With the $Al/C_{60}/OTiPc/ITO$ configuration, energy-conversion efficiencies (η) are high at low incident light intensities (Pc), due to the stable p–n junction between C_{60} and OTiPc, but drop with increase in light intensity. On the other hand, $ITO/C_{60}/OTiPc/CrAu$ cells are stable even under white-light illumination at high intensity. It was found that the insertion of an SiO_2 insulating layer between OTiPc and CrAu significantly enhances short-circuit current, open-circuit voltage, and efficiency η.[161]

The photovoltaic effect for a Pc complex with transition metals (MPc) doped with C_{60} on GaAs electrode was studied. The C_{60}–MPc adducts were mixed in organic solvents and a new absorption peak appeared in the UV-visible range (700 nm). Doping of MPc with C_{60} led to marked enhancement of photosensitivity and photovoltaic effects; voltage increased three times in comparison to the GaAs electrode alone.[162]

Composite $LiPc$–C_{60} films showed an absorption peak at 1.5 eV. The photocurrent of $CuPc$–C_{60} films showed an obvious phase shift with excitation of 1.6 eV.[163] $Au/H_2Pc/C_{60}/Au$ showed a photovoltaic effect with conversion efficiency of 0.02%.[164]

Photovoltaic properties of ITO (indium tin oxide)/$C_{60}/H_2Pc/Au$ sandwich PV cells were investigated. The cell prepared in a higher vacuum ($1 \cdot 10^{-6}$ torr) showed an open-circuit photovoltage of 0.18 V, a short-circuit photocurrent (I_{sc}) of 89 $\mu A/cm^2$, a fill factor of 0.25, and an energy-conversion yield of 0.03% when illuminated by white light with 12.5 mW/cm^2 intensity. The photocurrent action spectra of the cell revealed that the photocurrent is generated at the C_{60}–H_2Pc interface due to diffusion of C_{60} excitons because the excited state of C_{60} has a relatively long lifetime, while the cell prepared at a lower vacuum (3×10^{-5} torr) showed a smaller photocurrent ($I_{sc} = 1.4$ μA/cm^2). Oxygen in C_{60} acts as a carrier trap and increases the resistance of C_{60}.[165]

Photoconduction was increased by light-induced charge transfer in composites of C_{60} and conductive polymers in the presence of Pc.[166] Another approach to improving the spectral response of donor–acceptor photocells is inserting an excitonic middle layer between the donor–acceptor interface.[167] In the excitonic layer, light absorption produces bound electron–hole pairs (excitons), which migrate toward the interface and dissociate, injecting electrons in the acceptor layer and holes in the donor layer. The authors fabricated three-layered organic photocells utilizing poly (2,5-dioctyloxy-p-phenylenevinylene) (OOPPV) and octaethylporphine (OEP) C_{60}.

The photocurrent spectrum for the photocell illuminated from the Al electrode side is almost the same as that for the photocell with a single OEP thin film. Typical values of the short-circuit current I_{sc} using 600-nm excitation are I_{sc} (ITO side) = 1.4 μA/cm^2 and I_{sc} (Al side) = 1.1 μA/cm^2. Values of the open-circuit voltage are V_{oc} (ITO side) = 0.15V and V_{oc} (Al side) = 0.35 V, which correspond to an efficiency of photoconversion of η = 0.03%, for a typical fill factor of $FF = 0.3$ with radiation intensity of W = 400 μW/cm^2. The current-voltage (I–V) characteristics of this photocell, in the dark and under illumination with 600–nm-wavelength light, was investigated. The I–V characteristics obtained in the dark were almost symmetrical with positive and negative bias conditions, proving that there are only ohmic contacts with the metal electrodes.[167]

9.5.4 Composites of Conducting Polymers with Fullerenes

The combination of conducting polymers and fullerenes makes possible efficient phototransport and separation of charge carriers with long lifetimes. This increases the photoconductivity of the polymer, making it suitable for use in solar energy.[168] Usually the films of polymer–C_{60} composite are prepared by evaporation of the solvent (aromatic hydrocarbon) from a composition comprising a polymer and the fullerene whose content is 1 to 3% of the polymer weight.[169]

Under photoexcitation of the polymer, transfer of an electron in less than 10^{-12} sec from the photoexcited polymer to the C_{60} molecule occurs with formation of a polymer*–C_{60} complex in an excited state. A transition to free separated charge carriers takes place. In these composites, the fullerenes accept electrons under photoexcitation, with formation in the polymer of a vacancy (hole) in the lower free molecular orbital of the (HCMO)–polymer. Second, recombination of the photoexcited electrons with holes is effectively delayed due to their considerable spatial separation when a charge resides on the big fullerene molecule.[170]

Changes in the densities of electronic states of fluorinated fullerenes $C_{60}F_x$ as a function of fluorine content have been investigated by discrete-variation Hartree–Fock–Slater (X_α) MO calculations. The calculated densities of electronic states are in satisfactory agreement with observed electronic spectra. The band gap of $C_{60}F_x$ increases with increasing fluorine content.[171]

Because velocity of electron transfer is high and carrier recombination rate is relatively low, the quantum yield of the charge carrier formation in the presence of fullerenes is considerably higher. An essential drawback of fullerenes is their instability in air. One of the possible ways of solving this problem is to use a ternary system: organic cation or radical–donor such as neutral fullerene or halogenide–acceptor. In this case, a composition with high conductivity can be obtained. A large-sized organic donor sterically hampers the access of oxygen molecules to the anion-radical of the fullerene.

Delocalization of electrons in C_{60} under photoinitiated electron transfer accounts for the photoconductivity with emergence of a free charge carrier. Experimental studies of photoinduced absorption have been carried out on composites of C_{60} with conducting polymers such as poly[2-methoxy-5-(ethlyhexyloxy)-p-phenylene vinylene] (MEH–PPV), in order to understand the quenching of photoluminescence and increase of photoconduction caused by addition of small amounts of C_{60}. To account for the observed photoinduced absorption, it is first necessary to know where the C_{60} molecules sit relative to MEH–PPV chains. Using a Monte Carlo algorithm, the most likely position of C_{60} is in a channel formed 10 Å from the MEH–PPV backbone.[172]

Liess and coworkers found three absorption bands at 0.4, 1.18, and 1.65 eV, which they assigned to C_{60}^{-} ·PPV^{+} complex bond pairs.[173] The homogeneous nanostructured network of C_{60}–MEH–PPV 1:1 composites is a bulk heterojunction material that exhibits photovoltaic efficiency of more than two orders of magnitude higher than neat MEH–PPV devices. Photovoltaic characteristics of the fullerene–conjugated polymer thin films on two different substrates with large active areas were examined.[174] By optimizing the donor-to-acceptor ratio, conversion efficiency of 2.5% and collection efficiency of 26% have been achieved with fullerene OPVs. Under reverse bias, a photosensitivity of close to 0.26 A/W has been achieved at 430 nm.[175] For undoped MEH–PPV, which has an energy gap of about 2.4 eV, the maximum built-in potential is about 2.1 eV, whereas for C_{60}-doped MEH–PPV, the maximum built-in potential drops to 1.5 eV. The authors found that the electron acceptor energy level of C_{60} in MEH–PPV is about 1.7 eV above the hole polaron energy level of MEH–PPV.[176]

Numerous applications of conducting polymers such as photovoltaic cells and superconductors use C_{60} as a molecular dopant.[177] Fullerenes play the role of weak dopants in conducting polymers with highly extended π-electron systems in the main chain. They induce quenching of photoluminescense and enhancement of photoconduction. These results can be explained in terms of the photoinduced charge transfer between the conducting polymer and C_{60}; that is, photoexcited excitons or exciton polarons in the conducting polymer are effectively dissociated at C_{60} molecules. The result of photoexcitation of C_{60} is the transfer of holes from C_{60} to the conducting polymer. These novel C_{60} doping effects have been observed not only in conducting polymers with nondegenerate ground-state structure but in also those with degenerate ground-state structure such as di-substituted acetylene polymers with solitonic–electronic systems.

Highly effective photoinduced charge transfer has also been observed in a conducting polymer–C_{60} heterojunction, which is considered a donor–acceptor photocell. An organic photovoltaic cell, $Al/C_{60}/OEP/conducting\ polymer/ITO$, in which OEP is octaethylporphine as a light-absorbing antenna, was examined.[178] The addition of fullerene to π-conjugated polymers results in an electron transfer process from the conjugated polymer to the fullerene upon illumination. The photoinduced electron transfer in these composites is ultrafast with quantum efficiency approaching unity; the separated charges are metastable.[179]

OPVs with a semiconductor layer containing the fullerene and a conjugated polymer on the semiconductor are flexible and have high conversion efficiency.[154,180] The combination of outstanding properties of fullerenes as acceptors and conducting polymers as excellent donors can be exploited. Polyaniline was chosen due to the nucleophilic character of the NH groups. The investigations clearly showed a strong effect of solvent, i.e., N-methylpyrrolidone (NMP), on the characteristics of the doped films. The nucleophilic addition of NMP to C_{60} suggests complete charge transfer. The small increase of electrical conduction and almost unchanged ESR characteristics are indicative of small charge transfer due to steric hindrance caused by the large size of the fullerene molecules.[181]

Conductive polyurethane elastomers using fullerenols as a cross-linking center have been prepared. The change of the activation energy was correlated with the amount of the connecting polymer chains used in the network preparation.[182] The doping of conducting polymers by sulfonated fullerene derivatives and dendrimers with multiple $-(O)-SO_3H$ groups as proton acid dopants was done for the polyaniline–emeraldine base (PANI–EB). While doping of PANI–EB films with sulfonated dendrimers produced conductivity of up to ~10 S/cm, the hydrogensulfonated fullerenol-doped materials show metallic characteristics with room-temperature conductivities as high as 100 S/cm.[183] A review on the superconducting and magnetic properties of doped fullerenes such as critical fields and their characteristic lengths was made by Buntar and Weber.[184]

On a carbon paste electrode, C_{60} and methyl pyropheophorbide-a (chlorin) were successively cast to prepare a chlorine–fullerene-modified carbon paste electrode. Photocurrents were produced by irradiating with visible light (510 nm) while immersed in an aqueous solution of 0.05 M ethylenediaminetetraacetic acid and 0.1 M Na_2SO_4 at pH 6.7. Larger anodic photocurrents were induced by the chlorine–fullerene-modified electrode than by the carbon paste electrodes modified with the fullerene or the chlorine alone. In addition, the photocurrent was dependent upon the amount of the fullerene; photocurrent action spectra at 300 mV vs. Ag/Ag^+ showed that the photoinduced electron transfer occurred from the excited state of the chlorine to the fullerene or from the chlorine to the photoexcited fullerene. The electron of the fullerene anion radical produced was then shifted to the carbon paste. Upon irradiation with 375-nm lights, the anodic photocurrent was enhanced by increased illuminating light intensity and reached 0.03 mA/cm^2 in the present system.[185]

A living free-radical polymerization process was adopted to synthesize a narrow polydispersity fullerene-end-capped polystyrene (NPFECPS). The NPFECPS can be dissolved in a variety of solvents, such as THF (tetrahydrofuran), toluene, and trichloromethane. Good photoconductivity was obtained.[186] Sexithiophen–C_{60} blends prepared by co-evaporation were proposed. Preliminary results suggest a microphase morphology with domain size comparable to exciton diffusion length. A photovoltaic device with quantum efficiency of 1% has been demonstrated.[187]

9.5.5 Superconductive Systems

The alkali fullerides $(AC_{60})_n$ with A = K, Rb, or Cs are crystalline conducting polymers in which covalently bonded ions form parallel linear chains; their advantage is that the polymers are stable in air. At ambient temperatures, these polymers are conducting. The conduction of $(RbC_{60})_n$ and $(CsC_{60})_n$ slowly decreases with decreasing temperature. The $(KC_{60})_n$ remains metallic to low temperatures, while $(RbC_{60})_n$ has a metal insulator transition below 50 K.[188]

9.5.6 Methods of Producing Fullerenes

The USDOE National Renewable Energy Lab (NREL) has designed a solar concentration system that can be used to produce fullerenes.[189] (A detailed description of the syntheses of fullerenes has been given in recent French papers.[190]) Fullerenes can be produced by irradiation with high-energy beams in the presence of reactive metals on amorphous carbon in vacuum. Alternatively, fullerenes are produced by electron-beam irradiation of metal-oxide particles on amorphous carbon in vacuum. Other preparative methods for syntheses of C_{60}/C_{70} include cracking of naphthalene and organic synthesis.[191]

C_{60}, C_{70}, and other fullerenes are now produced in macroscopic quantities by an arc method. The mechanism of their formation is still unclear: polyhedral clusterization and the formation of fullerene structures are not well understood. Polymerization of fullerenes retaining the elements of their spheroidal framework (by stitching) is attracting interest because of the high electronic capacity projected for these polymers. The latter can make fullerenes useful in current sources and batteries; this is confirmed by widespread interest in donor–acceptor systems with conducting polymers and fullerenes as acceptors.[192]

9.6 CONDUCTING POLYMERS

Polymers based on conjugated and heteroaromatic systems comprise a huge class of materials that have received considerable attention due to their interesting electrical and electrochemical properties. Conducting polymers have seen application in photovoltaic cells, batteries, and active materials for electrochromic and light-emitting displays. The purpose of this literature survey of conductive polymers is to identify the main classes of conducting polymers and to estimate the relative value of their development for photovoltaic devices, relying on concepts of conduction mechanisms and physicochemical properties of polymer structures.

9.6.1 Conduction Mechanisms in Polymers

Conducting polymers are usually defined as long polymer chains with continuous conjugation along the chain. Significant organic polymer materials include poly(acetylene) poly(pyrrol), poly(thiophene), poly(aniline), poly-(phenylene-vinylene), poly(diacetylene), and other molecules. Electrical and physical characteristics of these materials are of the most interest when doped with donors or acceptors of electrons. High electrical conductivity of the doped polymers (at the level of ordinary metals) and unique electro-optical properties make these substances very promising for use in various fields of technology.

Real polymer chains are always split by defects into conjugated fragments, which can be very long but are always of finite length. Defects, such as methylene bridges and ring rotations interfering with conjugation along the chain, arise in the synthesis of a polymer material and their concentration depends on this process. The conducting polymer is, therefore, an aggregate of conjugated fragments of different and finite lengths. Successful synthesis resulting in fairly close packing and parallel orientation of the polymer chains and conjugated fragments yields a conducting polymer representing an anisotropic molecular crystal consisting of long fragments — molecules of different length.

The key theoretical problem in understanding the metallic transport and optical properties of conducting polymers is that their structural units are not metallic. As a result of doping, Coulomb impurity centers arise in the polymer. Merging shells of impurity centers into one large cluster filling the entire sample volume proceeds especially fast at the dopant concentration of 10^{-3} cm^{-1}. Above this dopant concentration, a drastic drop in the concentration of unpaired spins is observed because their absorption by the growing cluster causes pairing of the spins. The appearance of quasi-metallic regions with delocalized electrons, growth of these regions, and their merger into a large cluster with growing dopant concentration explain the disappearance of Curie paramagnetism and the abrupt rise in electrical conductivity.

In interpreting photophysical properties of conducting polymers, two important issues should be taken into consideration: first, maximum in the absorption spectra of all conducting polymers occurs at an energy of $E \approx 2$ eV due to excitation of electrons in a long conjugated fragment of the polymer chain. The formation energy E (+,–) of two charged fragments in an undoped conducting polymer is equal to about 2 eV.

The approximate equality $E_{max} \approx E(+,-)$ provides a natural explanation of the emergence of charged particles in a conducting polymer after absorption of just one quantum of 2 eV of energy. This equality also explains the subsequent creation of a localized exciton with charge transfer, the excited state of which mixes, with energy of 2 eV and according to quantum mechanical laws, with states in which an electron of the same fragment has been transferred to a neighboring fragment. The creation of the localized exciton changes the shape of the absorption spectrum compared with the spectrum of an isolated fragment and provides a clue to understanding kinetic and spectral features of the photoexcitation spectra of the conducting polymers.

Charge transfer between two cation fragments becomes very small and the material acquires metallic properties. This situation arises in polyaniline in the form of an emeraldine salt, where heavily charged fragments of polymer chains (chlorine anions) are present and metallic behavior is observed because of electron hopping between fragments. A similar situation arises in quasi-one-dimensional tetrathioful-valene(tetracyanoquinone)dimethane crystal (TTF-TCNQ), where positively charged stacks are split into fragments and the charge transfer is known to be related to the TTF stacks. The current is due to hopping of electrons between positively charged fragments of the TTF stacks.[193]

The nature of the doping process that induces the high conductivity in conjugated polymers is especially clear from consideration of the charge carriers (solitons, polarons, and bipolarons) produced. Dependence of electroconductivity of these polymers on various factors like doping level, conjugation length, temperature, and frequency has been studied.[194]

The transport properties of conducting polymers as a function of structure or microstructure were studied.[195] In strongly disordered conducting polymers, electron transport is dominated by hopping between conducting grains separated by insulating barriers. Although the nature of the metal–insulator transition in weakly disordered conducting polymers is still a controversial topic, several results indicate that heterogeneities play an important role. Thus, heterogeneous disorder seems to control the conductivity of most conducting polymers.

Competition between polaron and bipolaron in conjugated polymers with nondegenerate ground states was systematically studied in the extended Hubbard–Peierls model with the symmetry-breaking Brazovskii–Kirova term, using the matrix renormalization group method combined with lattice optimization in the adiabatic approximation. The relative stability of a bipolaron over two separated polarons depends significantly on on-site Hubbard U and the nearest-neighbor repulsion V. When U is much larger than V, the bipolaron state is more stable compared with mean field calculations.

The first conductive polymers appeared in the 1960s. These were poly(acetylenes) and poly(p-phenylenes). Initially the samples of these polymers possessed low electrical conductivity and chemostability to oxygen and other reagents. However, steady improvement is still being realized.[196,197]

9.6.2 Poly(acetylene)

Poly(acetylene) has been at center stage throughout the evolution of this field. Undoped electrical conductivity (10^{-5} S/cm) can be increased with doping to values comparable to those of copper and silver metal (10^5 S/cm). However, because this polymer is insoluble in any solvents, infusible, and unstable under ambient conditions, there are difficulties in characterizing by microstructure or morphology and in processing.

The most widely used procedure for polyacetylene synthesis is addition of polymerization of acetylene using soluble Ziegler–Natta catalysts with $Ti(OC_4H_9)_4$ and $(C_2H_5)_3Al$ co-catalysts. However, its polymerization activity is sensitive to catalyst concentration, Al/Ti ratio, solvent, aging conditions, acetylene pressure, and polymerization temperature, giving rise to different molecular weights, microstructure, morphology, defects, etc.[198] Substituted poly(acetylenes) with phenylene-yl-silylene side groups $-(C_6H_4\text{-}C\equiv C)_n Si\text{-}iPr_3$ ($n = 1, 2$) are described.[199] Shirakawa et al.[200] demonstrated that the conductivity of poly(acetylene) could be increased on doping with various electron acceptors (more than seven orders of magnitude in the case of I_2) or electron donors. Poly(acetylene), with the simplest conjugated structure, has served as the prototype for other conducting polymers.[201]

Although the simplified band theory seems to be useful for understanding changes in the electronic structure induced by the doping process, some conjugated polymers, such as poly(aniline), can also acquire high conductivities through the protonation of imine nitrogen atoms without any electron transfer between the polymer and dopants.

Most conducting polymers are unstable in air, insoluble, infusible, and brittle, so a number of techniques have been developed to overcome these problems. Relatively stable conducting polymers have been made through physically blending certain conjugated polymers with appropriate conventional macromolecules.[202,203] Synthesis of conjugated oligomers, such as sexithiophene,[204] and poly(aniline) oligomers,[205] has been described. Use of precursors is often preferred for device fabrication so that the conjugated polymer can be generated as a thin layer directly on the electrode.

9.6.3 Poly(p-phenylene) and Poly(p-phenylene vinylene)

Poly(p-phenylenes) attract much attention. This section aims to give examples of their application in various types of up-to-date PV devices. Preparation of highly conjugated poly(p-phenylene vinylene) via a halogen precursor route is described.[206] It was found that, when the surface of a nanocrystal is treated to remove the surface ligands, the polymer photoluminescence is quenched consistently, with rapid charge separation at the polymer/nanocrystal interface.

A simple photovoltaic device based on composite materials formed by mixing CdSe or CdS nanocrystals with the conjugated polymer poly(2-methoxy-5-(2'-ethyl)hexyloxy-p-phenylene vinylene (MEH–PPV) is known; its quantum efficiency was 12%.[207] Such devices are important light-emitting sources. The performance of a sandwich junction device based on doped 1,4-bis(2-ethylhexyloxy)poly-phenylene-vinylene (BEH-PPV) was described.[208] The power-conversion efficiency is dependent on the doping level, external load, and work function of the cathode. The increase in power-conversion efficiency ($\eta e = 0.8\%$) is due to efficient charge separation resulting from photoinduced electron transfer from the BEH–PPV donor to a perylenedicarboximide acceptor.

Synthesis and characterization of water-soluble poly(p-phenylene) (PPP) and preparation of self-assembled multilayer films were described.[209] Conducting (RuO_2 and ITO), semiconducting (Si wafer), and nonconducting (SiO_2) substrates were used in the preparation of self-assembled multilayers.

Pt foil electrodes coated with a conducting polymer were used as working electrodes to deposit a second conducting polymer layer by electrolysis. Poly(aniline), poly(vinylcarbazole), poly(pyrrole), and poly(pyrrole)/poly(aniline) films were synthesized on Pt foil electrodes by sequential electrolysis and the Raman spectra studied.[210]

In-situ external reflection FTIR measurements have been performed during cyclic voltametric polymerization of poly(p-phenylene) (PPP) film. The film was made in either 0.05 or 0.2 mol·dm^{-3} biphenyl solutions using acetonitrile as a solvent and tetrabuthylammonium tetrafluoroborate as the electrolyte salt.[211]

Synthetic poly(arylenevinylenes) are available commercially. They are obtained by ion irradiation of a corresponding polyelectrolyte precursor comprising sulfoniums. Pattern formation was achieved by patterned plasma polymerization of acetic acid onto perfluorinated ethylene–propylene copolymer (FEP) films, followed by selective adsorption of oligo(ethyleneoxide)–PPV from chloroform solutions.[212]

Preparation of conducting films of Ppy–PP oxides has been described.[213] Poly(2-benzoyl-1,4-phenylene) (PBP) and poly(2,5-dibenzoyl-1,4-phenylene) (PDBP) were synthesized by copolymerization of 2,5-dichlorobenzophenone and 2,5-dichloro-1,4-dibenzophenone using an Ni catalyst. These polymers are soluble in common organic solvents, have high thermal stability, and show bright blue photoluminescence. Light-emitting diodes fabricated with these polymers as the active layer emit blue electroluminescence with wavelength of 433 nm for PBP and 475 nm for PDBP.[214] The influence of molecular morphology of organic photovoltaic devices on power-conversion efficiency was shown.[215]

Single-layer light-emitting diodes (LEDs) were fabricated with poly(p-phenylene) derivatives as the emissive layer. All organic layers for the LEDs were deposited by spin coating from solution. The electrodes (Ca/Al) were deposited on top of the organic layers by thermal evaporation under a vacuum of $5 \cdot 10^{-7}$ mm Hg. The turn-on voltages for the LEDs (thickness ~80 nm) were about 12 and 20 V for PBP and PDBP devices, respectively.[216] Blue LEDs have been successfully fabricated utilizing soluble and thermally stable polymers PBP and PDBP.

OPVs made from spin-coated bilayer thin-film heterojunctions of poly(p-phenylene vinylene) and poly(benzimidazobenzophenanthroline ladder) showed power-conversion efficiency from 1.4% under sunlight illumination to 2% at peak wavelength.[217]

9.6.4 Poly(aniline)

Poly(aniline) (PANI) is an excellent example of a conjugated polymer that can be used for photovoltaic applications through the doping process. Polyaniline's conductivity increases with doping from the undoped insulating base form (N ≤ 10^{-10} S/cm) to the fully doped, conducting acid form (N ≥ 10^{10} S/cm).[218]

Depending on the preparation method, the properties of the PANI film can vary over a wide range (Table 9.2). This is explained by the fact that synthesis of PANI under any preparation procedure proceeds through a number of oxidation stages of aniline: formation of indamine from two aniline molecules, its conjugation into 4- and then 8-nucleus indamine with one quinoid ring (violet proto-emeraldine), oxidation in the latter of one more ring into a quinoid ring (blue emeraldine), oxidation of the third ring (dark-blue nigrosine) and the fourth ring (virescent black pernigraniline), and, ultimately, attachment of three aniline molecules (nonvirescent Aniline black).[219]

The polyamine film was synthesized by electrochemical method from aniline as monomer, a perchloric acid as dopant in aqueous solution, and stainless steel as electrode. The film had electrical conductivity of 6.67 S/cm, tensile strength of 0.34 N/mm^2, and elongation at break of 12.6%.[244]

Conducting poly(aniline) (PANI) films have been prepared via a new route comprising 2-acrylamido-2-methyl-1-propanesulfonic acid (AMPSA) as both protonating acid and solvating group and with dichloroacetic acid (DCA) as a solvent. The room temperature conduction along the stretch direction was increased to a maximum value of 670 ± 55 S cm^{-1} for a film drawn at 363 K, compared to 210 ± 20 S cm^{-1} for an undrawn sample.[245] Poly(aniline) (PANI) doped with dodecylbenzenesufonic acid (DBSA) is a new conducting polymer complex that combines high electric conductivity with processability. Molar ratio DBSA to PANI varies from 0.55 to 1.12. Conductivities on the order of 1 S/cm were achieved.[223]

By using dodecylbenzene sulfonic acid and water, polymer emulsion of poly(aniline) was carried out. By matching the poly(aniline) with electrical conductivity (0.01 to 10 S/cm) and absorption spectrum (200 to 500 nm and 800 to 1000 nm, respectively) with an n-semiconductor dye (absorption spectrum of 500 to 800 nm), high-performance photoelectric cells were fabricated. A p-poly(aniline-perylene) OPV had a short-circuit current of 10 μA/cm^2 and open-circuit voltage of 400 mV.[246]

An LED with emeraldine base polyaniline (PAn) as the emitting layer, indium-tin oxide-coated glass plate as the hole injector, and deposited aluminum (or magnesium) thin film as the electron injector was fabricated that can emit nearly white light covering the full range of visible light (380 to 750 nm).[247] It has been found that the white light is emitted from the phase with reduced repeat units (amine form), while the phase with oxidized repeat units (quinoid form) does not emit light. The turn-on voltage for eye-observable light intensity is 13 and 6 V for the LEDs with aluminum and magnesium electrodes, respectively, at a thickness of the emitting layer of 800 Å. The electroluminescence (EL) spectrum covering the range of 300 to 800 nm is much broader than that of the photoluminescence (PL) spectrum, 350 to 510 nm. LEDs were prepared by coating a thin layer (about 800 Å) of PANI (or substituted PANI) on an ITO glass plate through dipping in its solution with NMP, drying in a dust-free environment, and then placing under dynamic vacuum at a pressure below 10^{-1} torr to remove residual NMP. Finally, a layer of aluminum (or magnesium) was deposited on the surface of the PAn (or substituted PANI) film by thermal evaporation at a pressure of 10^{-5} torr.

These LEDs exhibit a lower turn-on voltage of 6 V and brightness of 0.3 cd/m^2 at 10 V. They also have a higher current density (820 mA/cm^2 at 8 V) than LEDs with poly(phenylene vinylenes) (39 mA/cm^2 at 8 V, film thickness about 1200 Å) and poly(thiophenes) (420 mA/cm^2 at 8 V, film thickness about 1000 Å) as the emitting layers, but brightness is much lower.[247]

The use of sulfonated fullerene derivatives and dendrimers with multiple sulfate groups as protonic acid dopants for polyaniline emeraldine base (PANI–EB) was described.[248] While doping of PANI–EB films with sulfonated dendrimers gives conductivities up to ca. 10 S/cm, hydrogensulfated fullerenol-doped materials show metallic characteristics with room-temperature conductivities as high as 100 S/cm — about six orders of magnitude higher than the typical value for fullerene-doped conducting polymers.

Photoinduced charge-transfer effects characteristic of C$_{60}$ fullerene and potential biomedical applications of dendrimers could make these highly conducting materials

Table 9.2   Poly(anilines)

Polymer	Dopant	S/cm	Ref.
PANI	Particular substituted compounds that simultaneously form H-bond and ring–ring interactions with NH-groups and 6-membered rings of conducting PANI		220
PANI prepared by oxidation polymerization in the presence $(NH_4)_2S_2O_8$	HCl, MeSO$_3$H	1.0	221
PANI prepared by mixing a solution of host polymer with colloidal PANI disperse Copolymer PANI – poly(vinyl alc)	Dodecyl benzene sulfonic acid in. Aq. Medium	2 – 3 at I = 50 :m	222
PANI	Dodecul benzene sulfonic acid, Molar ratio DBSA: PANI from 0.55 to 1.12	1.0	223
Oligomer: acetylene-terminated polyaniline	Bronsted acids	0.01	224
Poly[4-(2-thienyl)benzamine prepared] using FeCl$_3$ as oxidant at low t° (Emeraldine)	I$_2$	$5 \cdot 10^{-3}$	225
PANI Emeraldine +	Theoretical study of p, n-dopants		226
PANI	Lactone	Enhanced conductivity	227
PANI, Comprising dispersant	The dispersion is a key for making conducting polymers suitable for industrial application	—	228
PANI	Fluorinated alcohols 1,3,3,3-hexa-fluoro-2-propanol	Metallic type of conductivity to 200K	229
PANI + incorporated Me methacrylate		10^{-4}	230
PANI[a]			231
PANI	Methan sulfoacid[b]	$2 \cdot 10^{-3}$	232
PANI	Poly(methylmethacrylate)-co-3-(tri-methoxysilylpropylmeth-acryate) + camphosulfonic acid	High + good adhesion	233
PANI (emeraldine salts)			234
PANI (dihalogenated)	Introduction halogens in PANI are analogous to doping	The same as in doped PANI	235
PANI	H_2SO_4, HhlO$_4$, HNO$_3$	XANES investigation	236
PANI	CH$_3$SO$_3$H		237
PANI	Phenylphosphonic acid + plasticizer (dimethylphthate or dibuthylphthalate)	Good electrical properties	238
PANI	I$_2$	$2.64 \cdot 10^{-2}$	239

Table 9.2 Poly(anilines) (Continued)

Polymer	Dopant	S/cm	Ref.
PANI, (o-Me, o-MeO)	H_2SO_4, HCl, $HClO_4$		240
Copolymers aniline with poly(ethylene oxide)			241
PANI	Protonic acids with hydrophylic ethyleneoxide oligomer	10^{-3} to 10^{-2}	242
PANI prepared by emulsion polymerization	Dodecyl benzensulfoacid The cell Al/(PS-porous silicon)PANI(dodecylben-solsulfoacid)/Au	The characteristics are analogous to the doped PANI	243

[a] Ultra-thin vapor-deposited emeraldine exhibits conductivities several orders of magnitude higher than those of HCl-doped emeraldine films produced by wet-chemical deposition techniques.
[b] Vacuum vapor deposited.

useful in many advanced optoelectronic or biomedical devices. Emeraldine base is used as a primary and secondary dopant for poly(aniline). The self-secondary-doping effect was shown to induce significant enhancement in electrical properties. Direct-current conductivities of up to 100 S/cm, with a metallic conduction mechanism, have been obtained for poly(aniline) films doped by the sulfonated C_{60}.[248]

Thus, it should be emphasized that poly(aniline) (PANI) is the conducting polymer that has the best potential for industrial applications. The possibilities of the industrial transfer were assessed based on specific applications: high voltage, microwave absorption, distributed filtering, and electromagnetic shielding. The relative ease of synthesis, stable conductivity in air, and low cost make PANI of particular industrial interest.

9.6.5 Poly(pyrrole)

Poly(pyrroles) (PPy) are obtained by polymerization of pyrrole. Of all possible five-membered aromatic heterocycles, pyrrole is the primary biochemical building block for chlorophylls and hemoglobins as well as other vital supramolecular structures such as porphyrins, pródignosines, and vitamin Y_{12}. The parent polyconjugated polymer has conductivity from 10^{-10} to 10^{-5} S/cm; doping leads to metallic or semiconductor conductivity from 1 to 10^5 S/cm. Among different conducting polymers, PPy and their derivatives are of special interest because of their high conductivity, stability in the oxidized state, and interesting redox properties. They are also attractive because of simplicity and availability of the starting monomers — pyrroles — and can be produced by chemical or electrochemical methods.

PPy in the form of black powder are obtained by oxidizing the monomer in the presence of oxidants ($FeCl_3$, aqueous or anhydrous, other Fe^{+3} and Cu^{+2} salts, organic electron acceptors). When oxidized in the presence of $FeCl_3$, PPy are found doped with Cl^- ions.[249,250]

The highest conductivity of 190 S/cm is shown by PPy polymerized from methanol. By varying the ratio of $FeCl_3$ and $FeCl_2$[250] or making use of a binary solvent of acetonitrile with methanol,[251] conductivity can be increased to 220 S/cm in the first instance and to 328 S/cm in the second. Pyrrole slowly polymerizes in acetonitrile in the presence of $FeCl_3$ to yield a polymer of small particle size. The addition of water in trace amounts increases the polymerization rate about ten times and results in polypyrrole with larger particle size and higher concentration.[252]

The problem of processing PPy powder into a film was solved by deposition of PPy from the vapor phase by using $FeCl_3$ as an oxidant.[253] Using the method of chemical-oxidizing polymerization, a technique of layer-by-layer deposition of PPy and PANI films has been developed.[254] Polymers prepared in this way have a more ordered structure and higher crystallinity than polymers synthesized by electrochemical methods. A number of films are obtained by polymerization of complexes of pyrrole or N-methyl pyrrole with unsaturated carboxylic acids, e.g., polyacrylic or polymethacrylic acid on nonconducting substrates. The substrates are coated with aqueous or aqueous-alcohol solutions of a complex of pyrrole with unsaturated acid and then treated by oxidation agents, e.g., $(NH_4)_2S_2O_8$, and polymerized.[255]

The advantage of electrochemical polymerization of conducting polymers is the formation of an electrically active film on the electrode surface. Yield is close to 100%, giving films of specific weight and thickness. In 1979 Diaz and coworkers prepared flexible films with conductivity of 100 S/cm.[256] Electrochemical polymerization of PPy can be carried out in water and anhydrous media such as acetonitrile, propylene carbonate, and dichloromethane.[257-262]

Spectral characteristics of PPy films of thickness up to 1 μm depend on synthesis conditions and oxidation level. At higher oxidation levels, film color changes from yellow to blue and then black. Thin films have good adhesion to the substrate but at thickness above 10 μm come off the substrate rather easily.

The stability of PPy films is quite high. They are destroyed at temperatures above 150 to 350°C (depending on the anion-dopant). The thermal destruction usually starts from the anion.[263] Deprotonation is accompanied with formation in the polymer of an imine-like structure. Thermal degradation is attributed to release of HCl from the samples, which reduces the protonated conducting phase.

PPy films prepared by electrodeposition in the presence of inorganic anions are rather brittle. Mechanical properties of films with large organic anions are much better. It should be noted that PPy films are dense, have no voids, and have a microstructure that depends on the size of the anion-dopant. Films produced from aqueous solutions are more porous compared to those produced from acetonitrile.

Studies of the mechanism of electron conduction in PPy have shown that neutral undoped PPy is a dielectric with a band-gap of 4 eV. Oxidation shifts the band edges and band-gap width drops to 2.5 eV; i.e., PPy become semiconductors. Polarons and bipolarons usually straddle 3 to 4 links of the PPy polymer chain.[264] The correlation of parameters calculated from direct current and ESR supports the existence of Mott's variable range-hopping mechanism in poly(pyrrole).[265]

The choice of anion-dopant is determined by the required properties of the polymer. With hydrophobic anions, the sensitivity of PPy to moisture is reduced. With optically active anions such as (±)camphorsulfonic acid, an optically active polymer is obtained. To obtain flexible and smooth films, organic anions are used. Aromatic and surface-active sulfonate dopants improve conductivity and stability of a polymer.[264] Conductivities of PPy and their derivatives with different anions and

Table 9.3 Conductivity of PPy and PPy with Dopants

Compound	Dopant	Conductivity S/cm	Notes	Ref.
Copolymer PPy-cis-1,4-Polybutadiene composites		10^{-10} to 10^{-6}		271
Ppy in PMMA Matrix	$FeCl_3$	327	PPy (40%)	272
PPy	$(Pmo_{12}O_{40})$	=10		273
PE(poly)ethylene/PPy		10^{-11} to 1.0	PPy (0.25 to 17%)	274
Polyimid/PPy		1 to 10		275
PPy	Heavily doped by different dopants	10^{-2} to 45		276
PPy+polystyrene(porous)	$FeCl_3$	0.82		277
Polythiophene	Undoped	$2.3 \cdot 10^{-3}$	Band gap	278
PPy	I_2	$1.28 \cdot 10^{-1}$	2.4 eV	

additions of polyurethanes are in the range of 10^{-8} to 10^{-1} S/cm.[266-270] Selected data on the conductivity of PPy are given in Table 9.3.

Electrochemical oxidation of NiL_2 [LH = 2-(3-pyrrole-1-yl-propylimino-methyl) phenol] in MeCN produced a conducting film at the electrode surface.[279] Pyrrole was grafted on poly[(methylmethacrylate)-co-(2-N-pyrrolyl)ethylmethacrylate] [PMAA-co-PEMA] using potential electrodes. The thermal stability of PMAA-co-PEMA was improved as a result of electrochemical grafting with pyrrole.[280]

Among all the properties and applications, Ppy are especially important as electrode coatings. On inorganic semiconductors, oxidized electrodes are inoperative. PPy films covering the electrodes prevent passivation of their surfaces; simultaneously, holes that extend through the polymer develop in the film so that the electrochemical reaction takes place on the polymer surface. Systems of this sort have been designed with GaAs,[281] CdS, CdSe,[282] and PV cells with solid polyelectrolytes.[283]

9.6.6 Poly(thiophene)

Poly(thiophenes) (PTh) and their derivatives are narrow-band-gap-conducting polymers. It was believed that these materials might be inherent electronic conductors, that is, that they might be conducting without the need for doping, similar to metals with overlapping bands. If these materials were indeed like metals, they might also show the high electrical conductivity of metals. Upon doping, for example, by way of oxidation, new low-energy transitions were observed due to generation of bipolarons, which gave rise to two new states within the gap, while transitions involving these energy states increased in intensity.[284]

The first of the narrow-band-gap polymers to be prepared was poly(benzo-[c]thiophene), a blue-black insoluble material. Its band gap, with the edge at ~1100 nm, is about 1.0 to 1.2 eV.[284] When doped with chlorine ions, it has a conductivity of 2 to 7 S/cm; doping of film grown on four narrow Au electrodes with I_2 or perchloric acid gave conductivities of about 50 S/cm. Synthesized polymers with three condensed thiophene rings have a band gap of 0.58 eV; the corresponding pyrrole derivatives have a narrower band gap of 0.38 eV.[285]

Like poly(pyrroles), poly(thiophenes) are synthesized electrochemically.[284,286-291] A promising synthesis of conjugated donor–acceptor passivator polymers uses sequential electron-rich N,N′-dimethyl-3,4-diaminothiophene, electron-deficient 3,4-dinitrothiophene, and a repeat phenylene unit.[292]

Table 9.4 provides data on conductivity of some derivatives of poly(thiophenes). Attempts were made to obtain very narrow-band-gap-conducting polymers based on thiophene derivatives; the narrowest band gaps reported were about 0.5 eV. Scientists have also tried to make these polymers as transparent and colorless as possible in the conducting state. A recurring problem with many of the derivatives of poly-thiophenes is their poor stability, either doped or undoped, in ambient air. These polymers are not stable enough for long-term applications.

The electric characteristics of the junction obtained by depositing poly(4,4′-dipentoxy-2,2′-bithiophene) on n-type Si have been investigated in the dark and under white-light illumination. The dark current-voltage characteristics and imped-ance spectra suggest that the current is space-charge limited at forward bias > 0.2 V, whereas it has an exponential trend for very low forward voltages. A Schottky barrier formation at the polymer–Si interface has been demonstrated. The barrier height values obtained from the Mott-Schottky plot (0.80 eV) and from the dependence of open-circuit voltage on short-circuit current (0.77 eV) are in agreement with the polymer redox potential. The short-circuit current is a linear function of the incident light intensity.[301]

An OLED was constructed with an aluminum electrode as an electron collector and a poly(3,4-ethylenedioxythiophene)-polystyrenesulfonate)/ITO electrode as hole collector.[302] A hybrid organic–inorganic photovoltaic used a thin-film CdSe/poly(3-methylthiophene) junction fabricated by electrodeposition of poly(3-methyl-thiophene) (PMeTh) on polycrystalline CdSe layers deposited by chemical bath deposition (CBD) onto indium tin oxide (ITO) glass. Electrochemical pretreatment

Table 9.4 Conductivity of Undoped and Doped PTh

Compound	Dopant	Conductivity, S/cm	Notes	Ref.
Poly(ω-BrAlk)Th	I_2	1 to 8		293
	$FeCl_3$	7.5		
Poly(Th-1,3-dithione)	—	10^{-4}		294
	$NOBF_4$	10^{-1}		
Poly(4-dicyanomethylene-4,4-cyclopenta[2,1-b:3,4-b′]dith)	—	10^{-8}	Band gap 0.8 eV	295
Poly(3-methyl)Th			Electrodeposited on CdSe 5.6 mW/cm² max conversion efficiency	296, 297
Hydroisothianaphthalene	—	2.0		298
PmeTh/CIS, PmeTh/CdTe			η ~ 1%	299
Poly(3,4-ethylenedioxy)Th		10^{-4}		300

consisted of polarizing the ITO at a potential below the potential range of the Cd^+/Cd couple. The nonoptimized junction exhibited a photovoltaic energy-conversion efficiency of 1.3% under 56 mW/cm^2 white light from a xenon lamp.[303]

A novel family of thermally stable electron-transporting amorphous bithiophene or terthiophenes with dimethylsilylboryl substituents has been reported.[304] 3-(4-trimethylammonium phenyl)thiophene trifluoromethan sulfonate, bithiophene, and terthiophene analogs have attached anion exchange centers. Fast ion transport is demonstrated from the linear relationship between the scan rate and peak current. Polymer doping (both p- and n-) is accompanied solely by the movement of anions, which has implications for their use in electrochemical capacitors where fast ion transport is required for devices with high power.[305]

Charge transport in thin films of poly(phenylazomethinthiophene) (PPATh) using sandwich-type Al/PPATh/ITO and In/PPATh/ITO is explained in terms of the p-type semiconducting behavior of the PPATh film and by formation of a Schottky barrier with an In or Al electrode and an ohmic contact with ITO.[306] The electrical conductivity of such undoped polymers was less than 10^{-8} S/cm, while in doped polymers it was 10^{-7} to 10^{-6} S/cm. The low conductivity is attributed to the low degree of conjugation in the polymers caused by nonplanarity of the polymer chains. The thiophene-containing polymers lost sulfur moieties during polymerization, resulting in an interruption of conjugation in the polymer backbone.[307]

Conducting polymers with alternating 3,4-ethylene-dioxy-thiophene repeat units have low oxidation potential.[308] A series of electrically conducting polymers containing thionyl groups with very narrow band gap is reported.[309,310] Transport properties of the junction ITO/poly(4,4'-dipentoxy-2,2'-bithiophene)/aluminum were investigated in the dark and under white-light illumination for different values of the doping level of the polymer. Higher short-circuit photocurrents were achieved with less-oxidized polymer.[313]

The electrochemical polymerization of 3-alkylselenophenes[311] and bitelluro-phene[312] gave materials with conductivities of 10 and 10^{-6} to 10^{-9} S/cm, respectively.

9.6.7 Other Conducting Polymers

Chemical oxidative copolymerization of 1-naphthylamine with aniline, o-anisidine, and o-toluidine, as well as that of o-anisidine with N-methylaniline or N-ethylaniline with Fenton's reagent, yields copolymers with an electrical conductivity of 10^{-5} to 10^{-3} S/cm.[314] Cyclopolymerization of diethyl dipropargyl malonate and triethyl dipropargyl phosphonoacetate by $MoCl_5$ yielded conducting polymers with a conductivity of $2.5 \cdot 10^{-2}$ and $7.5 \cdot 10^{-2}$ Ω/cm when doped with I_2.[315]

The reaction of vinamidinium salts with p-substituted aromatic diamines gives rise to polymeric 2-arylmalonaldehyde dianils in good yields. The electric conductivity of these polymers after doping with I_2 or $FeCl_3$ is 0.03 to 50 S/cm.[316] This polymer type is expected to be stable in air, but commercial production would be expensive. An interesting type of polymer is poly(ferrocenophane); conductivity is from $5 \cdot 10^{-5}$ to $7.6 \cdot 10^{-4}$ S/cm after I_2 doping.[317] Poly(vinylpyrrolidones)[318] and Sb-containing bis-azomethines[319] are being studied.

9.6.8 Conducting Polymer Blends

Higher conductivity and stability of conducting polymers can be achieved not only by doping but also by preparing conducting blends, composites, and pastes. Depending on the basic material, bulk electrical resistance can be from 1 to 10^{-3} Ω/cm. These compositions are distinguished by low labor content, and the circuitry is simple and reliable. Examples of such materials are given next.

A polymer blend generally includes a matrix material, selected from thermoplastic polymers, monomers, polymer precursors, and their combinations, and nonpolymeric highly conducting additives dispersed in the polymer blend and having a conductivity greater than that of the blend matrix. This results in a blend having a conductivity greater than 2 to 5 S/cm.[320] Very often, siloxane polymers mixed with carbon black [e.g., acetylene black, lamp black, ISAF(N-234 black)] are used in conducting blends.[321-327] The electrical conductivity of such systems is explained by the formation of carbon black chain structures, which is confirmed by electron microscopic investigations.

Also, such compositions incorporating microflakes[328] or Zr compounds have been described.[329] The polymers obtained have an initial resistance of 0.52 and 0.84 Ω/cm at 25 and 150°C, respectively, or an initial resistance of $4.1 \cdot 10^{-4}$ Ω/cm and $4.3 \cdot 10^{-4}$ Ω/cm after shock thermal cycles.[328] Thermally conducting storage-stable polymer compositions contain metal powder having a surface layer of oxides or nitrides, preferably Al oxide or Al nitride[330]; often conducting paste contains epoxy resins.[331-334] Three classes of fillers are used in conducting polymer blends: polypyrrole, carbon black, and silver-coated copper.[335,336]

Conducting blends and conducting pastes provide vacuum-tight connections. The blends are fairly cheap if produced on a commercial scale. Protrusions from Si-organic rubbers with carbon black provide resistance between contact pads of 200 to 5000 Ω. The most promising with regard to long-term stable operation are conducting pastes incorporating epoxy resins in parts where high strength is important and siloxane rubber where elasticity is required.

9.7 DESIGN AND ENCAPSULATION OF PV DEVICES

From consideration of the major processes taking place in a PV cell where the light energy is converted into electricity, it becomes evident that the efficiency of each particular process is dependent on optical and electrophysical properties of the semiconductor (reflection from the surface, photoionization quantum yield, diffusion length of minority current carriers, spectral position of the main absorption band), characteristics of the p–n junction (mechanism of the reverse current, potential barrier height, width of the space-charge region), and the so-called geometrical factor (relationship between the diffusion length of charge carriers and the depth of the p–n junction).

Intense fluxes of particles, mainly free electrons and protons, constituting the so-called near-Earth radiation belts cause impairment of the electrical parameters of PV arrays that, for maximum use of the solar radiation, must be mounted on the outer surface of a spaceship or on special external panels. The most effective way of protecting PV arrays is the use of coatings of transparent and radiation-resistant materials. The effectiveness of a transparent protective coating depends on its ability to attenuate or block low-energy particles in Earth radiation belts from the semiconductor. These particles ruin PV arrays and reduce their efficiency.

The major obstacle to the practical solution of this problem is that, in addition to protecting the cell from damaging radiation, the optical coatings should have high anti-reflection and thermotaxic properties. They should reduce reflection in the operating spectral range and protect PV cells from excessive heat by virtue of a high integrated factor of thermal radiation from the surface. To resolve this problem, a system perspective on the current status of the technology, the changes already experienced, and the necessity for improvement in tomorrow's systems has been prepared.[337] Issues of reliability, efficiency, safety, and low cost are the principal concerns of the system approach.

Transparent elastic polymers such as ethylene vinyl acetate and fluoropolymer-impregnated glass fibers have been used as an encapsulant for PV modules. Cationic and anionic polymers are used as desiccants in cooling systems; they are regenerated at temperatures below 353 K.[338,339]

A light-scattering layer of Polyester (PET-E-5100), with a rough surface and a colored transparent resin (e.g., Vylon 200 containing Pc blue) and dispersed light-scattering agent (e.g., MBX-10), is used for the illumination devices.[340] Fluorinated ethylene monomers with OH, COOH, and COOR groups or epoxy groups (copolymers with different mol% of components) are preferred packing materials for PV cells.[341]

9.8 CONCLUSION

Intensive development in a number of research directions of organic PV cells has been reviewed. The most developed organic materials providing high photovoltaic-conversion potential are the dye-sensitized systems; considerable attention has been directed toward dye-sensitized nanocrystalline semiconductor films. Nanocrystalline semiconductor films are highly porous and have a large internal surface area. Recent works on dye-sensitized PV cells, centering on ruthenium–bipyridine complexes sensitizing nanocrystalline TiO_2 films, have proven to be highly efficient in proton-to-electron conversion. However, in order to study photoelectric conversion mechanisms and develop a more efficient dye-sensitized PV cell, it is necessary to probe sensitization with other kinds of dyes.

Among the other types of dyes, phthalocyanines (Pc) and pentacenes (Pn) are especially interesting. Phthalocyanines are the most promising materials because they are readily synthesized and nontoxic, and their electrical characteristics are widely researched. The use of unique electron acceptors such as C_{60} has greatly increased efficiency conversion of solar energy to electricity.

Conducting aromatic and heterocyclic polymers are attractive for OPV fabrication. Poly(p-phenylene-vinylene) (PPV), poly(aniline) (PANI), poly(pyrroles) (Ppy), and poly(thiophene) (PTh) are used most widely in prototyping OPVs. The fabrication of photoconducting layers based on these materials may be performed by varied methods. From this point of view, use of PPV and C_{60} as heterojunction materials between ITO and Al and blend system from PPV and C_{60} is most important. Efficiencies in PPV–C_{60} blends are about one order of magnitude lower than in PPV–C_{60} heterostructures, which indicates that charge-carrier transport in the blend systems is worse than in pure PPV and C_{60} layers. It is obvious that the architecture of the OPV has an influence upon efficiency conversion and that this is an important area for engineering development.

PANI is the most promising conducting polymer for industrial applications. Its possibilities are based on its specific characteristics: high voltage, microwave absorption, distributed filtering, and electromagnetic shielding. The relative ease of synthesis, stable conductivity in the air, and low cost make PANI of particular industrial interest. PPy is unique as an electrode coating. PPy films covering the electrodes prevent passivation of their surface; simultaneously, holes that extend into the polymer develop in the film so electrochemical reactions can proceed on the polymer surface. This kind of system has been demonstrated with GaAs, CdS, and CdSe. Passivation properties of PPy are attractive for the fabrication of photovoltaic cells using solid electrolytes.

Organic photoconducting materials offer high photosensitivity with low dark current in OPVs. They are fabricated with simple coating techniques that can be implemented at relatively low cost.

REFERENCES

1. Gratzel, M., Low cost and efficient photovoltaic conversion by nanocrystalline solar cells, *Proc. 6th Sede Boquer Symp. Sol. Electr. Prod.,* 32–55, 1994.
2. Ihara, M. et al., Enhancement of the absorption coefficient of cis-(NCS)$_2$ bis(2,2′-bipyridil-4,4′-dicarboxylate)ruthenium(II). Dye in dye-sensitized solar cell by a silver island film, *J. Phys. Chem. B,* 101, 26, 5133–5137, 1997.
3. Lindquist, S.E. et al., Electron transport properties in nanoporous TiO$_2$ from analysis of action spectra of dye sensitized electrodes, *Proc. SPIE-Int. Soc. Opt. Eng.* (Optical Materials Technology for Energy Efficiency and Solar Energy Conversion XIII), 2255, 803–810, 1994.
4. Nazeeruddin, M.K. et al., Conversion of light to electricity by cis-X2bis(2,2′-bipyridyl-4,4′-dicarboxylate)ruthenium(II) charge-transfer sensitizers (X = Cl⁻, Br⁻, I⁻, SCN⁻) on nanocrystalline TiO$_2$ electrodes, *J. Am. Chem. Soc.,* 115, 14, 6382–6390, 1993.
5. Dioczik, L. et al., Dynamic response of dye-sensitized nanocrystalline solar cells: characterization by intensity-modulated photocurrent spectroscopy, *J. Phys. Chem. B.,* 101, 49, 10281–10289, 1997.
6. Murakoshi, K. et al., Solid-state dye-sensitized TiO$_2$ solar cell with polypyrrole as hole transport layer, *Chem. Lett.,* 5, 471–472, 1997.
7. Gruenwald, R. and Tributsch, H., Mechanism of instability in Ru-based sensitization solar cells, *J. Phys. Chem. B.,* 101, 14, 2564–2575, 1997.
8. Frank, A.J., Gregg, B.A., and Gratzel, M., Photochemical solar cells based on dye sensitized of nanocrystalline TiO$_2$, in *AIP Conf. Proc.,* 404, 145–153, 1997.
9. Mirakoshi, K. et al., Fabrication of solid-state dye-sensitized TiO$_2$ solar cells combined with polypyrrole, *Sol. Energy Mater. Sol. Cells,* 55, 1–2, 113–125, 1998.

10. Ferrere, S., Zaban, A., and Gregg, B., Dye sensitization of nanocrystalline tin oxide by perylene derivatives, *J. Phys. Chem. B.*, 101, 23, 4490–4493, 1997.

11. Cao, F., Oskam, G., and Searson, P.S., A solid-state dye-sensitized photoelectrochemical cell, *J. Phys. Chem.*, 99, 47, 17071–17073, 1995.

12. Salafsky, J.S., Exciton dissociation, charge transport and recombination in ultra-thin conjugated polymer-TiO$_2$ nanocrystal intermixed composites, *Phys. Rev. B.: Condens. Mater. Phys.*, 99, 16, 10885–10894, 1999.

13. Bach, U. et al., Solid-state dye-sensitized mesoporous TiO$_2$ solar cells with high photon to electron conversion effeciency, *Nature (London)*, 583–585, 1998.

14. Arango, A.C., Carter, S.A., and Brock, P.J., Charge transfer in photovoltaic consisting of interpenetrating networks of conjugated polymer and TiO$_2$ nanoparticles, *Appl. Phys. Lett.*, 74, 12, 1698–1700, 1999.

15. Barbe, C.J. et al., Nanocrystalline TiO$_2$ electrodes for photovoltaic applications, *Amer. Cer. Soc.*, 80, 12, 3172–3180, 1997.

16. Liu, J. et al., Investigation of influence of redox species on the interfacial energetics of a dye–sensitized nanoporous TiO$_2$ solar cell, *Sol. Energy Mater. Sol. Cell.*, 55, 3, 267–281, 1998.

17. Bouzek, K. and Kavan, Z., Heat losses in Gratzel solar cells, *Sol. Energy Mater. Sol. Cell.*, 57, 4, 359–371, 1999.

18. Liu, D. et al., Picosecond dynamics of an IR sensitive squaraine dye. Role of singlet and triplet excited states in the photosensitization of TiO$_2$ nanoclusters, *J. Chem. Phys.*, 106, 15, 6406–6411, 1997.

19. Ruile, S. et al., Novel sensitizers for PV cells. Structural variation of Ru (II) complexes containing 2,6-bis(1-methyl-benzimidazol-2-yl)pyridine, *Inorg. Chim. Acta*, 261, 2, 129–140, 1997.

20. Alebbi, M. et al., The limiting role of iodide oxidation in cis-Os(dcb)$_2$(CN)$_2$/TiO$_2$ photoelectrochemical cells, *J. Phys. Chem. B.*, 102, 39, 7577–7581, 1998.

21. Papayeorgiou, N., Barbe, C., and Gratzel, M., Morphology and adsorbate dependence of ionic transport in dye sensitized mesoporous TiO$_2$ films, *J. Phys. Chem. B.*, 102, 21, 4156–4164, 1998.

22. Zakeeruddin, S.M. et al., Molecular engineering of photosensitizer nanocrystalline solar cells: synthesis and characterization of Ru dyes based on phosphonated terpyridines, *Inorg. Chem.*, 36, 25, 5937–5946, 1997.

23. Agrazzi, R. and Bignozzi, C.A., 4-phenylpyridine as ancillary ligand in ruthenium(II) polypyridyl complexes for sensitization of *n*-type TiO$_2$ electrodes, *J. Phochem. Photobiol., A.*, 115, 3, 239–242, 1998.

24. Mohammad, K. et al., Redox regulation in Ru(II) polypyridyl complexes and their application in solar energy conversion, *J. Chem. Soc., Dalton Trans.*, 23, 4571–4578, 1997.

25. Jing, B., Zhang, M., and Shen, T., Advances in dye-sensitized solar sell, *Chin. Sci. Bull.*, 42, 23, 1937–1948, 1997.

26. Franco, S.V. et al., Oxidative formation of electroactive film from poly(pyridinyl) complexes of Ru(II) containing 3-(pyrrol-1-ylmethyl)pyridine, *Synth. Met.*, 90, 2, 81–88, 1997.

27. Kohle, O. et al., The PV stability of bis(isorhiocyanato)-ruthenium(II)-bis-2,2'-bypyridine-4,4'-dicarboxylic acid and related sensitizers, *Adv. Mater.*, 9, 11, 904–906, 1997.

28. Nasr, C., Hotchandani, S., and Kamat, P.V., Role of iodide in photoelectrochemical solar cells electron transfer between iodide ions and ruthenium polypyridyl complex anchored on nanocrystalline, *J. Phys. Chem. B.*, 102, 25, 4944–4951, 1998.

29. Jing, B. et al., Ruthenium(II) thiocyanato complexes containing 4'(4-phosphonatophe-nyl)-2,2': 6',2''-terpyridine: synthesis, photophysics and photosensitization of nano-crystalline TiO$_2$ electrodes, *J. Mater. Chem.*, 8, 9, 2055–2060, 1998.
30. Argazzi, R. et al., Enhanced spectral sensitivity from Ru(II) polypyridyl based pho-tovoltaic devices, *Inorg. Chem.*, 33, 25, 5741–5749, 1994.
31. Tennakone, K. et al., A nanoporous solid-state PV cell sensitized with copper chlo-rophyllin, *J. Photochem. Photobiol. A*, 108, 2–3, 175–177, 1997.
32. Tennakone, K. et al., An efficient dye-sensitized photoelectrochemical solar cell made from oxides of tin and zinc, *Chem. Commun.*, 1, 15–16, 1999.
33. Fang, I. et al., PV study of nanocrystalline TiO$_2$ film modified with dye molecules, *J. Vac. Sci. Technol. B.*, 15, 4, 1468–1470, 1997.
34. Fang, J. et al., The PV study of co-sensitized microporous TiO$_2$ electrode with porphyrin and Pc, *Appl. Surf. Sci.*, 119, 3–4, 237–241, 1997.
35. Fang, J. et al., The photoresponse properties of nanocrystalline TiO$_2$ particulate films co-modified with dyes, *New J. Chem.*, 21, 6–7, 839–840, 1997.
36. Deng, H. et al., Fabrication, characterization and the photoelectric conversion of the nanostructured TiO$_2$ electrode, *J. Vac. Sci. Technol. B.*, 15, 4, 1460–1464, 1997.
37. Deng, H. et al., The mixed effect of Pc and porphyrin on the photoelectric conversion of nanostructured TiO$_2$ electrode, *Synth. Met.*, 92, 3, 269–274, 1998.
38. Cherepy, M.J. et al., Ultrafast electron injection: implications for a photoelectrochem-ical cell utilizing an anthocyanin dye-sensitized TiO$_2$ nanocrystalline electrode, *J. Phys. Chem. B.*, 101, 45, 9342–9351, 1997.
39. Honma, I., Sasabe, H., and Zhou, H.-S., Synthesis of self-assembled functional molecules in mesoporous materials, *Mater. Res. Soc. Symp. Proc.*, *(Nanophase and Nanocomposite Mater. II)*, 457, 525–531, 1997.
40. Tracey, S.M., Hodgson, S.N.B., and Ray, A.K., Sol-gel derived TiO$_2$/lead phthalocy-anine photovoltaic cells, *J. Sol-Gel. Technol.*, 13, 1–3, 219–222, 1998.
41. Wang, Z. et al., Photosensitization of ITO and nanocrystalline TiO$_2$ electrode with a hemicyanine derivative, *Synth. Met.*, 114, 201–207, 2000.
42. Pochtennji, A.E. et al., The jump conductivity in CuPc and its composite structures, *Fiz. Tverd. Tela.*, 38, 8, 2592–2600, 1996 [in Russian].
43. Dieing, R. et al., Soluble substituted μ-oxo(phthalocyaninato)iron (III) dimers, *Chem.*, B 128, 589, 1995.
44. Velichko, A.V., Sambrovsaya, M.A., and Moskvin, W.S., Russ. RU 2.050.359, 1995 [Isobreteniya, No. 35, 197–198, 1995].
45. Himeno, K., Hibara, T., and Sekioka, R., Appl. 94/69,139, 1994; JP 07 252,425 [95 252,425], 1995.
46. *Handbook of Conducting Polymers*, Skothiem, T.A., Elsenbaumer, R.L., and Reinolds, J.R., Eds., 2nd ed., Marcel Dekker, Inc., New York, 1998, 197.
47. Li, L., Yang, L., and Dai, H., Improved method for synthesis of oxotitanium Pc, *Huaxue Shiji.*, 19, 6, 379–380, 1997.
48. Sakamoto, K. and Ohno, E., Synthesis of CoPc derivatives and their cyclic voltam-mograms, *Dyes Pigm.*, 35, 4, 375–386, 1997.
49. Kondratenko, N.V. et al., The synthesis and properties of some polyfluoroalkoxy substituted Pc, *J. Porphyrins Phthalocyanines*, 1, 1, 341–347, 1997.
50. Katsube, H. and Tanaka, M., JP Appl. 94/11,661, 1994; JP 07 258,576 [95 258,576], 1995.
51. Iida, A., Japan Appl. 96/354,029, 1996; JP 10,175,978 [98,175,978], 1998.
52. Pope, M. and Svenberg, Ch. E., *Electronic Processes in Organic Crystals*, New York, vol. 2, 1982, 326.

53. Baruch, P. et al., On some thermodynamic aspects of photovoltaic solar energy conversion, *Sol. Energy Mater. Sol. Cells*, 36, 2, 201–222, 1995.
54. Simon, A. and Xavier, F., Conduction mechanism of metal-free phthalocyanine single crystals as a function of temperature, *Bull. Mater. Sci.*, 20, 3, 297–303, 1997.
55. Fornur, T. et al., Charge transfer dynamics of donor–acceptor systems in solutions and sol-gel matrixes, *J. Sol-Gel Sci.Technol.*, 2, 1–3, 737–740, 1994.
56. Popovich, Z.D. et al., Study of carrier generation in titanyl phthalocyanine (TiOPc) by electric field-induced quenching of integrated and time resolved fluorescence, *J. Phys. Chem. B.*, 102, 4, 657–663, 1998.
57. Zhou, X. and Qi, X., A study on the preparation of photoconductive film of X-type titanyl phthalocyanine and its performance in photoconduction, *Gongneng Caliao.*, 26 (Suppl.), 67–68, 1995.
58. Pan, J. et al., Steady-state photovoltaic and electroreflective spectra in Al/vanadyl phthalocyanine (VOPc in phase II) indium tin oxide (ITO) sandwich cell, *Thin Solid Films*, 324, 1(2), 303–213, 1998.
59. Tsizh, B.R., Potential barriers at organic semiconductor interfaces, *Zh. Fiz. Dosl.*, 1, 2, 276–278, 1997.
60. Kume, T. et al., Enhancement of photoelectric conversion efficiency in CuPc solar cell white light excitation of surface plasmon polariton, *Jpn. J. Appl. Phys. Part 1*, 34, 12A, 6448–6451, 1995.
61. Meier, H., Albrecht, W., and Zimmerhackl, E., Photoconductivity of polymeric copper phthalocyanine, *Polymer Bull.*, 13, 43, 1985.
62. Petritsch, K. et al., Liquid crystalline phthalocyanines in organic solar cells, *Synth. Met.*, 102, 1–3, 1776–1778, 1999.
63. Tsutsuçi, T., Nakashima, T., and Fujita, Y., Photovoltaic conversion efficiency in copper phthalocyanine/perylenetetracarboxylic acid benzimidazole heterojunction solar cell, *Synth. Met.*, 71, 1–3, 2281–2282, 1995.
64. Inigo, A.R., Xavier, F.P., and Goldsmith, G.J., CuPc as an efficient dopant in development of solar cells, *Mater. Res. Bull.*, 32, 5, 539–546, 1997.
65. Kunugi, T. et al., Molecular beam epitaxial growth of CuPc on Si(111), in *Jpn. IEMT Symp. Proc. Jpn. Int. Electron. Manuf. Technol. Symp.*, 1995, 187–190.
66. Pan, Y.L. et al., Transient photocurrent and charge-transfer excitation bands in a tin-phthalocyanine (SnPc) polycrystalline film, *Appl. Phys. A.: Mater. Sci. Process.*, A65, 4–5, 425–428, 1997.
67. Hooper, P.D. et al., Electrical properties of nickel Pc (NiPc) sandwich devices incorporating a tetracyanoquinonedimethan, *Semicond. Sci. Technol.*, 12, 4, 455–459, 1997.
68. Abdel-Malik, T.G., Abdel-Latif, R.M., and El-Samahy, A.E., Transport properties in NiPc thin films using gold electrodes, *Thin Solid Films*, 256, 1–2, 139–142, 1995.
69. Amar, N., Gould, R.D., and Salech, A.M., Space-charge-limited conductivity in evaporated K-form metal-free phthalocyanine thin films, *Vacuum*, 50, 1–2, 53–56, 1998.
70. Abdel-Malik, T.G. and Abdel-Latif, R.M., Ohmic and space–charge-limited conduction in CoPc thin films, *Thin Solid Films*, 305, 1(2), 336–340, 1997.
71. Abdel-Latif, R.M. and El-Samahy, A.E., Electrical properties of PbPc films, *Acta Phys. Pol. A.*, 90, 3, 557–564, 1996.
72. Dumm, M. et al., Charge transport in LiPc, *J. Chem. Phys.*, 104, 13, 5048–5053, 1996.
73. Ioannidis, A. and Dodelet, J.P., Hole and electron transport in chloroaluminum Phthalocyanine thin films, *J. Phys. Chem. B.*, 101, 26, 5100–5107, 1997.

74. Yanagida, S. and Mirakoshi, K., Novel solar cell for power generation: future of dye sensitized solar cell and its potential, *Seisan to Gijutsu.*, 50, 1, 31–36, 1998.
75. Goossens, A. and Schoonman, J., New generation of solar cells sees the light, *Chem. Mag.* (The Hague), 1997, 5, 186–188.
76. Morikawa, T. and Yamazaki, K., Appl. 94/51,011, 1994; JP 07 240,530 [95 240,530], 1995.
77. Shinohara, H. et al., JP Appl. 95/176,731, 1995; JP 09 69,642 [97 69,642], 1997.
78. De Haan, A., BE Appl. 96/704, 1996; PCT Int. Appl., WO 98, 08,084, 1998.
79. Kerp, H. R. and van Faassen, E.E., Photovoltaic yield from exciton dissociation in organic dye layers, *Phys. Chem. Chem. Phys.*, 1, 8, 1761–1763, 1999.
80. Azim-Araghi, M.E. and Krier, A., Optical charaterization of ClAlPc thin films, *Pure Appl. Opt.*, 6, 4, 443–453, 1997.
81. Honma, I. and Zhou, H.-S., Synthesis of self-assembled photosensitive molecules in mesostructured materials, *Chem. Mater.*, 10, 1, 103–108, 1998.
82. Signeski, R., Tarosz, G., and Godlevski, J., Photoelectric properties of heterojunctions formed from di-(pyridyl)-perylene tetracarboxylic diimide and copper phthalocyanine or pentacene, *Synth. Met.*, 94, 1, 135–137, 1998.
83. Rudiono, T., Improvement of photovoltaic performance in copper phthalocyanine binder layer solar cell, *Jpn. J. Appl. Phys. Part 2*, 36, 2, 127–129, 1997.
84. Sharma, G.D. and Sangodkar, S.C., Electrical and photovoltaic characteristics of a *p–n* junction organic device, in *Proc. SPIE – Int. Soc. Opt. Eng.*, Semiconductor Devices, 273, 3, 226–228, 1996.
85. Fujii, A. et al., Organic photovoltaic cell with donor–acceptor double heterojunction, *Jpn. J. Appl. Phys.*, 35, 11A, L.1438–1441, 1996.
86. Jutaka, H., Heishoku, A., and Josomiya, R., Photoconductive properties of TiO_2 film, prepared by sol-gel method and its application, *J. Mater. Sci.*, 32, 12, 3183–3188, 1997.
87. Kyokane J., Aojagi, R., and Joshino, K., Organic photoelectric devices using evaporated thin film by ion-beam-assisted method, *Kenkyo Kiyo-Nara. Kogio Koto Senmen Gakko.*, 33, 23–27, 1997.
88. Takahashi, K., Nakamura, J., and Yamaguchi, T., Enhanced photocurrent quantum yield by electronic interaction between zinc porphyrin and Rhodamine B molecules in Al/dye/Au sandwich-type solar cell, *J. Phys. Chem. B.*, 101, 6, 991–997, 1997.
89. Tsuzuki, T. et al., Photoelectrical conversion of *p–n* heterojunction devices using thin film of TiOPc and a perylene pigment, *Thin Solid Films*, 273, 1–2, 177–180, 1996.
90. Gregg, B.A., Bilayer molecular solar cells on spin-coated TiO_2 substrates, *Chem. Phys. Lett.*, 258, 3,4, 376–380, 1996.
91. Deng, H. et al., The liquid junction cell based on the nanostructured TiO_2 electrode sensitized with Zn tetrasulfonated phthalocyanine, *Chem. Phys.*, 221, 3, 323–331, 1997.
92. Woehrle, D., Kreienhoop, L., and Schnurpfeil, G., Investigation on *n–p* junction photovoltaic cells of perylene tetracarboxylic acid diimides and Pc, *J. Mater. Chem.*, 5, 11, 1819–1829, 1995.
93. Tsuzuki, T. et al., Effect of morphology on photovoltaic properties of titanyl phthalocyanine, *Jpn. J. Appl. Phys. Part 2*, 35, 4A, 447–450, 1996.
94. Meslenikov, S.V. and Fedorov, M.I., Organic semiconductor solar cells with a heterojunction, *Russ. Phys. J.*, 40, 1, 60–63, 1997 [in Russian].
95. Takanashi, K., Nakatani, S.-I., and Matsuda, T., Enhanced quantum yield in porhyrin solar cell with redox chain for electron transfer, *Chem. Lett.*, 11, 2001–2004, 1994.

96. Deng, H. et al., Improvement in photoelectric conversion on a nanostructured TiO_2 electrode cosensitized with Pc and porphyrin, *Jpn. J. Appl. Phys. Part 2.*, 37, 2A, L. 132–135, 1998.

97. Murata, K. et al., Long-lived excited state of C_{60} in C_{60}/phthalocyanine heterojunction solar cell, *Appl. Phys. Lett.*, 68, 3, 427–429, 1996.

98. Antohe, S. and Merticaru, A., Electrical properties of ITO/CuPc/CuPc+TPyP/ TPyP/Al cells, in *CAS '97. Proc. Int. Semicond. Conf., 20-st.*, 2, 501–504, 1997.

99. Antohe, S. et al., Electrical and photovoltaic properties of a three layered organic solar cell with an enlarged photoactive region of codeposited dyes, *Rom. Rep. Phys.*, 48, 7, 581–597, 1996.

100. Liang, Z. et al., Preparation and properties of PMPc/Si solar cells, *Congneng Cailiao*, 28, 4, 429–431, 1997.

101. Takahashi, K., Goda, T., and Yamaguchi, T., Enhanced photocurrent in Al/porphyrin Schottky barrier cell with heterodimer consisting of metal free porphyrin and Zn porphyrin, *J. Phys. Chem. B.*, 103, 23, 4868–4875, 1999.

102. Fedorov, M.I. et al., Physicochemical, optical and photoelectric properties of the organic semiconductor Bordeaux perylene, *Russ. J. Phys. Chem. (Engl. Transl.)*, 63, 11, 1688–1690, 1989.

103. Rudiono, K.F. and Takeuchi, M., Morphological characteristics of perylene-doped phthalocyanine thin films and their photovoltaic effect, *Appl. Surf. Sci.*, 142, 1–4, 598–602, 1999.

104. Yakushi, K. and Yonehara, Y., Pressure induced charge transfer in phthalocyanine conductors, *Koatsuryoku no Kagaku to Gijutsu*, 6, 3, 167–175, 1997.

105. Kajihaza, K., Tanaka, K., and Hirao, K., Photovoltaic effect in titanium dioxide/zinc phthalocyanine cell, *Jpn. J. Appl. Phys. Part 1.*, 35, 12A, 6110–6116, 1996.

106. Takahashi, K., Hashimoto, K., and Murata, K., Spectral cosensitization in organic solar cell with mixed film of zinc porphyrin and merocyanine, *Chem. Lett.*, 2, 269–272, 1994.

107. Brinkmann, M. et al., Electrodeposition of lithium phthalocyanine thin films: part 1. Structure and morphology, *J. Mater. Res.*, 14, 3, 2162–2172, 1999.

108. Ros, T.D. et al., Photoexcitation of a Zn tetra-phenyl-fullerene complex leads to electron transfer with very long lifetime of the charge separating pairs, *Chem. Commun.*, 7, 635–636, 1999.

109. Zhang, I. et al., Photovoltaic properties of porphyrin solid films with electric field induction, *Thin Solid Films*, 284, 569–599, 1996.

110. Komura, T., Takahashi, M., and Murata, K., Japan Appl. 93/36, 755, 1993; JP 06,252,379 [94,252,379], 1994.

111. Murata, K. and Ito, M., Appl. 95/165,993, 1995; JP 09,18,039 [97,18,039], 1997.

112. Takahashi, M., Murata, K., and Ito, M., Appl. 95/230,249, 1995; JP 09 74,217 [97 74,217], 1997.

113. Takahashi, K. et al., Contribution of electric-field-induced metal-free porphyrin dication to photocurrent in mixed solid of metal-free porphyrine and o-cloranil/Al Schottky-barrier cell, *J. Electrochem. Soc.*, 146, 5, 1717–1723, 1999.

114. Antohe, S. et al., Electrical and photovoltaic properties of ITO/chlorophyll-a/TPyP/Al *p–n* junction cell, *Phys. Status Solidi A*, 153, 2, 581–588.

115. Antohe, S., Tugulea, L., and Cheorgher, V., Photoelectrical properties of ITO/chlorophyll-a/TPyP/Al *p–n* heterojunction cell, *Rom. Rep. Phys.*, 48, 7–8, 571–580, 1996.

116. Videlot, Ch., Fichou, D., and Garnier, F., Photovoltaic solar cell based on rare-earth bisphthalocyanine complexes, *Mol. Cryst. Liq. Cryst. Sci. Technol. Sect. A.*, 322, 319–328, 1998.

117. Murrer, B.A., Graetzel, M., and Nazeeruddin, M.K., GB Appl. 97/14,905, 1997; PCT Int. Appl. WO 99, 03 868, 1999.

118. Tada, H. and Utsugi, K., Organic electroluminescent devices with moleculary doped polymers and anode interface, *Polym. Adv. Technol.*, 8, 7, 443–448, 1997.

119. Fujii, A., Ohmori, Y., and Yoshino, K., An organic infrared electroluminescent diode utilizing a Pc film, *IEEE Trans. Electron. Devices*, 44, 8, 1204–1207, 1997.

120. Mortimer, R.J., Electrochromic materials, *Chem. Soc. Rev.*, 26, 3, 147–156, 1997.

121. Oba, K. and Okada, Sh., JP Appl. 96/113,615, 1996; JP 10 26, 838 [98 26,838], 1998.

122. Kadoi, M. et al., Appl. 95/229,329, 1995; JP 09 73,182 [97 73,182], 1997.

123. Yanagida, Sh. et al., Appl. 96/22,814, 1996; JP 09 199,744 [97 199,744], 1997.

124. Hanack, M. and Lang, M., Conducting stacked metallophthalocyanines and related compounds, *Adv. Mater.*, 6, 11, 819–833, 1994.

125. Naarrmann, H., DE Appl. 4,242,676, 1992; Ger. Offen. DE 4, 242, 676, 1994.

126. Scherf, U. and Muellen, K., Design and synthesis of extended π-systems: monomers, oligomers, polymers, *Synthesis*, 2, 23–28, 1992.

127. Goldfinger, M.B. and Swager, T.W., Fused polycyclic aromatics via electrophile-induced cyclization reactions: application to the synthesis of graphite ribbons, *J. Amer. Chem. Soc.*, 116, 17, 7895–7896, 1994.

128. Clarkson, G.J. et al., Synthesis and characterization of mesogenic phthalocyanines containing a single poly(oxyethylene) side chain: an example of steric disturbance of the hexagonal columnar mesophase, *Macromolecules*, 29, 3, 913–917, 1996.

129. Han, A.M. et al., Effect of bridging group in poly(titanyloxo-phthalocyanine)son photocurrent generation, *Synth. Met.*, 101, 62–63, 1999.

130. Gangotri, K.M. and Regar, O.P., Role of azine dye as photosensitizer in solar cell: glucose–safranine system, *Heterocycl. Commun.*, 2, 6, 567–579, 1996.

131. Gangotri, K.M. and Regar, O.P., Use of azine dye as a photosensitizer in solar cell, *Int. J. Energy. Res.*, 21, 14, 1345–1350, 1997.

132. Gangotri, K.M. and Regar, O.P., Use of azine dye as a photosensitizer in solar cell: EDTA–safranine system, *Arabian J. Sci. Eng. Sect. A.*, 24, 1A, 67–71, 1999.

133. Yana, A.K. and Bhowmik, B.B., Enhancement in power output of solar cell consisting of mixed dyes, *J. Photochem. Photobiol. A,* 122, 1, 53–56, 1999.

134. Jang, X., Huang, S., and Wu, D., Studies on the photoelectric mutual conversion properties of a new perylene dicarboximide, *Huadong Ligong Daxue Xuebao.*, 22, 4, 495–498, 1996.

135. Higashino, K., Ishiguro, E., and Nakaya, T., Molecular structure and photovoltaic power of high efficiency and high-output azocompounds, *Senryoto Jakuhin.*, 42, 1, 21–26, 1997.

136. Van Slyke, S.A., Chen, C.H., and Tang, C.W., Organic electroluminescent devices with improved stability, *Appl. Phys. Lett.*, 69, 15, 2160–2162, 1996.

137. Videlot, C., Fichou, D., and Garnier, F., Organic semiconductor-based solar cells, *J. Chim. Phys-Chim. Biol.*, 95, 6, 1335–1338, 1998.

138. Harvey, P.P. and Fortin, D., Photoproperties of the polymeric $\{[M(dmb)_2]Y\}_n$ materials: photoinduced intrachain energy and intermolecular electron transfers, and design of photovoltaic cells, *Coord. Chem. Rev.*, 171, 351–354, 1998.

139. Salbeck, J. and Lupo, D., DE Appl. 19,711,568, 1997; PCT Int. Appl. WO 98, 42,715, 1998.

140. Salbeck, J. and Lupo, D., DE Appl. 19,711,714, 1997; PCT Int. Appl. WO 98, 42,655, 1998.

141. Boch, U. et al., DE Appl. 19,711,713, 1997; Ger. Offen. DE 19,711,713, 1998.

142. Lang, A.-D. et al., Studies of the photocurrent generation performances from a series of amphiphilic bis-chromophore zinc complexes and correlation between photocurrent generation performance and molecular structure, *Synth. Met.*, 99, 97–103, 1999.

143. Schon, J.H., Kloc, Ch., and Battlog, B., Efficient photovoltaic energy conversion in pentacene based heterojunctions, *Appl. Phys. Lett.*, 77, 16, 2473–2475, 2000.

144. Jarosz, G., Signerski, R., and Goldewski, J., Temperature and light dependence of photoenhanced current in thin organic layers, *Synth. Met.*, 109, 161–164, 2000.

145. Vertsimakha, Y. and Verbitsky, A., Oxygen effect on photovoltaic properties of pentacene-based barrier structure, *Synth. Met.*, 109, 291–294, 2000.

146. Koltun, M. et al,. Solar cells from carbon, *Sol. Energy Mater. Sol. Cells*, 44, 4, 485–491, 1996.

147. Konarev, D.V. and Lyubovskaya R.M., Donor–acceptor complexes and ion-radical salts based on fullerenes, *Uspekhi Khimii*, 68, 1, 23–44, 1999 [in Russian].

148. Munn, R.V., Calculation of charge-transfer states and electro-absorption for the fullerene C_{60} indicated a band-gap of 3.1 eV, in *NATO ASI Ser. 3*, 24 (Electrical and Related Properties of Org. Solid), 117–132, 1997.

149. Licht, S. et al., Photoactivation, temperature and O_2 depletion effects in Fu solar cells, *Sol. Energy Mater. Sol. Cells*, 56, 1, 45–46, 1998.

150. Babbery, J., Naides, R., and San, G., Temperature dependence of the phototransport properties in C_{60} films, *Fullerene Sci. Technol.*, 6, 1, 39–57, 1998.

151. Makarova, T.L. et al., Electrical conductivity of polymerized states of C_{60}, *Pis'ma Zh. Tech. Fiz.*, 22, 23, 75–81, 1996 [in Russian].

152. Nemchuk, N.I. et al., Surface and bulk effects in conductivity of fullerene films, *Mol. Cryst. Liq. Cryst. Sci. Tech. Sec. C.*, 7, 1–4, 183–186, 1996.

153. Al-Mohamad, A. and Allaf, A.W., Fullerene-60 thin films for electronic application, *Synth. Met.*, 104, 39–44, 1999.

154. Katz, E.A. et al., Photovoltaic properties of fullerene (C_{60}) thin films, in *Mater. Res. Soc. Symp. Proc.*, 485 (Thin Film Structures for Photovoltaic), 1998, 113–118.

155. Licht, S. et al., Fullerene photoelectrochemical solar cells, *Sol. Energy Mater. Sol.Cells*, 51, 1, 9–19, 1996.

156. Terukov, E.I., Davydov, V.Yu., and Kon'kov, O.I., Effect of photopolymerization on photoelectric properties of polycrystalline C_{60} films, *Pis'ma Zh. Tech. Fiz.*, 22, 5, 71–74, 1996 [in Russian].

157. Chen, K.M. et al., Rectification properties and interface state of heterojunction between solid C_{60} and *n*-type GaAs, *Appl. Phys. Lett.*, 69, 23, 3557–3559, 1996.

158. Murata, K. and Ito, M., Appl. 95/230,248, 1995; JP 09 74,216 [97 74,216], 1997.

159. Lin, Y.-Sh. et al., Synthesis and photoelectric properties for (η^2-h_{70})-Pd(PPh$_3$)$_2$, *Gaodeng Xuexiao Huanxue Xuebao*, 18, 4, 509–512, 1997.

160. Larina, L.L., Shevalevskii, O.I., and Chernozatonski, L.A., Structure organization of fullerene layers for photovoltaic devices, in *Diffus. Deffect. Data. Pt. B*, 51–52 (Polycrystalline Semiconductors IV), 553–558, 1996.

161. Yonehara, H. and Pac, C., Photoelectrical properties of double-layer organic solar cells using C_{60} and Pc, *Thin Solid Films*, 27, 1–2, 108–113, 1996.

162. Chen, Z. et al., PV effect of fullerene-doped phthalocyanine complexes of transition metals on a GaAs electrode, *Xiamen Daxue Xuebao Ziran Kexueban*, 35, 5, 745–749, 1996.

163. Matsuzaki, S., Kubota, H., and Nagato, M., Development of new functional materials by multilayered composite films containing fullerene, in *Sagava Senton Kagaku Gigutsu Shinko Zaidan Yosei Kenkyi Hakokusho*, 8th. ed., 1995, (publ. 1996), 25–33.

164. Hiromitsu, J., Kaimori, Y., and Ibo, T., Photovoltaic effect and electrically detected electron spin resonance of H_2-Pc/C_{60} heterojunction, *Solid State Commun.*, 104, 9, 511–515, 1997.

165. Murata, K. et al., Long-lived excited state of C_{60} in C_{60}/phthalocyanine heterojunction solar cell, *Appl. Phys. Lett.*, 68, 3, 427–429, 1996.

166. Schlebusch, C., Organic photoconductors and C_{60} - increase of the photoconduction by light-induced charge transfer, in *Schr. Forsch. Juelich. Mater. Mater.*, 1(Physik der Nanostrukturen), 1998, D 3.1–D 3.22.

167. Fujii, A. et al., Organic photovoltaic cell with donor-acceptor double heterojunctions, *Jpn. J. Appl. Phys.*, 35, 11A, L1438–L1441, 1996.

168. Wang, J., Photoconductivity of fullerene-doped polymers, *Nature* (London), 356, 585, 1992.

169. Sariftci, N.S. and Heeger, A.J., Photophysics of semiconducting polymer–C_{60} composites: a comparative study, *Synth. Met.*, 70, 1–3, 1349–1352, 1995.

170. Lee, C.H. et al., Sensitization of the photoconductivity of conducting polymers by fullerene C_{60}: photoinduced electron transfer, *Phys. Rev. B: Condens. Matter.*, 48, 20, 15425–15433, 1993.

171. Kawasaki, S. et al., Discrete-variational X_α calculations of $C_{60}F_x$ with x = 0.36 and 0.48, *Phys. Rev. B: Condens. Matter.*, 53, 24, 16652–16655, 1996.

172. Conwell, E.M., Mizes, H.A., and Perlstein, J., Photoinduced absorption of conducting polymer–C_{60} composites, in *Proc. SPIE — Int. Soc. Opt. Eng.*, 25–30, 1995, 87–98.

173. Liess, M. et al., Electro-modulated photoinduced absorption of C_{60}–doped MEH–PPV, *Synth. Met.*, 84, 1–3, 683–684, 1997.

174. Sariftci, N.S., Fullerenes as photoinduced electron acceptors: from photophysics to photovoltaics, in *Proc. Electrochem. Soc.*, 9–12 (Recent advances in the chemistry and physics of fullerenes and related materials), 1999, 279–285.

175. Gao, J., Hide, F., and Wang, H., Efficient photodetectors and photovoltaic cells from composites of fullerenes and conjugated polymers: photoinduced electron transfer, *Synth. Met.*, 84, 1, 979–980, 1997.

176. Heller, C.M. et al., Chemical potential pinning due to equilibrium electron transfer at metal/C_{60}-doped polymer interfaces, *J. Appl. Phys.*, 81, 7, 3227–3231, 1997.

177. Yosito, K. et al., Novel properties of a new type conducting and insulating polymers and their composites, *IEEE Trans. Dielectr. Electr. Insul.*, 3, 3, 331–334, 1996.

178. Yosito, K. et al., Charge transfer in fullerene-conducting polymer composite: electronic and excitonic properties, *Fullerene Sci. Technol.*, 5, 7, 1359–1386, 1997.

179. Brabec, C.J. et al., Plastic solar cells: from basic research to devices, *Chem. Ind.*, 80, 10, 1301–1307, 1998.

180. Miyamoto, H., Nishizawa, H., and Hosoya, M., JP Appl. 96/56,473, 1996; JP 09, 246,580 [97, 246,580], 1997.

181. Fracowiak, E. and Beguin, F., Intraction between electroconducting polymers and C_{60}, *J. Phys. Chem. Solid,* 57 (6-8 Proc. of the 8th Int. Symp. of Intercalation Compounds, 1995), 983–989, 1996.

182. Huong, C.Y., Lin, J.G., and Wang, L.J., Conductive behavior of interpenetrated polyanilines in fullerene-poly(urethane-ether) network, *Fullerene Sci. Technol.*, 5, 7, 1607–1614, 1997.

183. Dai, L. et al., Doping of conducting polymers by sulfonated fullerene derivatives and dendrimers, *J. Phys. Chem. B.*, 102, 21, 4049–4053, 1998.

184. Buntar, V. and Weber, H.W., Magnetic properties of fullerene superconductors, *Supercond. Sci. Technol.*, 9, 8, 589–595, 1996.

185. Kureishi, Y. et al., Photoinduced electron transfer from synthetic chlorophyll analogue to fullerene C_{60} on carbon paste electrode. Preparation of a novel solar cell, *Bioelectrochem. Bioenerg.*, 48, 1, 95–100, 1999.

186. Wang, Ch. et al., Synthesis and characterization of the narrow polydispersity fullerene-end-capped polysterene, *Polym. Bull.* (Berlin), 37, 3, 305–311, 1996.

187. Veenstra, S.C. et al., Preparation of photovoltaic cells from sexithiophene–C_{60} blend, in *Proc. SPIE-Int. Soc. Opt. Eng.*, 2852 (Nonlinear Optical Properties of Organic Materials IX), 1996, 277–285.

188. Janossy, A. et al., Conducting fulleride polymers, in *Springer Proc. Phys.*, 81 (Materials and Measurements in Molecular Electronics), 1996, 163–171.

189. Jenkins, D. et al., Uses of ultra high solar flux, in *Proc. Sol. 95: Am. Sol. Energy Soc. Annu. Conf.*, 130–135, 1995.

190. Lede, J. and Pharabod, F., Solar chemistry in the worldland in France, *Entropie*, 33, 204, 47–55, 1997.

191. Han, Sh., Jang, Zh., and Zheng, I., Preparative chemistry of C_{60}/C_{70}, *Hauxue Tongbao*, 5, 11, 11–16, 1996.

192. Buchachenko, A.L., Chemistry on the border of two centuries: achievements and prospects, *Uspekhi Khimii*, 68, 2, 99–118, 1999 [in Russian].

193. Misurkin, I.A., The theory of conducting polymers, *Khimicheskaya Fizika (Chem. Phys.)*, 15, 8, 110–115, 1996 [in Russian].

194. Sachdev, V.K. et al., Electrically conducting polymers. An overview, *Diffus. Defect Data., Pt. B, Semiconductor Materials and Technology*, 55, 104–109, 1997.

195. Travers, J.P., Transport mechanism in conducting polymers: do general behaviors exist? *J. Chim. Phys. Phis. — Chem. Biol.*, 95, 6, 1427–1432, 1998.

196. Hubner, E., DE Appl. 19,717,952, 1997; PCT Int. Appl. WO 98 49, 690, 1998.

197. Heinze, J., Tschunsky, P., and Smie, A., The oligomeric approach - the electrochemistry of conducting polymers in the light of recent research, *J. Solid State Electrochem.*, 2, 2, 102–109, 1998.

198. *Handbook of Conducting Polymers*, Skothiem, T.A., Elsenbaumer, R.L., and Reinolds, J.R., Eds., 2nd ed., Marcel Dekker, Inc., New York, 1998, 197.

199. Vohlidal, J. et al., New substituted polyacetylenes with phenylenethynylene side groups [–$(C_6H_4$–C≡h)n–Si–i–Pr_3 n=1,2]. Synthesis, characteristics, spectroscopy, photoelectronic properties, *Macromolecules*, 32, 20, 6439–6450, 1999.

200. Shirakawa, H. et al., Synthesis of electrically conducting organic polymers: halogen derivatives of polyacetylene $(CH)_x$, *J. Chem. Soc., Chem. Commun.*, 16, 578–580, 1997.

201. Dai, L., Conjugated and fullerene-containing polymers for electronic and photonic applications: advanced synthesis and microlithographic fabrications, *J. Macromol. Sci. — Rev. Macromol. Chem. Physics.*, C39, 2, 273–387, 1999.

202. Machado, J.M., Schlenoff, J.B., and Karasz, F.E., Morphology, doping and electrical properties of poly(p-phenylenevinylene)/poly(ethylene oxide) blends, *Macromolecules*, 22, 4, 1964–1973, 1989.

203. Heeger, A.G., in *Science and Applications of Conducting Polymers*, Salantck, W.R., Clark, D.T., and Samuelsen, E.J., Eds., IOP Publ., Bristol, U.K., 1990, 1–12.

204. Hadziioannou, G., Van Hutten, P.F., and Malliaras, G.G., Photonic polymers for the devices of the 21st century, *Macromol. Symp.*, 121, 27–34, 1997.

205. Zhang, W.J. et al., Synthesis of oligomeric anilines, *Synth. Met.*, 84, 1–3, 119–120, 1997.

206. Hsieh, B.R. et al., Synthesis of highly phenylated poly(p-phenylene-vinylenes) via a chlorin precursor route, *Macromolecules*, 31, 631, 1998.

207. Greenham, N.C., Peng, X., and Alivisator, A.P., A CdSe nanocrystal/MEH–PPV polymercomposite photovoltaic, in *AIP Conf. Proc. (Future Generation Photovoltaic Technology)*, AIP Press, 404, 295–301, 1997.

208. Angadi, M.A., Gosztola, D., and Wasielewski, M.R., Characterization of photovoltaic cells using poly(phenylenevinylene) doped with perylenediimide electron acceptors, *J. Appl. Phys.*, 83, 11, 6187–6189, 1998.

209. Shi, X. et al., Self-assembled multilayers and photoluminescence properties of a new water-soluble poly(para-phenylene), *Mater. Res. Soc. Symp. Proc. (Electrical, Optical, and Magnetic Properties of Org. Solid. State Materials IV)*, 488, 133–140, 1998.

210. Sacak, M., Akbulut, U., and Batchelder, D.N., Characterization of electrochemically produced two-component films of conducting polymers by Raman spectroscopy, *Polymer*, 39, 20, 4735–4739, 1998.

211. Damlin, P., Kvarnstroem, C., and Ivaska, A., *In situ* external reflection Fourier transform infrared spectroscopic study on the structure of the conducting polymer poly(para-phenylene), *Analyst* (Cambridge, U.K.), 121, 12, 1881–1884, 1999.

212. Winkler, B., Dai, L., and Mau, A.W.-H., Novel poly(p-phenylene vinylene). Derivatives with oligo(ethylene oxide) side chains: synthesis and pattern formation, *Chem. Mater.*, 11, 3, 704–711, 1999.

213. Polotskaya, G.A., Gladilin, S.V., and Sazanov, U.N., Synthesis and properties of poly(pyrrole)-poly(phenyleneoxide) electroconducting film, *Macromol. Compounds, A.*, 42, 1, 5–9, 2000 [in Russian].

214. Fu, H. et al., Blue photo- and electroluminescence based on poly(2-benzoyl-1,4-phenylene) and poly(2,5-dibenzoyl-1,4-phenylene), *Polym. Adv. Technol.*, 10, 5, 259–264, 1999.

215. Shaheen, S. E. et al., 2.5% efficient organic plastic solar cells, *Appl. Phys. Lett.*, 78, 6, 841–843, 2001.

216. Edwards, A. et al., Blue photo- and electroluminescence from poly-(benzoyl-1,4-phenylene), *Appl. Phys. Lett.*, 70, 3, 298, 1997.

217. Jenekhe, S.A. and Yi, Sh., Efficient photovoltaic cells from semiconducting polymer heterojunctions, *Appl. Phys. Lett.*, 77, 17, 2635–2637, 2000.

218. *Handbook of Conducting Polymers*, Skothiem, T.A., Elsenbaumer, R.L., and Reinolds, J.R., Eds., 2nd ed., Marcel Dekker, Inc., New York, 1998, 945.

219. Stepanov, B.I., *Introduction in the Chemistry and Technology of Organic Dyes*, 2nd ed., Chemistry, Moscow, 1977, chap. 8, 214 [in Russian].

220. Ikkala, O. et al., US Appl. 115,536, 1993; EP 643, 397, 1995.

221. Angelopoulos, M. and Liao, J.Sh., US Appl. 722,283 1996; JP 10 120,782 [98 120,782], 1998.

222. Gospodinova, N. et al., A new route to polyaniline composites, *Polymer*, 38, 3, 743–746, 1997.

223. Ahlskog, M. et al., Heat induced transition to the conducting state in polyaniline/dodecylbenzene sulfonic acid complex, *Synth. Met.*, 69, 1–3, 213–214, 1995.

224. Wilbur, J.M. et al., A representative of a new class of conducting oligomer: acetylene-terminated polyaniline, *Synth. Met.*, 82, 3, 175–181, 1996.

225. Ng, S.C., Xu, L.G., and Chan, H.S.O., A novel conductive polymer: poly[4-(2-thienyl)-benzamine], *J. Mater. Sci. Lett.*, 16, 21, 1738–1740, 1997.

226. Libert, J., Bredas, J.L., and Epstein, A.J., Theoretical study of p- and n-doping of the leucoemeraldine base form of polyaniline: evolution of the geometric and electronic structure, *Phys. Rev. B.: Condens. Matter.*, 51, 9, 5711–5724, 1995.

227. Kulkarni, V.G., Polyanilines — progress in processing and applications, *Polym. Prepr. (Am. Chem. Soc., Div. Polym. Chem.)*, 39, 1, 127, 1998.

228. Wessling, B., Dispersion as the link between basic research and commercial applications of conductive polymers (polyaniline), *Synth. Met.*, 93, 2, 143–154, 1998.

229. Rannou, P. et al., Spectroscopic, structural and transport properties of conductive PANI (polyanilines) processed from fluorinated alcohols, *Macromolecules*, 31, 9, 3007–3015, 1998.

230. Hupcey, M.A.Z. et al., Conducting electron beam resists based on polyaniline, in *Proc. SPIE – Int. Soc. Opt. Eng. (Emerging Lithographic Technologies II)*, 3331, 369–374, 1998.

231. Plank, R.V. et al., Characterization of highly conducting ultra-thin polyaniline films produced by evaporative deposition, *Chem. Phys. Lett.*, 263, 1, 2, 33–38, 1996.

232. Shi, W., Wey, H., and Huang, H., Appl. 98, 120, 067, 1998; Chine Pat. 1, 249, 315, 1999.

233. Wei, Y. et al., Composites of electronically conductive polyaniline with polyacrylate-silica hybrid sol-gel materials, *Chem. Mater.*, 7, 5, 969–974, 1995.

234. Kinlen, P.J., Liu, J., and Ding, Y., Emulsion polymerization process for organically soluble and electrically conducting polyaniline, *Macromolecules*, 31, 6, 1735–1744, 1998.

235. Diaz, F.R. et al., Synthesis, characterization and electrical properties of dihalogenated polyanilines, *Synth. Met.*, 92, 2, 99–106, 1998.

236. Hennig, C., Hallmeier, K.H., and Szargan, R., XANES investigation of chemical states of nitrogen in polyaniline, *Synth. Met.*, 92, 2, 161–166, 1998.

237. Sakthivel, S. et al., Structure, dielectric, a.c. and d.c. conduction properties of acid doped polyaniline films, *Eur. Polym. J.*, 33, 10–12, 1747–1752, 1997.

238. Pron, A. et al., FR Appl. 96/9,521, 1996; W.O. 98.05.040, 1998.

239. Zeng, X. et al., Structure and properties of iodine doped polyaniline, *Gaofenzi Cailiao Kexue Yu Gongcheng.*, 13 (Suppl.), 114–119, 1997.

240. Fabhane, V.A., Ranyar, A.D., and Dhakate, S., Electrical properties of the derivatives of substituted polyaniline, *Indian J. Pure Appl. Phys.*, 36, 6, 354–356, 1998.

241. Moon, H.-S. and Park, Y.-K., Synthesis and spectroscopic characterization of the copolymers of aniline and aniline derivatives with poly(ethylene oxide) chains at the 3-position, *Macromolecules*, 31, 19, 6461–6468, 1998.

242. Wang, X.H. et al., Preparation and properties of water-based conducting polyaniline, *Chem. Res. Chin. Univ.*, 14, 3, 309–314, 1998.

243. Li, J. and Wam, M., Preparation of doped polyaniline solution by emulsion polymerization-extraction method, *Congneng Gaofenzi Xuebao*, 11, 3, 337–342, 1998.

244. Xiao, S., Huang, Yi, and Tian, H., Synthesis and properties of electrical conductive free standing film of polyaniline by electrochemical method, *Huanan Ligong Daxue Xuebao, Ziran Kexueban*, 25, 8, 126–129, 1997.

245. Adams, P.N. et al., A new acid-processing route to polyaniline films which exhibit metallic conductivity and electrical transport strongly dependent upon intrachain molecular dynamics, *J. Phys.: Condens. Matter.*, 10, 37, 8293–8303, 1998.

246. Wei, F., Wei, W., and Wu, H., Study of photovoltaic properties of conducting polymers, in *Proc. IEEE. 6th Int. Conf. Conduct., Breakdown Solid Dielectr.*, 389–392, 1998.

247. Chen, S.-A. et al., White-light emission from electroluminescence diode with polyaniline as the emitting layer, *Synth. Met.*, 82, 3, 207–210, 1996.

248. Dai, L. et al., Doping of conducting polymers with sulfonated fullerene derivatives and dendrimers, *J. Phys. Chem. B*, 102, 4049–4053, 1998.

249. Pron, A. et al., Moessbauer spectroscopy studies of selected conducting polypyrroles, *J. Chem. Phys.*, 83, 11, 5923–5927, 1985.

250. Machida, S., Miata, S., and Techagumpuch, A., Chemical synthesis of highly electrically conductive polypyrrole, *Synth. Met.*, 31, 3, 311–318, 1989.
251. Whang, Y.E. et al., Polypyrroles prepared by chemical oxidative polymerization at different oxidation potentials, *Synth. Met.*, 45, 2, 151–161, 1991.
252. Stanker, D., Hallensleben, M.L., and Toppare, L., Oxidative polymerization of Py with $FeCl_3$ in CH_3NO_2, *Synth. Met*, 72, 2, 159–165, 1995.
253. Kurachi, K. and Kise, H., Preparation of polypyrroles/polyethylene composite films by vapor-phase oxidative polymerization, *Polym. J.*, 26, 12, 1325–1331, 1994.
254. Paul, A., Sarkar, D., and Misra, T.N., Electrical properties of organized assemblies of pyrrole-N-methylpyrrole copolymer in Langmuir –Blodgett films, *Solid State Commun.*, 89, 4, 363–367, 1994.
255. Dubitskii, Y.A., Zhubanov, V.A., and Maresch, G.G., Synthesis of polypyrroles in the presence of ferric tetrafluoroborate, *Synth. Met.*, 41, 1–2, 373–376, 1991.
256. Diaz, A.F., Kanazawa, K.K., and Gardin, G.P., Electrochemical polymerization of pyrrole, *J. Chem., Soc. Chem. Commun.*, 14, 635–636, 1979.
257. Diaz, A.F., Castillo, J.I., and Logan, J.A., Electrochemistry of conducting polypyrroles films, *J. Electroanal.Chem.*, *Interfacial Electrochem.*, 129, 1–2, 115–132, 1981.
258. Diaz, A.F. and Kanazawa, K.K., Polypyrrole: an electrochemical approach to conducting polymers, in *Extended Linear Chain Compounds*, Miler, J.S., Ed., Plenum Press, New York, 1983, 417–441.
259. Asavapiriyanont, S., Chandler, G.K., and Gunavardena, G.K., The electrodeposition of polypyrrole films from aqueous solutions, *J. Electroanal. Chem. Interfacial Electrochem.*, 177, 1–2, 229–244, 1984.
260. Scharifker, B.R., Garcia-Pastoriza, E., and Marino, W., The growth of polypyrrole films on electrodes, *J. Electroanal. Chem. Interfacial Electrochem.*, 300, 1–2, 85–89, 1991.
261. Jamaura, M. et al., Memory effect on electrical conductivity upon the counter-anione exchange of polypyrrole films, *Synth. Met.*, 48, 3, 337–354, 1992.
262. Downard, A.J. and Pletcher, D., The influence of water on the electrodeposition of polypyrrole in acetonitrile, *J. Electroanal. Chem. Interfacial Electrochem.*, 206, 1–2, 139–145, 1986.
263. Kang, E.T., Neoh, K.G., and Ong, Y.K., Thermal stability and degradation of some chemically synthesized polypyrrole complexes, *Termochim. Acta.*, 181, 57–70, 1991.
264. Vernitskaya, T.V. and Efimov, O.M., Polypyrrole: a conducting polymer (synthesis, properties and applications), *Uspechi Chimia.*, 66, 5, 489–505, 1997 [in Russian].
265. Singh, R., Narula, A.K., and Chandra, S., Mechanism of DC conduction and its correlation with electron spin resonance data of polypyrrole, in *Proc. Macromol.-New Front., Proc. IUPAC Int. Symp. Adv. Polym. Sci. Technol.*, 1, 367–370, 1998.
266. De Jesus, M.C., Weiss, R.A., and Chen, J., The development of conductive composite surfaces by diffusion-limited *in situ* polymerization of pyrrole in sulfonated polystyrene ionomers, *J. Polym. Sci. Part B. Polym. Phys.*, 35, 2, 347–357, 1997.
267. Weiss, R.A. et al., Development of conductive elastomer foams by *in situ* copolymerization of pyrrole and N-methyl-pyrrole, *Annu. Tech. Soc. Plast. Eng. 56th*, 2, 1370–1374, 1998.
268. Constantini, N. et al., Electrochemical synthesis of intrinsically conducting polymers of 3-alkylpyrroles, *Synth. Met.*, 92, 2, 139–147, 1998.
269. Seki, M. et al., Electrical properties of polypyrrole-poly(ether-urethane) composite films, *Polym. Polym. Comps*, 5, 5, 337–341, 1997.
270. Kim, I. and Elsenbaumer, R.L., Solution processible poly(1-alkyl-2,5-pyrrolylenevinylene): new low band gap, *Chem. Commun.*, 3, 327–328, 1998.

271. Yildirim, P. and Kucuyavuz, Z., Synthesis and characterization of conducting poly-pyrrole-cis-1,4-polybutadiene composites, *Synth. Met.*, 95, 1, 17–22, 1998.

272. Wang, C. et al., Preparation of polypyrrole conducting composite film, *Gongneng Gaofenzi Xuebao*, 11, 2, 167–171, 1998.

273. Wang, P. and Li, Y., Electrochemical and electrocatalytic properties of polypyrrole film doped with heteropolyanions, *J. Electroanal. Chem.*, 408, 1, 77–81, 1996.

274. Omastora, M. et al., Electrical properties and stability of polypyrrole containing conducting polymer composites, *Synth. Met.*, 81, 1, 49–57, 1996.

275. Selampinar, F., Akbult, U., and Toppare, L., Conducting polymer composites of polypyrrole and polyimide, *Macromol. Rep.*, A33 (Suppl. 586), 309–318, 1996.

276. Kaynak, A., Electromagnetic shielding effectiveness of galvanostatically synthesized conducting polypyrrole films, *Mater. Res. Bull.*, 31, 7, 845–860, 1996.

277. Park, J.-S., Preparation and characterization of conducting composites impregnated with thick polyheterocyclic polymers, *Kongop Hwanhak,* 9, 3, 342–347, 1998.

278. Parakka, J.P. et al., Synthesis and properties of two regular thienyl-pyrrole polymers, *Macromolecules*, 29, 6, 1928–1933, 1996.

279. Losada, J., del Peso, I., and Beyer, L., Redox and electrocatalytic properties of electrodes modified by films of polypyrrole nickel(II) Schiff-base complexes, *J. Electroanal. Chem.*, 447, 1–2, 147–154, 1998.

280. Balâi, N. et al., Electrically conductive polymer grafts prepared by electrochemical polymerization of pyrrole onto poly[(methylmethacrylate)-co-(2-(N-pyrrolyl)ethyl-methacrylate)] electrodes, *Turk. J. Chem.*, 22, 1, 73–80, 1998.

281. Noufi, R., Tench, D., and Warren, L.E., Photoelectrochemical evaluation of the *n*-cadmium selenide/methanol/ferro-ferricyanide system, *J. Electrochem. Soc.*, 128, 11, 2363–2366, 1981.

282. Frank, A.J. and Honda, K., *J. Photochem.*, 29, 195, 1989.

283. Skotheim, T.A., Feldberg, S.W., and Armand, M.B., Polypyrrole electrodes. Charge-transfer to aqueous and solid polymer electrolytes, *J. Phys. Collog.* (C 3 Conf. Int. Phys., Chim. Polym. Conduct., 1982), 615–620, 1983.

284. *Handbook of Conducting Polymers*, Skothiem, T.A., Elsenbaumer, R.L., and Reinolds, J.R., Eds., 2nd ed., Marcel Dekker, Inc., New York, 1998, 261.

285. Toussaint, I.M. and Bredas, J.L., Novel low bandgap polymers: polydicyanomethyl-ene cyclopenta-dithiophene and dipyrrole, *Synth. Met.*, 61, 1–2, 103–106, 1993.

286. King, G. and Higgins, S., Synthesis and characterization of novel substituted benzo (â) thiophenes and polybenzo(c)thiophenes: tuning the potential for *n*- and *p*-doping in transparent conducting polymers, *J. Mat. Chem.*, 5, 3, 447–455, 1995.

287. Rebart, E., Pepin-Donat, B., and Dinh, E., Fully conjugated conductive gels: synthesis and first characterizations, *J. Chim. Phys. Phys.–Chim. Biol.*, 92, 4, 775–778, 1995.

288. Bouachrine, M. and Zakhlifi, T., Optimization of the electrochemical synthesis of an organic polymer conductor: poly-3-*n*-hexylthiophene, *J. Chim. Phys. Phys.–Chim. Biol.*, 95, 5, 987–1000, 1998.

289. Irvin, J.A. and Reynolds, J.R., Low oxidation potential conducting polymers contain-ing substituted phenylene and 3,4-ethylenedioxythiophene units, *Polym. Prepr.* (Am. Chem. Soc; Div. Polym. Chem.), 37, 1, 682–683, 1996.

290. Iijima, Y., Electrochemical synthesis of polythiophene in magnetic fields, *Kino Zairyo*, 14, 8, 41–47, 1994.

291. Tanaka, S. and Yamashita, Y., Synthesis of narrow band gap heterocyclic copolymers of aromatic donor and quinonoid acceptor units, *Synth. Met.*, 69, 1–3, 599–600, 1995.

292. Devasagayaraj, A. and Tour, J.M., Synthesis of conjugated donor/acceptor passivator (DAP) polymer, *Macromolecules*, 32, 20, 6425–6430, 1999.

293. Pomerantz, M. and Liu, M. L., Synthesis and characterization of poly[3-(ω-bro-moalkyl)thiophenes], *Polym. Prepr.* (Am. Chem. Soc.; Div. Polym. Chem.), 39, 1, 151–152, 1998.

294. Van Asselt, R., Aagaard, O.M., and Froehling, P.E., BE Appl. 94/925, 1994; Eur. Pat. 707,023, 1996.

295. Huany, H. and Pickup, P.G., *In situ* conductivity of a low band-gap conducting polymer, *Acta Polym.*, 48, 10, 455–457, 1997.

296. Chartier, P., Nguyen, C.H., and Sene, C., Hybrid organic-inorganic photovoltaic junctions: case of the all thin-film CdSe/poly(3-methylthiophene) junction, *Sol. Energy Mater. Sol. Cells*, 52, 3–4, 413–421, 1998.

297. Sentein, C. et al., The influence of the stereoregularity of photodoped poly(3-alky-lthiophenes) on the thermal stability of their electrical conductivity, *J. Chim. Phys. Phys.–Chim. Biol.*, 92, 4, 983–986, 1995.

298. Saita, Y. and Ikenoue, Y., Appl. 96/282,525, 1996; JP 10 120,769 [98 120,769], 1998.

299. Gamboa S.A. et al., Photovoltaic structures based on polymer/semiconductor junctions, *Sol. Energy Mater. Sol. Cells*, 55, 1–2, 98–104, 1998.

300. Larmat, F. et al., Comparative reactivity of thiophene and 3,4-(ethylene-dioxy)thiophene as terminal electropolymerizable units in bis-heterocycle arylenes, *J. Polym. Sci. Part A. Polym. Chem.*, 35, 17, 3627–3636, 1997.

301. Camaioni, N. et al., Photovoltaic and transport properties of the heterojunction between poly(4,4'-dipentoxy-2,2'-bithiophene) and *n*-doped silicon, *Sol. Energy Mater. Soc. Cell.*, 53, 3–4, 217–227, 1998.

302. Roman, L.S. et al., High quantum efficiency polythiophene/C_{60} photodiodes, *Adv. Mater.*, 10, 10, 774–777, 1998.

303. Chartier, P., Nguyen, C.H., and Sene, C., Hybrid organic-inorganic photovoltaic junctions: case of the all thin-film CdSe/poly(3-methyl-thiophene) junction, *Sol. Energy Mater. Sol. Cells*, 52, 3–4, 413–421, 1998.

304. Noda, T. and Shirota, J., 5,5'-bis(dimesitylboryl)-2,2'-bithiophene and 5,5"-bis(dimesitylboryl)-2,2': 5',2"-terthiophene as a novel family of electron-transporting amorphous molecular materials, *J. Am. Chem. Soc.*, 120, 37, 9714–9715, 1998.

305. Loveday, V.C. et al., Characterization of self *n*-dopable polymers, *Polym. Prepr.*, 39, 1, 145–146, 1998.

306. Roy, M.S., Gupta, S.K., and Sharma, G.D., Photogeneration process and charge conduction mechanism in poly(phenylazomethinethiophene) thin film devices, in *Proc. SPIE — Int. Soc. Opt. Eng. (Physics of Semiconductor Devices. V. 2)*, 3316, 1243–1246, 1998.

307. Ng, S.C. et al., Novel heteroarylene polyazomethines: their synthesis and character-izations, *Polymer*, 39, 20, 4963–4968, 1998.

308. Irvin, J.A. and Reynolds, J.R., Low oxidation-potential conducting polymers: alter-nating substituted para-phenylene and 3,4-ethylenedioxythiophene repeat units, *Poly-mer*, 39, 11, 2339–2347, 1998.

309. Killan, J.G. et al., Electrochemical synthesis and characterization of series of fluoro-substituted phenylene-2-thienyl polymers, *Chem. Mater.*, 11, 4, 1075–1082, 1999.

310. Jounus, M. et al., Synthesis *n*-butylphosphane-substituted Pt(II) poly-ynes, *Angewandte Chemie. Int. Ed.*, 37, 21, 3036–3039, 1998.

311. Andrieu, C.G. et al., Chemical and electrochemical polymerization of 3-alkylsele-nophenes, *Sulfur Lett.*, 19, 6, 261–266, 1996.

312. Otsubo, T. et al., Synthesis, structure and polymerization 2,2'-bitellurophene, *Synth. Met.*, 69, 1, 537–538, 1995.

313. Camaioni, N., Casalbore–Miceli, G., and Beggiato, G., Influence of the doping of the polymer on the dark and photoelectrical properties in the junction ITO/poly(4,4'-dipentoxy-2,2'-bithiophene)/aluminum, *Synth. Met.*, 104, 169–173, 1999.

314. Maruthamuthu, M., Harichandran, G., and Jeyakumar, D., Conducting copolymers of aromatic amines, Macromol. New-Front, in *Proc. IUPAC Int. Symp. Adv. Polym. Sci. Technol.*, 1998, 1, 359–362.

315. Gal, Y.-S., Lee, W.-Ch., and Lee, H.-J., Cyclopolymerization of diethyl dipropargyl malonate and triethyl dipropargylphosphonoacetate by molibdenum pentachloride, *J. Macromol. Sci. Pure Appl. Chem.*, A34, 11, 2251–2267, 1997.

316. Gompper, R., Muller, Th.I.I., and Polborn, R., Polymeric malondialdehyde dianils — a novel type of electrically conducting polymers, *J. Mater. Chem.*, 8, 9, 2011–2018, 1998.

317. Neef, Ch. J., Glatzhofer, D.T., and Nicolas, K.M., Cyclopolymerization of 3-phenyl [5]ferrocenophane-1,5-dimethylene: synthesis and electronic properties of a polyferrocenophane, *J. Polym. Sci., Part A: Polym. Chem.*, 35, 16, 3365–3376, 1997.

318. Khore, P.K. et al., Electrical conduction mechanism in solution grown doped poly (vinyl pyrrolidone) films, *Bull. Mater. Sci.*, 21, 2, 139–147, 1998.

319. Karak, N., Maiti, S., and Sanigrahi, S. R., Antimony polymers (Part 4) – electrical properties of antimony polymers and blends, *Indian J. Chem. Technol.*, 5, 4, 217–221, 1998.

320. Takiguchi, T. et al., JP Appl. 90/16,811, 1990; EP 439170, 1991.

321. Princy, K.G., Josef, R., and Kartha, C.S., Studies on conductive silicone rubber compounds, *J. Appl. Polym. Sci.*, 69, 5, 1043–1050, 1998.

322. Heitmann, P. and Recharz, F., DE Appl. 19,651, 1996; PCT Int. Appl. WO 98 26,432, 1998.

323. Thielen, A., US Appl. 742,579, 1996; PCT Int. Appl. WO 98 19,312, 1998.

324. Oofuchi, Sh. et al., Appl. 96/282,020, 1996; JP 10 120, 887 [98 120,887], 1998.

325. Wartenberg, M.F., Lahlouh, J.G., and Toth, J., US Appl. 408,769 1995; Pat. Int. Appl. WO 96 30,443, 1996.

326. Haddadi–Asl, V. and Mohammadi, T., Effect of processing methods and on properties of conductive carbon-polyolefins composite, *Iran Polym. J.*, 5, 3, 153–164, 1996.

327. Meeker D.L. et al., Tailoring electrochromic properties using poly(N-vinylcarbazole) and poly(N-phenyl-2-(2'-thienyl)-5-(5''-vinyl-2''-thienyl)pyrrole) blends, *Macromolecules*, 31, 9, 2943–2946, 1998.

328. Mitani, O. et al., JP Appl. 96/305,670, 1996; Eur. Pat. Appl. EP 839 870, 1998.

329. Tobita, M., Japan Appl. 96/344,435, 1996; JP 10 168,186 [98 168,186], 1998.

330. Rikako, T., Kazumi, V., and Katsuotshi, M., Appl. 97/180,728, 1997; JP 11 12,481 [99 12,481], 1999.

331. Murayama, R., Matsuda, J., and Okuba, H., JP Appl. 96/278,101, 1996; JP 10 121,012 [98 121,012], 1998.

332. Nakamura, K., Japan Appl. 93/188,146, 1993; JP 07 40,498 [95 40,498], 1995.

333. Shizuhata, H., Appl. 96/88,763, 1996; JP 09 259,636 [97 259,636], 1997.

334. Hiroshiyi, J. and Itoh, K., PCT Int. Appl. 99/09101, 1999; WO 99 09,101, 1999.

335. Fournier, J. et al., Structure and properties of conductive epoxy composites. Application to PTC thermistors, in *Exstr. Abstr.-EUROFIZZERS. 97 Int. Conf. Filled Polym. Filler, 2nd* ed., 1997, 383–386.

336. Fournier, J. et al., Study of the PTC effect in conducting epoxy polymer composites, *Chim. Phys. Phys–Chim. Biol.*, 95, 6, 1510–1513, 1998.

337. Thomas, M.G., Post, H.N., and De Biasio, R., End-of-millenium review, *Prog. Photovoltaics*, 7, 1, 1–19, 1999.

338. Jorgensen, G. et al., Polymers for solar-energy devices, *Desk. Ref. Funct. Polym.*, 1997, 567–588.
339. Watanuke, Y., Appl. 95/227,597, 1995; JP 09 70,886 [97 70,886], 1997.
340. Arakawa, F., Appl. 95/271,515, 1995; JP 09 113,708 [97 113,708], 1997.
341. Araki, T. et al., Japan Appl. 97/59,599, 1997; JP. 10, 256,780 [98,256,780], 1997.

Radioisotope Microbattery
Commercialization

K.E. Bower, A.F. Rutkiewic, C.C. Bower, and S.M. Yousaf

CONTENTS

10.1 INTRODUCTION

Radioisotope power sources can revolutionize microelectronics by enabling emerging microelectromechanical systems (MEMs) and nanotechnology; distributed power supplies would last years, even decades. No power cords, rectifiers, or transformers will be needed for a new generation of microdevices. Safe, direct, long-life, temperature-stable electric power will be available from an energy source providing five orders of magnitude higher power density than chemical redox systems and more than two orders of magnitude higher power density than fossil fuel oxidation reactions.

Market analysis for a revolutionary technological development is wishful thinking. Without years of accumulated financial information regarding sales, only time can vindicate predictions. To avoid the overt appearance of these faults, this chapter sets the more modest goal of comparing the proposed technology to traditional "portable micropower" sources and their applications — low-power microcircuits and MEMs.

10.2 BATTERY, MICROELECTRONICS, AND MEMS MARKET PROJECTIONS

The chemical battery is the most widely used portable electrical energy storage technology. Of the world's battery production, 95% is for primary batteries, but the secondary (rechargeable) battery market is expected to increase rapidly.[1] The consumer market for portable electronics has grown by more than 10% a year, with similar growth expected in the future. The increased need for improved batteries in the consumer market is matched by increased expectations in the military and environmental sectors, where reliability, durability, and long life are primary considerations. As can be seen from Figure 10.1, chemical batteries make up the bulk of battery types.

The current battery market is more than $17 billion per year in the U.S. alone. "Aggregate growth reflects an unending proliferation of portable devices which drives demand not only for batteries, but for longer lasting, higher performance and hence more expensive batteries as well."[2] This high-value battery market segment is driving growth for secondary (rechargeable and storage) batteries at 7.9% per year to more than $9 billion in 2000. Business Communications Company estimates that the lithium battery, a rechargeable for portable electronics, will be worth $2 billion by 2003.[3] "The need for more expensive high-technology batteries will also drive

Battery Types

Chemical Batteries

Primary

Alkaline
Manganese
Silver oxide
Zinc air
Lithium manganese
Lithium thionylchloride
Organic

Secondary (Rechargeable)

Nickel cadmium
Nickel metal hydride
Lithium ion
Lead acid
Nickel zinc
Nickel iron
Organic

Other

Fuel cell
Biochemical cell

Physical Batteries

Solar Voltaic cell
Radioisotope cell
Flywheel
Water wheel
Windmill

Figure 10.1   Battery classification scheme.

industrial battery demand. The outlook for defense spending is an important factor in demand for specialized battery products..."[2] While radioisotope microbatteries might compete with chemical cells, particularly in hybrid configurations that take advantage of the high instantaneous power of the chemical battery and extraordinary energy density of the nuclear battery, the radioisotope battery is especially well suited for ultra-low-power devices.

Forbes ASAP, a popular business magazine, estimates that "the average U.S. middle-class home has 35 to 50 devices that contain microprocessors — and that number does not include the family computer. Along with its embedded software systems it's a $15.3 billion market that's on its way to $27 billion..." Numerous low-power circuits require less than 100 μW. CMOS backup, for example, needs less than 25 μW. Nuclear batteries can compete favorably with rechargeable batteries due to their long life. Ease of use, environmental stability, and power supply integration efficiencies are huge factors giving the nuclear battery competitive favor. Radioisotope batteries may find use in consumer applications such as security and weather sensors and transmitters and automobile computer memory backup. Industrial, military, and medical MEMs sensors will benefit most from the new technology; indeed, the nuclear battery will enable emerging MEMs technology.

"[The] U.S. MEMs-based sensors market is ... expected to reach $1.16 billion by 2005. Frost & Sullivan sees more accelerated growth in the outyears of this forecast due to pending technological innovations and development of MEMs-specific platforms."[4] Roger Grace, a MEMs market analyst, puts the market value at $5 billion by 2004, while Venture Development Corp. puts the market value at $3.6 billion in 2004.[5] Innovations in power supplies for these sensors and their transmitters are critical technology for this emerging market. "The defense role in MEMs development is mainly that of research and development and is the catalyst for several MEMs developments that have already entered or are slated to enter the commercial market."[4] Assuming the most conservative MEMs market estimate for 2005 ($1.16 billion), that 50% of MEMs machines are battery-powered, and that 20% of product value is the power supply, one can expect a new $116 million/year market well suited to the radioisotope battery.

10.3 ENERGY DENSITY COMPARISON OF RADIOISOTOPES WITH CHEMICAL BATTERIES, FOSSIL FUELS, FUEL CELLS, SOLAR CELLS, AND BIOCHEMICAL SOLAR SYSTEMS

Battery power density — The weight-to-power density of current battery technology is displayed in Figure 10.2. Table 10.1 provides another summary of energy densities based on various primary and secondary batteries. The energy density of chemical redox batteries is in the mW-h/g (W-h/k) range.

Fossil fuel — The amount of energy available in organic oxidation, as in all chemical reactions, is the difference of energy in chemical bonds, energy of state, and entropy of the starting material and ending material. The enthalpy component of the change for a given reaction can be calculated by subtracting enthalpies of

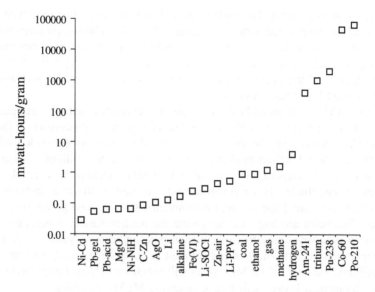

Figure 10.2 Battery energy density compared to fossil fuel and radioisotope decay energy.

Table 10.1 Energy Densities of Various Primary and Secondary Batteries

	Energy (kJ/g)	Energy (mW-h/g)
Primary		
Carbon/Zinc	0.314	87
Magnesium oxide	0.23	64
Silver oxide	0.383	106
Lithium	0.456	127
Alkaline	0.6	167
Iron (VI)	0.9	250
Zinc/Air	1.592	442
Secondary		
Rechargeable		
Nickel/Cadmium	0.1	28
Nickel hydride	0.234	65
Lead/Acid	0.223	62
Lead/Gel	0.192	53
Lithium powder	0.756	210
Li/thionyl chloride	1.08	300
Li / PPV	1.9	534

formation of the reactants from enthalpies of the products.[6] (The entropy component of the Gibbs free energy change will be neglected in this analysis.)

$$\Delta H^{\circ}_{\text{reaction}} = \Sigma \Delta H^{\circ}_{\text{f(products)}} - \Sigma \Delta H^{\circ}_{\text{f(reactants)}}$$

A simple approach to calculating enthalpy changes in chemical reactions is to sum the enthalpies of formation from the reactants to the elements in their standard state and add the enthalpies of formation from the elements to the products. Negative enthalpy changes represent exothermic reactions. The more negative the enthalpy change, the more energy is released. Based on this fundamental approach to energy calculations of oxidation of various fuels, Table 10.2 has been prepared. For the sake of consistency with other energy sources in this presentation, however, exothermicity is shown as positive in this table. This approach assumes that oxygen is available for combustion and that gaseous fragments do not escape. (The mass of the oxidant is included only in the thermite example.)

In comparison to the energy density of chemical batteries, the stored chemical energy in fossil fuels is substantial. The oxidation of these fuels is in the W-h/g range. On a weight basis, hydrogen is very attractive; on a volume basis, gasoline and diesel fuel are attractive. They are both plentiful and inexpensive; however, when used on a massive scale, fossil fuels will be exhausted, and product gases may cause long-term damage to the ecosystem through global warming.

The nitroglycerin example illustrates that the explosive power of organic molecules is not a result of high enthalpy changes only but also depends on the balanced internal oxygen source and gaseous products formed. The speed of a chemical reaction, reaction kinetics, is not the same as reaction thermodynamics. Explosives are not necessarily the most energetic materials by mass or volume.

Table 10.2 Heat of Reaction for Oxidation of Various Fuels

Fuel	Molar Mass (g/mol)	Energy (kJ/mol)	Energy (kJ/g)	W-h/g
Hydrogen	2	286	143	40
Coal	12	395	33	9
Acetylene	26	1,300	50	14
Methane	16	891	56	16
Methanol	32	765	24	7
Syngas	32	855	27	7.5
Propane	44	2,222	51	14
Ethanol	46	1,411	31	9
Gasoline	114	4,955	43	12
Glucose	180	2,805	16	4
Thermite	212	850	4	1
Nitroglycerin	227	1,870	8	2
Diesel	280	12,000	43	12
Cellulose	666	11,000	17	5

The efficiency of fuel use for electricity generation is especially relevant. Diesel and gasoline fuels in electric generators utilize several energy conversion steps. In the engine, chemical energy is converted to heat and gas to turn a shaft, creating mechanical energy. The turning shaft is connected to a generator to produce electricity inductively. Some arrangements use gas compressors, heat pumps, and the like. Even when the efficiency of the last conversion is 0.2, the overall efficiency for electric generation is generally less than 0.1.[7] Assuming 10% electrical-conversion efficiency, the diesel fuel for a generator is about 10 to 20 times more energy dense than a battery. It is this overall efficiency that is best compared when portable electrical power is the design goal.

Numerous hybrid configurations have merit with other design goals. For example, a thermophotovoltaic generator burning propane can achieve 2 W/cm² cell area, with overall fuel efficiency of 5%; the remainder is available as useful heat and light. These devices are commercially available from IX Crystal. Since the generator or turbine itself has significant mass and volume, there is a barrier to miniaturization for use in distributed power applications. One good answer to power plant expense and miniaturization limits is development of fuel cells.

Fuel cell — The fuel cell is the most important emerging energy-conversion technology with widespread consumer market potential. It does not produce greenhouse gas and shows excellent power-to-mass scaling as the engineered energy requirement increases. Each percent of the global vehicle market is worth $2 billion in sales alone.[8]

A fuel cell consists of two electrodes separated by an electrolyte. Hydrogen fuel is fed into the anode of the fuel cell, and oxygen (or air) enters the fuel cell through the cathode. The hydrogen is split into proton and hydride over a heterogeneous platinum catalyst, which take different paths to the cathode. The proton passes through the electrolyte, and electrons from the hydride pass through a load before returning to the cathode to be reunited with protons and oxygen, producing water.[9] The fuel cell may gradually transform a carbon-based energy industrial economy to a hydrogen basis, perhaps using solar-powered water electrolysis as the primary reverse reaction.

The efficiency of fuel cell operation approaches 35% in the laboratory. The fundamental fuel cell reduction and oxidation reactions are

$$\text{Anode:} \quad H_2 \rightarrow 2H^+ + 2e^- \qquad \varepsilon^\circ = 0 \text{ volt}$$
$$\text{Cathode:} \quad O_2 + 4e^- + 4H^+ \rightarrow 2H_2O \quad \varepsilon^\circ = 1.23 \text{ volt} \, (25°C)$$

Commercial fuel-cell power plants currently cost about $3000 per kilowatt. At that price the units are competitive only in high-value "niche" markets and in areas where electricity prices are high and natural gas prices are low. Many companies are competing in this area, so investment capital is currently plentiful. One of the more successful research firms has been Dais Corporation, whose proton exchange membrane (PEM), a sulfonated triblock polymer, is expected to reduce fuel cell cost greatly. In an ambient environment with no forced air, the Dais MEA produces ca. 50 mW/cm² at ½ volt. With controlled oxygen flows, heat (60°C), and increased pressure (2 atm), the Dais 585 membrane yields ca. 400 mW/cm² at ½ volt.

An 11-mW unit was demonstrated in Japan, and more than 100 200-kW units have been installed worldwide. High-energy designs such as carbonate fuel cells and solid oxide fuel cells are the focus of major electric utility companies' efforts to bring this technology to the market. Full-sized (commercial) cells and full-height stacks have been successfully demonstrated for the carbonate fuel cell design.

The fuel cell can use different hydrogen sources, including methanol, landfill gas, natural gas, and other waste products, to make electricity. Primarily, the fabrication cost of catalytic converters of other fuel sources into an onboard hydrogen supply has limited commercial introduction of the portable fuel cell. The possibility of hydrogen fuel storage in carbon nanotubes at 40% hydrogen by weight[10] would be a revolutionary advance for fuel cell use. This hydrogen storage media would be much more hydrogen dense than hydrides or liquid hydrogen. At 35% efficiency, with 40% hydrogen by fuel-source mass and 80% hydrogen recovery, the fuel cell could deliver 4.5 kW-h per kilogram of fuel.

Solar energy — Solar voltaic technology has been touted for decades as a primary solution to overreliance on fossil fuel oxidation for electricity generation. Despite intense government and private investment, costs still greatly exceed that of fossil fuel–generated electricity. Solar voltaic sales have grown at an annual rate of 20%, with 152 mW worldwide in 1998[11] — only a small fraction of world energy consumption. Rooftop PV programs in the U.S., Japan, and Europe increased interest in PV building integration and growth of on-grid PV markets. High-value, off-grid PV markets such as space, telecommunications, and other remote power applications remain robust. PV modularity, reliability, and environmental value provide its competitive edge.

Crystalline silicon continues to dominate the market, but thin-film PV is expanding. National Renewable Energy Laboratory (NREL) has verified the 16.6% efficiency of an AstroPower thin-film silicon laboratory cell.[12] Multijunction III-V solar cells are the most efficient PV technology, at 30.9% efficiency.[13] While III-V multijunction cells have been attractive to the space industry due to their high efficiency, low temperature coefficients, and good radiation hardness, multijunction cells in concentrator modules hold promise for terrestrial manufacturers, as well. The multijunction approach is also productive for hydrogenated amorphous silicon (a-Si:H) solar cells. With the support of NREL, United Solar Systems Corp. was able to develop a triple-junction a-Si:H cell with a commercial efficiency of 12%.[14] This cell is fabricated on a flexible stainless-steel substrate in a continuous roll-to-roll process that substantially reduces manufacturing costs.

Siemens Solar achieved $CuInSe_2$ (CIS) module efficiencies as high as 11.8% and is shipping 10 mW of commercially produced CIS modules annually.[15] NREL achieved an 18.8%-efficient $Cu(InGa)Se_2$ (CIGS) laboratory cell.[16] Efficiency of 12.5% has been achieved on a CIS module, and 10.5% efficiency has been achieved on a CdTe module.[17]

Bioenergy — The energy basis for life is visible sunlight; nuclear fusion in the sun is the ultimate source of photons that promote endothermic chemical reactions on the earth for all biological activity. Because of the distance between the sun and the earth, the fraction of the sun's emitted energy that falls on the earth[18] is less than 10^{-17}. Carbon fixation by photosynthesis in forests accounts for 10^{10} tons/year, on cultivated land 4.3×10^9 tons/year, on grassland 10^9 tons/year, on the desert 3×10^8

tons/year, and in the ocean 1.2×10^{11} tons/year.[19] The amount of photosynthetic activity on land, which covers roughly two-sevenths of the earth surface, is about one-seventh of that in the ocean. The oceans can be viewed as nearly continuous forests of photosynthetic growth. A deceptively simple overall photosynthetic reaction is the conversion of carbon dioxide and water to glucose and oxygen:

$$6CO_2 + 6H_2O \rightarrow C_6H_{12}O_6 + 6O_2 \qquad \Delta H^0_{reaction} = 2870 \ kJ/mol$$

This equation says nothing about the mechanism of the reaction. There are more than a hundred chemical steps in the photosynthetic process, each requiring a special catalyst in the right place at the right time. Of course, other important carbon compounds are also produced in various catabolic pathways driven by the same process.

Based on the overlapping spectra of chlorophyll *a* light absorption and photosynthesis efficiency, chlorophyll is known to be the primary initiator of photosynthesis. Numerous accessory pigments such as carotene are often involved in absorbing light energy in the cell, but they are temporary traps for the energy. The efficient electron transport from photoexcited chlorophyll to do catabolic work is at the heart of biochemistry. While the overall efficiency of photosynthesis appears low — carbon is fixed in a cornfield during the growing season with 2% efficiency, for example — the process can be made 35% efficient in the laboratory.[20] This critical biochemical system is a source of inspiration for new materials and engineered systems for portable energy storage and conversion.

Living organisms have evolved other interesting energy-conversion methods.

The electric eel can deliver a shock of several hundred volts of electric potential, which is derived from chemical energy. The flashes of the firefly are brought about by the conversion of chemical energy into light. The bombardier beetle can fire its startling cannon by converting the chemical energy of hydrogen peroxide into the pressure-volume energy of expanding oxygen gas.[21]

The conversion of biochemical energy to electricity is worthy of greater study. As the electric eel demonstrates, the conversion can be done more compactly and efficiently to turn a turbine or generator, than burning fuel to boil water or compress gas.

Radioisotope decay energy — An assembly of plutonium-238 heat sources was recently sent to Saturn on the NASA Cassini Space Probe. The plutonium heat source was designed to pipe heat to mechanical joints and through a thermoelectric process, in order to generate the necessary electricity for spacecraft electronics. Thermoelectric batteries made by Union Carbide in the 1960s with 2.14 g of plutonium-238 are still functioning at 2.22 V and 48 mA.*[22] The U.S. Atomic Energy Commission once had a very active program using radioactive isotopes as battery sources:

---

* Batteries were inspected by the author 6/14/2000 at Picatinny Arsenal, NJ. Efficiency of 3 to 4% is currently achieved using BiTe alloys at Hi-Z Technology, 7606 Miramar Road, Suite 7400, San Diego, CA 92126. Using tritium in titanium, direct conversion is expected to reach 2% steady-state efficiency, while indirect conversion will reach about 1.5%.

It was shortly recognized that certain isotopes — both the "refuse" from the fission reaction, and specific isotopes produced by irradiation, had just the right combination of power density, favorable half-lives, and chemical properties to provide a long lived, low weight, reliable heat source whose energy could be ultimately converted to [electric] power. On January 16, 1959, a device that transformed the heat from radioactivity into electricity was demonstrated on the desk of the President of the United States. The device was the size of a grapefruit. It weighed four pounds and was capable of delivering 11,600 watt-hours of electricity for about 280 days. This is equivalent to the energy produced by nearly 700 pounds of nickel-cadmium batteries. It was fueled with polonium-210 and designated SNAP-3.[22]

Milliwatt-to-watt systems were built and successfully tested. The milliwatt system typically contained 200 mg of plutonium, while kilowatt systems were designed to contain hundreds of thousands of curies of stontium-90. Wider application of the nuclear thermoelectric battery was limited primarily by very low efficiency and few low-power applications.[23] The efficiencies are increasing with use of new materials. Numerous ultra-low-power applications are being created.

There is great potential in renewed research and development where emphasis is placed on efficiency, proper packaging, and microcircuit integration.

Nuclear isotopes offer the most potential for Army applications. The energy density of radioisotopes is enormous, and they are extremely cost effective per watt-hour delivered...Isotope systems are extremely reliable and have been the system of choice for deep space probes. (Indeed, they have made them possible.)[25]

The potential energy in radioisotope decay is over a thousand times higher on a mass basis than that of chemical batteries, and the electrical conversion technology can be conceptually scaled to microscopic dimensions. After 25 years of research stagnation, recent papers and patents by NASA,[25] Sandia National Laboratory,[26] Ioffe Institute[27] Ontario Hydro,[28] TRACE,[29] and others[30] show that conversion of radioisotope decay energy to electricity is an area of revitalized progress as new materials are investigated.

Tremendous versatility, stability, and very long life are theoretically possible by matching alphavoltaic, betavoltaic, photovoltaic, scintillation, and radioactive materials in a battery or capacitor configuration. An advantage of the nuclear battery over a chemical battery is the design range possible in power allocation between the voltage and amperage components through engineered wiring of many-layered arrays. The nuclear battery can provide the power of two-dimensional solar cells (mW/cm^2) in an autonomous three-dimensional (mW/cm^3) package. The radioisotope power supplied is slow, steady, and independent of load because it is dependent on the rate of radioactive decay.

For long-lived radioisotopes (half-life greater than 300 years) in secular equilibrium with daughters, the number of decay events among a large population will not significantly change over a human lifetime. The engineered life of such a nuclear battery set arbitrarily at one-half the radioactive half-life of the source could readily exceed 100 years if radiation-robust materials with suitable heat-transfer properties are chosen. This battery concept is fundamentally different from nuclear power sources based on thermionic or fission principles.

The energy of radioisotopes is a function of the radioisotope source and its purity, half-life, and energy and decay. A variety of alpha and beta parent isotopes is listed in Table 10.3.[31] These isotopes have been selected partly on the basis of half-life (10 to 200 years) and useful energy. For direct conversion, beta emitters appear to be adequate to produce useful power. Alpha emitters are included with half-lives short enough to have enough useful power in combination with their daughters but long enough to provide a long continuous life under otherwise favorable conditions. For simplicity, the daughters have not been included, but they would contribute to nuclear converter performance.

The total decay energy, where t is the time in years and λ is the decay constant, is given by

$$\int_0^t P_o e^{-\lambda t}\, dt = P_o \left.\frac{\left(e^{-\lambda t} - 1\right)}{-\lambda}\right|_{t=0}^{t}$$

Based on this integral, the last two columns of Table 10.3 have been prepared. One can see that 1 g of polonium-210 could provide over 50 kW-h over a 4-year period if converted at 8% efficiency while being converted to a stable product. While energy density of radioisotopes is 100 to 100,000 times higher than other systems considered in this review, instantaneous power is relatively low. The radioisotope converter would be optimally designed for long autonomous use of low-power circuits such as pulsed radio frequency tags and MEM devices.

Direct-conversion — The efficiency of energy conversion is initially high through the use of direct alpha- and beta radiation exposure to semiconductor junctions. However, the direct-conversion scheme fails because crystalline semiconductors are damaged by high-energy particle bombardment. TRACE Photonics, Inc. has made progress in solving this direct-conversion problem by using amorphous Si:H *nip* drift junctions and polycrystalline IIB-VIA semiconductors with self-annealing properties. Stable performance in air has been achieved at 1.5% conversion efficiency using amorphous silicon and 4.4% conversion efficiency using CdTe/CdS semiconductors with direct exposure to [147]Pm.

Indirect-conversion — An alternative approach is to convert radioisotope energy to light with phosphors that emit in a region of high photovoltaic efficiency. The technical challenge is that plastic and glass scintillators transparent to the radioluminescent light have efficiencies of less than 1%, while highly light-self-absorbing inorganic crystals have efficiencies up to 25%. While the indirect scheme protects the photovoltaic cell from ionizing radiation, the phosphor is subject to radiation degradation. Research emphasis has been to construct scintillation glass using sol gel processes with high light-conversion efficiency under radiation exposure. This has resulted in an overall conversion efficiency of less than 0.5%. While these results are not impressive in comparison to radioisotope thermoelectric generator efficiencies of 3%, the lightweight components of photonic and direct-conversion can be better integrated into microelectronics, and alpha-radiation can be used over the long term with more confidence, using indirect-conversion.

Table 10.3 Specific Power and Work for Various Radioisotopes

Radioisotope	Half-Life (Year)	Energy (MeV)	Specific Power (W/g)	Ci/g	Work (W-h/4year/Ci)	Work (kW-h/4year/g)
Es-252	1.3	6.6	4.3E+01	1,098	559.2	614.0
Cf-252	2.6	6.1	1.9E+01	536	777.9	417.2
Cf-251	898.4	5.7	5.3E-02	2	1168.0	1.9
Cf-250	13.1	6	3.9E+00	109	1112.5	121.6
Cf-248	1.0	6.25	5.3E+01	1,579	397.2	627.2
Bk-249	1.0	5.4	45.5	1,639	329.3	539.7
Cm-248	3.4E+05	4.65	0.0	0	953.2	0.0
Cm-243	29.0	5.8	1.7	52	1092.4	56.4
Cm-244	18.0	5.8	2.7	81	1090.9	88.3
Am-241	435.0	5.5	0.1	3	1122.6	3.8
Np-235	1.0	5	46.1	1,402	389.3	545.7
Pu-241	14.0	4.9	3.2	103	994.9	102.5
Pu-238	86.4	5.6	0.6	17	1151.9	19.7
Pu-236	2.9	5.8	17.8	531	755.8	401.6
U-232	70.0	5.3	0.7	22	1033.3	23.1
Ac-227	21.0	0.0679	0.0	72	13.1	0.9
Th-228	1.9	5.4	26.2	820	588.8	482.5
Po-210	0.4	5.2	137.2	4,493	146.2	656.9
Ir-192	0.2	0.8137	44.0	9,211	12.2	112.5
Tm-170	0.4	0.00542	0.2	5973	0.1	0.8
Pm-147	2.6	0.062	3.4E-01	927	7.9	7.3
Cs-137	30.2	0.094	4.8E-02	87	18.4	1.6
Sr-90	28	0.9	7.5E-01	139	180.0	25.0
Tc-99	2.14E+05	0.085	8.4E-06	0	17.4	0.0
Mo-99	7.50E-03	0.0176	49.8	480,000	0.0	4.7
Co-60	5	2.50442	16.6	1,130	400.9	453.2
H-3	12	0.0057	0.3	9,664	1.1	10.3

Another important consideration is radioisotope transportation requirements that reflect public perception of handling and human safety hazard. Table 10.4 has been prepared to compare the available energy and transportation limits of various radioisotopes based on 49CFR173.423-435. This constraint, which is a practical limit for any conceivable consumer application, limits the radioisotope power over 4 years to 600 W-h based on Am-241 and 100 W-h based on tritium. The instantaneous power limits are 0.02 W and 0.36 mW for these two isotopes, respectively.

If one assumes that the technology used converts 10% of this energy into electricity, consumer products are limited to 2 mW. Even this limit assumes a technological solution to several difficult problems concerning material degradation from alpha radiation. TRACE Photonics has proven the stability of amorphous silicon and IIB-VIA semiconductors to moderate beta radiation but has not yet tested IIB-VIA converters with actinides. Use of radioisotope trickle-chargers for secondary chemical cells or direct capacitor rechargers could expand this limit significantly.

Isotope cost, safety, security, and recycle options — The cost of feed materials, user and manufacturing safety and security, regulatory approval, transportation controls, disposal or recycling, and quality assurance must be considered before commercialization of this technology. Distributed radioisotope micropower can provide

Table 10.4 Radioisotope Decay Energy and U.S. Department of Transportation Limits for Special Form (Sealed) Isotopes

Radioisotope	Specific Power (W/g)	Ci/g	Energy (kW-h/4year/g)	U.S. DOT Limits (mg)	DOT Limit (W-h/4 year)
Am-241	0.11	3	3.4	180	600
Pu-238	0.57	17	20	32	640
Ac-227	0.03	72	0.9	150	135
Th-228	26	820	483	170	82,000
Po-210	137	4493	656	2.4	1,550
Pm-147	0.34	927	73	10.8	790
Cs-137	0.047	87	1.6	11.6	19
Sr-90	0.75	139	25	0.4	10
H-3	0.325	9664	99	1.1	109

high-value-added products from recycled (processed) spent fission fuel, thereby mitigating difficult disposition issues. Modern technological progress has been characterized by increasing electronic functionality with declining power needs. The portable nuclear battery is a very long-life, low instantaneous power source. New materials are being considered that can give the nuclear battery greater durability and higher efficiency.

Outstanding features of radioisotope microbatteries include their long life (based on material stability and radioisotope half-life), continuous power because the source is always on, low weight compared to chemical batteries, potentially miniature size, very high voltage and efficiency using the direct capacitor charging scheme, temperature insensitivity (unlike chemical batteries that are very sensitive to temperature for the rate of the chemical reaction), and shock and acceleration stability since the device is tiny and solid.

Cost and availability of isotopes are critically important considerations. Tritium, cobalt-60, strontium-90, polonium-210, americium-241, and cesium-137 are available for direct purchase. Promethium-147 was the isotope of choice in the second-generation, direct-conversion atomic battery,[32] but it is difficult to obtain in quantity. Strontium-90 was found to be too damaging to electronic devices in early direct-conversion schemes,[33] but it might be used with the new generation polycrystalline and amorphous semiconductors.

An important economic and ecological feature of radioluminescent, thermoelectric, and direct-energy-conversion devices is that purified radioisotopes are not required. The mixed fission fragment of spent nuclear fuel may make a cheap and plentiful source. Fission products can be energy storage systems, rather than waste at Yucca Mountain, Nevada, at a cost of over $10,000/ft³. Estimates of the composition of spent nuclear fuel are provided in Table 10.5. These are based on reactors and burn rates from early commercial reactors. When long-lived uranium and plutonium are removed and recycled, fission products are estimated to have an activity of approximately 1 Ci/g.[34]

With the cheapest purified radionuclides costing $0.50/Ci (tritium) and promethium costing over $1000/Ci, it is economically unlikely that tens of millions of on-circuit power supplies could be built using millions of curies of these radionuclides. When

Table 10.5 Spent Commercial Nuclear Reactor Fuel Composition (t)

Isotope	Grams	Activity in Curies
U-238	998,000	0.335
Pu-239	800	496
Cs-137	110	13,100
Sr-90	40	41,500
Ba	40	4,200
Y	20	51,000
La	40	0
Ce	100	174,000
Pr	155	15,000
Zr	115	112,000
Nb	5	203,000
Mo	85	0
Tc	25	0
Ru	55	37,000
Rh	12	0
Other	100	2,000
Total Power (W/Ci)		**0.03**
Total Power (W/g)		**0.02**

infrastructure development costs for material handling and recycling are included, the estimated $116 million total radioisotope power supply market cannot accommodate radioisotope liability without significant government assistance. A partnership between government nuclear scientists and private radioisotope microbattery manufacturers, with oversight by government legislators and regulators, could create an industrial reuse option for high-level nuclear waste. This win–win partnership would greatly reduce overall government cost for high-level waste disposition while creating an important new high-technology commercial enterprise.

Radiothermoelectric generation will scale up with source concentration, while direct conversion on semiconductors and indirect conversion with scintillation glass and semiconductors will productively scale down to ultra-low power, long-life electronic and electromechanical devices.

Future work will include conversion-efficiency studies with mixed sources using indirect-conversion scintillation glass, direct conversion using amorphous silicon and II-VI semiconductors, and integration into useful microcircuits.

10.4 ULTRA-LOW-POWER MICROELECTRONICS APPLICATIONS

The radioisotope converter would be suited for long autonomous electronic circuits such as GPS and sensor data transmission. It might trickle-charge secondary cells or be used in pulse mode to provide much greater instantaneous power. Hybrid radioisotope-chemical battery systems have been proposed for development. This energy source is also suitable for high-voltage ultra-capacitor charging. This approach has been demonstrated to be over 65% efficient with the Linder cell. Table 10.6 gives a sampling of low-power device applications.

Table 10.6 Ultra-Low-Power Device Sampler

Category	Device Name, Supplier, or Manufacturer	Nominal Power Consump.	Back-Up Power Needs	Volt.	Current	Standby Current	Comments and Ref.
4-bit processor	S-13L40AF Seiko	.3 mW	6 µW	1.2 V	.25 mA	5 µA	
16-bit processor	MSP430F11x Texas Instruments	1.1 mW	1.8 µW	2.2 V	1.6 µA	0.8 µA	www.ti.com
Processor	STW/TUD	1 µW					www.dimes.tudelft.nl/oldwww/19 96/SSC/nodel12
Wideband transmitter	PulsOn, Time Domain Inc	50 µW					www.time-domain.com
Active-pixel-sensor digital camera	research	20 mW	40 µW				NASA Tech Briefs Vol. 22, No.10, page 44
Pressure sensor	Lucas NovaSensor	0.2 mW		0.1 V	2 mA		
Light sensor		50 µW	NA	5 V	10 µA	NA	
Salt sensor					70 µA		
Battery low					90 µA	1mA	
Short detector					100 µA		
Thermopneumatic valve	CalTech	61 mW					www.mems.caltech.edu/home/publications/xing/hh98
VFC Conv.		15 µW			9 µA		
Crystal oscillator	HA7210 Harris Semicond.			2–7 V	5 µA @ 32 kHz		32.768 kHz clock at 3V
Blood vessel monitor	NASA	80 mW		3.2 V	2–25 µA		Bob Ricks, NASA
Fetal monitor	NASA	30 µW		3.2 V	9 µA		Bob Ricks, NASA
Heart pacemaker	Wilson Greatbatch Inc.			2.8 V	.4 msec pulses		
Heart defibrillator	Wilson Greatbatch Inc.		1 mA	600 V	30 J pulse 1 msec		
UHF transceiver	BiM-418-F Radiometrix	45 mW		5 V	15 mA		www.radiometrix.co.uk
IR transceiver	Hummingbird HP	100 mW	1.35 µW	2.7 V	25 mA	0.5 mA	11/3/97 press release HP
VLF receiver	Micropower Concorde	.5 mW		5 V	100 µA		www.cms-asic.com
Serial transceiver	DS276 Dallas Semicond.	70 mA	70 mA	2.7 V		25 mA	www.dalsemi.com

Frequency synthesizer	LMX2335L National Semicond.			5 V	4 mA		www.national.com
D/A converter	LTC1329-10	.14 mW	.5 µW	2.5 V	55 µA	0.2 µA	Linear Technology
Op. Amp.	LT1097 Linear Technology	3.2 mW		9 V	350 µA		
Fixed gain amplifier	LT1101 Linear Technology	0.4 mW		5 V	75 µA		
Quad comparator	LP339 National Semicond.	120 µW		2 V	60 µA per comp.		www.national.com
Radar transmitter	MIR LLNL			1.5 V			www.lasers.llnl.gov
Amplifier	AD627 Analog Devices	0.2 mW		2.2 V	85 µA		www.analog.com
High gain optocoupler	HCPL-4701 HP	0.1 mW		1.6 V	50 µA		AlGaAs LED
HM5165405F 64M DRAM	Hitachi	468 mW		3.3 V	140 mA	300 µA	L-Version
HB288032C5 32M flash	Hitachi			3.3 V		300 µA	
AM29SL800 flash	Advanced Micro Devices Inc.	1.8 mW					www.news.cnet.com/0,10000, 0-1003-200-324816,00
UPPD45D128 164 DRAM	NEC	1 W		2.5 V	100 mA	25 mA	
UPD431000A	NEC			3.5 V	35 mA	500 µA	
MB81N643289-50 FCRAM	Fujitsu	2 W		2.5 V	230 mA	25 mA	
EM6603 Multi I/O MCU	EM Microelectronics	0.36 µW					www.emmarin.com/Line.asp
EM6617 MCU w/ADC	EM Microelectronics	1.2 µW					www.emmarin.com/Line.asp
EM6629 MCU w/LCD driver	EM Microelectronics	0.6 µW					www.emmarin.com/Line.asp
CMOS/VGA	EM Microelectronics	12 mW					www.emmarin.com/Line.asp

REFERENCES

1. Wada, M. and Kabuki, K., Batteries, in *Polymeric Materials Encyclopedia*, Joseph C. Salamone, Ed., CRC Press, Boca Raton, FL, 1996.
2. The Freedonia Group Inc., *Batteries to 2000*, company report, Cleveland, OH, 1996.
3. *www.becresearch.com/editors/RGB-210.html*
4. Frost & Sullivan, *U.S. MEMs-Based Sensor Markets*, 5999-32, 1999, company report, NY, 1-1.
5. *www.ebnews.com/story/OEG20011012S0060*
6. Zumdahl, S.S., *Chemistry*, 2nd ed., Heath, Lexington, MA, 1989, 221–250.
7. Sharpe, G.J., *Applied Thermodynamics and Energy Conversion*, Longman Scientific & Technical Press, New York, 1987, 72–90.
8. Arthur D. Little, Inc., *Fuel Cell Market Study*, Cambridge, MA, 1999.
9. Fuel cells heading for sale, C & E News, *Am. Chem. Soc.*, June 14, 1999, 31–37.
10. Fuel cells heading for sale, C & E News, *Am. Chem. Soc.*, June 14, 1999, 36.
11. Maycock, P., *PV News*, 18(2), 1999.
12. Mitchell, R.L. et al., Progress update on he U.S. Photovoltaic Manufacturing Technology Project, *Proc. 26th IEEE Photovoltaic Specialists Conf.*, 1997.
13. Masuda, K., *Proc. 11th Int. Photovoltaic Sci. Eng. Conf.*, 1999.
14. Maycock, P., *PV News*, 17(9), 1998.
15. 100 Most Significant Technical Products of the Year, *R&D Mag.*, Sept. 1999.
16. Maycock, P., *PV News*, 17(11), 1998.
17. Delahoy, A.E. et al., CIS Photovoltaic Technology Final Report, NICH Rpt. No. SR-520-25713, April, 1998.
18. Personal communication, Patrick Coulton, Eastern Illinois University.
19. Lehninger, A.L., *Bioenergetics*, 2nd ed., Benjamin Cummings, Menlo Park, CA, 1973, 11.
20. Lehninger, A.L., *Bioenergetics*, 2nd ed., Benjamin Cummings, Menlo Park, CA, 1973, 117–118.
21. Lehninger, A.L., *Bioenergetics*, 2nd ed., Benjamin Cummings, Menlo Park, CA, 1973, 13.
22. Seiken, S.J., Terrestrial isotope power systems in perspective, in *Proc. Symp. Mater. Radioisotope Heat Sources*, in *Nuclear Metallurgy*, Vol. 14, American Institute of Mining, Metallurgical, and Petroleum Engineers, 1969, 29.
23. Seiken, S.J., Terrestrial isotope power systems in perspective, in *Proc. Symp. Mater. Radioisotope Heat Sources*, in *Nuclear Metallurgy*, Vol. 14, American Institute of Mining, Metallurgical, and Petroleum Engineers, 1969, 41.
24. *Energy-Efficient Technologies for the Dismounted Soldier*, National Academy Press, Washington D.C., 1997, 38.
25. Patel, J., Generating electric power from alpha-particle sources, *Electron. Tech. Briefs*, December, 2000.
26. Ashley, C.S. et al., US Patent 5,240,647, 1993; Walko, R.J. et al., Tritium-fueled beta cells, *IEEE Proc. Intersoc. Energy Convers. Eng. Conf.*, 6, 1997, 135–140.
27. Andreev, V.M. et al., Betavoltaic cells based on tritium and AlGaAs–GaAs heterojunctions, *Tech. Dig. PVSEC-11*, 1999, 817–818; Andreev, V.M. et al., Tritium-powered beta cells based on $Al_xGa_{1-x}As$, *28th IEEE PVSC*, Anchorage, September, 2000.

28. Kherani, N.P. et al., US Patent 5,606,213, 1997; Mannik, L. et al., US Patent 5,859,484, 1999; Kherani, N.P. et al., Tritiated amorphous silicon for micropower application, *Fusion Technol.*, 28, 1609; Kosteski, T. et al., Tritiated amorphous silicon films and devices, *J. Vac. Sci. Technol. A*, 16(2), 1998, 893; Ruda, H.E. et al., Semiconductor-based light emitters powered by tritium, *Appl. Phys. Lett.*, 71(18), 1997, 2644.

29. Deus, S., Alpha- and betavoltaic cells based on amorphous silicon, *16th EPSEC*, 2000; Deus, S., Tritium-powered betavoltaic cells based on amorphous silicon, *28th IEEE PVSC*, Anchorage, September, 2000.

30. Cots, A.O. and Reed, J.J., US Patent 5,082,505, 1992; Shanks, H., US Patent 5,721,462, 1998; Olsen, L.C., Review of betavoltaic energy conversion, *Space Photovoltaic Res. Technol. Conf. XII*, NASA Conference Publication 3210, 1993, 256–267.

31. Browne, E. and Firestone, R.B., *Table of Radioactive Isotopes*, Wiley Interscience, New York, 1986.

32. Flicker, H., Loferski, J.J., and Elleman, T.S., Construction of a Promethium-147 atomic battery, *IEEE Trans. Electron Devices*, 1964, 2–8.

33. Rappaport, P., The electron-voltaic effect in *p–n* junctions induced by beta particle bombardment, *Phys. Rev.*, 93, 1953, 246–247; Pfann, W.G. and van Roosbroeck, W., Radioactive and photoelectric *p–n* junction power sources, *J. Appl. Phys.*, 25, 1954, 1422–1429.

34. Howells, G.R. et al., The chemical processing of irradiated fuels from thermal reactors, *Peaceful Uses of Atomic Energy*, 1958, 3.

Index

cathode ray–induced degradation of
   description of, 193, 197
   thin silicone oxide layer for preventing, 201
coactivators for, 122
europium in, 133

green-emitting, 133–136
luminescence spectra of, 122–124, 132
synthesis of, 121
yellow-emitting, 142
Zinc tetrasulfophthalocyanine, 393